ROBERT M. GROVES, University of Michigan and Joint Program in Survey Methodology, College Park, MD

ALISON HALE, Statistics Canada, Ottawa, ON

LON HOFMAN, Statistics Netherlands, Voorburg, Netherlands

EDWIN HUNT, National Opinion Research Center, Chicago, IL

DONNA M. JEWELL, Research Triangle Institute, Research Triangle Park, NC

JOHN KENNEDY, Indiana University, Bloomington, IN

SUSAN H. KINSEY, Research Triangle Institute, Research Triangle Park, NC

JAMES M. LEPKOWSKI, University of Michigan, Ann Arbor, MI

LARS LYBERG, Statistics Sweden, Stockholm, Sweden

TONY MANNERS, Office for National Statistics, London, U.K.

JEAN MARTIN, Office for National Statistics, London, U.K.

SYLVIE MICHAUD, Statistics Canada, Ottawa, ON

HEATHER G. MILLER, Research Triangle Institute, Research Triangle Park, NC

KENT J. MILLER, Washington State University, Pullman, WA

WILLIAM L. NICHOLLS II, U.S. Bureau of the Census, Washington, DC

COLM O'MUIRCHEARTAIGH, London School of Economics and Political Science, London, U.K.

JAMES M. O'REILLY, Research Triangle Institute, Research Triangle Park, NC

THOMAS PIAZZA, University of California, Berkeley, CA

MARK PIERZCHALA, Westat, Rockville, MD

D. E. B. POTTER, Agency for Health Care Policy and Research, Rockville, MD

MAGDALENA RAMOS, U.S. Bureau of the Census, Washington, DC

MICHAEL RHOADS, Westat, Rockville, MD

SUSAN M. ROGERS, Research Triangle Institute, Research Triangle Park, NC

SALLY ANN SADOSKY, Intel Corporation, Santa Clara, CA

WILLEM E. SARIS, University of Amsterdam, Amsterdam, Netherlands

DAVID SCUDDER, Boise State University, Boise, ID

mputer Assisted Survey
formation Collection

Computer Assisted Survey Information Collection

Edited by

MICK P. COUPER
University of Michigan

REGINALD P. BAKER
Market Strategies, Inc.

JELKE BETHLEHEM
Statistics Netherlands

CYNTHIA Z. F. CLARK
U.S. Bureau of the Census

JEAN MARTIN
Office for National Statistics, U.K.

WILLIAM L. NICHOLLS II
U.S. Bureau of the Census

JAMES M. O'REILLY
Research Triangle Institute

A Wiley-Interscience Publication
JOHN WILEY & SONS, INC.
New York · Chichester · Weinheim · Brisbane · Singapore · Toronto

Contributors

REGINALD P. BAKER, Market Strategies, Inc., Southfield, MI

NANCY BATES, U.S. Bureau of the Census, Washington, DC

JELKE BETHLEHEM, Statistics Netherlands, Voorburg, Netherlands

EVERT BLOM, Statistics Sweden, Örebro, Sweden

BILL BLYTH, Taylor Nelson Sofres, London, UK

JOHN BOSLEY, U.S. Bureau of Labor Statistics, Washington, DC

ANN BROWN, Statistics Canada, Ottawa, ON

PAUL BUCKLEY, Abt Associates, Inc., Chicago, IL

CYNTHIA Z. F. CLARK, U.S. Bureau of the Census, Washington, DC

RICHARD L. CLAYTON, U.S. Bureau of Labor Statistics, Washington, DC

MARTIN COLLINS, City University Business School, London, England

WILLIAM E. CONNETT, University of Michigan, Ann Arbor, MI

FREDERICK G. CONRAD, U.S. Bureau of Labor Statistics, Washington, DC

PHILLIP C. COOLEY, Research Triangle Institute, Research Triangle Park, NC

MICK P. COUPER, University of Michigan and Joint Program in Survey Methodology, Ann Arbor, MI

DON A. DILLMAN, Washington State University, Pullman, WA

RICHARD DULANEY, Westat, Rockville, MD

BRAD EDWARDS, Westat, Rockville, MD

TERESA PARSLEY EDWARDS, Research Triangle Institute, Research Triangle Park, NC

BARBARA H. FORSYTH, Westat, Rockville, MD

JAMES GRAY, Office for National Statistics, U.K.

ROBERT M. GROVES, University of Michigan and Joint Program in Survey Methodology, College Park, MD

ALISON HALE, Statistics Canada, Ottawa, ON

LON HOFMAN, Statistics Netherlands, Voorburg, Netherlands

EDWIN HUNT, National Opinion Research Center, Chicago, IL

DONNA M. JEWELL, Research Triangle Institute, Research Triangle Park, NC

JOHN KENNEDY, Indiana University, Bloomington, IN

SUSAN H. KINSEY, Research Triangle Institute, Research Triangle Park, NC

JAMES M. LEPKOWSKI, University of Michigan, Ann Arbor, MI

LARS LYBERG, Statistics Sweden, Stockholm, Sweden

TONY MANNERS, Office for National Statistics, London, U.K.

JEAN MARTIN, Office for National Statistics, London, U.K.

SYLVIE MICHAUD, Statistics Canada, Ottawa, ON

HEATHER G. MILLER, Research Triangle Institute, Research Triangle Park, NC

KENT J. MILLER, Washington State University, Pullman, WA

WILLIAM L. NICHOLLS II, U.S. Bureau of the Census, Washington, DC

COLM O'MUIRCHEARTAIGH, London School of Economics and Political Science, London, U.K.

JAMES M. O'REILLY, Research Triangle Institute, Research Triangle Park, NC

THOMAS PIAZZA, University of California, Berkeley, CA

MARK PIERZCHALA, Westat, Rockville, MD

D. E. B. POTTER, Agency for Health Care Policy and Research, Rockville, MD

MAGDALENA RAMOS, U.S. Bureau of the Census, Washington, DC

MICHAEL RHOADS, Westat, Rockville, MD

SUSAN M. ROGERS, Research Triangle Institute, Research Triangle Park, NC

SALLY ANN SADOSKY, Intel Corporation, Santa Clara, CA

WILLEM E. SARIS, University of Amsterdam, Amsterdam, Netherlands

DAVID SCUDDER, Boise State University, Boise, ID

BARBARA M. SEDIVI, U.S. Bureau of the Census, Washington, DC

JANE SHEPHERD, Westat, Rockville, MD

DIRK SIKKEL, CentERdata, Tilburg, Netherlands

JAMES SMITH, Westat, Rockville, MD

TIMOTHY K. SMITH, Research Triangle Institute, Research Triangle Park, NC

TOM W. SMITH, National Opinion Research Center, Chicago, IL

PAUL M. SNIDERMAN, Stanford University, Stanford, CA

HOWARD SPEIZER, National Opinion Research Center, Chicago, IL

SANDRA SPERRY, Westat, Rockville, MD

R. SURESH, Research Triangle Institute, Research Triangle Park, NC

ELIZABETH M. SWEET, U.S. Bureau of the Census, Washington, DC

WENDY SYKES, City University Business School, London, U.K.

JOHN TARNAI, Washington State University, Pullman, WA

ROBERT D. TORTORA, The Gallup Organization, Alexandria, VA

ROGER TOURANGEAU, National Opinion Research Center, Washington, DC

CHARLES F. TURNER, Research Triangle Institute, Washington, DC

DAVID UGLOW, U.S. Bureau of Labor Statistics, Washington, DC

FRANK VAN DE POL, Statistics Netherlands, Voorburg, Netherlands

MICHAEL F. WEEKS, Research Triangle Institute, Research Triangle Park, NC

PAUL S. WEISS, Trilogy Consulting Corporation, Smyrna, GA

GEORGE S. WERKING, U.S. Bureau of Labor Statistics, Washington, DC

MARK S. WOJCIK, Abt Associates, Inc., Chicago, IL

Preface

The phrase "computer assisted survey information collection" (CASIC) is gaining widespread acceptance as denoting the use of computers for survey data collection, data capture, and data preparation and the activities that support those tasks. A variety of CASIC technologies have already had a major effect on the survey industry, and this trend is likely to accelerate in the future. While some CASIC methods, such as computer assisted telephone interviewing, have now been in use for decades, the last few years have witnessed a rapid evolution of CASIC methods and an accelerating replacement of traditional (paper and pencil) survey procedures with CASIC alternatives.

This monograph reflects, as did the conference on which it is based, a watershed in the development and evaluation of these methods. In part this volume serves as a historical record of past accomplishments, providing an occasion to reflect on progress made in implementing CASIC methods and the lessons learned from both prior missteps and triumphs. At the same time this monograph provides an opportunity to look to the future, to anticipate new directions and improvements, and to identify the work needed to further advance survey research through computer assisted methods.

The conference was the first devoted exclusively to the topic of computer assisted survey information, and we suspect it may also be the last. CASIC has become an integral part of survey research and survey methodology; its influence is pervasive and widespread. An international conference and edited monograph are a timely contribution and testimony that these issues have come of age.

In the summer of 1993 Judith Lessler, then incoming chair of the Survey Research Methods Section of the American Statistical Association, approached Mick Couper with the idea of organizing a conference on the topic of computer assisted survey information collection. With Couper as chair, an organizing committee was formed. In addition to the editors of the monograph, the committee consisted of William Connett (University of Michigan), Lee Decker (American Statistical Association), Tony Manners (Office for National Statistics, U.K.), and Alan Tupek (National Science Foundation).

The committee decided at the outset to follow the example of previous conferences in this series by preparing an edited monograph of invited conference papers. In 1995 abstracts were solicited for invited conference papers with the understanding that they would become monograph chapters, and the development of the monograph proceeded in parallel with planning for the conference.

The conference was organized under the scientific sponsorship of the following professional associations: the Survey Research Methods Section of the American Statistical Association (ASA), the American Association for Public Opinion Research (AAPOR), and the International Association of Survey Statisticians (IASS) of the International Statistical Institute (ISI). Each of these associations also provided funds to support the planning activities of the committee and the development of the conference and monograph. Additional fundings was provided by the following research and professional organizations.

Abt Associates, Inc.

Australian Bureau of Statistics

Center for Survey Methods, University of California, Berkeley

Computers for Marketing Corporation

The Gallup Organization

International Field Directors and Technologies Conference

Joint Program in Survey Methodology

National Opinion Research Center

Office for National Statistics, U.K.

Quantime Corporation

Research Triangle Institute

Response Analysis Corporation

Sawtooth Technologies, Inc.

Statistics Finland

Statistics Netherlands

Statistics New Zealand

Statistics Sweden

Survey Research Center, University of Michigan

Surveycraft, Inc.

U.S. Bureau of the Census

U.S. Bureau of Labor Statistics

U.S. Department of Agriculture, National Agricultural Statistics Service

U.S. National Center for Educational Statistics

U.S. Department of Transportation, Bureau of Transportation Statistics

U.S. Department of Energy, Energy Information Administration

U.S. Centers for Disease Control and Prevention, National Center for Health Statistics

Westat

Without the generous support of these organizations, neither the conference nor the monograph would have been possible.

The International Conference on Computer Assisted Survey Information Collection (InterCASIC) was held in San Antonio, Texas, from December 11 to 14, 1996. The conference was attended by over 360 persons from 21 countries. Fully a quarter of the attendees came from outside the United States. Two half-day workshops were held on the first day of the conference. A workshop on "Total Survey Error Considerations in CASIC" was presented by Paul J. Lavrakas of Ohio State University. A workshop on "A Practical Approach to Usability Testing" was presented by Janice C. Redish of Redish & Associates. More than a fourth of the conference attendees enrolled in one or both of these workshops. The conference was formally opened with a provocative keynote address on "User Interfaces for Survey Data Collection" by Ben Shneiderman, Professor of Computer Science and head of the Human Computer Interaction Laboratory at the University of Maryland.

The technical program consisted of 48 sessions over a two and a half day period. In total, 131 invited and contributed papers, software demonstrations, and panel discussions were presented. An exhibit hall for new CASIC hardware and software and for institutional exhibits was well attended throughout the conference.

The goal of the monograph editors went far beyond that of producing a conference proceedings volume. From the outset our mission was to guide and encourage a volume that would become an authoritative and comprehensive review of the CASIC field. No book, including this one, can claim a fully exhaustive coverage of its subject matter, but this volume does encompass a broad range of CASIC methods and issues extant at the time of its preparation.

The monograph, like the conference, looks both backward and forward. It documents past and present CASIC developments and consolidates our knowledge of the field. At the same time it points to the future by considering likely directions for survey research in the information age and identifying challenges and opportunities for further research and development. We appreciate the forbearance of the chapter authors in allowing us to mold their papers into an organic and coherent volume. It is our belief that this book documents the state of the art of CASIC at the time of writing.

The chapters in this book represent a variety of different evidentiary styles, reflecting the different disciplines of the authors and the current state of knowledge on various CASIC methods. Some authors present the results of empirical experiments on alternative approaches. Others review the existing literature on a particular topic or canvass a number of survey organizations to describe typical strategies or approaches that are used. Still others describe procedures used at their own organizations, offer advice based on their own experiences, or speculate on likely future directions of the field. This variety of approaches is appropriate for a book on a rapidly evolving field such as CASIC. However, we urge the reader to exercise caution in generalizing from the experiences described in these chapters.

The book is organized into eight sections. Chapter 1 (Section A) begins with a historical overview and introduction to CASIC and its many forms. The remaining sections and their editors are as follows:

Section B, Chapters 2–5: Transition to CASIC, Jean Martin

Section C, Chapters 6–8: Instrument Design, James O'Reilly

Section D, Chapters 9–12: Issues in Survey Design, Cynthia Clark

Section E, Chapters 13–16: Case Management, Jelke Bethlehem

Section F, Chapters 17–19: Interviewers as Users of CASIC, Mick Couper

Section G, Chapters 20–24: Self-Administered Surveys, William Nicholls

Section H, Chapters 25–29: Emerging Technologies in CASIC, Reginald Baker

Each editor also had responsibilities as the secondary editor for another section. Chapter authors were also called on to review other monograph chapters. This was done in an effort to minimize redundancies across chapters and ensure consistency of terminology, while attempting not to stifle individual style. Because of the field's predilection for acronyms, we sought to improve readability by appending a brief glossary of key CASIC terms and concepts to the end of the book. This volume is intended both for graduate students in survey methods and for survey research practitioners. As editors, we hope this monograph will lay a solid foundation for future CASIC research and development.

We have several others to thank for their contributions to the conference and monograph. Christie Nader of the Joint Program in Survey Methodology (JPSM) developed and maintained the conference web site. The JPSM staff provided assistance with bulk mailing and numerous other administrative tasks. We thank Lee Decker and her conference staff at ASA for planning and site activities related to the conference. Susan Ohm provided local assistance in San Antonio. The staff of the Survey Methodology Program in the Survey Research Center at the University of Michigan also provided valuable administrative support during preparation of the manuscript. Pat Dean Brick had the unenviable task of editing and reviewing all chapters before delivery to the publisher. Her contributions went beyond that of copyediting. Steve Quigley of John Wiley & Sons ensured a smooth production process.

Last, but not least, we thank our employers for providing us with the time and other support needed to organize the conference and edit the monograph.

MICK P. COUPER
REGINALD P. BAKER
JELKE BETHLEHEM
CYNTHIA Z. F. CLARK
JEAN MARTIN
WILLIAM L. NICHOLLS II
JAMES M. O'REILLY

Computer Assisted Survey Information Collection

reported of centralized telephone interviewing and CATI before the early 1980s (Collins, 1983; Porst, Schneid, and van Brouwershaven, 1994), but both have grown rapidly with telephone subscribership since that time.

In the 1980s government statistical offices in Canada, the Netherlands, Sweden, the United Kingdom, and other countries came under increasing legislative scrutiny and demands for budgetary reductions. These external pressures often prompted a close internal examination of data collection and processing methods and a readiness to consider computerized alternatives. Government statistical bureaus in these countries were among the leading developers of interactive editing and CAPI (see Clark, Martin, and Bates, Chapter 4). In the United States, federal government expenditures for surveys began to level off in the 1980s, but U.S. statistical agencies were not subject to the intense budgetary pressures experienced in Canada and Europe until the mid-1990s. Even then, the gross revenues of major U.S. survey organizations in the private sector continued to grow (Rudolph and Greenberg, 1994). CASIC emerged during a period of growing demand, initial affluence, innovation, and budgetary pressures for change in survey research.

As public and private budgetary scrutiny increased, business philosophies were developed emphasizing organizational renewal and downsizing under the rubrics of total quality management (TQM), business process re-engineering (BPR), and "re-inventing government." The success of these movements in achieving their stated objectives and avoiding unintended consequences may be debated. At a minimum, they contributed to a climate fostering change within which CASIC could be advanced as one means of achieving the promised objectives of increased productivity and quality improvement. At best, they provided a conceptual framework for the redesign of survey processes making maximum uses of computer systems and automated methods (Bethlehem, 1997).

1.2.2   Trends in Computer Technology

The develoment of CASIC methods must also be viewed as part of the broader trend of computerization of industrialized societies. The growth and development of CASIC has been, and will continue to be, highly dependent on advances in computer technology itself.

This is most evident for computer hardware. When CATI was invented in 1971, computers generally were large mainframes affordable by only large organizations. Survey research generally had to purchase computer time for sampling, tabulation, and analysis on shared mainframes dominated by other uses. Usually mainframes were not only too expensive but too slow or too committed to batch activities to permit interactive data collection with CATI. The spread of lower-cost minicomputers beginning in the 1970s permitted medium-sized survey research units to have their own computers that could be used for data collection as well as for other survey activities. The later growth of stand-alone and networked personal computers in the 1980s made many of these computerized survey functions available to small survey agencies (Rudolph and Greenberg, 1994). The production of portable laptop or note-book microcomputers in the mid-1980s permitted the transfer of the CATI technology from centralized telephone interviewing to CAPI face-to-face inter-viewing by geographically distributed field interviewers. With each major advance in computer hardware, at least one survey research agency (and typically two or more) rushed to test it for survey data collection within a year or two of its release.

The rapid growth in the speed, power, and memory of computers was equally important to survey research. Throughout much of the past quarter century, computers have reputedly doubled their speed and memory capacity for the same price every 18 to 24 months. In the 1970s and early 1980s, the response time of CATI survey items, the number of questions permitted in a CATI questionnaire, and the number of checks and edits it contained were frequently capped by affordable computer resources. These limits have largely vanished, while computing power has increased so dramatically that relatively small computers can now undertake highly complex tasks, such as voice recognition, handwritten character recognition, and multimedia interviewing at reasonably affordable costs. These trends are expected to continue.

Developments in computer operating systems and software also have had their effect on the survey field. Early use of mainframe computers and minicomputers often meant reliance on proprietary operating systems and languages. This encouraged the proliferation of home-grown CATI systems and divided survey agencies into groups by the hardware they possessed. The spread of networked, microcomputers running DOS and Novell in the 1980s spurred a wide distribution of general-purpose CATI systems, facilitated the transfer of computer assisted methods from office-based CATI to distributed CAPI, and encouraged a rapid expansion of CASIC survey functions sup-ported by general-purpose systems.

Despite these trends the CASIC field has always experienced a tension between reliance on general-purpose systems and organizational pressures to write special-purpose or home-grown systems to meet apparently unfulfilled needs. This tension is especially strong when a new generation of hardware, new operating systems, and new communication methods, such as the Internet, gain acceptance. None of the major general-purpose CAI systems is supported by a staff of more than half a dozen full-time systems analysts and pro-grammers (and some only by one or two) who must attend to discovered problems and answer user questions as well as undertake future development. When they cannot easily keep up with broad changes in the computing field, survey organizations turn away from general-purpose CAI systems to develop their own special-purpose systems.

For survey research, as for many industries in society, the future develop-ment of computing hardware, software, and communications is largely un-predictable and beyond control. New products in the production pipeline can be anticipated from computing trade journals, but as Popper (1957) observed, when the main dynamic in history is the growth of human knowledge, the future is by definition unknowable. We cannot predict with confidence what computer technology will be in 20 or 50 years, only that it will be different. Moreover the future computer technology available to survey research will depend on market forces to which the survey research industry is an insignificant element. Computing hardware, operating systems, programming languages, and electronic communications will be developed to meet military, government, business, scientific, and entertainment needs, and so survey research must choose its tools from products produced for those purposes.

The growth of computer technology also has affected survey research through the new perspectives it has brought to the work flow of survey activities. Traditional survey research was very much a batch-oriented activity. Questionnaires or interview guides were reproduced in batches, address labels or telephone numbers were assigned in batches, returned questionnaires were reviewed and key-entered in batches, and the data were then computer-edited and imputed in batches. Interactive computing made it possible to distribute cases to interviewers sequentially or on demand, to move the edit and reconciliation process from survey offices to the interview itself, and to consolidate collection, capture, and editing into fewer steps. For some, computer assisted methods can be viewed not so much as a change in survey field operations but as an extension of survey processing to the point of data collection or capture.

The successful growth of computing technology in the past quarter century produced an even more important change in perspective. It persuaded many decision makers, including those in survey research, that the computerization of paper and pencil clerical tasks typically resulted in faster, more accurate, and more efficient performance. This has been historically true not only for banking and commercial air travel but also for a variety of survey research activities, such as sampling frame construction, sample selection, data tabulation, and statistical analysis. This growing recognition of the power and value of computer technology over the last quarter century undoubtedly raised expectations that survey data collection should reap similar benefits. Faith in this belief undoubtedly motivated the growth of computer-based survey data collection, even when the anticipated benefits could not always be demonstrated in the short run.

1.3 HISTORICAL REVIEW OF KEY CASIC AREAS

1.3.1 Computer Assisted Telephone Interviewing (CATI)

CATI was the first form of computer-based survey data collection to gain broad usage and was largely developed in the United States. The concept was proposed by R. M. Gryb of AT&T, and the first CATI survey was conducted by Chilton Research Services in 1971 (Fink, 1981; Nicholls and Groves, 1986).

The phrase "computer assisted telephone interviewing" and its acronym "CATI" were coined (and copyrighted) by Gerald Glasser (1978, 1996, personal communication), who with Gale Metzger developed the second CATI system at Statistical Research, Inc. Glasser did not enforce the copyright, and "CATI" rapidly became common parlance in the field, establishing the model for subsequent CASIC acronyms. During the remainder of the 1970s and early 1980s additional market research agencies developed their own CATI systems or leased systems from others (Shanks, 1983). The first CATI systems appeared in Europe in the early 1980s and grew with European telephone subscribership and telephone interviewing (Collins, 1983; Porst, Schneid, and van Brouwershaven, 1994).

Potential key benefits of CATI, especially in market research, were identified from the start. Based on the first use of CATI, Nelson, Peyton, and Bortner (1972) reported that it reduced costs, increased timeliness, and improved data quality relative to paper and pencil telephone surveys. They did not support these claims with detailed comparative data, and CATI's potential benefits have been debated ever since (see Nicholls and Groves, 1986). CATI generally has been recognized as greatly reducing postinterview processing time by combining data collection with data capture and editing during the interview. Whether CATI speeds or slows survey design and setup and the actual pace of interviewing apparently depends on the system, the application, and the complexity of the survey. For many market research studies, CATI apparently does contribute to faster setup, making the 24-hour survey possible. The ability of CATI to reduce skip pattern errors and postinterview batch edits through computer-controlled branching and checks built into the interview has been repeatedly demonstrated (Nicholls, Baker, and Martin, 1997). Few studies have attempted to compare the overall costs of CATI with P&P interviews, but in Chapter 3, Groves and Tortora report that U.S. survey organizations only infrequently gave cost savings as a primary motive for adoption of CATI. Most chose CATI to enhance data quality or to keep pace with the survey field.

While the early market research advocates of CATI generally acknowledged its potential to increase timeliness and reduce costs, they typically emphasized that its major contribution was to allow surveys to do things that were not easily possible with P&P methods. These included the effective management of very large telephone samples, especially those with precise quotas, sequential sampling, randomization of question and response category order, on-line and automatic arithmetic, access to on-line databases, and complex forms of trade-off analyses (Dutka and Frankel, 1980; Fink, 1983; Porst, Schneid, and van Brouwershaven, 1994; Roshwald, 1984; Smith and Smith, 1980).

University survey research centers began their largely independent development of CATI in the mid-1970s (Nicholls, 1978; Palit and Sharp, 1980; Shure and Meeker, 1978). The first large-scale production use of CATI outside the commercial sector was for the 1978 California Disability Survey conducted by the Survey Research Centers at Berkeley and UCLA (Shanks, Nicholls, and Freeman, 1981). In the United States and Canada, the number of academic and nonprofit survey organizations using CATI grew from 3 in 1978 to 16 in 1987 to at least 29 in 1990 with 10 more planning to move to CATI within two years. In Europe, the first university survey organizations to use CATI were the State University of Utrecht in the Netherlands (Dekker and Dorn, 1984) and Social and Community Planning Research (SCPR) in the United Kingdom (Sykes and Collins, 1987).

University survey research centers brought their own unique perspectives to enhance the field. First, to attain the typically higher response rates expected of academic and government surveys, they focused more thoroughly on effective call scheduling and efficient callbacks of sampled telephone cases (see Edwards, Suresh, and Weeks, Chapter 15). Second, greater flexibility was added to questionnaire construction methods to accommodate the wide variety of questionnaire styles employed in academic research. Third, the range of permissible CATI interviewer actions was expanded to approximate more closely those in academic opinion surveys. Fourth, an academic survey research center, the Wisconsin Survey Research Laboratory, was among the first to experiment with the use of microcomputers for CATI interviewing (Palit and Sharp, 1983). In time, the use of stand-alone and networked microcomputers made CATI feasible for even the smallest survey organizations. Alternative approaches to CATI hardware, call scheduling, and software in small survey organizations are reviewed by Tarnai, Kennedy, and Scudder (Chapter 5).

University developers of CATI also played a major role in introducing computer assisted interviewing to the broader survey research and statistical community through professional meetings and journals. They emphasized somewhat different advantages and objectives for CATI data collection. Academic survey research centers seem to have been less successful than their market research counterparts in using CATI to reduce costs or speed the design and setup of telephone surveys. They emphasized the advantages of CATI for efficient clustered random digit dialing (RDD), for standardizing survey practice across interviewers, time, and survey organizations, and for survey documentation. They also sought to study interviewer behavior through CATI's video and audio monitoring capabilities and survey performance through automation of supervisory records. And they emphasized the importance of this emerging technology as a new tool to study and advance methodology (Freeman, 1983; Groves, 1983; Shanks, 1983). Some of the gains they hoped to achieve are now being realized by comparative studies of interviewing described by Lepkowski, Sadosky, and Weiss (Chapter 19) and by building experiments directly into CATI surveys as described by Piazza and Sniderman (Chapter 9).

While U.S. government agencies showed an early interest in the potential of CATI and supported its development with grants and contracts, they did not develop their own capabilities until the early 1980s. The first government users of CATI were the Census Bureau and the National Agricultural Statistics Service (NASS) in the United States, and Statistics Netherlands, all in 1982 (Nicholls, 1983; Tortora, 1985; Mokken, 1987). Even then, CATI remained a relatively small part of government data collection. The late entry of government statistical bureaus into CATI often reflected their commitment to geographically dispersed field interviewers (sometime organized by regional or state offices) conducting face-to-face household interviews or telephone calls from their homes. In government, CATI was often used as part of a mixed-mode strategy (e.g., Woltman, Turner, and Bushery, 1980).

By the early 1990s CATI was a well-established method of data collection in market research, university, and government research organizations in most of the developed world. Saris (1990) estimated the worldwide number of CATI installations exceeded 1000. However, CATI has not fully replaced P&P telephone interviewing, especially among the smaller survey organizations. In a 1992 study of full-service research organizations in the United States, Rudolf and Greenberg (1994) found that although fully 92 percent of these survey organizations conducted telephone interviews, only 60 percent used CATI and only 10 percent relied exclusively on CATI for their telephone surveys. Groves and Tortora (Chapter 3) report that U.S. survey research organizations still typically conduct both CATI and P&P telephone interviews. In Germany, Schneid (1991) found that one in three market research companies conducted CATI interviews in 1990.

1.3.2 Computer Assisted Personal Interviewing (CAPI)

While the development of CATI began in market research, the implementation of CAPI began in governmental statistical agencies, perhaps because government agencies were more committed to face-to-face interviews conducted by local field interviewers and because only government agencies could afford the resources to equip a large field staff with portable microcomputers.

The concept of conducting face-to-face household interviews with portable computers, and the acronym "CAPI," were well understood in the late 1970s (see Shanks, 1983). The difficulty was finding appropriate hardware to perform the task. Efforts began at both ends of the hardware spectrum. In the United States, Standford Research Institute and Response Analysis asked interviewers to carry 20 to 25 pound (9 to 11 kg) computer terminals to respondents' homes, connect them to a telephone line, and dial in a connection to a headquarter's mainframe computer (Rothschild and Wilson, 1987). Demands on the interviewers were burdensome, response times between items were unacceptable, and connections often were lost before the interview was completed.

In Europe, small handheld computers were initially tested for CAPI, first by Statistics Sweden in 1982 and then by Statistics Netherlands in 1984 (Danielsson and Marstad, 1982; Bemelmans-Spork and Sikkel, 1985). Although handheld computers were easily portable and had sufficient battery power to operate for long periods, they did not have sufficient speed, memory, and display areas at that time for more than the simplest data collection tasks, such as recording the prices of goods in stores. Statistics Sweden contracted for the development of a microcomputer specifically designed for a decentralized field staff which "should be handy, reliable, light-weight, and ergonomically sound,... able to accommodate four or five surveys simultaneously and to store at least one day's field work without recharging" (Lyberg, 1985). This device was to be used for face-to-face interviewing, telephone interviews from interviewers' homes, and data entry tasks in the field. Although a prototype was produced, this effort was superseded by the emergence of commercially available portable laptop microcomputers in the mid-1980s. While not meeting all of Statistics Sweden's requirements, off-the-shelf alternatives made independent manufacture for survey research difficult to justify.

The first national survey to use CAPI, the Dutch Labor Force Survey, began in January 1987 using a small portable computer running under CP/M (van Bastelaer, Kerssemakers, and Sikkel, 1988). In the same year the first U.S. national survey to employ CAPI for at least part of its data collection, the Nationwide Food Consumption Survey, was conducted by National Analysts (Rothschild and Wilson, 1988). A variety of CAPI field tests also were conducted in that year for the French Labor Force Survey (Bernard, 1990; Heller, 1993), the U.S. National Health Interview Survey (Nicholls, 1988), and by Research Triangle Institute (RTI) (Sebestik et al., 1988). The laptop computers available for CAPI in these early applications were often so limited by weight (up to 15 lb or 7 kg), screen visibility, capacity, and battery life that they presented sufficient operational challenges to raise concerns about the health of interviewers. These problems were solved by applied research, improved hardware, and experience (Couper and Groves, 1992).

From 1988 to 1994 the use of CAPI grew rapidly. After periods of testing, the British, Canadian, French, Swedish, and U.S. government labor force surveys were moved to CAPI (or a mixture of CAPI and CATI) while various organizations in the private sector in the United States and United Kingdom undertook their first major CAPI surveys. Most of these large organizations are now committed to moving all their face-to-face interview surveys to CAPI or have already done so. In the broader survey community, Rudolf and Greenberg (1994) found that only 23 percent of U.S. full-service survey organizations were using CAPI as one of their data collection methods in 1992. This was less than a third of those full-service survey organizations who conducted face-to-face interviews.

The benefits anticipated in moving face-to-face surveys from P&P to CAPI have varied from organization to organization. Cost savings were important motives and reportedly realized by Statistics Netherlands and the U.K. Office for National Statistics (ONS) (Martin and Manners, 1995). A reduction in the postinterview processing burden was critical for several surveys (Rothschild and Wilson, 1987; Baker, Bradburn, and Johnston, 1995; Sperry, Bittner, and Branden, 1991; Heller, 1993). Initial CAPI hardware and training costs were at least partially offset with savings in processing. The opportunity CAPI provided to employ dependent interviewing, more complex questionnaire routing, and other advanced interviewing techniques to improve data quality were key objectives for the U.S. Current Population Survey and for the Canadian Survey of Labour and Income Dynamics (Dippo et al., 1992; Grondin and Michaud, 1994). And all users anticipated time savings between the end of field work and the release of estimates or the next round of panel interviewing. In their study of U.S. academic and commercial survey research organizations, Groves and Tortora (Chapter 3) find that data quality goals were apparently more important in CAPI adoption than cost-saving goals and more likely to be realized.

In the first CAPI surveys, the interview software was preloaded on the CAPI laptops at headquarters or sent to the field staff on floppy disks. Completed interview data were returned on floppy disks, usually by mail. Interviewing assignments and reports of progress were handled by P&P forms, telephone calls, or disk exchange. Efforts to move all these transactions between interviewers and headquarters to modem-based telecommunications were pioneered by RTI (Sebestik et al., 1988; O'Reilly et al., 1989), and refined at other organizations (e.g., Spiezer and Doughterty, 1991; Hofman and Keller, 1991; Nicholls and Kindel, 1993).

Once decentralized field interviewers were equipped with laptop or notebook CAPI microcomputers, they could use the same equipment for telephone interviews conducted from their homes (see Bergman et al., 1994). Generalized CASIC case management systems have been designed to move cases between CAPI and CATI (both centralized and decentralized) for multimode surveys or as data collection needs and resources demand. Recent developments and future prospects for CASIC case management and communications systems are discussed further by Connett (Chapter 13), Hofman and Gray (Chapter 14), and Smith, Rhoads, and Shepherd (Chapter 16).

While the laptop or notebook microcomputer has currently become the hardware standard for CAPI surveys conducted by government statistical bureaus, alternative hardware may be more suitable for specialized applications. The U.S. Bureau of Labor Statistics (BLS) plans to use pen-based (rather than keyboard) microcomputers for CAPI implementation of its Consumer Price Index Survey (see Bosley, Conrad, and Uglow, Chapter 26). At A. C. Neilsen, microcomputers are equipped with bar code scanners to read both the UPC codes of products purchased by sampled houlseholds and individually bar-coded response categories to survey questions. New uses also are being found for advanced handheld microcomputers, designed as "personal digital assistants" or PDAs. These have smaller screens and less computing power than a laptop, but since they are much lighter, more durable, less expensive, and have a much longer battery life, they seem especially well suited for relatively simple surveys conducted in difficult field circumstances. PDAs have been used for data collection on land use, soil characteristics, and similar natural resource data (Nusser, Thompson, and DeLozier, 1996) and have been recommended for population and health studies in developing nations (Foster and Snow, 1995).

1.3.3   Computerized Self-Administered Data Collection

In self-administered forms of data collection, respondents read the survey questions and record their answers by themselves. Traditionally paper questionnaires or forms are used, which are mailed to respondents and returned in similar fashion. Paper questionnaires also may be administered to groups of respondents in an office or classroom setting or handed to respondents during face-to-face interviews. Many computerized forms of self-administered data collection have evolved, and they are known by a variety of overlapping names and acronyms. Their variety reflects the many uses of paper questionnaires in survey research and the multiple origins of their computerized alternatives.

The *electronic questionnaire* has a long history, beginning with the use of computers to obtain medical histories and administer psychological tests. The concept of having respondents read questions from a computer display and enter their own answers on a keyboard was used in university medical settings even before CATI was invented (Evan and Miller, 1969). Major uses in survey research began in the early 1980s both in the United Kingdom (Duffy and Waterton, 1984) and in the United States (Kiesler and Sproull, 1986). In these early studies the electronic questionnaire was sometimes seen as an alternative to a P&P face-to-face interview and sometimes as an alternative to a mailed paper questinnaire.

Ramos, Sedivi, and Sweet (Chapter 20) describe the history, status, and future prospects of *computerized self-administered questionnaires* (CSAQ). This technology is the most direct analog of the mail-out, mail-back P&P questionnaire. In CSAQ the electronic questionnaire may be mailed to the respondent on a floppy disk for a *disk-by-mail* (DBM) survey. Or it may be transmitted by modem or e-mail to the respondent's computer or terminal for an *electronic mail survey* (EMS). The respondent then installs the questionnaire on his or her computer, answers the questions, and returns the completed answers to the research organization by disk or modem. The CSAQ approach has been used both in academic and commercial surveys of professional populations and in government surveys of industrial and business establishments. Production use in government began at the U.S., Energy Information Administration (EIA) in 1988. The special advantages and difficulties of extending the CSAQ concept to the Internet and World Wide Web are also described by Clayton and Werking (Chapter 27).

A major weakness of CSAQ data collection is that it can be used only with respondents who have access to a PC with the appropriate configuration and operating system. This greatly inhibits its use for general population samples. One way to overcome this coverage bias is to select a sample in the normal way and then provide a computer to those respondents who do not have one (see Saris, Chapter 21). The computer also serves as an inducement for the respondent to answer questions downloaded at regular intervals. This approach has been effectively implemented with European consumer panels, beginning with the Netherlands Gallup Poll in 1986.

Another form of computerized self-interviewing represents an extension of CAPI. During a CAPI interview, the laptop can be turned around to face the respondent who then reads the survey questions from the screen and keys in his or her own answers. This is equivalent to the self-administered question-naire (SAQ) sometimes used in P&P face-to-face interviews to ask sensitive questions the respondent may feel reluctant to answer to the interviewer or within hearing of other household members. This data collection method has been called *computer assisted self-interviewing* or CASI.

In an important variant of CASI, called *audio-CASI* or ACASI, the respondent wears an audio headset, and the survey questions are not only displayed on the computer screen but read to the respondent through the headset in a computer-controlled digitized voice. Each answer entered on the keyboard triggers the next question to be displayed and read. (CASI without the audio component is sometimes called *video-CASI* or VCASI.) ACASI provides maximum privacy to the respondent for sensitive topics and also helps overcome respondent illiteracy. These methods are described in more detail by Tourangeau and Smith (Chapter 22) and Turner et al. (Chapter 23). A telephone version of ACASI, called T-ACASI, also is discussed in Chapter 23.

All forms of computer assisted self-interviewing described thus far ask the respondent to operate a computer. *Touchtone data entry* (TDE) only requires respondent access to a touchtone telephone. In common applications, respon-dents call a toll free number at the data collection organization. The system reads voice-digitized survey questions to the respondent who enters his/her answers using the telephone keypad (Clayton and Harrell, 1989). First produc-tion use of TDE was for the BLS Current Employment Statistics (CES) survey in 1987. It has since been employed in other business and establishment surveys. TDE also has been tested for surveys of the general public (Frankovic, Ramnath, and Arnedt, 1994; McKay and Robinson, 1994). TDE is functionally equivalent to telephone ACASI described above, except that in TDE the respondent calls the survey organization while in T-ACASI a survey inter-viewer makes contact with the respondent before turning the interview over to the computer.

Voice recognition entry (VRE or VR) is similar to TDE but accepts spoken rather than keyed responses. A computer reads voice-digitized survey questions to the respondent, analyzes the spoken answer, and records its recognized meaning. Recognition failure may prompt a repeat of the question or a related probe. Depending on the sophistication of the system, respodents may be asked to answer with a simple Yes or No, single digits, continuous numbers, spoken words, or spelled names. Small vocabulary voice recognition (Yes, No, and digits) may serve as a backup or replacement for TDE in call-in, panel surveys of business establishments. It has been used for this purpose in the CES since 1992 (Clayton and Winter, 1992). Large vocabulary VR may become an alternative to call-in CATI interviewing for consumer panels (Blyth and Piper, 1994; Blyth, 1997). It also was tested as one of several means of answering the year 2000 U.S. Decennial Census (Appel and Cole, 1994).

1.3.4 Modern Data Capture and Data Preparation

In traditional survey processing, the answers to survey questions are recorded on paper forms which then are clerically reviewed, key-entered, and batch-edited to identify errors, omissions, and inconsistencies. Problem fields are resolved by clerical reconciliation to the original paper forms, by recalls to respondents, or by referral to content experts. The corrections are then rekeyed and the process repeated until all editing criteria are met. Alternately, at some stage batch imputation programs are used to assign acceptable values to problem fields.

Modern computing methods provide alternative ways to perform these tasks. Data collection, data capture, and at least some data editing can be combined ito a single process before the data leave the field. Survey data entered on paper forms may use *computer assisted data input* (CADI), also called *computer assisted data entry* (CADE), to capture and edit the data in a single integrated step, case by case, much as a CATI or CAPI interview does.

One round of data capture and editing is sufficient for some surveys. For others, the data are passed through additional stages of review, editing, and preparation prior to analysis. This may be accomplished either with batch programs or case by case in an interactive environment using *computer assisted data editing* (also CADE). CADE systems may be used by clerks or by content experts. The distinction between CADI and the two meanings of CADE is primarily one of degree and whether the data have already been captured. All use essentially the same software employed for CATI and CAPI but may employ more edits and reference databases. These new approaches to data capture and editing were pioneered for survey research by Statistics Netherlands in the mid-1980s (Denteneer et al., 1987).

The difference between traditional and CASIC approaches to data capture/editing has diminished with time. Standard key-entry software now has features to permit interactive, case-by-case editing, while CADI and CADE programs can now typically operate in batch mode with batch controls. The optimal present strategy is to find the best combination of key entry, CADI, CADE, and batch editing and imputation to meet a survey's needs (see Bethlehem and Van de Pol, Chapter 11).

Survey coding is the process of assigning standard alphanumeric codes to the text answers of survey questions on such topics as city of residence, occupation and industry of employment, and the products of a business or industry. Traditional coding has been performed by clerks who review each handwritten answer and choose the appropriate code based on rules and examples provided in a large classification volume. Automation of this process has been encouraged by the capture of text answers in machine-readable form with CASIC methods. *Automated coding* uses batch computer programs to compare survey responses with the entries of computerized reference dictionaries. *Computer assisted coding* visually displays text answers on a clerk's computer screen while providing on-line access to reference materials and suggested codes. These two methods often are used sequentially. These topics are further discussed by Spiezer and Buckley (Chapter 12).

Optical scanning of self-administered paper forms is a relatively old technology. It "reads" answers on a form with a moving electronic beam and light sensitive devices. The U.S. Census Bureau used *optical mark sensing* of filled circles for the 1960 through 1990 decennial censuses. Education and psychological testing services have employed mark sensing to score standardized tests for decades. Although scanning and mark sensing have generally been less expensive than data keying for mass data capture, survey applications have traditionally been limited by their volume requirements for cost efficiency, by the long lead times for design, by the stringent printing tolerances of scannable forms, and by the limitations scanning placed on questionnaire length, formatting, and folding for mail distribution.

In the last decade advances in scanning, in *document imaging*, and in *optical character recognition* (OCR) have removed many of the restrictions of mark sense data capture and increased the accuracy with which both printed and handwritten characters can be scanned and recognized. They also have developed improved methods of detecting and repairing instances of failed recognition. Blom and Lyberg (Chapter 25) describe new uses of scanning and OCR by government statistical bureaus, while Dillman and Miller (Chapter 24) consider the advantages and disadvantages of this technology for academic surveys of small to moderate size.

Electronic data interchange (EDI) is one of the most challenging data capture strategies developed in recent years. Its goal is to obtain economic data directly from the computer records of sampled businesses by an automatic electronic transfer of information to a government statistical bureau, eliminating the need for a survey. The Census Bureau has approached this goal by enhancing the American Standard (X12) EDI standard messaging format already in use for transfers of information between business partners to include the transfer of statistical data to government agencies (Ambler and Mesenbourg, 1992; Nicholls and Appel, 1994). Statistics Netherlands is approaching the same goal by encouraging national standards of administrative and statistical concepts and by seeking to obtain the needed information through intermediaries, such as the producers of accounting packages and business service bureaus (Keller, 1994). This strategy, if successful, may make many traditional business surveys obsolete.

1.3.5 CASIC: A Summary Definition

The acronym CASIC was coined in 1988 by the self-named CASIC Subcommittee of the U.S. Statistical Policy Office's Federal Committee on Statistical Methodology (1990b). They defined CASIC as "those information gathering activities using computers as a major feature of the collection of data from respondents, and in transmitting of data to other sites for postcollection processing." Their intention was to introduce CASIC as an umbrella term encompassing both the familiar collection technologies of computer assisted telephone and personal interviewing (CATI and CAPI) and various forms of computer-based self-administered data collection (CSAQ, TDE, VRE, etc.).

The CASIC Subcommittee also coined the acronym CASI to cover all forms of self-administered data collection using computers. This was something of a misnomer since many of these technologies were developed as alternatives to P&P questionnaires rather than to interviewing. The CASI acronym now has at least three common meanings in the literature: (1) as any computer-based form of self-administered data collection, including TDE and VRE, (2) as any form of self-administered data collection where the respondent operates a computer, and (3) as a specific type of data collection where the respondent operates a computer with an interviewer present. CASI remains the most ambiguous and inconsistently used acronym in the field.

Although the CASIC Subcommittee emphasized CASIC as a term applicable to survey data collection, they recognized that it encompassed the use of computers "in transmitting of data to other sites for postcollection processing" and "in the conversion of data to proper formats." Later use of the same term by the U.S. Census Bureau and others have expanded its coverage: (1) to all computer-based methods of data capture and data preparation as described above, (2) to activities preparing for, supporting, and managing computer-based data collection and capture, and (3) to the tasks of building interfaces to their later steps in the survey process.

For this volume, CASIC is defined as the use of computers for survey data collection, data capture, and data preparation and for related activities preparing for, supporting, managing, and coordinating those tasks with each other and with later stages of the survey process.

Since the CASIC acronym originated in the U.S. government, it is best known in North America. In Europe, the phrase *computer assisted data collection* (and its acronym CADAC) is more familiar (Lyberg, 1985; Saris, 1990). Manners (1992) has proposed the phrase *computer assisted survey methodology* (CASM) as a more general rubric for all computer assisted steps in the survey process. Unfortunately, this acronym is confused with *cognitive aspects of survey methodology* (also CASM; see Jabine et al., 1984), predating Manners's use of the term. We thus use the term CASIC in this volume, following the definitions given above.

1.4 CONSEQUENCES OF CASIC FOR SURVEY RESEARCH

The final section of the chapter briefly explores the consequences of CASIC for survey research in two areas: (1) changes in the performance characteristics of surveys using CASIC and (2) changes in the survey process and in survey organizations.

1.4.1 Changes in Survey Performance Characteristics

When assessing the performance characteristics of CASIC surveys relative to those of P&P methods, three considerations must be borne in mind.

First, the effects of CASIC methods for individual surveys will partly depend on what their survey managers hope to achieve with them. Some managers are primarily interested in CASIC's potential to reduce the costs or increase the timeliness of surveys over traditional P&P methods. Some may choose CASIC methods to enhance survey data quality in any of several ways, with costs as a secondary consideration. Some may turn to CASIC as a means of collecting data of a range, detail, depth, or type not possible with P&P methods. And some may have other objectives in mind, such as a standardization of survey practice or the ability to accommodate surveys of greater complexity. Saris (1990) observed that CASIC has the power to reduce costs, increase timeliness, and improve quality but rarely all three at the same time. The same principle applies when a longer list of potential CASIC benefits is used. One cannot expect to maximize all potential benefits at once.

Second, the advantages and disadvantages of CASIC will vary among the diverse technologies encompassed within this broad umbrella term. For example, Weeks (1992) concluded that computer assisted interviewing when appropriately used often enhanced survey data quality in comparison with P&P methods but did not necessarily reduce survey costs. By contrast, Nicholls, Baker, and Martin (1997) concluded that in busines surveys the substitution of TDE for P&P questionnaires or telephone interviews often reduced costs without affecting data quality.

Third, other elements of the survey design and the type of application also may change CASIC's relative performance characteristics. Dillman and Miller (Chapter 24) suggests that scanning and OCR may save money relative to traditional data keying in large surveys but not at the present time in small ones. Because survey applications in academic, commercial, and government research often differ greatly in their designs and objectives, results of a move to CASIC in one of these areas may not apply to the others. Partly for this reason, generalizations about CATI and CAPI presented in Section 1.3 were set in the historical development of these separate areas.

Even with these cautions in mind, we can offer a few generalizations about CASIC performance characteristics that appear to have broad applicability.

First, CASIC methods generally decrease the time between data collection and data tabulation. Although few controlled studies are available to demonstrate the point, the anecdotal evidence is compelling. Nicholls and De Leeuw (1996) found that increased timeliness was the most frequently given reason for adopting CATI and CAPI for specific studies mentioned in the literature. Gains in postinterview timeliness in some application areas may come at the expense of increased pre-interview preparation time.

Second, cost savings are not a common outcome when P&P interview methods are replaced by CATI or CAPI, although reduced costs have been demonstrated with a sufficient volume of usage or appropriate organizational restructuring (see Palit and Sharp, 1983; Weeks, 1992; De Leeuw and Hox, 1995; Martin and Manners, 1995). Groves and Tortora (Chapter 3) report that cost savings typically were not a prime motive for survey organizations to move to CAI, and most organizations did not think that cost savings were realized.

Third, the belief that appropriately used CAI methods improve survey data quality is both widely held and supported by a substantial body of research (Groves and Nicholls, 1986; Groves and Tortora, Chapter 3; Weeks, 1992; De Leeuw and Hox, 1995). However, as Nicholls, Baker, and Martin (1997) have observed, while "… promised data quality gains have now been demonstrated by major reductions in item nonresponse and postinterview edit failures," with a few significant exceptions other data quality gains remain "unsubstantiated by controlled studies," CASIC data quality gains, especially reductions in measurement error, have been elusive and not as large as early CASIC proponents anticipated.

Uses of CASIC methods that go beyond emulation of P&P surveys to employ survey procedures not easily possible with P&P methods offer greater promise for the future. They include (1) the improved reporting of sensitive information and the reduction of interviewer contributions to variance through the use of audio-CASI, as reported in Chapters 22 and 23, (2) the reduction of respondent memory errors in panel surveys through the use of controlled dependent interviewing as described by Dippo et al. (1992) and by Brown, Hale, and Michaud (Chapter 10), (3) the reduction or elimination of common order effect biases in questionnaire design through the use of randomized item and response category order (e.g., Piazza and Sniderman, Chapter 9), and (4) reductions in survey biases arising from researcher preconceptions through the use of individualized interviewing techniques as reported by Sikkel (Chapter 8). Whether such advances presage broader gains against common forms of survey bias and error through CASIC methods remain to be seen (see Nicholls, 1997).

1.4.2 Changes in Survey Process and Organization

The effect of CASIC on the broader survey field may be at a fairly early stage. The consequences of technologically driven change, as Schnaars (1989) has emphasized, are rarely predictable in advance and typically occur more slowly than anticipated. Different CASIC technologies also may bring different, or even opposing, changes to the field. The introduction of CATI encouraged the spread of centralized telephone interviewing and some decentralization of the survey industry because CATI telephone operations could be set up at relatively low cost with few survey professionals. The spread of the newer and more costly forms of CASIC, especially CAPI and ACASI, may have the opposite effect, especially occurring at a time of declining telephone response rates and the growth of multinational surveys (see Blyth, Chapter 28). Only the largest, or best capitalized, survey organizations may have the financial resources to establish and maintain these forms of data collection over large geographic areas. Survey organizations with fewer resources may not survive the transition or may become specialists in the least expensive forms of CASIC, such as surveys conducted through the Internet and World Wide Web.

The evolution of competitive CASIC technologies may also complicate the choice of basic survey collection modes. Advances in dependent interviewing, use of on-line and remote administrative databases, more complex branching and customization, growing opportunities for randomization, and the increasingly rapid turnaround of panel surveys suggest a strengthening and growth of traditional interviewing, especially for complex topics where the assistance of trained interviewers is critical. But concurrent advances in self-administered forms of CASIC, including the special benefits of ACASI and EDI and the prospects of low-cost collection through the Internet may suggest to some a decline in the survey interviewer's role and perhaps in survey interviewing. While both these trends imply an obsolescence of P&P questionnaires, whether completed by interviewers or respondents, advances in scanning and OCR may well extend the life of P&P methods indefinitely into the future. Which modes and CASIC technologies will dominate the future will probably depend both on their technological capabilities and on their ability to satisfy the traditional performance criteria of survey research: costs, timeliness, coverage, unit and item response rates, and measurement accuracy.

Whatever specific forms of CASIC prove most successful in the future, CASIC is clearly bringing fundamental changes to the field of survey research. To date progress has been uneven across technologies and organizations. While much progress has been made on technology issues (hardware and software), we have been hard pressed to keep up with the change on the human side of the transition. This means changing the skills, tasks, and tools of users to make optimal use of the new technologies. While technological developments in CASIC are likely to continue apace, the human side of the process is demanding increasing attention.

One central belief of the CASIC perspective is acceptance of computer technology as an integral part of the entire survey process. This is not just data collection (where we have made much progress in recent years) but also processing, dissemination, management, and the like. We are coming to realize that CASIC facilitates an integrated view of data collection and processing. Previous distinctions between different stages of the survey process are now fading. For example, editing can now be done during data collection rather than after capture or keying. Decisions about data set construction and documentation (metadata issues) are made during instrument design rather than at the end of the process. As a result of the adoption of new processes and procedures, the management of the survey process changes. We are developing new approaches to training interviewers and to supervising and evaluating their work. The ability of CASIC methods to provide a wealth of real-time data on the progress of a study needs to be harnessed and managed to serve the goals of efficiency and quality. This will require organizational changes, retraining, new hardware and software systems, and so on.

We are coming to the increasing realization that CASIC represents a whole new way of doing business. Within the next few years we expect to see the other side of the change occurring. This stage in the development of CASIC will be characterized by some of the following features: (1) increasing integration of the entire survey process, (2) organizational restructuring to adapt to the demands of the new systems, (3) a greater focus on usability or interface issues, (4) development of tools to aid the work of designers, testers, and others in implementing CASIC surveys, and (5) development of management tools designed to make use of information produced by CASIC systems and operations.

However, it is apparent that while we need to consolidate our gains and our knowledge and identify key gaps for further work, developments both internal and external to the survey industry will produce pressures for even faster change. Developments in computer hardware and software, for example, are occurring so rapidly that survey organizations barely have a chance to stabilize one platform or system before upgrades are needed. At the same time, demands from users and producers of survey data are increasing as the potential of the new methods becomes apparent. We believe that managing this change will be one of the key challenges facing the survey industry over the next few years. Merely keeping pace with technology change may demand an increasing proportion of scarce resources.

The chapters in this volume demonstrate that we have come a long way in the last few years. We are witnessing a fundamental transformation in the process of survey data collection. This monograph also makes clear that there is much we do not know, for example, (1) about how best to apply the technologies to the survey endeavor, (2) about what changes (organization, roles, skills, tools, methods, training, etc.) to make to accommodate the technological advances we are witnessing, and (3) about the effect of these changes on all aspects of the survey world, including the societal contexts within which we conduct our work. While we must press forward to explore new methods and develop innovative applications of new technologies, we must temper this with carefully considered research and evaluation, and ferret out the implications of these changes on all aspects of the survey process. These are truly exciting times for survey research.

CHAPTER 2

Diffusion of Technological Innovation: Computer Assisted Data Collection in the U.K.

Martin Collins and Wendy Sykes
City University Business School

Colm O'Muircheartaigh
London School of Economics and Political Science

2.1 INTRODUCTION

In this chapter we chart the adoption in the United Kingdom of computer technology as an aid to the collection of survey data in interviews carried out face to face (CAPI), by telephone (CATI), and through self-completion interviews (CASI). Our aim is broad: to describe changes over time in the pattern of investment in and applications of facilities for computer assisted interviewing (CAI) — both in the U.K. research industry as a whole and within organizations; and to explore the processes underlying these developments.

Our analysis borrows from theory arising from research into the diffusion of technological innovation. This, we find, provides a useful framework for thinking about the changes that have taken place within the U.K. in respect of CAI, as well as those that are yet to come.

2.2 CAI INNOVATION AND THE U.K. CONTEXT

The use of computers on any scale as part of the survey interview process is a relatively new phenomenon in the U.K. It postdates by some years both

Computer Assisted Survey Information Collection, Edited by Mick P. Couper, Reginald P. Baker, Jelke Bethlehem, Cynthia Z. F. Clark, Jean Martin, William L. Nicholls II, and James M. O'Reilly.
ISBN 0-471-17848-9 © 1998 John Wiley & Sons, Inc.

recognition by U.K. researchers of the potential of CAI and the development of practical systems for realizing that potential. Examples of limited applications of CAI — in all three forms with which we are concerned — could be found in the U.K. in the 1970s. But it was the 1980s that saw the first significant breakthroughs in routine use on a large-scale, associated with twin revolutions in technology and industry outlook.

The broad background to our research is the study of the diffusion of technological innovation: the way in which new technologies are taken up within a social system — an industry or a society — over time. Such study has been concerned with ways of measuring adoption, with general patterns over time and, centrally, with the factors that foster or inhibit progress. In an industrial context, diffusion research may be undertaken to build basic understanding, to inform the marketing and promotion of new products, or to assist in predicting the success and likely effects of a new technology (Mahajan and Peterson, 1985).

A key conception that informs our consideration of diffusion is *resistance*. Those involved intimately with new developments tend often to assume that all that is needed is an innovation, and its diffusion will be not only inevitable but swift. In practice, however, a number of barriers must be removed for an innovation to be accepted.

Survey research management draws on the standard management and organization theories common to the commercial and industrial sectors. One of the most important of these is *scientific management*, the implementation of which still features the principles expressed in the seminal work *Principles of Scientific Management* (Taylor, 1911). Characteristically these involve (often extreme) *specialization* of tasks, the definition of a *best way* of carrying out a task, and *division* of work and responsibility such that management is responsible for planning work methods and principles, and workers are responsible for executing the work accordingly. Of these principles, potentially the most restrictive is the last — the separation of thinking from doing. Innovations, by contrast, usually involve the redistribution and recomposition of tasks, either of which may threaten the stability or even the viability of an existing organization.

The second barrier to innovation is functional dependency of thinking. That is to say, those who work with a particular methodology tend to view the activity in terms of the functions as they are currently carried out. An innovation may challenge that structure and therefore threaten to isolate or render redundant those currently in positions and functions of importance. A third issue that complicates the picture is the change that occurs in the innovation itself as it is being diffused; a successful innovation adapts to the environment. Thus what we now consider standard CAI is very different in effectiveness, power, and cost from earlier manifestations.

These factors will play a greater or lesser role in encouraging or suppressing innovation to the extent that the status quo is affected by the innovation. In plotting and explaining the diffusion of CAI in the U.K., we therefore find it necessary to look separately at CATI, CAPI, and CASI as well as to explore the common ground. Thus, while "cyberhobia" is a general theme in the history of CAI, other factors have operated differentially as incentives or barriers to adoption of particular technologies.

Especially marked is the distinction between CATI, on one hand, and CAPI/CASI, on the other. With CATI, early adoption in the U.K. faced the barrier of distrust of telephone interviewing itself (Collins, 1983), and the injection of computer technology probably helped to lower that barrier. CAPI and CASI, by contrast, represent the addition of computer technology to familiar and trusted modes of data collection; barriers relate to the technology, not to the mode.

Here we see a marked difference with the United States where it was CATI that called for major changes to an accepted methodology (see Groves and Tortora, Chapter 3; Tarnai, Kennedy, and Scudder, Chapter 5), while the adoption of CAPI has meant a return to face-to-face interviewing. This and other "cultural" differences between Europe and the United States are discussed further by Blyth (Chapter 28).

One key distinction made in diffusion research is between *inter*organization diffusion, where the focus is on the pattern of adoption among the different organizations making up an industry, and *intra*-organization diffusion, where the focus is on the way in which the use of a new technology spreads through individual organizations. An organization may adopt a technology but not necessarily use it for all possible applications. In a study of "no-till" agriculture in the United States, for instance, Carlson and Dillman (1988) found that whereas volume of mechanical activity on a farm (essentially size) was the primary influence on introduction of the practice (interorganization diffusion), the level of mechanical sophistication of the farmer was more important for adoption (intra-organization diffusion). Our findings suggest that both phenomena have to be considered in respect of CAI and that they cannot be treated as independent.

A very basic area in which this distinction arises is that of measuring the current level of adoption of a given new technology in a given industrial context. The level of diffusion may be measured in several ways. One is simply the proportion of organizations that use the new technology at all, which is the most obvious measure of *inter*organization diffusion. An alternative measure is the share of industry output made with the aid of the new technology. This will usually confound inter- and intra-organization diffusion. Our findings on CAI suggest that this confounding is not useful and rather that the "volume" measure used must be broken down into two component parts: the proportion using the technology (or penetration of technology) and their average output volume (or intensity of use).

Perhaps the most obvious illustration of the need for distinguishing between the two forms of diffusion is a basic conclusion derived from studies of the diffusion of technological innovation: that the level of adoption over time tends to assume a rising sigmoid (S-shaped) curve, namely a curve that starts slowly, accelerates into rapid growth, and levels off as the process of diffusion nears completion. Much research is characterized by attempts to understand the influences that give rise to this S-shaped diffusion curve (e.g., see Rogers and Shoemaker, 1971).

Indeed, if we look at the simple measure of technological penetration — interorganization diffusion — we find some empirical evidence of the early stages of an S-shaped curve, with all three CAI modes showing a point of inflexion — the acceleration of diffusion — some five years after mainstream introduction. But now, when we might expect to see further acceleration, we find quite steady upward trends. As we will discuss in Section 2.4, we believe that a measure of volume of use (not available to us) will show further acceleration, arising from increased use in adopting organizations — intra-organization diffusion. (Clearly we will have to wait some time to see the second point of inflexion as adoption of the technology approaches saturation.)

Looking at the current position of inter- and intra-organization diffusion, we find some differences between these modes. With CAPI and CASI, we see quite low penetration of the technology *and* quite low intensity of use in the average organization using the technology. Our respondents in such organizations expect rapid increase in their intensity of use and point to a number of economic and organizational factors to justify that expectation (which is also supported by past performance in government-sponsored survey research). Here is some evidence of a technology in the early stages of adoption, with a few brave experimenters and even fewer totally committed users. We can expect in the near future new users and more intensive use by existing users, that is, a distinct acceleration in any measure of volume of use.

The more mature technology of CATI has already seen a fairly steady rise in penetration as earlier adopters greatly increased their intensity of use of the technology. From this juncture further growth is limited. We can expect penetration growth to continue but without parallel growth in intensity of use, so the growth in volume will slow down. In other words, the final phase of the S-shaped curve will be more imminent.

Historically, as Nabseth and Ray (1974) show, the early stages of diffusion of a new technology has tended to be dominated by interorganization diffusion. Intra-organization diffusion becomes important later as organizations that have made the transition begin to acquire new capacity to increase production, further replace existing technology, and extend their use of the new technology to other areas. We also note in passing a close parallel with the work of Ehrenberg (1969) and others in research into buyer behavior, where some insight has been gained by similarly breaking down sales volume into the two components of penetration and average weight or frequency of purchase per buyer. The most pertinent conclusion from such research is that big brands (cf. well used technologies) are differentiated from small brands (cf. little used technologies) by the number of people who buy them and by the average frequency of purchase of those buyers. A second, related conclusion is that some specialist "niche" brands can prosper because of an exceptionally intensive use by a rather small number of buyers (Collins, 1971). This may also be true of some specialist applications of technology in survey data collection (e.g., speech recognition; see Blyth, 1997).

Despite the generalizable features of the diffusion curve, substantial variation in its shape has been observed for different innovations and social systems (see Rogers, 1962, and later editions). Research — conducted both retrospectively and prospectively — includes attempts to describe and explain this variation, for example, across a range of innovations in a given context and or within different social systems (e.g., countries).

Description of the diffusion of new technologies centers on features such as the rate of adoption, the characteristics of adopters and nonadopters at different stages of the process, and the ceiling to which the diffusion curve tends. *Explanators* (Mahajan and Peterson, 1985) may be looked for in terms of three basic sets of factors associated with the innovation itself, the industry or culture to which it is introduced, or the ways in which information about the innovation is communicated. We are only following in these footsteps.

2.3 DATA SOURCES

2.3.1 Yearbook Estimates

In our analysis of the diffusion of CAI in the U.K. we have used three sources. The first is the *1996 Yearbook of the Market Research Society*, which includes a listing of commercial survey research suppliers. Among the details submitted by entrants to this listing is their ability to offer CATI and CAPI data collection. Not all of these claims, especially among smaller suppliers, reflect the existence of in-house facilities; many will be subcontracting to larger suppliers or specialist fieldwork organizations. Nevertheless, we see these claims as valid indicators of the level of enthusiasm — adoption in a qualitative sense — in the U.K. market research industry.

In the yearbook we identified 200 commercial organizations, which operated from and in the U.K., carried out quantitative surveys, and had their own interviewing resources. To these we added two unlisted suppliers of survey research — Social and Community Planning Research (SCPR) and the Social Survey Division of the government Office for National Statistics (ONS) — the only major noncommercial suppliers of nationwide survey research in the U.K. (We should draw attention to this pattern of concentration of activity in commercial market research companies, general in Europe but different from the pattern of activity in the United States discussed later by Groves and Tortora, Chapter 3.)

2.3.2 Survey of Suppliers

In order to probe beneath these surface claims, we carried out a mail survey of research suppliers. This was directed at the same population of 202 organizations described above. Questionnaires, which were addressed to the most senior identifiable individual, were sent to a sample of 86 organizations:

- All 24 of the large organizations, with annual turnover or fee income of £5 million (approx. $8) or more.

- 34 (1 in 2) of the 67 medium sized organizations, with annual turnover of at least £1 million (approx. $1.6m) but under £5 million.

- 28 (1 in 4) of the small organizations, with annual turnover of under £1 million.

Six of the replies we received indicated that the organizations were not eligible. Two carried out only qualitative research, one operated only overseas, and three used interviewing resources within the same group of companies. After a second questionnaire mailing and a telephone follow-up, completed questionnaires were received from 53 of the 80 organizations remaining eligible, representing a response rate of 66 percent.

The questionnaire was largely focused on collecting factual information on CAI: the existence of *in-house* facilities for CATI, CAPI, and CASI; the year of installation and source of capital funding; the use of the facilities, in terms of both clients and survey types; the hardware and software in use; and the number of interviewers trained in its use. We also asked respondents to report in similar terms on their first year of operation of the facility.

The only nonfactual questions asked respondents to categorize themselves and their organization in terms of openness to innovation and about the level of commitment at the time of installation — was the organization committed from the beginning to full-scale implementation, was it confident of such implementation but still testing plans, or was the installation only a test of feasibility? Where the organization reported no CAI facilities of any kind, questions were asked about the use of subcontract facilities and broad intentions for the future.

2.3.3 Depth Interviews

Respondents to our survey were asked if they would be willing to take part in a follow-up interview to discuss in greater depth their decision on investment in CAI. The two-thirds who consented supplied us with the name of a key individual to contact. In-depth interviews were conducted either face to face or by telephone with 15 organizations which were selected to include both adopters and nonadopters, and to reflect in other ways the diversity of the U.K. research industry.

2.4 ADOPTION OF CAI

2.4.1 Current Levels

In looking at the current level of adoption of computer assisted interviewing, we have concentrated on the yearbook estimates described above. Of the 202 organizations surveyed, 43 percent claimed to offer CATI, CAPI, or both. CATI (offered by 38%) was far more common than CAPI (offered by 22%).

These general figures conceal very large differences by size of organization, as measured by annual turnover, which is shown in Table 2.1. Among the smallest suppliers (turnover under £1m per year) only 1 in 4 offered any CAI services and only 1 in 10 offered CAPI. At the other end of the scale (turnover £5m plus per year), as many as 88 percent offered some CAI services and 71 percent offered CAPI as well as CATI.

A second available source was our own small survey of *in-house* facilities. The adoption estimates differed greatly, and the differences were not altogether easy to understand. Our survey estimate of CATI adoption, for example, is as high as 51 percent, well above the yearbook-based estimate. While that could be due to some underreporting in the yearbook entries, we believe that the more likely reason for the large variation is differential nonresponse in our survey. The vast majority of the organizations offering CAI include CATI. Conversely, those without CATI often do not have other CAI facilties; such people are almost certainly less likely to respond to a questionnaire about CAI.

For CAPI, on the other hand, the survey estimate of 16 percent adoption is substantially below the 22 percent found in the yearbook analysis. Here we think that any effect due to the differential nonresponse was outweighed by the difference between offering the service and having in-house facilities. On balance, we prefer to take the incidence rates derived from the yearbook as the more valid estimates of technology acceptance in the U.K. Unfortunately, the yearbook does not list the availability of CASI, so our survey estimate of adoption, 14 percent, is the best available estimate.

Table 2.1. Cell Incidence Rates Estimated from the MRS Yearbook (in %)

	Annual Turnover		
	Under £1m	£1m, under £5m	£5m plus
Any CAI claimed	23	63	88
CATI only	13	36	17
CAPI only	5	8	—
Both CATI and CAPI	5	19	71
Total claiming CATI	18	55	88
Total claiming CAPI	10	27	71
Base (n)	(111)	(67)	(24)

2.4.2 Year of Adoption

Another important limitation of the yearbook estimates is that they are only available for the last couple of years, since computer assisted interviewing was not identified separately until then. To see the pattern of adoption over time, we could use only our survey estimates. While our estimates may overestimate the incidence of in-house facilities, we have no reason to believe that the pattern over time will have been disturbed.

Figure 2.1 plots the cumulative penetration of the three forms of CASIC. The figure shows a dramatic increase in adoption of CATI over the decade of the 1980s; it also suggests that we are well short of saturation: The first six years of the 1990s show a further doubling of the level of adoption of CATI. In simple terms of penetration or interorganization diffusion, we see the second phase of the widely quoted S-shaped curve—a steady upward trend.

CAPI is of course a much more recent phenomenon, and although the incidence has quadrupled in the 1990s, the current level of adoption is still little higher than that seen for CATI 10 years ago. As will be seen in later sections, this result is hardly surprising. The incentives and pressure for adoption have been far less for CAPI than they were for CATI; the investment required is much greater. Perhaps most important, while CATI boosted a relatively new and widely distrusted data collection method—telephone interviewing—CAPI represents a greater intrusion into a key area of experience and tradition, face-to-face interviewing. Nevertheless, the steady upward trend shows no sign of leveling off.

The use of computers in self-completion surveys (or self-completion components of interview surveys, CASI) tends to be an extension of CAPI. As Figure

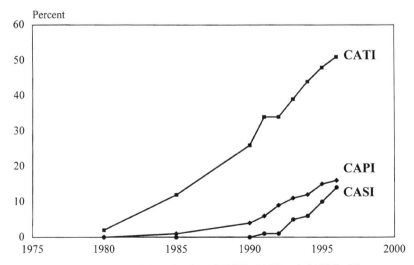

Figure 2.1. Cumulative adoption of CATI, CAPI, and CASI (in %).

2.1 shows, there has been recent and very rapid growth, but there is too little information to allow a confident forecast of future development. We can only expect this application of CASIC to continue to track the adoption of CAPI.

2.4.3 Adopters: Company Size

Our survey shows that in-house CAPI facilities are mainly confined to the largest survey suppliers, with annual turnover exceeding £5 million. In this case our survey estimate is almost exactly in line with the yearbook analysis, showing 73 percent adoption. Elsewhere, the survey estimates are lower than the claims made in the yearbook, with 10 percent adoption among medium-sized organizations (compared with 27%) and only 5 percent among the smallest, with annual turnover of under £1 million (compared with 10%). This reinforces our belief that smaller organizations offering CAPI services tend to be dependent on subcontracting to the larger organizations. As we will discuss later, several observers see this as a rising trend toward concentration of face-to-face interviewing resources.

In the case of CATI, the bias toward larger suppliers is still visible but less dramatic. Approximate adoption levels are 80 percent among large suppliers, 60 percent among medium-sized suppliers, and 40 percent among small suppliers. (As explained above, these are almost certainly all overestimates, but the pattern is probably accurate.)

More marked differences emerged in terms of *when* CATI was adopted. Among the large suppliers, adoption occurred earlier. Thus, about half of all CATI installations were made before 1991 — these suppliers we will treat as early adopters. Among large suppliers, as many as 75 percent of adopters did so early. In the medium-sized group the proportion of early adopters was just under 50 percent; among the small companies about 40 percent were early adopters. This is a classic pattern of diffusion for a field requiring substantial investment. We might look for the same pattern to evolve in the future in respect of CAPI, but as we will show below, this is far from certain.

2.4.4 Adopters: Attitudes toward Innovation

Our survey respondents were asked a series of questions allowing us to categorize them as individuals and by the organization for which they work, on a scale measuring openness to technological innovation. This was done by asking how closely each of the following fitted as a description:

1	Adventurous	Eager to try new ideas — risk taker
2	Respectable innovator	"The man to check with" on the sanity of new ideas.
3	Deliberator	Motto: "Be not the last to lay the old aside, nor the first by which the new is tried"

| 4 | Cautious sceptic | Responds to economic necessity and increasing peer pressure |
| 5 | Traditionalist | Past as a key point of reference |

As shown in Table 2.2, about half of our respondents were scored 1 or 2 — innovation friendly — on this scale, with 14 percent scored at point 1. There was, however, a marked correlation with organization size; those working in large organizations were far more likely to be scored at the innovative end of the scale. Given all that we hear about bureaucracy in large organizations and dynamism in small businesses, this pattern came as a surprise. It may, however, be explained by factors such as the capacity of large organizations to fund more investment and spread risk, or their involvement in larger surveys that demand or at least can justify the investment. Over time it seems inevitable that the higher rate of successful CAI adoptions in larger organizations would foster enthusiasm among their staff.

Perceptions of the organization's attitude toward innovation tended to be close to self-perceptions, with only a slight tendency for organizations to be seen as somewhat more cautious than the reporting individuals. Again, the relationship with organization size was clear, with large organizations markedly more likely to be seen as being receptive to technological innovation.

Not surprisingly, attitudes toward technological innovation are more positive among respondents in organizations that have adopted CAI, especially in those that have adopted CASI or CAPI (see Table 2.3). While it seems that greater receptivity is a factor leading to adoption, we should not reject the possibility that it is the experience of adoption that fuels enthusiasm. Some support for this alternative explanation arises from the fact that early adopters of CATI do not differ in this respect from all adopters of that technology.

2.4.5   Initial Investment

Many installations of both CATI and CAPI were initially stimulated by the pressures of particular surveys or particular clients. They were, however,

Table 2.2. Individuals' Attitude to Innovation (in %)

	Attitude Score				
	Adventurous				Traditionalist
	1	2	3	4	5
Large	27	53	13	0	7
Medium	14	43	29	14	0
Small	11	28	22	28	11
Overall	14	36	23	20	7

Table 2.3. Attitudes Among Individuals in Adopting Organizations (in %)

Adopted System	Attitude Score				
	Adventurous 1	2	3	4	Traditionalist 5
CASI	28	61	11	0	0
CAPI	21	63	11	0	5
CATI	20	46	21	12	1

almost all funded from operating revenue rather than from grants or other ear-marked funding.

At the time of adoption, the level of commitment to CATI was high: Fifty-six percent were committed to full-scale implementation (full steam ahead) and another 25 percent were at least confident that they would move to full implementation. With CAPI the approach seems to be much more cautious. Only five organizations (one in four of those using CAPI) claim to have been fully committed from the start. Most of the rest are confident that their implementation will become full scale.

2.4.6   Using the Technology

As Table 2.4 shows, where CATI has been adopted, it has generally become established as the normal operating mode. It is used for a high percentage of all telephone interview surveys and clients and by most interviewers. CAPI, in contrast, is used much more selectively by a smaller percentage of surveys and clients, and by a smaller percentage of interviewers. In terms of survey type, the distinction is that most CAPI facilities are used only for questionnaires involving complex routing.

We can, however expect the situation of CAPI to change, even quite dramatically. When first installed, CATI systems were used far less intensively

Table 2.4. Current Use of CATI and CAPI (in %)

	CATI	CAPI
Used for more than 75% of telephone/face-to-face surveys	62	11
Used for less than 10% of telephone/face-to-face surveys	21	53
Used by all or almost all clients	43	11
Used by only one or two clients	11	32
More than 75% of relevant interviewers trained	79	16
Fewer than 25% trained	12	53

than now. In the first year, fewer than one in five were used for more than 75 percent of surveys or by all or almost all clients. As noted earlier, it is the subsequent increase in intensity for use — intra-organization diffusion — that we believe brings our case into line with the generally accepted S-shaped diffusion curve. CAPI has already become the norm in two or three organizations. We can expect in the relatively near future, organizations that have made the investment in CAPI to use their facilities intensively. Thus we should see rapid growth in CAPI volume, with both new entrants and more intensive use by earlier adopters. Given that the leaders, suppliers as well as their customers, are mostly large and influential, we can expect them to have profound effects on the rest of the survey industry, as we will see later.

2.4.7   Nonadopters

The vast majority of U.K. organizations with in-house CAPI facilities also have CATI; indeed favorable experience with the latter has often affected decisions on the former. But CAPI adoption is still only true of a minority of organizations, even those with CATI experience. We will discuss in Section 2.5 the barriers to adoption of CAI and especially CAPI, based on our qualitative interviews.

Within our survey, we directed specific questions only at organizations with *no* adoption of CAI. Among organizations that have not adopted CAI in any form, there are major signs of interest. One in four have already used CAI through subcontractors or have collaborated in research projects that used other CAI facilities. Most have given some consideration to investment, and 30 percent are currently considering the possibility. Only 30 percent of nonadopters, representing 12 percent of the total sample, have not considered investment.

2.5   FACTORS SHAPING THE DIFFUSION OF CAI IN THE U.K.

2.5.1   Introduction

This section examines the factors that have governed the diffusion of CAI within the U.K. It is based on in-depth interviews, carried out with 15 of the organizations that responded to our survey. These organizations were purposively selected to include nonadopters and adopters, and to reflect in other ways the diversity of the U.K. research industry.

The stories told to us by various organizations about their transition to CAI, or not, are very different. The range of factors conditioning decisions at any point in time, and the complexity of their interplay, do not yield to any simple analysis, so we cannot yet tell tales of "typical" organizations. However, we believe we have gone some way toward identifying some key themes, and these are presented below under three headings, in line with other research into diffusion: factors relating to the technology, CAI; factors relating to the industry, U.K. research organizations (largely in the commercial sector); and factors to do with communication and the spread of knowledge.

2.5.2 CAI Technology

Attitudes toward Computerization of Interviewing

The acceptance of CAI into the U.K. research industry has relied, and to some extent still does, fundamentally on the lowering of resistance to the idea not just of changes but of *computerized* changes to aspects of the survey process outside the familiar territory of data processing and analysis. In earlier work on CATI in the U.K., Collins (1983) described a research culture in which levels of such resistance were high. Key components of this included concerns about limitations that CAI might impose on the type and quality of data that could be collected, and loss of control of aspects of the survey process such as questionnaire design to computer "nerds" with a historically poor reputation for delivering their promises on time. To these we may add doubts about the reception by respondents of CAI, professional fears about coping with the technical transition, and a host of other, by now familiar, worries.

It is clear that such "cyberphobia" has now abated to a great extent. In the period of nearly one and a half decades that have elapsed since that was written, computers have become a feature of everyday life, both at home and at work. Researchers in most organizations are equipped with personal computers and increasingly user friendly software continues to close the gap between computing personnel and research staff. Concerns about the reactions of respondents are less pressing in the face of increasing computer literacy and familiarity with computerized processes among the general public.

Feeding this decline in computer resistance has been a burgeoning array of large-scale, high-profile applications of CAI conducted by major agencies across all sectors, as well as influential research addressing in a systematic way the methodological questions raised. In general, organizations that made an early transition to CAI were more preoccupied with the basic anxieties surrounding computerization of the interview process than later adopters and current nonadopters. And, as might be expected, organizations that have already invested in one type of CAI (typically CATI) tend to be more phlegmatic about such issues when contemplating investment in other types of CAI than are CAI virgins.

We do, however, find some exceptions to the general lowering of resistance to computerization of the interview process. Although in some cases Luddite forces are undoubtedly still at work, more typically such resistance centers on special types of fieldwork. For example, two of the nonadopters we spoke to (both small organizations working in specialist fields) were concerned about the possible effects of computerization on data quality: the depth and feel of information in one case (advertising) and the accurate capture of difficult brand names in the other (pharmaceuticals). There were also worries about the effect of computerization (specifically CAPI) on the morale and stability of a specialist fieldforce (medical research), and the ability of certain (more disadvantaged) categories of respondent to cope with CASI (social research).

Pace of Technological Change and Development

Aside from factors arising directly from the involvement of computers in CAI, other aspects of the technology have impinged on its diffusion in the U.K. Key among these have been changes and anticipation of changes in successive generations of CAI systems. This is currently well exemplified in the diffusion of CAPI, perceived as a rapidly evolving technology. Organizations that we interviewed mentioned some or all of the following developments or awaited developments in describing the timing or postponements of their entry into the market: smaller (even) more portable machines; interviewing software not linked proprietorially to specific analysis packages; pen-based systems that can deal with complex questionnaires and routing; less expensive or less restrictive software licensing; Windows based systems facilitating the writing of scripts, especially by research executives; and of course cheaper systems.

While the pace of change in CAI technology appears to be a significant factor affecting investment decisions in the U.K. research industry, its effect on different organizations has continued to be a variable one. For example, early adopters in the U.K. tend to include organizations that can amortize their investments in a relatively short period of time, carry out large volumes of work compatible with the early generations of CAI technology (e.g., in-home or central location CAPI interviews as opposed to street interviews), and are keen to establish themselves as market leaders. On the other hand, our sample included a number of smaller organizations who stressed their reluctance to enter too soon a rapidly changing field for fear of finding themselves stuck with redundant or old fashioned technology.

Demand for CAI Products

Suppliers who were early adopters of CAI in the U.K. have had to fight quite hard to sell their technology, especially in the case of CAPI. Their success is now reflected in that most of the organizations we interviewed commented on the growing demand among clients for CAI surveys. The existence of CAI and awareness of its existence may have as much to do with this growth in demand, and informed understanding of its capabilities and some element of fashion consciousness must be suspected. Nevertheless, suppliers who have not yet invested in CAI may well experience some pressure to bring forward their transition and to seek ways creatively to manage the attendant economic risks or other perceived barriers to investment. As customers become more sophisticated and discriminating purchasers of CAI, we may also expect to see growing pressure on organizations to update undepreciated and still serviceable systems, creating yet another set of problems.

Although demand for CAI generally is seen to have increased in recent years, there are variations between sectors and across the different forms of

CAI. For example, while CATI has never been markedly successful in the social research sector in the U.K., a number of those to whom we spoke mentioned a recent surge in demand for CAPI, particularly among government clients and other main buyers of large-scale social research. Organizations serving this sector were well aware of the need to be able to meet client demand, and the dangers associated with not doing so. On the other hand, market research clients were relatively quick in taking to CATI, perceived as enhancing the possibilities of telephone interviewing, which was already well established at the start of the 1980s and remains a growth market.

2.5.3 U.K. Research Organizations

Attitudes toward Innovation

As was noted in Section 2.4.4, there is variation between U.K. research organizations in their attitudes toward innovation. This is a clear factor affecting the spread of CAI within the industry, with larger organizations demonstrating a generally more bullish attitude toward change than smaller organizations. This may be accounted for in a variety of ways. For example, larger organizations may be more outward looking, mindful of the need to keep abreast of or to anticipate developments on which their continued success might depend, more concerned with their responsibilities to shareholders whose expectations are for continued growth in profits, and better able to attract ideas people in relevant fields. Larger organizations may also be among the first to feel the pressure of changing demand, in this case for CAI by virtue of their perhaps more sophisticated client base, as well as by being generally better placed to incorporate new technology at an early stage. Both these factors may condition attitudes toward change.

This general trend notwithstanding, our sample included a number of smaller organizations with pronounced forward-looking attitudes. Often such attitude arise with and filter down from the most senior levels, typically the rugged individualist heading the organization. The story of the diffusion of CAI in the U.K. is not simply the conversion of traditional survey organizations. In particular, a number of CATI suppliers, including, if not dominated by, smaller outfits, were established for that purpose and led often by entrepreneurial figures who broke away from more traditional organizations in order to catch early the rising tide of change.

Ability to Invest

Perhaps the most important factor affecting adoption of CAI in the U.K. has been the ability of organizations to invest in the technology. This appears to be increasingly the issue as CAI is accepted into the industry. Two important components of the ability to invest are the availability of outlay capital and the speed with which such outlay may be recovered. In the U.K. the typical early investor, in both CATI and CAPI, has been the organization able to attract suitable large-scale surveys on which to "hang" their investment. In some cases clients for such surveys have offered or been persuaded to share some of the investment costs.

As the pressure to invest in CAI increases, for example, through client demand, we believe that more organizations are feeling the need to examine creatively solutions to these twin problems. For example, our sample of a small firm working in a specialist area (agriculture) had recently purchased an organization with both CATI facilities and the (wider based) business to support it. This enabled it to offer CATI to its traditional clients while making full use of newly acquired facilities. In other cases risks have been taken or are being considered on the basis that the investment in CAI facilities will, in the light of increasing demand, attract new business or prevent the erosion of existing business. For example, the loss of a job specified for CATI spurred one of our sample to introduce computerized telephone interviewing facilities shortly thereafter.

The decision to invest in CAPI appears on the whole to have been more difficult for U.K. organizations than investment in CATI. A number of factors account for this that impinge on outlay costs and their recovery. The appearance of viable CATI systems coincided in this country with the emergence of telephone interviewing, and particularly centralized telephone interviewing, as an important method of data collection. Decisions by many organizations to invest in CATI have tended therefore to be bound up with decisions to invest in centralized telephone interviewing facilities. Where that step has been taken, the costs of installing CATI have appeared as marginal costs. Moreover the existence of a fixed production point for CATI surveys has meant that—demand permitting—facilities can be employed around the clock, ensuring rapid recovery of outlay. CAPI, on the other hand, is in most cases being grafted on to existing face-to-face fieldwork arrangements. The costs of outlay on equipment and the training of interviewers tend to be viewed more starkly, since that which CAPI seeks to replace "ain't broke." CAPI machines tend also to be allocated to individual interviewers, meaning both that a larger number is required than for the equivalent volume of CATI work and that their use at any one time is restricted.

This state of affairs is exacerbated by the nature of the typical U.K. interviewer fieldforce and its relationship with employers. Most face-to-face interviewers are not—and do not wish to be—employed on a full-time basis. For the most part they are individuals who value their jobs because they are flexible and part-time, and these features enable them to tolerate the typically low rates of pay. As one of our respondents commented, interviewers may work as little as two days per week, meaning that machines may lie idle for more than two-thirds of the time.

Among the leaders in CAPI adoption, ONS had the singular advantage of a different relationship with interviewers, seeking to be their sole employers and to use them for at least three days a week. In time, we must expect to see the spread of such arrangements between interviewers and research organizations in order to address the problem of underutilization of capital. This is an issue to which we will return below.

2.5.4   Communications and the Spread of Influence

Although several very successful CAI systems have been developed in the U.K., for the most part the main impetus for change originated abroad. Taking together the countries of Western Europe and North America, we must regard ourselves as followers rather than as early adopters of CAI — but we are probably not laggards.

The spread of influence from other countries to the U.K. is impossible for us to chart in any detail without (and perhaps even with) the aid of a study set up specifically to address the question. However, an important part of the tale, undoubtedly, is the story of the enthusiast. Martin (1995), discussing the adoption of CAPI by ONS in the late 1980s, mentions "key individuals with vision and drive" and "trail blazers" within ONS, in contact with organizations in Europe (e.g., Statistics Netherlands), in the United States (e.g., the Bureau of the Census), and in Canada (Statistics Canada).

Within the U.K. we believe that an important part of the process of diffusion of CAI has been the trickle down of influence from early adopters, especially larger organizations conducting high-profile, well-publicized CAI surveys. These organizations have provided others within the industry with working examples of the application of CAI, with inspiration to explore the potential of the technology, and in some cases with good documentation of the advantages and pitfalls associated with the new methods. They have also exerted influence by stimulating demand among research clients generally.

2.6   EFFECTS OF CAI ON ORGANIZATIONS AND THE INDUSTRY

2.6.1   Effects on Organization Structure and Procedures

The introduction of CAI technology to U.K. research organization has in many cases undermined the logic governing the traditional structure and functioning of the organizations. Here we discuss some of the changes that appear to be taking place as well as those that are widely expected. In part, this is a description of the experiences to date of adopter organizations, but several observers foresee other changes affecting both adopters and nonadopters as the proportion of organizations using CAI increases. It is necessarily an account written at a time of major transition. Many adopter organizations in the U.K. are still wrestling to identify those areas where the shoe no longer fits and are making piecemeal changes where it pinches most. Even where efforts have been made to engineer more radical changes to an organizational structure, there is little sense that a final solution has been achieved — only the first significant steps toward a new type of survey organization.

The best report of the changes in a U.K. survey organization arising from the introduction of CAI is that provided by Martin (1995) of the effect of CAPI introduction on the Social Survey Division of ONS. Many changes are reported, not all of them easily managed. For example, the transition implied changes to survey timetables, with a heavier workload prior to fieldwork than for equivalent surveys using PAPI, a change not always recognized by customers. Interviewers had to learn new skills (see De Leeuw and Collins, 1997), research staff needed more computing skills, and role changes were needed throughout virtually the entire organization.

Martin's account is of wholesale change in a government survey agency, encountering problems that would be recognized easily by other such agencies (see Clark, Martin, and Bates, Chapter 4). Commercial agencies can be very different, and in many respects the problems can be even greater. Most are far more reliant on one-off (ad hoc) surveys, often with small samples and carried out within very tight timetables. They may be less able to influence client demand and have to be more reactive to the demands of customers who may have had less exposure to the benefits of CAI. They will expect their research staff to spend as much time generating business as to survey design and management. Their existing relationships with interviewers, as we have already mentioned, are very different, characterized by ease of access to a large pool of self-employed interviewers who in turn may work for several agencies.

A very few large commercial agencies have undergone change similar to that experienced by ONS and other government agencies. They tend to recognize the same issues, although they may attach different priorities to them and often react differently. Elsewhere, as we have seen, even CAI adopters have not made a complete transition but are using the new technology for only a small part of their work. This may represent an opportunity in that change in organizational practices can evolve gently to meet the new challenges. But more often it is seen as an added problem insofar as CAI technology demands major organizational reengineering that is not easily achieved through such evolution. At the extreme the contrasts between the requirements of the part of the workload transferred to CAI and the part still using traditional methods can create new problems and tensions.

For all the potential differences, we do find some common strands running both through the account of change in government agencies and our own interviews with commercial suppliers, both adopters and nonadopters of CAI. We attempt below to identify some of the key driving forces for change associated with CAI (particularly CAPI) and to analyze responses to those forces.

Increased Demand for Computing Skills
Automation of most parts of the survey process has led to a generally higher computer literacy profile for all categories of staff in the adopter organizations. This has affected employment and training policies at all levels.

The main issue here concerns the traditional separation between research and computer divisions in U.K. agencies. Typically researchers are not expected to possess much competence in computer progamming — even at the level of using standard packages to analyze their own survey data. This has been the province of computer staff operating in a black box late in the production line: research staff outline their requirements and wait patiently for the computing division to deliver.

The introduction of CAI with more friendly technology may help take down these barriers. Computer expertise is needed from the outset of a CAI survey and as such must be combined with the traditional skills of the survey researcher. Organizations vary greatly in their approach here. The most radical vision being pursued in a few organizations is to create a team of researchers who have the computing skills to manage the whole survey process, from the writing of the questionnaire to the analysis of the data.

Less ambitious approaches encourage research staff to acquire just enough programming skills to participate in the writing of computer-based versions of their questionnaires. They are assisted by computing specialists who are closely integrated into the survey process working at all stages with the research team.

The tradition of researcher and computing specialist doing separate tasks does, however, survive in some organizations. So the involvement of researchers in programming questionnaries is kept to the absolute minimum. The main concern in such organizations seems to be that of efficient division of labor. Clearly the diversion of the energies of researchers into programming, best handled by specialists, would be inefficient, There is further recognition that such an approach does not erode the position of the specialists, probably originally employed to handle the analysis end of the computer packages — the "spec writers" who translate the researchers' analysis requirements into computer code. The model also perhaps fits best in organizations seeking to maintain a structure that can deal simultaneously with CAI and non-CAI surveys. The main concern about CAI in such organizations is that researchers may lose control of the survey process at a critical stage, a concern heightened by the lack of a hard copy version of the final instrument in many packages. This concern has brought organizations following this practice to recognize the need of involving computing specialists earlier in the survey process so that they can understand the objectives behind the paper questionnaire submitted by the researcher.

Merging of Parts of the Survey Process

Automation of the survey process has led to the merging of previously separate tasks such as interviewing and data entry. In some organizations the introduction of CAPI has caused job losses among data entry and editing clerks, the most affected group. More generally, the negative effects have been mitigated by other factors: the conversion of only part of the total survey activity, substantial real growth in that total activity, and rapid growth in the use of scanning and self-completion questionnaires. Since we can expect that CAI will over time lead to leaner organizations, the transitions now in place will ease the pain of future unemployment.

Relationships among organizations have also been affected. As new contracts are being generated for contracting out occasional data entry to specialists, others are being severed. The introduction of CAI will demand more self-sufficiency in agencies that previously depended on outside specialists. The net effects on the industry cannot yet be envisaged clearly, but we can expect to see, for example, considerable impact on specialist suppliers of nonintegrated processes such as data entry, data processing, and tabulation. Such organizations will not necessarily disappear, but those that remain will have to diversify or merge in order to meet the demands of a more integrated survey process — for example, they could provide all the necessary computing services including questionnaire translation, to agencies that do not wish or cannot afford to become self-sufficient.

Interviewers

From our contacts with commercial organizations it is clear that the greatest effects of growth in CAPI will be on interviewers and their relationship with survey organizations. CAPI interviewers must be able to work with the new technology, and the need for literacy in this technology will change the kind of people attracted or recruited in this field. More dramatic changes are expected to arise in terms of relationships and employment contracts — government agencies already have contracts with interviewers quite different from the norm in the commercial world.

Adopter organizations will seek ways of ensuring a return on their substantial investment in having to provide an interviewer with equipment. They will demand more intensive use rather than depend on nonintensive use of a large pool of part-timers looking for pin money. They will seek a smaller body of interviewers committed to the task as a career. Changes already reported by the earliest adopters show that the people recruited must be prepared to commit to both greater mobility and at least a three-day week. But changes will not be only one-sided, career-oriented interviewers will have very different expectations in terms of their conditions of employment.

As in the case of the clerical staff, the changes should be gradual so as not to cause pain to individuals. The present pace of adoption of CAPI and the continuing growth in the industry may well allow change to be absorbed relatively painlessly. Of course we have no guarantee of this: CAPI adopters are large and powerful, and buyers demanding CAPI control very large budgets. Both may be expected to be aggressive in forcing through change in the rest of the industry.

2.6.2   Effects on Industry Structure

The effects on industry structure which will be the most dramatic are also the most difficult to predict and the most dependent on the actions of key players.

So far the growth of CATI has had little effect on the industry, since it has been growing sufficiently fast to absorb the growth of telephone interviewing, which now accounts for around 25 percent of all data collection in the U.K. As the use of CAPI spreads, more visible effects can be expected.

In general, most observers agree that there will be greater concentration in the industry — within the U.K., on the continent, or trans-Atlantic. Large organizations have the capacity to handle change, so they will likely prosper provided that they can increase demand for the higher data quality offered by CAI. Most small and specialist organizations are expected to survive. They may even prosper if the growth of the heavily capitalized research machine leads to reduced creativity, flexibility, or responsiveness in the large organizations, since such organizations often do not maintain their own fieldforces. Even if CAI becomes the norm, small organizations can survive if they have access to outside facilities — mostly now offered by specialist fieldwork organizations but increasingly available from large full-service agencies eager to make full use of their resources.

Medium-sized organizations — too large to specialize and too small to carry the investment — are seen to be most at risk. Some will disappear, others will merge, acquire or be acquired; this change is already apparent. The other group seen to be at high risk is that of specialist fieldwork suppliers. Many such organizations highly depend on the existing pattern of interviewer use. They have access to very large pools of part-time interviewers, quickly and flexibly available to all comers. If this pattern of use does not survive, nor will those dependent upon it. To survive, they will have to re-engineer themselves to embrace change — establishing clearer and continuing relationships with their interviewers, making the investment in a CAPI capacity, or merging with the other threatened group of data-processing specialists.

There is further the possibility that the whole technology of CAPI, even face-to-face interviewing itself, may be overtaken by other new technologies, especially those linked to self-completion surveys. At this point the future is unpredictable. It will be determined by the big players in both supply and demand, with their vested interest in CAPI technology. Will new players find creative ways of using that technology, or will exciting new technologies overtake us all?

CHAPTER 3

Integrating CASIC into Existing Designs and Organizations: A Survey of the Field

Robert M. Groves
University of Michigan and Joint Program in Survey Methodology

Robert D. Tortora
The Gallup Organization

3.1 INTRODUCTION

Collins, Sykes, and O'Muircheartaigh (Chapter 2) use theories of the diffusion of innovation to contrast interorganizational with intra-organizational diffusion of CASIC. The present chapter focuses on the intra-organizational effects of the adoption of CASIC within U.S. academic and commercial survey organizations. The chapter was stimulated by set of observations in the field:

- Some adoption of CASIC innovations appeared to be the result of a search for higher quality or lower cost of survey data collection; others seemed to be motivated by an uninformed sense that a "state-of-the-art" survey required its use.
- After adoption of the technology, some organizations appeared to have altered their personnel mix, while others retained their earlier staffing patterns.
- Many organizations faced issues of whether to move all surveys to CASIC or operate in dual or multiple modes.
- It appeared that some organizations were more satisfied with their CASIC systems than others.

Computer Assisted Survey Information Collection, Edited by Mick P. Couper, Reginald P. Baker, Jelke Bethlehem, Cynthia Z. F. Clark, Jean Martin, William L. Nicholls II, and James M. O'Reilly. ISBN 0-471-17848-9 © 1998 John Wiley & Sons, Inc.

Our impression was that the existing CASIC research literature did not cover these issues well, and we were interested in whether the CASIC research literature fully reflected the experience of survey managers using the technology on a daily basis. Thus we undertook a review of the literature with a keen interest in findings that would lend themselves to comparison with self-reports of active CASIC users. We used a review of that literature to motivate the design of a survey of CASIC users.

3.2 LITERATURE REVIEW ON CASIC

Our efforts to characterize the CASIC literature were based on computer literature searches. We also accessed the extensive CASIC literature archives of other researchers, followed up on the references of those articles, and examined the proceedings of various professional organizations involved in CASIC. As with all such literature reviews, the vast majority of the papers were published in journals or conference proceedings, and we suspect this weakens our insight into the performance of CASIC in practice. However, in this case, it is precisely the contrast between the view of CASIC in the literature and the view of CASIC practitioners that we wanted to highlight.

The abstract of each article was read, and the article was classified according to (1) whether it was a qualitative review of CASIC use, an observational study reporting on a CASIC use, or an experimental evaluation of CASIC using a contrast group, (2) what CASIC technologies were examined in the paper, (3) what survey was used as the vehicle to study the method, and (4) whether household or establishment survey use was being examined. First we classify the literature according to these aspects, and then we offer some commentary on the key questions addressed by the literature.

We assembled over 70 articles on the quality of CASIC data, of which the majority are dated between 1991 and 1995. The literature is therefore a relatively recent one. This five-year period witnessed a fourfold increase in papers in the field over the prior two five-year periods (1980–85, 1986–90). The majority of the articles are experimental in nature, typically comparing the results of some CASIC technology with some paper-based analogue or comparing different types of CASIC designs. The most dramatic change in the literature in the last five years is the waning number of qualitative papers that describe some use of CASIC without formal evaluation. There is also a larger proportion of papers of an observational nature, reporting empirical data on CASIC uses without a contrast methodology. This may reflect a move from research and development to production work, as might be expected from an emerging technology.

Most of the papers concern the use of CASIC in household surveys versus establishment surveys. Most of the CASIC papers on household surveys concern CATI, whereas establishment surveys employ other CASIC technologies, especially touchtone data entry (TDE). This reflects, we believe, the dominance of household surveys in telephone surveys and the common use of self-administered measurement in establishment surveys.

The qualitative articles mostly take a nonscientific perspective. Many of these assume an evangelical tone which befits sales better than science. For example, Freeman (1983, p. 144) wrote "CATI, over the long run, will result in a qualitatively different set of norms and expectations regarding the survey: optimistically, we should be able, through CATI, to improve markedly the art of data collection."

There appear to be several consistent quality-related themes in the literature. A very common evaluative tool is the measurement of interviewer attitudes toward the technology (e.g., Baker, 1992; Perron, Bethelot, and Blakeney, 1991; Couper and Burt, 1994). We deduce that this flows from the perspective that interviewer discomfort would lead to behavior affecting either question presentation or respondent comprehension, and thus lead to measurement errors. In almost all papers interviewers tend to report that they like the technology. Because almost all the studies examine first (or early) uses of a method, it is easy to question whether these responses reflect the enthusiasm of the investigators and trainers more than the well-considered reactions of interviewers. This piqued our curiosity about whether experienced users of CASIC would make similar evaluations.

Another common focus of the literature is a set of properties of the questionnaire—item missing data and skip errors (Groves and Mathiowetz, 1984), "don't know" answers (Olsen, 1992), effects of contingent branching on responses (Catlin and Ingram, 1988; Baker, Bradburn, and Johnson, 1995), segmentation effects on interviewers (ability of interviewers to navigate throughout the questionnaire) (Groves and Nicholls, 1986), and responses to open questions (Harlow, Rosenthal, and Ziegler, 1985). In general, other than the harmful effects of segmentation, the evidence leans toward improved performance of CASIC over paper methods, although responses to open questions may be sensitive to system and interviewer attributes. We see these studies as examination of those functional attributes of CASIC that are visibly different from those of paper and pencil methods. We were interested in whether these benefits of CASIC were exploited in routine use of the technologies.

Another focus, especially with ACASI but also with other modes, is the susceptibility of the method to social desirability effects. This probably stems from a perspective that the computer (and sometimes the medium of communication it uses) creates greater social distance between the measurer (interviewer, collection organization) and the respondent. With greater distance, the argument goes, comes improved reporting of socially undesirable attributes. There are many examples of such studies (e.g., O'Reilly et al., 1994; Kiesler and Sproull, 1986), and the findings either show no differences between CASIC and paper modes or higher reporting of socially undesirable attributes on CASIC.

A relatively new area of inquiry focuses on the human-computer interface, sometimes importing concepts from human-machine interaction studies. These study interviewer behavior when the interviewer is the operator of the CASIC technology (e.g., Couper, Hansen, and Sadosky, 1997) or respondent behavior in ACASI or TDE (e.g., Phipps and Tupek, 1991). They often take the perspective that the specific screen format, keystroke protocol, and sequencing design of CASIC affect operator ease of comprehension. Through that comprehension step the investigators, at least implicitly, are interested in error rates or length of time to complete the response task. In some sense this branch of inquiry might be characterized as seeking the causal mechanisms of interviewer variance in CATI and CAPI use, a theme that existed in prior studies. We were interested in whether CASIC users have enlarged the role of programmers to address these user interface issues.

The later CASIC literature sometimes addresses integration of CASIC into a broader information technologies framework. It describes issues of combining CASIC methods across an organization and of the use of multiple CASIC systems in a single survey. It discusses the desirable organizational locus for staffs to develop CASIC, how large the staffs should be, and what skills they should have in order to improve CASIC application development. Finally some papers address how traditional information technology paradigms and methods might be applied to improve the implementation of CASIC methods. This part of the literature tends to have been written by those involved in CASIC survey production. We suspect that the perspectives taken would more closely be mirrored by reports in a survey of CASIC users.

In short, the CASIC literature seems to have gradually moved its focus from issues related to its feasibility to ones that arise after a survey production process is committed to the technologies. These latter questions involve how the technologies' use might be optimized or at least improved, and not whether the technology is better than the prior alternative.

Along with data quality, the cost of introducing or using CASIC methods has been discussed in several papers. Particularly in early papers the discussion often took the form of statements about costs, without much documentation. For example, Fink (1983, p. 154) states, "CATI provides increased interviewing efficiency and the virtual elimination of many operations." Palit and Sharp (1983, p. 169) state, "When a CATI system is operational ... the total cost of any survey should be less ... but the cost may not be 'substantially' less."

Cost discussions in later papers are not radically different, but the debate becomes more focused. Tinari (1988, p. 304) noted that "Without good cost information, one cannot make any judgements as to allocation of resources or optimal workloads." Tinari continues by pointing out that "you are talking about replacing clerical staff with programmers and technicians ... [that] are difficult to hire and costly to train." However, by 1992 the debate on cost, as well as on data quality, began to change. Nicholls and Matchett (1992, p. 378) noted that a panel of experts used by the Census Bureau, many from outside government, stated that "the survey organizations represented on the panel were not awaiting the results of formal research studies on CASIC costs, timeliness, and data quality before proceeding to implement these methods where judged to be appropriate." Image and client demand became important in adopting CASIC technology.

In a recent review of 22 papers discussing organizations' motivations for adopting computer assisted interviewing systems, Nicholls and deLeeuw (1997) note that the most common reports are a desire for increased timeliness and improved efficiency of sample management. Cost savings are less often mentioned.

Given the themes in the CASIC literature reviewed above, we thought that a comparison of that literature with the reports of CASIC practitioner would help assess the role of the technologies in the survey field. The chapter examines reported motivations for adopting CASIC, levels of use within the organization, the effect on staff skill needs, the evaluation of the relative quality of CASIC surveys, and an evaluation of their relative costs.

3.3   SURVEY OF U.S. COMMERCIAL AND ACADEMIC ORGANIZATIONS USING CASIC

A sampling frame of U.S. survey and research organizations was assembled by including all organizations in the current list of academic survey centers maintained by *Survey Research*, a newsletter written and distributed by the University of Illinois Survey Laboratory. In addition commercial survey firms, listed as members in the directory of the Council of American Survey Research Organization (CASRO), were included, for a total of 244 organizations. CASRO is the trade organization of the commercial survey industry. CASRO members are full-service survey research organizations in the United States, with an overrepresentation of the larger firms.

The frame was stratified into three size categories, and a systematic sample of 1 in 5 of the small academic organizations was taken, and 1 in 4 of the small commercial organizations, and complete census of the medium and large organizations in both sectors.

A screening survey was conducted on the resulting 105 organizations. The screening interview verified that the organization conducted its own data collection and obtained year of start of the organization and number of full-time employees. The survey inquired about what types of CASIC technologies were used and whether paper and pencil equivalents were also used. Finally the survey sought the name of the person who would be knowledgeable about the effect of computer assisted data collection on the organization. This included "the organization and management of computer assisted data collection applications and staff, as well as some technical aspects of computer assisted data collection." If the respondent offered several names, the interviewer was instructed to choose that person who was most familiar with the survey administrative side of CASIC use, versus the hardware, software, or systems side of CASIC.

Not all of the 105 organizations were eligible for the survey. The 25 who did not use CASIC in-house or were out of business were eliminated from further examination; in addition 4 refused, and 2 were not contacted. The remaining 74 (2 of which we later ruled out as ineligible) were sent a questionnaire that was tailored in two ways: CATI and/or CAPI use, and measures on the transition to CATI and/or CAPI for those organizations formerly using paper-only methods.

The overall response rate for the survey was 75 percent (ratio of completed returns to sample units conducting some CASIC surveys). The rate for commercial firms (70%) was much lower than that for the academic organizations (89%). Those organizations using both CATI and CAPI were sent a longer questionnaire; they tended to have lower response rates (66%). Organizations with 0–9 employees had a 100 percent response rate; those with 10–99, 57 percent; and those with 100 or more employees, 83 percent. The analyses of the data that follow attempt to be sensitive to these differences in rates of measurement.

A note about inferential rules in the analysis of data is in order. This chapter deals with a small pool of survey organizations. The pool represents, however, the vast majority of survey activity ongoing in the academic and commercial sector because of the sample inclusion with certainty of large survey organizations. Smaller survey organizations were sampled in both the academic and commercial domains. We present estimates that weight the sampled organizations by the reciprocal of the sampling probability. The weighted counts appearing in the table are thus sample-based estimates of the total number of organizations on the sampling frame, in a given category. Our inference is to the entire set of academic and commercial organizations using a least one CASIC mode, covered by the CASRO and *Survey Research* sampling frame.

3.4   REPORTED MOTIVATION FOR ADOPTING CASIC

The survey asked the informant, "As you can best recall, what was the *most important reason* that your organization first decided to use" CATI and/or CAPI, depending on what methods the organization used. The question provided some response alternatives and allowed the respondents to provide a response in their own words. We collapsed the answers into three categories, with a fairly even split between hopes for better quality data (47%) or for an improved image of the organization (37%) as with "keeping up with technology in the survey field," for example. Less frequently reported (by 16%) is the hope of reduced costs. Judging from our review of the literature, where hopes for reduced costs were common, these responses seem a small mismatch with impressions suggested by the history. We suspect that the recall of initial motivation by respondents is heavily colored by their subsequent experiences with CATI.

The reasons for adopting CAPI among the 14 organizations using CAPI yield a different distribution, with the hope for an improved image of the organization (57%), proportionally fewer hoping for better data quality (35%), and a smaller portion (9%) expecting to save money. It is likely that the contrast between CATI and CAPI is based on the fact that the fixed costs of CAPI reflect a one to one correspondence of computers and interviewers (versus the computer sharing in centralized facilities), that CAPI requires added software support for telecommunication, and that CAPI has relatively unsophisticated sample administration functions. Since all of the CAPI organizations are also CATI organizations, lower expectation of CAPI cost savings might also be based on the experience with CATI cost structures.

3.5 EXTENT AND CHARACTER OF USE OF CASIC AND NON-CASIC MODES

By design, all of the respondent organizations use at least one CASIC method. CATI use predominates — all but one of the 54 organizations use CATI (there is one that uses computerized self-administrated questionnaires, or CSAQ, and disk-by-mail only). As Table 3.1 shows, the next most prevalent CASIC mode in the organizations are CAPI, disk-by-mail, and FAX, but none of these have a large penetration among the survey organizations.

About half of the organizations use multiple CASIC methods simultaneously, if we include all nine CASIC modes covered in the survey. The most prevalent combinations are CATI and CAPI (26%) and CATI and FAX

Table 3.1. Percentage of Organizations Using Different CASIC Technologies, Number of Studies Conducted for Major CASIC and Non-CASIC Modes

Technology	Number of Studies in 1995 among Users						
	0	1	2–5	6–19	20–49	50+	Total
Major CASIC modes							
CATI	5	0	12	17	18	48	100
CAPI	73	10	8	5	1	3	100
CSAQ	90	1	3	1	0	5	100
FAX	82	3	7	7	2	0	100
Other CASIC modes							
Disk-by-mail	75	···················25····················					100
TDE	94	···················· 6····················					100
Voice recognition	97	···················· 3····················					100
Internet	89	···················11····················					100
Audio-CASI	97	···················· 3····················					100
Major non-CASIC modes							
PAPI, face-to-face	43	1	14	10	5	27	100
PAPI, telephone	37	2	12	21	10	18	100
Mail	13	6	36	21	8	16	100

(18%). As might be expected, the larger organizations tend to use multiple technologies (79% for those with 100 or more employees; 53% for those with 10–99 employees and 36% for those with 0–9 employees).

Almost none of the organizations are total CASIC shops; all use some mix of CASIC and paper methods. (One organization that does no paper surveys is a CATI-only operation that did two to five surveys in 1995.) The most prevalent paper and pencil mode is mailed questionnaires, used by 86 percent of all organizations.

Perhaps more significant than whether there are all-CASIC survey organizations is whether CASIC dominates production volume. Recall that Collins, Sykes, and O'Muircheartaigh (Chapter 2) found that CASIC had not yet dominated within adopting firms in the United Kingdom, although the authors forecast such dominance in the future. Table 3.2 shows the number of CASIC surveys conducted in 1995 by the number of non-CASIC surveys in the same year. Since there is no CASIC alternative for most mailed questionnaire surveys, we excluded counts of surveys done in that mode. The table shows that organizations doing a small number of CASIC surveys tend to do no (or much fewer) non-CASIC surveys. For example, 42 percent of those doing 2–5 CASIC surveys did no non-CASIC survey; 73 percent of those doing 6–19 CASIC surveys did no non-CASIC survey in 1995. Organizations with higher volumes of CASIC surveys also tend to do more non-CASIC work. One gets the impression that larger organizations have maintained the offering of both CASIC and non-CASIC modes, relative to smaller organizations. This furthers the evidence that CASIC adds another tool to the existing set available to survey organizations, one that does not necessarily lead to abandonment of paper-based methods. Larger-volume organizations can perhaps afford supporting CASIC and non-CASIC modes more easily.

Table 3.2. Percent Distribution of Estimated Number of CASIC Surveys by Number of Non-CASIC Surveys in 1995[a]

CASIC Surveys[c]	Non-CASIC Surveys[b]						Total	(Weighted n)
	0	1	2–5	6–19	20–49	50+		
1	0	0	0	0	0	0		(0)
2–5	42	0	42	17	0	0	100	(12)
6–19	73	0	0	27	0	0	100	(15)
20–49	26	0	32	32	11	0	100	(19)
50 or more	2	0	9	11	11	68	100	(56)

[a]Estimated total number based on using midpoints for reported ranges of studies 2–5 (3.5), 6–19 (12.5), 20–49 (34.5), 50 or more (75).
[b]Includes paper and pencil interviews (by telephone or in person); excludes mailed paper and pencil surveys.
[c]Includes CATI, CAPI, CSAQ, and FAX surveys.

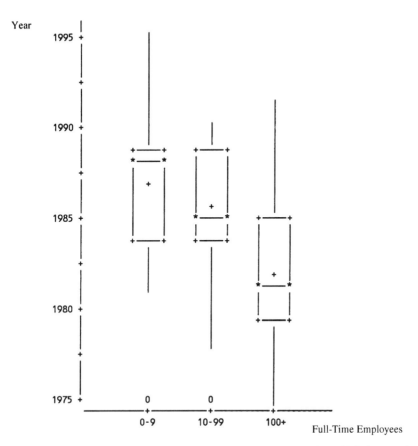

Figure 3.1. Box and whisker plots for year of first CATI survey by number of full-time employees in organization.

The move to CASIC requires an investment in hardware and software that might be easier to make with higher levels of cash flow through the survey organization. As Figure 3.1 shows, currently larger survey organizations were the first to move to CATI. The median year of the first CATI survey is about 1981 for the organizations with more than 100 current full-time employees; 1985, for those with 10–99 employees; 1988, for those with 0–9 employees. This relationship of speed of adoption of CATI and organization size is similar to that found by Collins, Sykes, and O'Muircheartaigh (Chapter 2) for the United Kingdom.

There is also some evidence of inertial factors affecting the date of first adoption. Among survey organizations created in the decade of the 1980s, about 75 percent moved immediately to CATI, in contrast to a much slower movement among those organizations created before 1980. This may be an example that newer organizations, which face fresh decisions on the administrative, methodological, and financial structure without prior commitments, adopt new technologies with relatively greater ease.

3.6   EFFECTS OF CASIC ON STAFF SKILL NEEDS

Large survey organizations, especially those created in the 1930–1950 period, tended to take on an organizational structure resembling a manufacturing plant. Individual units were charged with the completion of one component of the full survey product and passing their completed component to the unit next in the production flow. Thus they tend to have sample design and execution units, questionnaire development and testing units, field administrative units, editing and coding units, and data-processing units. CASIC challenges this traditional structure because it forces an integration of the components of a survey much earlier than non-CASIC surveys (see Dillman, 1996; Groves, 1995).

The most frequently observed example of this is the demand that editing specifications be completed at the time of questionnaire development and integrated into the CASIC questionnaire application. However, forced integration occurs with other traditional specialties — sample units records are often coextensive with interview records, interviewing result codes and sample administrative reporting are part of the same software application, and so on. These observations led to early speculations (Sudman, 1983; Groves, 1983) that some broadening of the skill mix of the traditional staff might improve the relative efficiency of CASIC implementation.

Organizations using CATI were asked about perceived needs for cross training of three types of employees — computer programmers, interviewer supervisors, and interviewers. The categories of skills in question were "general computer literacy, computer programming, data set structure and setup, questionnaire design, administration of the sample, interviewing, and interviewing staff management." We created an awareness measure by coding whether the organization recognized the need for training outside the traditional expertise of the role (e.g., for programmers, training in questionnaire design, administration of the sample, interviewer, or interviewing staff management). The recognition of the need for cross training is widespread, ranging from 59% to 91% of the sample organizations. Furthermore there seems to be no systematic variation by size of organization on this recognition.

For those organizations that experienced the transition from non-CATI to CATI, we asked whether job responsibilities indeed were broadened since CATI began being used. Broadening of responsibilities was most common for technical and computer support staff (85%), interviewer management (68%) and data-processing staff (63%). It is rarer for broadening to occur for administrative staff (43%) and for interviewers (49%). It is interesting to note similar findings in Tarnai, Kennedy, and Scudder (Chapter 5) for smaller organizations.

As Table 3.3 shows, the broadening of responsibilities is most common in larger organizations, those having 100 or more full-time employees, than in smaller organizations. This fits the speculations of 15 years ago that staff in smaller survey organizations generally have broader responsibilities than those

Table 3.3. Percentage Reporting CATI Produced Broadening of Responsibilities by Job Type and Size of Organization

Job Type	Number of Full-time Staff			All Organizations
	0–9	10–99	100+	
Administrative staff	52	26	50	43
Field management	58	78	68	68
Interviewers	48	41	62	49
Project management	71	52	67	63
Data processors	35	74	86	63
Technical support	65	100	95	85
Average over five job types	55	62	71	62

in larger organizations. Their ability to adapt to CASIC demands requires smaller changes in the operations. In support of this, Tarnai, Kennedy, and Scudder (Chapter 5) note that 36 percent of the smaller survey organizations reported that *all* of their professional staff have CATI programming skills.

3.7 JUDGMENTS ABOUT ASPECTS OF CATI QUALITY

Not all the traditional data quality aspects — accuracy, precision, relevance, and timeliness — are necessarily affected by CATI or CAPI. The literature devotes some attention to rates of skip errors, out of range responses, completeness of open responses, item missing data rates, and interviewer variance, but there is little on improved question comprehension through change of structure, levels of irrecoverable errors in application software, or other features affecting quality that are unique to CASIC.

The questionnaire asked the informant to provide several different types of judgments about the quality of CATI and CAPI data. Some of these were more global indicators. For example, the vast majority of respondents report beliefs that data from CATI (95%) and CAPI (100%) are of "better quality" than those from comparable paper and pencil surveys. Regarding interviewer "comfort" with using CATI or CAPI, they report either "most" or "all" interviewers are comfortable, relative to "some," "few," or "none."

We asked informants to use a "satisfied–dissatisfied" scale to rate four attributes of their CATI system — ease of programming questionnaires, ease of discovering errors in questionnaires, the quality of information available on the sample from the CATI system, and the call scheduling system. We constructed a simple summation of scores on these four items, ranging from 0 for "very dissatisfied" with all four attributes to 13 for "very satisfied" with all four. About half (51%) have scores of 10 or higher, reflecting tendencies toward reports of "very satisfied."

There is one hint in the data that satisfaction with CATI is a function of the features of the particular system used. We asked whether CATI offered sample management, call scheduling, automatic dialing, remote monitoring of interviewer screens, productivity statistics, postsurvey coding, and postsurvey editing. A count of these seven functions was created. About half of the organizations used five or fewer of these functions. Table 3.4 shows a tendency for those using five or fewer functions to report lower satisfaction ratings compared to those using more functions.

Informants also reported more specific quality related information concerning CATI and CAPI. These provide a less rosy view of the methods. About half report "often" or "sometimes" discovering CATI programming errors after the interviewing has begun. This is somewhat lower than the 62 percent who reported to Tarnai, Kennedy, and Scudder (Chapter 5) that they have had to recontact respondents because of programming errors. Similar results pertain to CAPI (43%). The problem of debugging CASIC application software is a long-recognized weakness of the field (see Kinsey and Jewell, Chapter 6).

The CATI debugging problem seems more commonly recognized in commercial organizations (61%) than in academic (28%). This might be associated with the shorter cycles of surveys in commercial environments. Relatively fewer small organizations report this problem (24%) than the large organizations of 100 or more employees (65%). We suspect that this result is related to the greater challenge of coordinating specialized staff in large organizations and, perhaps, a tendency for larger organizations to conduct relatively more complex surveys.

Specific CASIC features have the potential of increasing data quality. Indeed, without using some CASIC features, there is little hope of improving

Table 3.4. Percentage with Different Satisfaction Scores with CATI Features by Number of CATI System Features Used in Practice

Satisfaction Rating[b]	CATI Features Used of Seven Mentioned[a]		Total
	Less than 6	6 or 7	
<8	24	6	14
8–9	51	24	35
10–11	3	36	22
12–13	22	34	29
Total	100	100	100
(Weighted n)	(37)	(50)	(87)

[a]CATI features include sample management, call scheduling, automatic dialing, monitoring of interviewer screens, productivity statistics, postsurvey coding, postsurvey editing.
[b]Equally weighted sum of four different four-point satisfaction items, on ease of programming, ease of discovering errors in questionnaires, quality of information about the sample, call scheduling.

measurement error features of the method. Most notable of these are tailoring of question structure and wording to individual respondents (through fills and contingent skips). For organizations that experienced the transition from non-CATI to CATI, we asked whether they increased the use of these techniques, when changing a "previously done paper and pencil telephone questionnaire to a CATI questionnaire." The answers are mixed — about half taking advantage of the opportunity and half (55%) not tailoring questions to previous answers, and 46 percent using more complex branching. There appear to be no large differences across sectors or organization size in these behaviors.

Finally on none of the quality indicators did reports of the need to broaden responsibilities of roles or the need for training in nontraditional areas affect quality ratings.

3.8 JUDGMENTS ABOUT CASIC COSTS

The survey included several measures of cost and quality of CATI and CAPI surveys, sometimes comparing them to their paper and pencil equivalents. We asked the informant to compare CATI and/or CAPI fixed costs and variable costs to those of PAPI. By "fixed costs" was meant "facility and business costs, hardware, core staff, and software maintenance." By "variable costs" was meant "study specific costs that vary based on the number of interviews per study." The modal response categories for both CATI (51%) and CAPI (43%) were higher fixed costs and lower variable costs for CASIC. This implies that these CASIC technologies can be more expensive under low production volumes.

For each of the two CASIC modes, Table 3.5 shows a combination of the fixed cost and variable cost judgments. A minority thought CATI to be less costly in all survey designs (35%). Proportionally fewer (24%) believe the same for CAPI. This is in line with the relatively higher CAPI hardware costs for each active interviewer than would be true for CATI, combined with the

Table 3.5. Cost Comparison Reports for CATI and CAPI, among Those Using the Method (%)

Response	CASIC Method	
	CATI	CAPI
CASIC costs always lower	35	24
CASIC costs higher only with low volume	51	43
CASIC costs higher only with high volume	0	0
CASIC costs always same or higher	14	33
Total	100	100
(Weighted *n*)	(95)	(21)

tendencies to have smaller workloads in face-to-face surveys than telephone surveys.

The reports of higher fixed costs probably reflect the real investment in hardware, software and technical staff support for both CATI and CAPI, which is not demanded by paper and pencil operations. There was some evidence that commercial and larger organizations were more convinced of higher fixed costs and lower variable costs for CASIC surveys relative to non-CASIC equivalents.

The questionnaire asked more specific information about the components of cost, forcing the informant to compare CATI and/or CAPI on various steps in survey production. Most informants (70%) thought PAPI required more time for sample preparation than did CATI; most thought CAPI offered no advantage over a similar PAPI survey on this point. Conversely, most thought CATI required more time for questionnaire development (74%) than an equivalent PAPI survey. Most reported that data editing and processing took longer on PAPI (95%), probably because of the need for a separate data-entry step.

The questionnaire asked the informant to report the number of person hours and different persons required to program "a national RDD survey with 50 questions, 10 of which involve skip patterns based on the respondent's answers, with a sample size of 1000 cases." For the organizations using CAPI, a similar question was asked pertaining to a face-to-face survey. Few organizations (27%) answered that it could be done in less than one day or by a single staff member (Table 3.6). The modal response for CATI was one to two days, using two staff; for CAPI, one to two days, using three staff. We suspect that sample assignment, administrative data, and telecommunication issues increase the staff size for CAPI versus CATI.

Table 3.6. Reported Number of Person Hours and Number of Persons Required to Program 50-Question Questionnaire, with 10 Skip Questions, by CASIC Method (%)

Number of Days and Persons	CASIC Method	
	CATI	CAPI
<1 day	27	5
1–2 days	44	77
>2 days	29	18
Total	100	100
(Weighted *n*)	(96)	(22)
1 person	14	32
2 people	37	9
3 people	24	45
>3 people	25	14
Total	100	100
(Weighted *n*)	(92)	(22)

Are there systematic differences across the organizations in the responses on the cost performance of CASIC? Do some organizations appear to enjoy advantages that others do not? Organization size appears relevant. There was some tendency for the largest organizations to report using three or more people to program a questionnaire (61%) versus the smaller organizations (45% for small- and medium-sized organizations).

Is there any evidence that CATI efficiency is greater in organizations that recognize needs to broaden the training or responsibilities of staff? The reader will recall that larger organizations were more likely to recognize this need. There is no evidence of increased efficiency ratings in firms concerned about cross training; if any, there is a tendency to report slower completion of the application and use of more persons to program the application. We have no explanation for this result, except that larger organizations seem to be sensitive to the training issue and that they tend to use more persons for the programming task.

There appear to be no systematic differences in cost judgments by reasons why CATI was adopted originally, but those who adopted it for a "better image" seem to use more persons to program an application than those seeking better data quality or lower costs. In addition there do seem to be effects of experience with CATI (or use of systems incorporated more recently). For example, fewer of the organizations with over 15 years of CATI experience believe that CATI *always* offers lower costs than those with less experience. They are more likely to report higher fixed costs and lower variable costs with CASIC than paper surveys. Consistent with this they report using more people to construct a CATI instrument and requiring more days to do it than those adopting the technology more recently. We know that these organizations tend to be larger and are more likely to be commercial than those adopting the method more recently; we could thus be seeing combined effects of size, complexity of the surveys conducted, age of CATI system, and amount of experience with using CATI.

Taking all of these results into account, we suspect combined effects in high-volume, larger organizations who adopted CATI early in its existence, and who tend to use larger staffs to do their work. Finally we obviously cannot discern from the survey data whether the different informants are making accurate judgments about the time and staffing requirements to complete the instrument application.

3.9 SUMMARY AND CONCLUSIONS

The literature on data quality in CASIC has focused on interviewer reactions to the technology, the performance of specific CASIC functions on measurement error, CASIC effects on social desirability influences on response, and various issues in human-computer interaction. The literature tends to show clear reductions in indicators of measurement error when specific CASIC functions addressing those errors are used in the application, general satisfaction among interviewers, and mixed reactions on reduction of social desirability effects.

There are many fewer studies of the relative costs of CASIC than of the relative data quality. Early promoters espoused the view that CATI would reduce survey costs. Dramatic cost savings are possible when mode switches accompany CASIC adoption (e.g., the move from telephone surveys to TDE). But past cost comparisons vary on how the fixed costs of hardware and software acquisition are factored into the cost comparisons. It is difficult to gain a coherent picture of relative costs. The hope of lower costs with CASIC did not appear to be matched by actual evidence of lower costs.

This chapter was motivated by the possibilities of a mismatch between the CASIC literature and the experiences of CASIC users. In general, we found correspondence between overall evaluations of CASIC in the literature and among users. We found frequent gaps, however, between the potential performance capabilities of CASIC and actual practices.

Among all CASIC technologies CATI is most prevalent. Other CASIC modes appear to serve particular clientele (e.g., disk-by-mail) or be limited in measurement functions (e.g., TDE). In the United States, CAPI penetration will be limited merely by the increasing rarity of face-to-face interviewing.

Since CASIC, by definition, requires some investment in hardware devices, organizations with more discretionary resources tend to adopt CASIC first. These tend to be larger organizations or start-up organizations, which incorporate the CASIC technology by judgment of the state-of-the-art.

There is little evidence that organizations have moved to all-CASIC shops (particularly because of the popularity of mailed self-administered surveys). Thus CASIC organizations are also non-CASIC organizations, maintaining any requisite infrastructure for paper instrumentation and processing as well as that for CASIC. The exception to the might be a set of small CATI-only units. For most other organizations, especially large ones, we have noted that CASIC might be better viewed as an *additional* tool for survey researchers instead of a *replacement* tool.

There is variation in what functions of CATI are used in practice. In that sense CATI means different things in different contexts. Our survey showed us that this influences how satisfied the organization is with key performance features of the system. Those using more functions were more satisfied.

CATI has prompted a belief in the need to expand the skills of programmers working in survey organizations and to a lesser extent those of interviewing supervisors and interviewers. Large organizations exhibit this belief slightly more often than smaller organizations, probably reflecting CASIC's forced integration of formerly disparate components of a survey product. Staffs in smaller organizations, we suspect, possessed breadth of skills better suited to that demanded by CATI. We would speculate that new organizations created after CATI may organize themselves differently about the technology.

Informants in the organizations express near total enthusiasm for the relative data quality of CASIC and for levels of interviewer comfort. Relative to that enthusiasm, however, common reports of finding programming errors after production start and some loss of cases to error are surprising. In short, global ratings of the methods seem more favorable than specific ones.

Regarding costs, the dominant belief about CATI and CAPI is that they entail higher fixed but lower variable costs. Those with more experience with the technologies tend to give this answer more frequently than the inexperienced (who believe CATI or CAPI is always cheaper or always more expensive). We suspect that a similar survey conducted in the mid-1980s would have generated the judgment of consistently lower CASIC costs.

The ubiquity of the PC in all survey organizations has reduced the marginal fixed costs for CATI (not for CAPI), but questionnaire development, especially for complex instruments, appear to remain a cost disadvantage for CATI and CAPI. Although CASIC generally provides more functionality in reducing some nonsampling errors (range checks, skip facilities, question-tailoring abilities), most systems do not *demand* that they be used. We obtained reports that this usage is quite mixed. We did not survey CASIC questionnaire designers, but we suspect that the familiarity of CASIC capabilities is lower in this group than among interviewing, sampling, and editing groups. If this is true, full potential gains of CASIC on data quality will require a broadening of the skills of questionnaire designers. This will entail a move away from the conception of a questionnaire as pieces of paper to one more compatible with CASIC.

ACKNOWLEDGMENTS

The authors express their appreciation to Stacey Erth for literature search, work on the annotated bibliography, and questionnaire design, to Stephen Hanway for data analysis, and to the U.S. Bureau of the Census, the Joint Program in Survey Methodology, and The Gallup Organization for staff support.

CHAPTER 4

Development and Implementation of CASIC in Government Statistical Agencies

Cynthia Z. F. Clark
U.S. Bureau of the Census

Jean Martin
Office for National Statistics, U.K.

Nancy Bates
U.S. Bureau of the Census

4.1 INTRODUCTION

Government agencies have long played an important role in developing and implementing computer assisted survey information collection (CASIC) technologies for household and business surveys. They supported the development of CATI with grants and contracts in the 1970s; they have been at the forefront in the development of CAPI, TDE, CSAQ, and other technologies since the 1980s. Their motivation for adopting CASIC may be similar to that of institutions working in the private sector (e.g., reduced costs, better data quality, improved timeliness, and keeping up with the technology; see Groves and Tortora, Chapter 3), but private and public sector organizations differ in structure, in the nature of surveys they conduct, and in the types of data they collect. These differences cause their strategies for implementing CASIC to diverge.

Government agencies typically conduct surveys using larger sample sizes than their private sector counterparts do. Consequently, adoption of CASIC technology requires a substantial investment in new equipment and training.

Computer Assisted Survey Information Collection, Edited by Mick P. Couper, Reginald P. Baker, Jelke Bethlehem, Cynthia Z. F. Clark, Jean Martin, William L. Nicholls II, and James M. O'Reilly. ISBN 0-471-17848-9 © 1998 John Wiley & Sons, Inc.

This in turn causes ripple effects within many levels of the organization. Surveys conducted by public agencies are often longer and more complex than those carried out in the private sector. These complexities stretch the limits of CASIC capabilities, largely because they demand highly skilled personnel to create, program, and test a contemplated survey. Many government agencies also conduct continuing surveys used to produce data that underlie policy making. Once the decision is made to switch a continuing survey from PAPI to CATI or CAPI, the agency must be aware of the way changes in data collection will affect the estimates. Adoption of CASIC may require special plans for phasing in the change over time, carefully measuring any changes in the estimates, and developing calibration techniques to adjust estimates between the old and new methods. Finally, these methodological innovations must often be implemented in the context of densely bureaucratic organizational structures. As noted by Dillman (1996) and Groves (1994), such innovations are difficult for government agencies to effect because of dysfunctional organizational structures and archaic budgeting practices.

The CASIC technologies implemented by government agencies have covered the full scope of technologies available, including CATI, CAPI, pen-based computers, TDE, VR, facsimile, scanning, Internet, and EDI. Agencies' experiences have also covered virtually the full scope of possible outcomes, reflecting their organizational structures, their mixes of staff skills, their organizational financing, and, not least, the types of surveys conducted.

This chapter describes the historical transitions to CASIC within four major government agencies, two in North America and two in Europe. These four agencies provide excellent comparative case studies because they exhibit a differing scope of CASIC technology implementation, differences in software use and development, and starkly diverse funding arrangements. The experiences of several other agencies are cited as contrasts to the case studies, often because of their size or specialization.

Four factors affecting the nature and speed of technology implementation are discussed. A critical initial factor is the ability to finance CASIC technologies. In our introductory summary for each agency, we discuss briefly how each met the financial resource requirements brought on by the advent of new technologies.

A second factor is the dramatic effect that CASIC technologies have had on government survey operations. Our second section reviews the CAI's effect on survey timetables, instrument development, case management, data entry, edit and processing, interface with analysis software, and on computer hardware platforms.

Demand for particular human resources is a third factor that has changed as CASIC technologies have been implemented. In our third section we contrast the various approaches taken in addressing these personnel requirements. Finally a fourth factor is the organization-wide effect felt when CASIC technologies have been implemented. We conclude by examining the choices agencies have made concerning the phase-in of these technologies and the effect these strategies have had on the agencies' organizational structures.

4.2   HISTORY OF TECHNOLOGY RESEARCH AND IMPLEMENTATION IN SPECIFIC GOVERNMENT AGENCIES

This section discusses the history of technology implementation in the four agencies selected as examples: the U.S. Census Bureau, Statistics Canada, Statistics Netherlands, and the Office for National Statistics (ONS) in Great Britain. The section concludes by providing both supporting and contrasting information from seven additional government organizations across the world, namely: Australian Bureau of Statistics (ABS), Institut National de Statistiques et des Etudes Economiques (INSEE) in France, Statistics Finland, Statistics Sweden, Statistics New Zealand, the U.S. National Agricultural Statistics Service (NASS), and the U.S. Bureau of Labor Statistics (BLS).

4.2.1   United States Bureau of the Census

The Census Bureau's initial impetus flowed from university influences and followed a slow-paced research model until the bureau was pushed into action by its customers and a new director was appointed in 1989. Even then it moved slowly. The Census Bureau was the first of the four agencies into CASIC, but the last to get its labor force survey moved over. The change was accompanied, however, by extensive evaluation followed by broad changes in the surveys. Often survey questionnaire content and field procedures were altered to improve quality. The bureau did not expand beyond CATI and CAPI until a centrally led effort took hold in the 1990s. It modeled its initial software on university software, later adopting the CASES software as its standard. In the mid-1980s, the Economic Directorate developed a separate system on a microcomputer platform for telephone follow-up to mail nonresponse and edit failures on retail and wholesale trade surveys. Additionally, for nonresponse follow-up to the 1992 Census of Agriculture, the directorate undertook the largest-scale CATI data collection ever done at the Census Bureau. Using a highly diverse set of CASIC options, the bureau is in the process of automating its household and business surveys.

   Capital outlays needed for CASIC implementation came (1) from overhead accounts and (2) from customers willing to provide increased funds to further technology development (Bureau of Justice Statistics, National Center for Health Statistics, and BLS). Budget initiatives were proposed for CATI, and later CASIC, but were not successful until 1996, when partial funding for maintaining a technologies research staff and purchasing new technologies was received from Congress. This line item funding is used for research, development, and initial implementation of new CASIC technologies. The funding for CATI and CAPI equipment now comes from survey funds.

4.2.2   Statistics Canada

Statistics Canada started into CASIC later; its initial impetus was to follow the Census Bureau's software and research approach. It had no external clients pushing it. It chose to move into CATI and CAPI to achieve expected internal gains in quality and reduced costs. Statistics Canada established corporate direction and monitoring to ensure that these goals would be met. It used CASIC for both household and business surveys in ways very similar to those employed by the Census Bureau, but tested and used a wide variety of hardware and software to accomplish its aims, survey by survey. It moved its household surveys (and many of its business surveys, including all agricultural surveys) to CASIC rather quickly, surpassing the Census Bureau in this regard.

A business case was prepared for the transition of the Labour Force Survey to a fully automated CAI system using both CATI and CAPI. Such analyses demonstrate the potential to repay an initial investment through savings made during a specified time period. This specific case projected savings of 700,000 Canadian dollars (approximately 500,000 U.S. dollars at 1997 exchange rates) during the fourth year of operation, and then 1.2 million Canadian dollars savings per year thereafter. This underlay an up-front loan from the corporation to pay for communications hardware and the acquisition of 1200 laptop computers. The projected cost saving was based on an assumption that CAPI would be used by a number of surveys. The proposal considered the need to refresh the technologies every three years.

4.2.3   Statistics Netherlands

Statistics Netherlands followed a different path in 1987 as it converted its labor force survey to CAPI. Although it used U.S. commercial software for its first CATI interviews, its major impetus for CASIC came from the need to improve survey timeliness. This it achieved through improved editing, a broad movement from mainframe to mini- to microcomputers, and a restructuring of the entire agency. In part, these changes were necessitated by overall agency budget reductions. Its solution was to adopt changes both in cost management and staffing (in part, by removing mainframe computers and data preparation departments). This enabled it to stay at the forefront of technological change and build its own internal software (Blaise). Research on altered survey methods was minimal, as were changes in survey design that might arguably have enhanced quality. A strict policy of hardware and software standardization was implemented, resulting in large gains in efficiency while simultaneously reducing operational costs. The Blaise software was so good that it had broad appeal to other statistical agencies.

Unlike Statistics Canada and the Census Bureau, the change to new technologies did not require additional funds at Statistics Netherlands. Replacing mainframes and dumb terminals with personal computers and laptops reduced costs. The automation budget was 25 million guilders (approximately

12.7 million U.S. dollars at 1997 exchange rates) in 1987, 20 million guilders in 1990, but only 10 million guilders in 1996. Because there was no phase-in of the new technologies, Statistics Netherlands did not face the cost of simultaneously maintaining new and old operating systems.

4.2.4 Office for National Statistics (Great Britain)

In Britain during the 1980s, the Social Survey Division (now part of ONS) was initially motivated to use CAPI for its Labour Force Survey (LFS) as a cost efficient means of achieving a fourfold increase in sample size within the same delivery timetable. Because the LFS questions undergo minor changes every year, there was no need to plan a major redesign to coincide with the move to CAI. Additionally, since the LFS is less high profile than labor force surveys in some other countries (e.g., the U.S. Current Population Survey—CPS), there was less emphasis on methods research to assess the effect of the switch. This approach concentrated on cost control and minimized software development costs by using the Blaise software developed at Statistics Netherlands. (A similar approach to CASIC implementation was also followed by INSEE, Statistics Finland, the ABS, and Statistics New Zealand.) Exploratory work on the feasibility of CAI for the LFS was initially funded by the Employment Department, then the sponsor for the LFS. Subsequently, full development costs for the LFS and other surveys were recovered from survey customers in response to a business case.

As do other business survey organizations, the Central Statistics Office (CSO), now part of ONS, primarily conducts mail surveys with telephone follow-up of nonrespondents. In July 1995 business statistics in Britain were centralized as responsibility for the collection of employment and earnings statistics was transferred from the Employment Department to the CSO. The CSO had pioneered early use of OCR for data capture in the 1993 Census of Employment. Investigaton of CATI for nonresponse follow-up to mail surveys, TDE, and EDI (using statistical modules as part of widely used accounting software) was initiated in 1996 by CSO.

For the most part there was little centralized coordination in the implementation of CASIC methods. Each program area (social surveys, censuses, business surveys) developed separately and secured CASIC costs either from customer funds based on a business case for investment finance or from core funding.

4.2.5 Commonalities and Differences across Organizations

We found that time frames for CAI development varied greatly across the agencies surveyed. At both the Census Bureau and Statistics Sweden, there was a long research and development period. By contrast, Statistics Netherlands conducted several small-scale research projects over a short time period with small size teams (see Brakenhoff, Remmerswaal, and Sikkel, 1987). Both

Statistics Canada and ONS adopted software developed elsewhere and then went quickly to an implementation stage. NASS also spent a long time doing CATI research, effecting mere partial implementation of CATI during the 1980s. All of its state offices effected full implementation during the early 1990s.

Within the four organizations cited, also at INSEE and Statistics Sweden, the labor force survey was the first survey to be fully automated. A few countries (e.g., United States and Britain) used a combination of CAPI and CATI. Labor force surveys were continuous and sufficiently large to justify the technology investment without the complexity of other household surveys. Other cross-sectional and longitudinal household surveys, including surveys of income, expenditures, program participation, health, and crime, were automated soon thereafter.

Business surveys have not gone to CASIC survey automation completely. At several agencies, CATI was first used for nonresponse to mail business surveys. Next came the research and development of TDE and VR, with CATI nonresponse follow-up. Many agencies have moved to testing and development of scanning technologies (OCR, OMR) to replace data entry and the previous use of microfiche (or paper retrieval). At Statistics New Zealand and several other agencies, the development of EDI for business surveys and pen-based computing for price surveys followed. Currently, both BLS and the Census Bureau are testing CSAQ by diskette and Internet, with BLS also using a prototype Web site for data reporting (see Ramos, Sedivi, and Sweet, Chapter 20; Clayton and Werking, Chapter 27).

Three funding strategies are illustrated in the four organizations represented. The Census Bureau initially funded its CASIC initiatives with research and development funds. Now, however, the first stage of a CASIC technology is a line item in the budget. CASIC survey implementation was based on business cases at Statistics Netherlands, Statistics Canada, and the ONS (for social surveys). At Statistics Netherlands the case was based on total organization automation on a client-server networked computer platform; at the U.S. Census Bureau (and initially at ONS for the LFS) cases were built on individual surveys. Statistics Canada based its case on implementation of the LFS with planned supplementation of laptops for all existing surveys. ONS and INSEE based their business cases on laptops for all personal interview surveys (Dussert and Luciani, 1995). Agency overhead or infrastructure funding contributed to CATI implementation at Statistics Canada and NASS, where calling facilities were a by-product of office automation. These were also the funding sources for CAPI at the Census Bureau before such costs came to be factored into reimbursable survey costs.

Agencies have faced different problems as they have planned for CASIC technology adoption. Continuing changes in hardware and software pose a recurring problem of technology implementation. This affects adoption or procurement decisions, and often stretches both planning and funding capabilities. For CAPI, neither NASS nor Statistics New Zealand has been able to show benefits commensurate with costs. The ABS did a similar cost-benefit analysis for CAPI with reference to its LFS but found it more costly to implement than originally envisioned. The cost effective implementation of CAPI depends heavily on the ratio of the number of survey interviews to the number of laptop computers required. In these orgnaizations CAPI did not compete well with other infrastruture investments such as wide area networks at NASS.

4.3   EFFECT OF CASIC ON SURVEY OPERATIONS

The introduction of CASIC technologies into survey operations may affect survey processes in two major ways: first, by changing the way existing PAPI surveys have traditionally been carried out; second, by facilitating surveys with new features and greater complexity than PAPI methods could previously support. Not all government statistical agencies initially recognized the potential organizational effect of these changes, but many have now re-engineered their survey processes to accommodate these new technologies, or are planning to do so.

4.3.1   Survey Timetables

Survey schedules have had to be adjusted to new processes. With CASIC implementation, once-distinct processes may be integrated, or may need to be carried out earlier in the survey cycle. CASIC accomplishes data collection and entry simultaneously and may include interactive editing (either as part of the interview or interactively after the interview). Field case-management systems need to be integrated with the instrument. The development and operational schedules for CASIC may differ vastly from those associated with previous nonautomated systems.

There was a general expectation that more time would be needed to develop CASIC instruments. The extent to which CASIC has lengthened the questionnaire design process seems to vary by survey, by agency, and by the CASIC software used. In the commercial world, simple survey instruments can be developed quickly using software suited to the purpose. Government surveys are generally more complex and use highly sophisticated CASIC systems. Because of their complexity they require considerable time for development and testing.

Adding edit checks to the CASIC instrument (rather than programming them separately for a later editing stage) changes the timing of this part of the work and, in many cases, places it on a critical path whereby completion is required before data collection begins. Time is also needed to set up the data-processing systems interfacing with the instrument. The effect of a lengthy questionnairc development period is less of a problem when the same questionnaire is used repeatedly. The extra time for development is then viewed as a fixed cost.

The CAI development time is more of a problem for agencies that carry out many new surveys each year or have continuous surveys with frequent changes in their questionnaires. A number of agencies mention problems completing the additional tasks before interviewing starts and the difficulty of coping with last minute requests for changes in an environment lacking the constraint of printing deadlines.

Statistics Netherlands reports that development time can be reduced and efficiency increased for some surveys by using a modular approach to building large and complex questionnaires from simpler subquestionnaires. Tested and well-developed subquestionnaires are then used in other surveys. One example is the "household-box" survey component that records the composition of a household. This piece is used in all household surveys and has the added advantage of ensuring consistency between surveys. NASS uses a similar approach in designing its agricultural surveys. The surveys have a common beginning and ending, and they use a library of survey questions. ONS is also developing both standard blocks of code for questions asked on most surveys and standard templates for common types of questions (which can then be customized for particular surveys) (see Pierzchala and Manners, Chapter 7).

Agencies that have moved to CAI for their labor force surveys all report savings in delivery time. However, different countries have different survey procedures, so it is difficult to make direct comparisons. The Census Bureau reports a time savings of two days (down from 15 to 13 days) for data collection to file delivery; ONS reports saving 29 days (down from 38 to 9 days). Additionally, ONS was able to quadruple the sample size of its LFS without affecting the delivery timetable. INSEE reports that automation reduced the data delivery time schedule for the survey by two months. Statistics Netherlands reports a change in the data collection and processing period from two years to two months. Statistics Canada did not reduce the data collection and processing period.

4.3.2 Questionnaire Development Processes

Several agencies report that the questionnaire development process for CATI/CAPI is very different from that for PAPI. In some agencies the decision to automate brings with it a conscious decision to redesign a survey "from scratch," including a re-evaluation of question wordings, question ordering, layouts, and formats. This is very different from simply trying to turn a paper instrument into a computerized instrument. In the United States, the BLS adopted the "redesign" decision model as it planned and implemented the automation of the Current Population Survey (see Polivka and Rothgeb, 1993; Rothgeb et al., 1991). Statistics Canada, on the other hand, approached the redesign and automation tasks in separate stages. The first CAPI version of the Canadian LFS used questions taken from the old paper instrument. In the second version of the CATI instrument, changes in question wordings and other design elements were incorporated into the finished product (see Gambino, 1996). ONS did not carry out any major redesigns at the time of automation but is doing so years later to harmonize questions across the major surveys and to take advantage of features available in a new version of the Blaise software.

The process for CAI instrument development varies from agency to agency. NASS, for example, starts with a paper questionnaire from which the CAI instrument is programmed. Others start with a comprehensive paper specification drawn up by the survey or subject matter experts, which they pass on to programmers to put into CAI code. At the Census Bureau the development of a CASIC instrument is a time-consuming cycle of events starting with paper specifications painstakingly prepared by a questionnaire designer (i.e., the "specs") (see Kinsey and Jewell, Chapter 6). The specs provide a blueprint for the CAI author in the form of a screen-by-screen delineation of the questionnaire, including question wording, data edits, branching routes, valid values, and variable names. Designers are responsible for question wording, while authors write the computer code. The specs are programmed into an instrument prototype that is tested and refined by the designer. The ONS and Statistics Netherlands do not maintain such a sharp distinction between questionnaire designers and authors as the same staff usually perform both tasks. Consequently, formal specifications are not required.

A few innovations are being developed to help bridge the gap between designers and CAI authors when these are different. All major software systems are redesigning to utilize a Windows environment for authoring, and designers are beginning to produce tools to help nonprogrammers more quickly and easily produce specifications mapping out CAI data items, response categories, flowcharts, and skip patterns (Katz, 1996). The ABS has developed a questionnaire development tool that does not require specific familiarity with the CAI software (see Colledge, Wensing, and Brinkley, 1996). It provides information including metadata from the survey processing system for designing the instrument.

Another aspect of instrument design mentioned by several agencies is the problem of obtaining adequate documentation of the instrument on paper (Martin, 1995; de Heer, 1991). This lack of paper documentation makes it difficult to discuss questionnaires with customers, sponsors, and government regulators of information collection (e.g., the Office of Management and Budget in the United States). It also makes the task of path testing more difficult, time-consuming, and error prone. This is particularly critical when the instrument designer and author are one and the same and the specification-writing step is circumvented. The quality of documentation produced automatically as a part of the instrument output varies among the major CAI software packages, but new capabilities are now being introduced into these systems to remedy the problem.

4.3.3   Effect of Automated Case Management on the Work Flow/Data Collection Process

Agencies vary in the extent to which case managment has been automated, but all agree that greater automation gives management better information and more control over the work flow. Typically, case management is viewed as having two components: field case management (transmitting to headquarters, scheduling cases, selecting addresses, recording outcomes) and general case management (linking field processes with sampling, office coding, editing, and analysis). This section addresses field case management; subsequent sections relate field case management to other survey processes.

Many agencies have very sophisticated case management systems for both CATI and CAPI. These systems record the status of each case, deal with re-issued cases, control the progress of further processing, and produce management reports relating to various stages. Some systems also deal with interviewer pay claims and other administrative tasks. At ONS, for example, the centralized case management system integrates field and centralized processing, bringing together the sample address list, interviewing, field management, pay claims, coding, and editing. Some agencies have incorporated e-mail or other arrangements for sending messages to interviewers.

All agencies using CAPI provide their interviewers with modems to transmit completed interviews to headquarters. Most send work out to interviewers by the same means, but some (e.g., ONS) still send new instruments out on diskettes because the files are so large. Most agencies provide laptops configured in such a way that interviewers can select from a menu of survey tasks and work on more than one survey at a time.

With respect to CAPI, agencies employing regional fieldwork organizations (as do the Census Bureau and Statistics Canada) tend to have more complicated systems. The ABS has found CAPI case management to be a challenge because of the potential for case duplication, but it has recently developed a solution for its system (Henden et al., 1997). Automation of field case management complicates the survey process because it is often developed by programmers and analysts not based in the field, and is integrated with the physical production instrument, though usually it is supervised by the field organizational unit.

4.3.4   Effect of CASIC on Data Processing

Under PAPI, survey questionnaires were checked in, coded, keyed, and edited — regardless whether data collection was conducted by mail, telephone, or in-person. Under PAPI, most agencies adopted a two-stage approach to editing, at least on the more complex surveys or those that required coding. Paper questionnaires returned to the office were passed to clerks prior to data entry; these clerks coded text answers and carried out key edit checks. Once the data were in the computer, batch editing was carried out as clerks consulted the paper questonnaires to resolve problems. CASIC eliminates the separate data check-in and data-entry stages by incorporating many of the edit checks and coding tasks into the interview.

As a result several agencies (ONS, NASS, Statistics Netherlands) report that their data preparation units have been closed. (ONS does report an occasional need to contract out data keying of mail household surveys.) The Census Bureau closed its regional office data preparation units when collection for the Survey of Income and Program Participation (SIPP) was transferred to CAPI, moving any remaining demographic survey data keying to the Data Preparation Division. Additionally, with the conversion of the CPS to CAI, the Census Bureau closed its FOSDIC processing unit.

Some data are still collected on paper—diary, multimode, and business mail surveys providing the most common examples. In this circumstance, varied arrangements for data entry from paper have been adopted, sometimes within the same agency. Some agencies use fast heads-down keying, particularly for business surveys. Other agencies (Statistics Canada, NASS) have interviewers enter data from mail questionnaires. ONS and Statistics Netherlands use the same staff to enter and code data from expenditure diaries—in the case of ONS these staff are also telephone interviewers.

Business surveys and decennial censuses in Sweden, Britain, and the United States (both at the Census Bureau and BLS) have undertaken automated data capture to produce electronic images of paper documents. By using computers to recognize respondent marks and convert them into computer readable files, technologies such as OMR and OCR eliminate or greatly reduce the need for manual keying. These technologies also eliminate the need for microfilming and paper archiving of forms.

Edit checks carried out during the interview on less complex surveys avoid the need for an office editing stage. ONS and Statistics Netherlands do virtually no further editing of their labor force surveys once the data are transmitted by the interviewers. The Census Bureau has retained batch editing on the CPS, although the number of edit failures has been reduced. Most agencies have found that there is still a need for staff to run additional edit checks on complex surveys.

Despite the availability of computer assisted coding facilities, no agency has entirely eliminated manual coding. In some cases the coding staff are now interviewers working at home rather than clerical staff at headquarters. For example, ONS uses computer assisted coding in the interview for simpler questions, followed at home by postinterview clerical coding of occupation and industry, before completed cases are transmitted electronically (see Bushnell, 1995). Computer assisted coding and automated coding are widely used by agencies in connection with population censuses because of the high volume of coding work (see Speizer and Buckley, Chapter 12).

4.3.5 Interfaces with Other Data Processing Systems

Output files from data collection become input files for subsequent data processing, including sample weighting and estimation, tabulation, and data·

analysis. Various CASIC software packages provide different output file formats and interface with different exogenous survey software. The success of CASIC implementation is also affected by the agency computer platform and operating systems—DOS for CASIC may not interface well with client-server architecture using Windows (95 or NT) for data processing and telecommunication. Agencies' experiences are affected by the CAI software they are using—and by its output data features. The need for complex reformats depends on the data requirements of the summary and analysis system receiving the CAI data. The ONS Social Survey Division works in a customer-contractor environment wherein customers, specifying the required output formats, have been more or less amenable to changing their processing and analysis systems in response to ONS's move to CAI.

The Census Bureau found interfaces with processing software to be a major problem. Each survey (or group of surveys) had its own proprietary processing, tabulation, and analysis system maintained by its own group of programmers. Because of interface problems in connection with demographic surveys, general methods had to be developed for taking CAI output an routing it to individual processing systems as some surveys were moved to a SAS standard. For business surveys the growing use of CASIC methods prompted the development of a generalized processing system to cover all business censuses and surveys. At the Census Bureau the pressures of CASIC (technological and organizational) began breaking down the barriers created by the prior existence of separate processing systems for each survey.

To facilitate subsequent data processing, CASIC systems may need to receive both data and metadata from other systems and from output files. Thus, interfaces at both input and output ends are important. Many agencies have separate sampling systems passing details of the selected sample to the CAI system for assignment of cases to interviewers. On longitudinal surveys, CASIC methods facilitate the feeding forward of information about cases from one wave to the next. These methods facilitate the design of survey instruments that allow previously reported information to be used for edit checks, probing, and reconciliation. Such capabilities are used by several agencies in their labor force surveys and by the Census Bureau and ONS for interviewer follow-up of census data collection.

4.4   CHANGES IN STAFF SKILLS

The use of new technologies often requires different, generally undeveloped, staff skills. Existing staff resources and skills are no longer needed; extensive training for existing staff is required, failing which, new staff skills and expertise will have to be acquired. This section addresses the changing staff resources and skills required when CAI technologies are implemented.

4.4.1   Effect of CASIC on Interviewers and Their Supervisors

CAI technologies have their most obvious organizational effect on the interviewers. Somewhat surprisingly, most agencies report that CAI implementation has required no dramatic shift in their techniques of recruiting and training interviewers. Typing skills are now a big plus in most agencies, but previous computer experience does not seem to be critical. Interviewer training for CAPI may now take longer—ONS reported a one-day increase (Martin, 1995). CAPI interviewer training now generally includes a computer-based tutorial component emphasizing the technical aspects of the data collection process (how to use function keys, how to transmit data, how to recover from errors, etc.). The Census Bureau reported that the need for increased computer training has resulted in some transfer of training related to questionnaire content and basic interviewing skills from the classroom to home study.

4.4.2   Changes in the Responsibilities of Interviewer Supervisors

While the interviewer's responsibility and job description have remained fairly stable during the shift from PAPI to CAI, the same cannot be said of CAPI interviewer supervisors at all agencies. Traditionally, field interviewer supervisors or other clerical staff have reviewed questionnaires as a quality check on completeness and consistency and for evidence of falsification. The introduction of CAPI has made supervisory clerical review of completed interviews neither possible nor necessary. At the Census Bureau this case-by-case supervisory review is being replaced by analysis of keystroke files and statistical analysis of internal flags programmed to identify cases evidencing poor data quality or possible falsification. In more recent surveys, quality assurance procedures are integrated into the instrument so that field supervisors can quickly isolate and re-assign suspect cases. In the case of CATI, supervisors have long been able to monitor banks of interviewers, concentrating, as necessary, either on individual interviewers or on troublesome sections of the instrument.

A recent query of the Census Bureau's regional offices revealed several common observations about the organizational effects of CAPI on field survey management. One relates to the increased technical responsibilities of the supervisors. Supervisors may now have to troubleshoot hardware/software problems. In some cases they have become much more dependent on headquarters staff to solve these instrument problems. In the PAPI environment an incorrectly printed skip pattern can be fixed by a manual correction; a similar mistake in a CAPI instrument poses a much more daunting technical problem. This change in responsibility is echoed by Statistics Netherlands, where supervisors report an increasing need to answer questions of a more technical nature. In ONS many of these supervisory functions are carried out by headquarters staff. Interviewer supervisors have a greater training role, and their responsibilities have changed less with the move to CAPI than in other organizations. Both ONS and ABS have help desks at headquarters for technical queries; many other queries are dealt with by the headquarters field management staff.

4.4.3 Needs for Instrument Authoring

Implementation of CAI has made the role of the questionnaire designer rather murky. In a paper environment, questionnaire design experts have easy access to and control over the layout, wording, and design of the paper document. Skip patterns are obvious and easy for the interviewer to follow using general questionnaire design conventions (arrows, skip instructions, italics, etc.). This and other PAPI questionnaire design norms quickly changed with the advent of CAI.

In an automated environment one confronts a need for instrument authors, namely, persons knowledgeable in CAI programming languages. Our review revealed that for some agencies the questionnaire designers and authors are one and the same; for others, the tasks remain separate. In Statistics Sweden simple instruments are developed by the survey design staff with little specialist help. This arrangement is carried further at ONS. There, even on complex surveys, the norm is for survey designers to do the bulk of the programming with some help from computer experts on more difficult aspects. At INSEE both programmers and survey designers have developed instruments. Because of turnover problems among the design and analysis staff, INSEE has now decided to develop a team of specialist programmers. Statistics Netherlands mentions that there is no sharp division between the authoring knowledge of survey designers and that of computer specialists.

At the Census Bureau different arrangements for authoring CAI instruments have developed over time and on different surveys. In general, separate from the survey design staff, teams of authoring specialists have done their work. Survey designers specify the wording and layout of demographic CATI/CAPI questionnaires; staff proficient in CASES bear responsibility for authoring the instrument. Because CASES training at the bureau is not very accessible to questionnaire designers, few, if any, are capable of using it. The Census Bureau notes that many of the CASIC authors were not originally trained or hired as programmers. Rather, they were retrained as such. The authoring process is similar for the Census Bureau's many business CATI surveys; instrument development is primarily the function of programmers within the Technologies Management Office (TMO). For other CAI modes, such as business CSAQ surveys, the bureau has contracted out some of its instrument development.

When more widely accepted instrument software languages do not fit the needs of specific survey applications, some organizations have chosen to hard code the survey instrument in a programming language. Statistics Canada, BLS, and the U.S. National Center for Health Statistics have used this approach with Visual Basic, C, Clipper, or other languages. This approach requires authors to be familiar both with programming languages and with systems development.

4.4.4 Building New Staff Skills

CAI technologies engender a need both to retrain staff to use a new method and to hire or train additional technicians to develop, maintain, and support that new method. Most of the agencies represented in this chapter incline toward retraining rather than rehiring—possibly because they lack flexibility in restructuring the organization's workforce.

Martin (1995) reports from ONS that automation has meant more than just a change from doing one task to doing another. The challenge was to retrain staff to perform new functions. Most staff members found themselves dealing with a wider range of tasks: Survey designers had to learn to program in Blaise and become familiar with interviewer-friendly design principles; field managers had to become aware of the capabilities of CAI and also become more technically oriented; programmers had to support the whole of the survey environment rather than just carrying out one step in the survey processing chain.

Similar trends are noted in the Netherlands and at NASS. Open training opportunities encourage employees to become "generalists" rather than "specialists." If employees are to be successful, they must be cross-trained in information technologies, survey methodology, and subject matter areas. Within the Census Bureau and Statistics Canada, the teamwork approach is now more used to implement CAI methods. This management style asks technical and subject matter experts to work side by side, thus increasing each other's knowledge of diverse parts of the operation and fostering communication among professional staff. Note, however, that this management style may also increase the number of staff required to do the work, thus increasing the costs of the operation.

4.5 EFFECTS ON ORGANIZATIONAL STRUCTURE

Statistical agencies, like other organizations, are structured according to the functions they perform. As technologies are implemented, tasks and functions change. These internal and external changes have significant effects on the organizational structure. This section discusses working relationships between organizational units and the process of restructuring an organization to achieve the appropriate mix of staff resources in a CASIC environment.

4.5.1 Effect on Working Relationships between Organizational Units

As organizations attempt to put new technological systems in place, they confront a need both to restructure organizational units and to redefine the working relationships among those units. Pre-CASIC organizational work relationships are not always compatible with the technical resource requirements of CASIC technologies. Both the Census Bureau and Statistics Canada are large complex bureaucracies where different parts of the organization bear responsibility for different groups of applications and different aspects of the survey process. Methods and systems for adopting CASIC (and the specific CASIC technologies adopted) varied widely across the Census Bureau's major program directorates—Demographic, Economic, and Decennial—so also between regional and headquarters offices within Statistics Canada.

At the Census Bureau early development activities were autonomous within directorates; sometimes even within divisions and surveys. This situation changed in 1990 when the new Census Bureau director identified implementation of CASIC methods as one of four major goals for the Census Bureau. Toward this end a bureau-wide CASIC Office was established to stimulate and coordinate CASIC procedures and activities.

The operational implementation of established CASIC technologies was later passed from the CASIC Office to the Technologies Management Office (TMO). This office became responsible for automated questionnaire authoring, centralized CATI, and case management automation. The adoption of CASIC changed the functions of the division responsible for data preparation; some were diminished or eliminated (e.g., keying, scanning, mail preparation, forms check-in), but others were created (CATI and CADE). Inevitably, the creation of the CASIC Office, and later the TMO, caused changes in the working relationships between the directorates. One result was to increase the number of players involved in a typical survey project.

At Statistics Canada, CATI (and later CAPI) developed independently in the regional and central offices. Even today these two organizational levels within Statistics Canada exhibit different CAI environments. Both conduct surveys dealing with economic and social topics, but the regional offices now have a steering committee for case management systems that cuts across the organization. Additionally, the CAI Development Center serving the regional offices provides programmers for teams working to develop CAI instruments. It also establishes CAI authoring standards and specifies generic CAI applications.

Statistics Netherlands reports that working relationships have become more decentralized under CAI. More of the survey processing activities reside within the subject matter departments. Consequently, the Automation Department no longer processes data; instead, it markets itself as a "partner in innovation." This unit now offers information system services to the subject matter divisions. These services include software and hardware tools, training, and support.

The Social Survey Division at ONS has always operated using matrix management. Project teams for the different surveys have been formed from staff in different functional areas (research, field operations, coding, computing, etc.). Many staff members work on more than one survey, moving regularly between the major continuing surveys. This management model applies even now, but the move to CAI resulted in a substantial relative increase in the number of staff members who provide support functions across all surveys. It was soon recognized that integrated computing support had to be provided for all parts of the survey process. This led ONS to move the computing support function for surveys from the Information Technology Division to the Social Survey Division.

4.5.2 Rescaling and Refurbishing an Organization to Achieve an Appropriate Mix of Resources

Perhaps the most obvious organizational change occasioned by the adoption of new technologies has been the formation of new organizational units and the dissolution of others. All of the agencies surveyed in this chapter reported a great reduction (and in some cases, the elimination) in the need for data entry, coding, and editing staffs. At ONS the separate coding and editing branch has been eliminated; the small remaining staff was merged into the Field Branch. Neither Statistics Netherlands nor ONS supports a separate data-entry unit. While Statistics Canada still supports a mixed-mode data processing system (DC2), the implementation of CAI has caused this system to be used much less than originally anticipated. The decreased utilization of the DC2 system stems directly from the decreased need for heads-down data entry.

While some units will decrease or be eliminated, others will be formed or built up. The increase in centralized CATI operations has resulted in the reorganization of interviewer units within several agencies. At Statistics Sweden, in 1994, the central telephone group was increased to 50 workstations, with an ability to expand up to 80 (Blom, 1994). Similarly, the Census Bureau has now expanded its centralized CATI facilities to three sites throughout the United States for a total of 250 dedicated CATI workstations. Implementation of CATI at NASS was completed in 1992, with CATI operating in 42 of the 45 State Offices. There telephone interviewers use regular office space for nighttime calling. The Census Bureau also established two new units (TMO and the CASIC Office) to develop and manage CASIC technologies throughout the organization.

4.6   COMPARISON OF ORGANIZATIONAL STRATEGIES FOR TECHNOLOGY IMPLEMENTATION

The preceding sections have reviewed the effects of funding strategies, survey operations, human resource needs, and organizational structure on the process of implementing CASIC methods in government statistical agencies. In this section organizational strategies are compared as we seek to identify the factors that contribute to success. But there are many different strategies for implementing CASIC methods, and this organizational diversity makes it difficult to generalize.

Few agencies reported that they had a comprehensive organizational strategy for accommodating CAI. In all cases the technology was implemented first; organizational changes followed later. Some agencies have yet to make the organizational changes. This suggests that they may not have achieved the optimal organizational structure needed to fully harness CASIC's potential.

Although most of the organizations represented here have successfully moved to CASIC methods, they differed in the timing and extent of changes. It is important to recognize that some agencies, such as Statistics New Zealand and NASS, have not yet been able to implement CAPI. Likewise, the ABS encountered higher costs than expected to implement CAI and are reconsidering their plans. These organizations have chosen to postpone CAPI technology because no financial argument could be made that would justify the cost of initial investments. In these organizations the survey business process did not optimize the number of survey contacts for each laptop.

4.6.1   Organizational Differences Prior to CASIC

Part of the reason that different agencies adopt different strategies is simply that they exhibited crucial differences before any move to CASIC was made. Two differences—size and degree of centralization—have led agencies to adopt different strategies.

The Census Bureau and Statistics Canada are very large organizations compared with Statistics Netherlands or ONS (and prior to 1996 ONS was two much smaller agencies). Large agencies find it more difficult to effect major changes; on the other hand, they may have more money available for research and development work to speed the introduction of CASIC methods.

The majority of countries have central statistical agencies responsible for all government surveys and statistics. The United States and Britain are clear exceptions: They maintain separate statistical agencies and also support statistical units in other government departments—units that are the clients for many of the major government statistical surveys. These client agencies have a profound influence on technological development within the statistical agencies. In Britain some government surveys are contracted out to private sector survey organizations. These contractors may have a positive influence on technological developments, but they too may have to be persuaded that it would be beneficial to use new technologies.

As a variant on both of these issues, agencies differ in their degree of bureaucratization. Bureaucracies are better at maintaining the status quo than in implementing major changes, particularly changes that affect so many parts of a large and complex organization. If a large organization is subdivided into relatively autonomous smaller units empowered to make changes independently, size and bureauacracy are less of a barrier to overall change. This was essentially the situation at ONS when the Social Survey Division functioned autonomously within a much larger government survey organization. On the other hand, the lack of coordination on CASIC activities at the Census Bureau resulted in many duplicative efforts across divisions and directorates.

Organizational strategies will differ when one agency has a coordinated approach to both household and business surveys while another handles them separately. This affects in turn the degree of centralization one might expect: In Britain, with its decentralized statistical system, different government departments were responsible for household and business surveys until April 1996. Although now in one organization, functionally they continue to be separate. Within the more centralized system prevailing in the Netherlands, there is a greater appreciation of technology issues that recur across different types of surveys.

Even for household surveys, agencies differ in their approach to assignment of survey tasks. The Field Directorate at the Census Bureau provides services for each of the program directorates—Decennial, Demographic, and Economic. The ONS, in contrast, is structured to provide, within each division, all of the services needed to conduct its surveys, although less so in the case of business surveys (as compared with social surveys). This gives the divisions greater autonomy for innovation and technology implementation. Furthermore, staff members may work broadly for a specific survey, performing many functions therein, or they may work on a more narrowly specialized function for a number of surveys. The latter type of assignment structure tends to favor an organization-wide perspective over a survey specific one. It is the organization-wide perspective that is most likely to embrace and advance technological innovation.

4.6.2 Facilitating Change

Baseline differences among agencies clearly shape their approaches to technological development—and these differences can only be magnified by the varied pressures agencies experience. For those statistical agencies that provide a service to other government agencies (as in the United States), pressures to adopt CASIC methods can come from outside the agency. Government survey organizations (e.g., Statistics Netherlands, Statistics Sweden, and ONS) tend to be innovators in their own right, as Couper and Nicholls describe in Chapter 1.

If CAI technologies are to be successfully adopted by a government agency, its strategic action plan must enjoy strong support from management at all levels. As we have suggested in this chapter, CAI changes an organization's staff requirements, functions, and tasks. More to the point, a change in organizational structure may be required to achieve cost savings. At the very least, the adoption of CAI will disrupt "business as usual." To minimize this disruption and gain cooperation across the organization, it is crucial to publicize CAI as a priority and make it highly visible within the organization.

Support for CAI technologies was elevated to the executive level at the Census Bureau in 1990 when the new director, Barbara Everitt Bryant, established the executive position of CASIC Manager and placed it on the Director's staff. This position became the focal point of leadership and accountability for CAI efforts at the bureau, subsuming initiatives previously developed within divisions or directorates on an autonomous basis. This high-level placement of the appointment made it easier to gain multidivisional acceptance of a CAI strategic plan. It also aided efforts to first obtain (and continue to receive) funding for research and development, and for establishment and maintenance of the new systems.

At ONS, CAI was first introduced by the relatively autonomous Social Survey Division, with little effect on the rest of the agency. Management commitment and strategic direction occurred at the divisional level, and was therefore much easier to control. The organizational climate was quite different, initially placing more emphasis on innovation and "trailblazing"; less on formal strategic plans until CAI development was well underway. CASIC initiatives on business surveys and the decennial census have also taken place largely at the divisional level. There is no coordination of CASIC activities across the whole of ONS.

In the early phases of CASIC development at Statistics Netherlands, Wouter Keller (then head of the Statistical Methodology Department) was a key instigator of client-server computers, a new technology automation policy, and standardization of hardware and software. The first CATI experiments were carried out in his department, and his vision for automation of the survey process had a dramatic effect on Statistics Netherlands and other government survey organizations that followed the lead of Statistics Netherlands. The development of the Blaise software greatly advanced the capabilities of CAI systems.

After some initial false steps Statistics Canada recognized that some corporate mechanism was required to set and monitor cost goals for CASIC development and implementation. Units of the organization requesting funds for CASIC were required to submit to a senior management committee a plan and timetable demonstrating how funds would be repaid through savings within three years. The progress of CASIC implementation efforts was then monitored by that committee.

If change is to be made, funding arrangements are clearly crucial. Agencies need development funds in the first instance and may, for implementation, need an investment in equipment. A separate funding initiative will then be required unless, perchance, a stock of underutilized personal computers is available at an existing CATI site. CAPI is more of a problem in this regard, given that each interviewer needs a laptop computer. But government agencies tend to find themselves in a more favorable financial position than run-of-the-mill private sector survey organizations do. Government agencies carry out large continuous surveys, wherein initial investment costs can be amortized over a number of years and spread over a large volume of survey activity.

Management monitoring of technology implementation also plays a key role in achieving organizational objectives for CASIC, not the least of which is cost containment (or cost savings). Statistics Canada has a committee that disburses development money and the proceeds of repayment agreements; at the Social Survey Division of ONS cost-accounting mechanisms restrained CASIC development costs within identified funding sources. At the Census

Bureau the lack of corporate mechanisms able to monitor CASIC implementation has affected negatively the organization's ability to realize cost savings.

4.6.3 Diverse Organizational Strategies

The government organizational opportunity to spread total costs over a longer time with a large volume of survey activity argues for moving to CASIC methods as quickly as possible, not only to share costs but also to minimize the period during which two different modes must be supported. But few agencies have moved all of their household interview surveys to CAI. Both ONS and Statistics Netherlands have one remaining PAPI survey each (in both cases, diary surveys), even though they were among the first to adopt CAPI for face-to-face interviewing.

A factor that seems to have slowed progress at the Census Bureau was the decision to undertake major revisions of surveys at the very time when automation was introduced. In theory this strategy allowed the agency to more fully utilize the novel instrument design features of CAI, but such a strategy also occasioned a much longer development and testing phase than was experienced at ONS and Statistics Canada, where little redevelopment work was undertaken in conjunction with automation.

Because CAI requires an integrated type of survey processing, tasks can no longer be managed and processed in a linear sequence. But not all agencies have changed their structures to efficiently accommodate these integrated processes. The Social Survey Division of ONS was restructured significantly when surveys were moved to CAPI. Statistics Netherlands recently reorganized the entire agency to better reflect the resulting changes in its functions after automation of its data collection and survey processes. Other agencies have adopted less radical approaches.

In view of the differing pre-CASIC organizational structures and cultures that one will encounter—to say nothing of diverse political pressures—it is virtually impossible to predict which strategy of organizational change will work best. One can, however (after the fact), readily determine whether an agency has managed to realize the promised advantages of CASIC: timeliness, improvements in data quality, and cost reduction. It is apparent that agencies have experienced improvements in data quality at the point of collection. There is consequently a lesser need for complex editing later on. Even though the total time required to carry out a continuous survey might not have been reduced (a consideration that looms large in the experience of most governmental statistical agencies), the time elapsed from initiation of data collection to availability of results has been much improved. Moreover, some agencies (not all) report cost savings. This is more often true with CATI than with CAPI. Cost savings are brought about through planned resource management.

Most significantly, there is some indication that agencies working under extreme pressure to reduce costs tend to adopt a minimalist stategy (e.g.,

Statistics Canada and ONS), but it is precisely this pressure that is most likely to achieve cost savings. In summary, the more pressure there is to achieve greater timeliness, improve data quality, and reduce costs, the greater is the likelihood that a successful transition to CASIC will be effected— whatever organizational strategy be adopted.

ACKNOWLEDGMENTS

The authors wish to thank the following contributors from the various statistical agencies represented in this chapter: U.S. Bureau of the Census (Shirin Ahmed, Richard Blass, Chester Bowie, William Nicholls II, Greg Russell), Statistics Canada (Gregg Alexander, Rolly Jamieson, Sylvie Michaud, Brian Williams), Statistics Netherlands (Jelke Bethlehem), Office for National Statistics (Tony Manners, Graeme Walker), Australian Bureau of Statistics (Fred Wensing), Statistics Finland (Vesa Kuusela), Institut National de Statistiques et des Etudes Economiques (Francoise Dussert), Statistics New Zealand (David Archer), Statistics Sweden (Evert Blom), U.S. Bureau of Labor Statistics (Catherine Dippo, Richard Clayton), and the U.S. National Agricultural Statistics Service (Jack Nealon). The views expressed are those of the authors and not of their employing organizations.

CHAPTER 5

Organizational Effects of CATI in Small to Medium Survey Centers

John Tarnai
Washington State University

John Kennedy
Indiana University

David Scudder
Boise State University

5.1 INTRODUCTION

This chapter describes the experiences of three small- to medium-sized survey organizations that have been conducting computer assisted telephone interview (CATI) surveys for the past decade. Since these are our organizations, we are familiar with them and with the way CATI has affected organizational aspects. We focus on CATI technology because we believe that this is the dominant CASIC technique used by small to medium survey organizations, and that CATI is the motivating force behind many of the other changes in survey organizations (Brent and Anderson, 1990). Some of our discussion pertains to computer assisted personal interviewing (CAPI) as well; however, CAPI is only rarely used by small- to medium-sized survey organizations (see Collins, O'Muircheartaigh, and Sykes, Chapter 2; Groves and Tortora, Chapter 3). We hope that a description of our experiences with CATI and the organizational issues that we have had to confront will be useful to others engaged in similar work.

An organization is made up of people, resources, work processes, and an organizational structure. When we discuss the organizational effects of CATI,

Computer Assisted Survey Information Collection, Edited by Mick P. Couper, Reginald P. Baker, Jelke Bethlehem, Cynthia Z. F. Clark, Jean Martin, William L. Nicholls II, and James M. O'Reilly.
ISBN 0-471-17848-9 © 1998 John Wiley & Sons, Inc.

we are referring to the effects that CATI has on all of the elements that comprise an organization. The general approach taken in this chapter is to outline the major organizational changes or issues that arise when organizations undertake CATI surveys. The issues identified are derived from a review of the CASIC literature, from the experiences of our own survey centers, and from the results of a small survey of CATI organizations. We first summarize the most prevalent issues identified in the CASIC literature. The next section describes and compares the implementation of CATI by each of the three survey organizations and the organizational changes that resulted. Finally we summarize the major organizational issues associated with the use of CATI and describe the results of an informal survey of CATI organizations.

5.2 LITERATURE REVIEW

While there has been relatively little research on organizational effects of CASIC technology, a number of organizational issues associated with CATI are mentioned in the literature on telephone surveys and CATI technology. We identified five major categories: (1) managing hardware and software, (2) human resource/staffing changes, (3) changes in designing and testing questionnaires, (4) changes in sample administration; and (5) issues in managing survey data.

Discussions about the kinds of hardware and software considerations that organizations must deal with in computer assisted interviewing are common (Carpenter, 1988; Curry, 1987; Frey, 1989; Henne, 1993; Saris, 1991). Articles generally include considerations of selecting the right hardware and software, maintaining and financing systems, dealing with computer and software upgrades, evaluating the features of different CATI systems, selecting appropriate software, and estimating the costs of establishing a small- to medium-sized CATI facility.

A number of authors have recognized the human resource changes involved with implementing or using CATI (Berry and O'Rourke, 1988; Curry, 1987; Martin and Manners, 1995; Weeks, 1992). Not only are new, more qualified personnel required to manage CATI technology, but the recruitment and training of interviewing staff is affected as well. In a small survey of telephone centers conducted by Berry and O'Rourke (1988), the principal changes reported by survey respondents due to implementing CATI include recruitment of telephone center staff, need for retraining of supervisory staff, selection and training of interviewers, and changes in salary and career patterns of telephone center staff. A review of CASIC methods (Weeks, 1992) concludes that CATI reduces interviewer supervisor time but requires more interviewer training. Shanks and Tortora (1985) find that many survey organizations have changed their division of labor as a result of CATI, and they expect this trend to continue as the technology for integration of separate tasks improves. Among organizations responding to a survey in 1979 and again in 1987 (Spaeth, 1987), the most frequent problems mentioned in converting to CATI deal with staff and interviewer acceptance and retraining needs. Groves (1983) notes that there is an increased need for coordination of work with CATI because the hardware and software now become the central work distribution system for the organization. He also suggests that administration of CATI may be less problematic in smaller organizations because employees can assume multiple roles, whereas in larger organizations, entire units are often involved in adapting to new systems and procedures.

It has been mentioned frequently that CATI requires more up-front time in designing and pretesting questionnaires (Curry, 1987; Driessen et al., 1987; House, 1985; House and Nicholls, 1988; Saris, 1991). Designing a CATI questionnaire is a complex process involving more precise specification of valid response values and skip logic than usually occurs with paper and pencil instruments. Most authors conclude that the net effect of this increased time spent on questionnaire authoring is to decrease back-end data management time (Weeks, 1992).

Changes to sample administration and call scheduling processes as a result of CATI are also mentioned frequently in the literature (Baker and Lefes, 1988; Klehn, 1993; Palit and Sharp, 1983; Weeks, 1988). CATI systems have the capability of providing substantial data about sample disposition and the status of survey samples, thereby enabling quicker control over sample allocations. Managing CATI survey data has been discussed both in terms of the types of data produced by CATI and its effects on survey operations (Carpenter, 1988; Bethlehem, 1997). Dealing with open-ended text that is in machine readable form rather than paper is one issue that has been recognized but rarely addressed (see Groves, 1983).

A variety of data management issues occur with CATI, involving primarily the creation, editing, and finalization of survey data (Baker and Lefes, 1988); the importance of managing system failures and backup of data files (Curry, 1987); and the effects of peak staffing problems; staffing for research and development; systems design, programming, and maintenance; training; and emergency procedures (Federal Committee on Statistical Methodology, 1990b). One question is to what extent manual procedures are needed to accompany CATI procedures in case of hardware or software failures (Shangraw, 1986).

The relevance of these five issues is confirmed in the experiences with CATI that our own survey centers, have had, and in the data obtained in an informal survey of CATI sites that we conducted. We have added three additional organizational issues which we believe are relevant to current CATI organizations: (6) changes due to networks (7) reporting survey results, and (8) administrative information.

5.3   EXPERIENCES WITH USING CATI IN THREE SURVEY CENTERS

This section presents a brief summary of the major organizational issues faced by each of our three organizations in conducting CATI surveys. The three organizations have different degrees of experience with CATI technology, and thus they represent a broad range of CATI experiences.

The Social and Economic Sciences Research Center (SESRC) at Washington State University uses CATI at its most basic level. The Center developed its own in-house CATI system which it has used for the past sixteen years. The interviewing stations in the SESRC consist of stand-alone computers that are not networked. This organization is currently making a transition to a networked computing environment.

The Social Science Research Center (SSRC) at Boise State University recently made the transition from stand-alone computers to a networked CATI system, and it uses Ci3 CATI software. The SSRC has also automated its sample administration procedures and many of its accounting procedures as a result.

The Center for Survey Research (CSR) at Indiana University is the most highly computerized of the three. It has a networked CATI system running the CASES software on a Novell network linking a number of other users.

5.3.1   Social and Economic Sciences Research Center, Washington State University

The SESRC conducts an average of 45 telephone and mail surveys each year and has a professional staff of 15 to 20 people, depending on the number of active projects. It has had a 14-station telephone survey facility since 1971, and decided in 1982 to develop its own in-house CATI system, at a time when there were few commercial systems available for small survey centers. The initial system was developed in Pascal for Apple microcomputers but was converted in 1988 to a DOS-based system. The SESRC has continued to add interview stations and now has 36 interview stations located in three separate rooms. All stations are equipped with 386 or 486 PCs running DOS as stand-alone units that are not networked. Thus all sample administration and call record management continue to be accomplished by paper and pencil, and collection of daily survey data files is accomplished by going to individual stations and collecting floppy diskettes. These files are then merged into one master data file and uploaded to a mainframe computer for data cleaning and analysis.

CATI Software
The SESRC's CATI system is maintained and upgraded by one programmer, who is assisted by other data management staff when feasible. The system is also designed for use in data entry of mail or self-administered questionnaires, since it has a double-entry verification feature. Having an in-house developed CATI system provides great flexibility in being able to make adjustments in

the software when a new feature is desired. This seems to happen regularly as new problems are encountered with new survey projects, requiring new capabilities for the CATI system. However, each new version invariably has a bug and must be thoroughly tested before it can be used for production interviewing. The CATI system is not currently set up for use on a network; thus it is limited for sample administration and call record management. The system is primarily used for questionnaire authoring, administration of the questionnaire on the computer screen, and recording of respondent answers into a survey database. For these reasons the SESRC is evaluating new CATI software for use with a new network that will allow it to link all 36 computers in the three separate interviewing rooms.

Key Organizational Issues
The most important issues that the SESRC confronts with its CATI system include (1) determining whether to continue to develop its in-house software or to purchase or lease new CATI software and, if the latter, what kind of system and network to purchase; (2) identifying a source of funds to purchase networking equipment, cables, and CATI software; (3) finding a way to make the transition without disrupting ongoing surveys; (4) finding the time and funds to train staff in network software and a new CATI system; and (5) developing the procedural changes required to convert from a paper call record system to a computerized one.

5.3.2 Social Science Research Center, Boise State University

The SSRC was established in 1991 to provide survey research services to university departments and state agency clients. The SSRC has between four and ten temporary staff, some of whom are part-time and some full-time, depending on the number of active projects. Initially telephone surveys were conducted by paper and pencil, but with several commercially available CATI systems in existence, a decision was made in 1993 to purchase ten stand-alone PCs, and Ci2 CATI. At the time the limited capabilities of the computers and the software meant that there was no archive of survey results, no ability to account for survey costs, and no ability to administer samples. With funding from SSRC reserves and university grants, the SSRC upgraded its equipment and CATI software in 1995. The SSRC now has 13 networked interview stations also linked to the university-wide network. The network provides the SSRC with the ability to use other software (dBase, Excel, and MS Project) for survey samples, budgeting, and planning and to have an in-house accounting system that is compatible with the requirements of the Federal Acquisition Regulation (FAR).

CATI Software
The SSRC uses Ci3 CATI, and Ci3 screen writer for questionnaire authoring, administration of the interview, and collection of survey responses. In addition

the system provides for proactive dialing through a modem, call scheduling, and sample management. Software for statistical analysis and variance estimation is separate from the CATI system, as are programs used to prepare survey reports. The instructions for managing call scheduling are much simpler using the CATI system than they were under the previous paper and pencil method. Interviewer turnover is relatively high, and few interviewers still with the center were trained to conduct paper and pencil interviews or trained under earlier versions of the software. However, there was no difficulty in training interviewers on how to use the system. The CATI system permits a number of very useful administrative reports about interviewer productivity and call disposition management. Changes in reporting and analysis of survey results are minimal, although the CATI system has made it easy to export data files into several different formats for data analysis. While the CATI system has increased the overall efficiency of the SSRC and the quality of survey data provided to clients, it has also resulted in increased costs of hiring and retaining telephone supervisors.

Key Organizational Issues
The issues faced by the SSRC in making the conversion to a networked system and to an upgrade in its CATI software were substantial. One was simply the amount of time required to make the transition—13 months from the initiation of planning to the first survey under the new system, and two years before all problems with the system were resolved. A second problem was finding that some hardware was incompatible with the network software. A third problem was the difficulty of having to conduct surveys while the new CATI software was still being installed and debugged. A fourth was the need for new staff with the technological ability to maintain the network and CATI system. The fifth challenge is to find sufficient paying survey projects to fund the cost of maintaining and upgrading the CATI facility.

One of the main changes that occurred as a result of the networked CATI system was in the role of telephone supervisors, who now assume explicit responsibilities for programming and computer maintenance. This required increasing their salaries, and providing training in use of CATI and screen writing software. New telephone supervisors with more advanced computer skills were recruited and hired as needed for funded projects. To compensate for the increased time that supervisors had to spend on programming and managing the computer system, their responsibilities for hiring interviewers were transferred to other staff. With the sample management features of a networked CATI system, telephone supervisors could also spend more time monitoring the progress of a survey rather than dealing with sample management issues.

5.3.3   Center for Survey Research, Indiana University

The CSR at Indiana University was established in 1981 as part of the Institute of Social Research (ISR) which is the research home for the Department of

Sociology. The CSR from its inception has nearly all of its survey, training, cost accounting, and management procedures computerized and shares computing resources including hardware, software, and support staff with the ISR and Sociology. This arrangement allows all three organizations (CSR, ISR, Sociology) to acquire and support a higher level of technology than any of the three could do alone. The CSR conducts all of its telephone surveys by CATI. The survey facility initially consisted of eight interviewing stations connected by serial cable to a minicomputer, which also served faculty, staff, and students in ISR and the Department of Sociology.

CATI Software

The first CATI system used by CSR was CASES, version 1, which was used principally for interview administration and sample management. In 1988 the CSR upgraded its CATI system to a Novell network and the then current verison of CASES (3.3). There were substantial differences between the earlier and this later version, which required many changes to CSR survey and management procedures as well. The current CATI system uses CASES 4.1 on a network run under Novell. There are approximately 200 users on the network, and approximately 150 computers connected to the network; only about one-fifth are CSR staff. The network supports video monitoring of production interviewing. The CSR has already decided that in the near future it will change its operating system again in order to take advantage of the improvements that are occurring in microcomputer software and hardware.

Key Organizational Issues

The main issues confronted by CSR as it has made the transition from earlier systems to later CATI systems include the following: Like the experience of the SSRC, a first challenge was the amount of time required to upgrade to a new CATI system, which averaged about two years. A second issue was that upgrades tended to involve learning a new network system, learning a new CASES upgrade, and learning new computerized survey functions, all at the same time. A third problem was having to put survey operations on hold for ten weeks while the new network and CATI software were being installed and tested. A fourth challenge was realizing that additional support staff would be needed to maintain the network, to track and solve hardware and network problems, manage backups, manage software changes, and set up new computers. A fifth issue is the continuing concern over how to fund maintenance upgrades of the network, computer equipment, and CATI software.

5.4 ORGANIZATIONAL ISSUES IN MANAGING CATI PROCESSES

Our three survey organizations have encountered and dealt with many organizational issues associated with using CATI during the past decade. A number of these issues are also confronted by large survey organizations as demonstrated by other chapters. However, smaller organizations may have an advantage in dealing with these issues because employees are more likely to perform multiple functions (Groves, 1983) and because the administrative process for changing survey procedures may be more flexible (Dillman, 1996). This seems to be true for our organizations, since job descriptions for our employees keep changing depending on the current critical needs for providing assistance in programming CATI questionnaires, managing CATI data, dealing with upgrades and maintenance of software and hardware, and testing and debugging CATI questionnaires. We have often found that better results are achieved when interviewing staff are involved in multiple other responsibilities associated with the CATI system. We were interested in knowing whether our experience with CATI was true of other similar sized survey organizations as well.

5.4.1   Results of a Survey of Small Survey Organizations

To identify some of the organizational issues faced by smaller CATI survey centers, we designed an informal mail survey of all survey organizations listed in the 1996–97 AAPOR/WAPOR Blue Book, including those with international addresses. Membership in the American Association for Public Opinion Research (AAPOR) and the World Association for Public Opinion Research (WAPOR) includes survey representatives from academic institutions, commercial organizations, government agencies, and nonprofit organizations.

The questionnaire was mailed to all 204 organizations listed in the Blue Book, including 148 U.S. and 56 international organizations. After follow-up telephone calls, we obtained a 64 percent response rate. Twenty-four responding organizations identified themselves as large, and 63 did not conduct surveys or did not have a CATI capability. This left 44 who identified themselves as small or medium survey centers with CATI facilities. In the next section we present the results of this survey in the context of the eight organizational issues presented earlier.

5.4.2   Managing Hardware/Software

A fundamental organizational change with CATI is having to spend staff time and organizational funds on computer hardware and software, and on maintaining equipment and software upgrades. In the early days of microcomputers, making decisions about the kind of system to get, whether to purchase or lease a system, and how to maintain it could involve substantial effort and anxiety because of the rapid pace of technological change. A common recommendation then was that CATI software should be selected first and hardware should be matched to the requirements of the software (Saris, 1991); otherwise, incompatibility problems could surface. However, recent advances in standardization of operating systems, hardware configurations, and software make this issue less important than others.

Cost

The substantial cost of CATI hardware/software is one of the first issues faced by an organization. Initial costs of establishing a CATI facility can be substantial (Frey, 1989). Because of this, our own organizations have adopted an incremental approach to purchasing, starting out often with fewer than ten interview stations and then adding or upgrading equipment as funds become available. The funds to pay for CATI systems come from a variety of sources, but the two main sources identified in our survey of organizations are accumulated cash reserves (59%) and fees charged to survey projects (51%). These sources correspond to the way our own centers are funding CATI.

Type of Software

A number of different CATI systems are on the market and deciding which software to purchase can be a major undertaking for survey staff. For our own survey organizations this decision was based mainly on cost, ease of use, whether the software would work within the existing computer environment, and what features the system had. Once a decision is made about a CATI system, it becomes difficult to change to another system because of the amount of time and resources spent learning a particular CATI system. Our own organizations demonstrate this principle; all of us have remained with the initial CATI system (whether purchased or developed in-house). In our survey 46 percent of survey organizations responded that CATI features were the most important factor in selecting their particular CATI software, compared with 18 percent reporting cost, and 15 percent ease of use as the most important factor.

Type of Hardware

Deciding on hardware is often a matter of cost and match with existing equipment. The main factor identified by the survey was cost, with 28 percent of organizations selecting equipment on the basis of this factor. Another 23 percent reported that the main factor was the match with the organization's existing computer hardware. Only 18 percent of organizations considered CATI features as the most important reason for selecting the specific equipment.

Maintenance

Maintaining hardware is an ongoing issue because computer equipment has to be set up, installed, made ready for users, tested, and repaired if things break. Some organizations have dedicated staff for these kinds of tasks; others have people in the organization who are identified as the technology experts but who normally have other job responsibilities. The SESRC, for instance, has one supervisor and one data manager, both of whom are also very knowledgeable about computers and who are therefore called on to assist in solving technological problems. In our survey 87 percent of organizations reported that they have in-house technical programming staff who maintain their CATI hardware and software. Fifty-six percent of organizations indicated that other in-house survey staff also deal with these issues. A related issue has to do with equipment replacement and how to fund new equipment. Our own organizations charge survey projects a computer use fee, which pays for new or replacement computers each year. About 50 percent of the respondents to our survey indicated that they charged a similar free for CATI surveys, which was used to pay for software upgrades and hardware maintenance.

Upgrades

Upgrading and maintaining software for CATI is also an issue for both in-house and purchased or leased CATI systems. Even new programs may have "bugs" that are not encountered until production interviewing has begun. Software upgrades or fixes must then be made to a network or to individual interview stations. All three of our organizations have upgraded our CATI software several times. The SESRC has upgraded its progams at least once a year over the last 15 years; having a system that was developed in-house made it easier to upgrade frequently. What we have learned through these frequent upgrades is that many problems are created when different versions of CATI software are operating at the same time. Failing to upgrade all interview stations and computers can lead to further problems later with data sets having data in different formats. This can be a data management nightmare and reinforces the idea that upgrades and fixes should happen between surveys rather than during a survey, if possible. Upgrades to commercial CATI systems can also occur fairly often. In our survey only 15 percent of organizations indicated that they have not experienced any upgrades to their CATI software. On the other hand, over 38 percent reported having had more than two upgrades to their systems.

An additional advantage of microcomputers is that with the increasing variety of available software, they become valuable for purposes other than telephone interviewing, when production interviewing is not in progress. Most organizations use their computers for other CASIC applications, including computer assisted data entry (CADE), computer assisted self-administered questionnaires (CSAQ), computer-aided instruction, word processing, and data management/analysis. Of the organizations responding to the survey, 85 percent use their CATI system for other survey related purposes. About 25 percent indicated using CATI computers for instructional purposes, 25 percent reported using it for self-administered questionnaires, and 50 percent reported using it for data management and statistical analysis. The dominant secondary use of CATI systems was for data entry purposes, reported by approximately 75 percent of respondents. Because of the many mail surveys undertaken by the SESRC, its CATI system was specifically programmed for double entry verification of self-administered questionnaires. A side benefit of using a CATI system for data entry work is that it can also be a good way to train telephone interviewers on the basics of interacting with a CATI system, which is what frequently happens at the SESRC.

5.4.3 Human Resource Changes

Managing computer equipment, knowing how to maintain a network, dealing with software problems, and knowing how to install and repair computers become major administrative issues for organizations with CATI systems (see Clark, Martin, and Bates, Chapter 4, for the large organization view of this issue). A variety of ways have evolved to deal with these kinds of administrative issues, with the more technologically oriented centers hiring dedicated computer experts and the less computerized environments emphasizing training of existing staff. Among our own organizations the CSR is the most computerized, and it hires dedicated programmers to maintain its CATI system. In comparison the SESRC trains existing staff to maintain its system. The SSRC also trains its own staff but has had to hire more experienced people to manage its CATI system. Our survey indicated that 87 percent of survey centers rely on their own in-house technical programming staff for maintenance of CATI software and hardware, and 57 percent also use other survey staff for this purpose. Twenty-eight percent report contracting for these services, and 36 percent indicate that CATI vendors provide maintenance services.

Interviewers

Interviewing skills are somewhat more demanding with CATI. The basic skills required of interviewers include how to log on to the CATI system, how to interact with the computer through a keyboard and mouse, and sufficient typing skill to enter respondent comments and other text into a computer. All of these skills are trainable, although it is generally easier to find people who already have typing skills than to train people in this skill. Some organizations recruit and hire only interviewers with computer skills. This is not too difficult in university environments where students have become accustomed to working with computers. Interviewers are on the whole positive about using computers to conduct interviews (Couper and Burt, 1994), and we have encountered no reports of negative experiences, except for initial fears of dealing with computers. Our own observations are that once they are taught how to use the computer and overcome their initial fear that they will "break" something, interviewers prefer CATI to paper and pencil surveys. Two-thirds of our survey respondents reported that the recruitment and training of telephone interviewers changed when CATI surveys where started. Twenty-six percent of centers now require their interviewers to have typing skills when they hire them. Another 56 percent say this is a desired skill but not required (19 percent say that typing skills are not needed at all). Some additional time is needed to train interviewers in the use of CATI systems, however: Average training time reported in the survey was four hours or less for 56 percent of centers. Only 15 percent of organizations reported eight or more hours of CATI training for interviewers.

Computerizing the interviewer training process can have advantages, as the CSR has learned. The CSR has a computerized self-paced training program for training new telephone interviewers and data entry clerks, using CASES and Powerpoint software. In the past, training was done by a lecturer in a classroom and with close supervision by shift supervisors in interviewing carrels. Now the information is presented on computers, using graphics and color. The trainee controls the pace that she or he moves through each training module. At the end of each module, a CASES instrument is used to test the concepts taught in the module. If the new interviewer does not pass the test, she or he repeats the module. This new training allows more flexibility so that interview carrels are not lost during prime interviewing time, and new inter-viewers can complete their training at any time of the day. This new training also significantly reduces training costs because there is less need for manage-ment staff to be involved.

Programmers

Programming responsibilities may be distributed among several organization staff, or consolidated in one or more individuals dedicated to CATI functions, depending on the size and complexity of the system. In small- to medium-sized organizations where the same people often do multiple tasks, the CATI software is usually learned by other noninterview staff. Thirty-six percent of respondents to the survey reported that all or most of their professional staff are trained in programming CATI questionnaires. Among the three survey organizations represented in this chapter, authoring of CATI questionnaires is typically accomplished by project managers, and sometimes by principal investigators. However, for longer and more complex questionnaires, program-ming staff are more likely to assume responsibilty for authoring.

5.4.4   Designing and Testing Questionnaires

CATI allows for complex questionnaire design but then requires substantial testing to ensure that the questionnaire works as intended. There seems to be general agreement that CATI questionnaires require more careful design and planning than conventional questionnaires (Nicholls and Groves, 1986). More careful planning is required because all aspects of the questionnaire, including display of questions, branching patterns, valid response categories, edit checks, must be specified before the questionnaire can be properly installed on the computer. Inadequately tested CATI questionnaires can lead to some ques-tions being skipped inappropriately or not being asked at all in the main survey. When this happens respondents may need to be recontacted and asked the skipped questions, thereby increasing survey costs, or item-nonresponse rates or both.

To minimize these kinds of errors, a critical part of programming CATI questionnaires at the CSR is maintaining programming standards. All ques-tionnaires are reviewed by the field director or center director to ensure that programming standards are maintained. These standards allow any CSR management person to understand quickly any programming that was done by another staff member. For archival purposes the standardized programming allows current staff to find programming and questions used in previous studies and reuse them for new surveys if appropriate. Standardization has saved significant amounts of programming and review time. At the SESRC, CATI questionnaires are archived on floppy diskettes contained in the codebooks created for each project so that they can be readily retrieved and used in developing new surveys.

Despite this kind of standardization, pretesting questionnaires continues to be one of the major unsolved problems with CATI surveys. A typical CATI questionnaire has so many aspects that must be checked for accuracy, including question text, response values, fills, computed values, branching and skip logic, and missing value codes, that it becomes very difficult to identify all errors in advance. A related problem is the potential for new errors whenever one version of a CATI questionnaire has been changed. Consequently most survey organizations (62 in the survey) have had to recontact survey respondents because of errors found in CATI questionnaires. When asked whether they had written procedures for testing and debugging CATI questionnaires, only 56 percent of organizations in the survey reported having them. Some organizations (13%) prefer to pretest a paper and pencil version of the questionnaire, instead of the CATI version, because of difficulties in making modifications or corrections to a CATI questionnaire. Most respondents in our survey (87%) indicated that they usually use the CATI version of the questionnaire when conducting pretests.

Some CATI systems can accept a word processing version of the questionnaire file, or an ASCII file created from the word processor, and use this to create the CATI version of the questionnaire. This capability seems critical because there is always the issue of the correspondence between the paper version of a questionnaire, seen by the client, and the one actually used in CATI interviews. The likelihood of differences between the two versions of the questionnaire is always greater if the same files cannot be used for both the word processing version and the CATI system version. This ability to use word processing files for CATI also makes it more likely that nonprogramming staff can develop final versions of the questionnaires that are used in the CATI system. Eighty-four percent of our survey responents indicated that their CATI system can import questionnaires created by word processing software.

5.4.5 Sample Administration

Besides questionnaire control, the greatest benefit of CATI is the ease of case and sample management offered by networked CATI systems. The immediate access to status reports on the progress of the study is critical for effective survey managment. The ability to track the progress of a study permits shift supervisors to assign interviewing staff efficiently, which is critical when there are multiple studies in the field. Computerized case and sample management reduces the need to process paper forms and questionnaires which can get lost or misplaced. Automated call scheduling is another advantage of CATI systems. The ability to automatically assign cases to interviewers based on standards set by the study manager permits efficient calling.

Most CATI organizations rely on these features to manage their samples, as 82 percent of our survey respondents said that they regularly use the sample management and call scheduling features of their CATI systems. The remaining 18 percent of organizations that do not use or do not have these features presumably still rely on paper call records and other ways of managing sample databases.

Organizations without a network must deal separately with sample administration and call scheduling. The SESRC, for instance, has two approaches for sample administration. The first is entirely dependent on paper call records which are prepared for every telephone number in the sample. These are released to interviewers who use them to initiate telephone contacts with respondents, and then to record information about callbacks and result codes. Following each interview shift, call records are sorted according to result code and a summary disposition table is prepared. The second approach accomplishes the same thing by computer. This involves creating floppy diskettes containing sample telephone numbers which are released to interviewers. The CATI system reads the information from the floppy diskettes to initiate interviews. Result codes and interview data are returned to the floppy diskettes, which are then collected by supervisors for merging into a daily master database. Summary statistics and dispositions are then run on the daily databases. Both of these approaches are significantly more labor intensive without the call scheduling advantages of a networked CATI system.

5.4.6   Managing Survey Data

About half of the organizations surveyed (51%) reported that their CATI systems generate final survey data sets. The remainder reported that final survey data sets are created by a separate data management process. In many respects data management with CATI does not differ much from that of paper and pencil interviewing, since at some point data must be in electronic form under both conditions. However, one major difference between paper and CATI interviews lies in the final form of the "raw" data from the interview. For paper and pencil interviews, this is the completed paper questionnaire. For a CATI interview, this is an electronic file that resides on some magnetic medium such as a floppy diskette, a hard drive, or a tape backup system.

One of the most imperative requirements for any CATI system is having a procedure for backing up survey data to protect against the inadvertent loss of data. Conventional paper and pencil surveys do not usually worry about losing data because the paper questionnaire is physically visible and serves as a backup after the data have been entered into a computer database. Should something happen to damage or lose the computer database, it would always be possible to recover the data by re-entering the questionnaire responses into a computer database again. CATI surveys, however, have no paper question-naires; data are generally stored on floppy diskettes, as files on a hard drive, or a file server computer. If a floppy diskette is damaged or if a hard drive "crashes," survey data may be irretrievably lost. Without an appropriate backup system, lost data could cost an organization not only thousands of dollars of lost interviews but also its credibility as a survey research facility. Adequate protection from data loss is provided by some very basic backup procedures. In our experience backup files are rarely needed. The small investment in data storage, however, is worth the peace of mind and insurance that is provided against data loss. In our survey 53 percent of organizations reported that they had lost interview data collected by CATI because of a hardware or software failure. Most organizations (90%) reported having specific procedures for backing up interview data collected by CATI.

5.4.7 Networks

Connecting interviewing computers to a local area network (LAN) is a desirable feature of a CATI system, offering advantages to a number of survey tasks. Two of the three survey organizations represented in this chapter have advanced networks for their CATI systems, while the third organization is currently installing a network. Most survey organizations are using networked CATI systems. More than 87 percent of the organizations responding to our survey indicated that their interviewing stations are networked.

For survey organizations probably the greatest advantage of having a networked system is that this allows sample management and call scheduling to be conducted by the CATI system rather than by manual means. It is not surprising then that 82 percent of organizations in the survey responded that they use their CATI system's sample management features and that 80 percent responded that they use call scheduling features. When accomplished manually, sample management and call scheduling are highly labor intensive, especially for interviewing staff. CATI systems reduce the labor involved but require more skilled staff to prepare the sample for input to the program. Consolidation of questionnaire data files and creation of survey data sets can also be facilitated by networked systems. However, only 51 percent of survey respondents indicated that their CATI software creates final survey data sets. For the remainder these are created by separate data management processes.

The organizational issue posed by networks is that this changes how samples are administered, how data are collected, and who in the organization is responsible for overseeing these tasks. For the SSRC this has meant having to hire more experienced supervisors with skills in working with networked systems. For the CSR this has meant hiring a dedicated LAN administrator to maintain the system. The CSR, like many organizations, is now totally dependent on a networked interactive computing environment that generates and records substantial amounts of shared information. Without the network many survey functions would not be possible or would be drastically altered.

5.4.8   Reporting Survey Results

CATI facilitates the production of preliminary survey data sets and survey codebooks. Survey clients appreciate this, since they often request survey results to assess the progress of a survey. Some CATI software provides the capability to generate a codebook that describes the variables in the survey database. Approximately 80 percent of survey respondents indicated that their CATI system creates a codebook of survey variables. However, only 41 percent of respondents indicated that their CATI software had the capability to conduct basic statistical analyses. Most organizations reported using other software for analysis of survey data.

At the three organizations described here, simple standardized procedures have been developed to generate preliminary data as needed from active studies. Frequency distributions are analyzed during the survey field period to check skip patterns, to provide preliminary data to clients, and to provide feedback to staff on survey questions that might be troublesome. At CSR, the CASES program used for CATI interviewing has a companion program, Conversational Survey Analysis (CSA) that is used for reporting survey results. CSA is a set of programs that generates simple statistics and frequency distributions. CSA uses the CASES questionnaire to create a file that contains all the question text, precodes, and some questionnaire flow information. When this file is merged with the raw survey data, using another CSA program, a codebook of frequency distributions is created. This codebook is generally sent to clients along with their survey data. Other CSA programs use the questionnaire text to generate SPSS and SAS program files that can read the data in these formats.

5.4.9   Administrative Information

Some CATI systems provide a wealth of additional information about the interviewing process, including data about length of interviews (or parts), completion rate statistics for each interviewer, complete keystroke sequences (for editing), and the like. Some of this information is quite useful, particularly for budgeting and tracking of survey expenses and to support payroll, billing, and accounting systems. Eighty-seven percent of respondents to our survey reported that their CATI systems routinely provide this kind of information.

All hourly employees at the CSR have the number of hours and minutes they work automatically recorded in a computerized database. After they log on the network, employees respond to a few simple questions programmed in a CASES questionnaire. At the end of each shift or workday, another CASES questionnaire provides employees with the number of hours worked and asks them to account for their time by translating it to tasks and projects. The CASES questionnaire is programmed to ensure that all time worked is charged to a project and task. This information is linked to a payroll and wage database which generates project expenses and payroll reports. This accounting information is also linked to other information produced by CASES, such as the interview data file. By linking this information, it is possible to track productivity differences in studies and interviewers. This provides information for future cost estimates and is useful for evaluation of interviewers. It also becomes possible to link daily attendance reports, tallies from interviewers' monitoring sheets, and other information stored in center databases. The SESRC's CATI system is used for data entry of mail or self-administered questionnaires, which are keyed in twice to ensure accurate data entry. The CATI system records information about the sources of error when they are encountered during data entry, providing information about interviewer performance, question difficulty, and rates of data entry errors for different kinds of surveys.

5.5   CONCLUSIONS

This chapter has presented eight organizational issues faced by small or medium survey centers when implementing or using CATI. As other chapters show, some of the same issues confront larger survey organizations as well. Several themes emerge from our data.

One theme concerns the effects of introducing new technology and work processes into a stable organization. Making the transition from paper and pencil interviews to CATI interviewing presents numerous problems to organizations. Staff may have little experience with computers and relatively little knowledge of CATI systems and requirements. Problems with incompatible hardware, software bugs, and network details can cause delays in getting a CATI system functioning. Installing a new CATI system can also disrupt ongoing surveys and will cause changes in survey procedures causing frustration and additional work for staff. The more complex the CATI software and the network are, the longer the period required for installation and training of staff.

Both the survey data and the experiences of our own survey centers suggest that maintenance and upgrade issues become a major ongoing concern after CATI implementation. All three of the organizations represented in this chapter have had negative experiences with upgrading hardware and software during production interviewing. Our conclusion leans toward upgrading in stages and testing all elements of an upgrade completely before conducting field interviews with the new CATI system. Hardware upgrades can be costly if all interview stations must be upgraded at the same time. Software upgrades are relatively frequent as the survey data indicate. In addition they are disruptive and take staff time to accomplish because someone in the organization must take responsibility for upgrading interviewing stations and training staff in software changes.

Another major theme is the effect of CATI on survey staff. CATI processes require continual training and upgrading of staff skills. In smaller survey organizations, where most staff have multiple responsibilities, usually everyone from interviewers to management must learn aspects of the CATI system. Rapid advances in computer technology and improvements in CATI software require continual retraining and mastering of new systems. There is some evidence that survey staff must be more skilled in computer technology and that interviewers with typing or computing skills are preferred over others.

The research literature indicates that CATI can improve the quality of survey data collected (Nicholls, Baker, and Martin, 1997). However, there is no evidence that it lowers survey costs (see Groves and Tortora, Chapter 3, for a discussion of fixed and variable costs). CATI incurs new costs for an organization because of maintenance and training costs introduced by the CATI system itself. In addition costs are shifted among various survey tasks so that, for instance, data entry costs are virtually eliminated, but questionnaire development and testing costs increase. Organizations will still find investment in CATI systems to be worthwhile, since both data quality and timeliness can improve.

Our findings indicate that most CATI facilities are networked and that most are making extensive use of the network's capabilities for sample administration and call scheduling. Networks offer substantial advantages for sharing daily useful information among survey staff. The rapid developments in computer technology now make it possible for even the very small survey organizations to invest in a network for its interviewing stations. A network is essential for taking advantage of the sample administration and call scheduling features of CATI. A network also makes it possible for survey staff to share files, to communicate with one another by e-mail, and to collect administrative data useful for payroll and budgeting applications. Small- to medium-sized organizations can take advantage of their CATI systems by using the computers for other purposes than simply telephone interviewing. The hardware and software can also be used for computer assisted data entry, computer assisted self-administered questionnaires, project accounting, billing and payroll, preparation of reports and presentations, budgeting and forecasting, instruction and training, and e-mail and communications. This spreads the cost of investing in CATI technology over a number of organizational functions. Future developments in CASIC technology will most likely change the relative significance of some of these organizational issues (see Baker, Chapter 29). Baker (1994) and others (Shanks and Tortora, 1985; Dillman and Tarnai, 1988) argue that increasing use of mixed mode surveys, and integration of different survey functions are more likely in the near future and that these have organizational effects that are unanticipated as yet. Small to medium survey centers have an advantage here over larger organizations, which are generally slower to change and adapt to new ideas and technology (Dillman, 1997). It seems clear that computers and technology are changing our organizations and how we conduct surveys, and this trend will most likely accelerate into the next century (Van der Spiegel, 1995).

ACKNOWLEDGMENTS

The authors would like to thank the administrative, research, and interviewing staffs of each one of their organizations for assistance in identifying and, in some instances, creating these organizational issues.

CHAPTER 6

A Systematic Approach to Instrument Development in CAI

Susan H. Kinsey and Donna M. Jewell
Research Triangle Institute

6.1 INTRODUCTION

Development of a quality computer assisted interviewing (CAI) system requires instrument design, programming, testing, and debugging. Organizations address each of these using a range of methods, procedures, and systems to produce an interviewing application. Underlying this diversity of tools and techniques is a set of common features and elements. This chapter describes these common elements and processes which, we believe, comprise a systematic general approach to development of CAI systems.

We studied current CAI development practices to determine how different organizations approach the CAI development process. Our goals were to

- identify key steps in the CAI development process and the issues associated with each,
- determine methods and tools currently used to develop CAI instruments,
- assess advantages and disadvantages of these methods and tools for instruments of varying scale, and
- explore the roles of question authors, survey staff, and programmers in the development process.

Our investigation included a review of the literature on CAI questionnaire design, programming, testing, and debugging. We also conducted discussions by mail and telephone with CAI developers at 11 survey and marketing

Computer Assisted Survey Information Collection, Edited by Mick P. Couper, Reginald P. Baker, Jelke Bethlehem, Cynthia Z. F. Clark, Jean Martin, William L. Nicholls II, and James M. O'Reilly.
ISBN 0-471-17848-9 © 1998 John Wiley & Sons, Inc.

research organizations in the United States and Europe. The participating organizations were: Research Triangle Institute (RTI), the U.S. Bureau of the Census, Abt Associates, National Opinion Research Center (NORC), Office for National Statistics (ONS), U.K., National Agricultural Statistics Service (NASS), Battelle/SRA, Survey Research Center (SRC) at the University of Michigan, The Gallup Organization, General Accounting Office (GAO), and one anonymous contributor. This chapter thus reflects practices and procedures of a relatively narrow sector of the survey industry, and predominantly in the United States.

This qualitative investigation makes clear that no single CAI development method works for every organization or survey. Opinions and practices vary among and within these organizations based on organizational structure, defined or perceived roles and responsibilities of CAI developers, the size and complexity of instruments, the quality and format of initial instruments, the capabilities of the CAI systems, and the schedule for various development steps. Nonetheless, these organizations practice a common, systematic approach to CAI development, involving content development and testing, specifications development, programming, and testing and debugging.

The chapter discusses each of these areas in turn, examining the CAI development methods and tools that are best suited to specific instruments. A key theme is that a team approach to development is a critical mechanism for improving instrument quality and expediting CAI development at each of these steps.

On the other hand, this chapter is not intended to be prescriptive of a particular approach or of a single best way to do a job. Very different strategies and methods are being used successfully in different organizations. Perhaps the most fundamental difference among firms in CAI development is whether the same or different people are responsible for content development and CAI programming.

In some organizations these two roles are merged, for instance, content specialists and statisticians program CAI instruments. Programmers may become involved for more demanding development issues, but significant portions of the instruments are programmed by nonprogrammers. In other organizations these roles are usually separated, with content specialists developing paper specifications and programmers writing the CAI code. This type of division increases the need for methods and tools for communication and interaction among content and programming specialists. For the most part, our discussion is based on some sort of division of labor between content and programming specialists. However, most of the techniques and tools discussed can be applied with only slight modification in less complex situations.

6.2   CONTENT DEVELOPMENT AND TESTING

Content development for paper and pencil (PAPI) and CAI surveys entails identifying information needs and translating these into a questionnaire

(Platek, 1985). Careful consideration must be given to the study objectives and requirements prior to question authoring. There is a valuable body of literature on the content development process, the rules for authoring questions, and the relationship between question characteristics and survey error (e.g., Fowler, 1995), as well as cognitive aspects of question writing (Sudman, Bradburn, and Schwarz, 1996; Schwarz and Sudman, 1996; Forsyth and Lessler, 1991).

It is also important to consider the attributes of CASIC instruments (see Nicholls, 1988; Weeks, 1992) and the way these apply to new questionnaires and PAPI questionnaires that move to CAI administration. These attributes include:

- more complicated routing patterns;
- on-line range, consistency, and data completeness checks;
- automated prompting and standardized prompts for error resolution;
- on-line assistance;
- customized wording of complex questions based on preloaded data or answers to previous questions;
- accessing external data sources for table lookup routines and across wave consistency checks in longitudinal studies;
- randomization of question order and response options for methodological research; and
- flexible display of items.

Questionnaire authors can build these features into new CAI applications. Often, however, instruments are written without considering the administration mode or involving programmers and other key staff. Saris (1991) notes that "simple translations of paper and pencil interviews to computer assisted procedures will not necessarily lead to better quality data." Question authors need to consider the overall instrument content and design to maximize the advantages of a CAI system.

A new type of questionnaire content that is becoming increasingly important in surveys is nontextual content, particularly multimedia. Audio, graphics, and video are beginning to be used in questionnaires. Multimedia in surveys can be expected to burgeon as the supporting software and hardware for general survey applications become widely available, and as these technologies demonstrate their capacity to improve the quality of research results (see Tourangeau and Smith, Chapter 22; Turner et al., Chapter 23). Multimedia questionnaires naturally make content development more complex and demanding.

As with content development, there is a considerable literature on various content-testing techniques (e.g., Forsyth and Lessler, 1991; DeMaio and Rothgeb, 1996). Content tests usually focus on traditional design issues, such as minimizing respondent burden and identifying questions that are not well-written or that fail to elicit an adequate response. On the other hand, with the growing complexity of today's CAI instruments, content tests can also identify items that may benefit from customized wording, internal consistency checks, the use of fills, help screens, and other features available in a CAI environment.

The decision to test the initial content of the instrument is usually driven by the programming and testing schedule (e.g., adequate time to content test given the production schedule) and the origin of the questionnaire (whether the content has been tested previously). Content testing can be conducted before or after programming. There are both advantages and disadvantage of testing prior to programming.

The primary disadvantage of testing prior to programming is the inability to test the CAI program and simulate the CAI interview process. Later in this chapter we offer a separate discussion on pretesting the CAI program. Overall, however, testing early in the process is usually advisable because many important questionnaire issues can be addressed using a paper version.

Regardless of the methods used to test the initial instrument content, the goals of these tests should be to improve instrument and data quality, and reduce content changes after programming begins. Section 6.6 discusses the advantages of a team approach to content development and testing.

6.3 SPECIFICATIONS DEVELOPMENT

6.3.1 Overview

Specifications embody the instructions for writing the CAI code and implementing various design features. They include not only the wording for question and answer choices but also the routing instructions and other attributes of the instrument. As such they are a critical communication tool between designers and programmers (Katz, Conrad, and Stinson, 1996). In our discussions with CAI developers, we found no consensus in methods and tools for specifications development. Moreover organizations address the specifications process in a variety of ways for different surveys. For some surveys, detailed specifications are developed as a separate, preprogramming task to guide the programming and testing operations — usually for complex CAI surveys that are subject to extensive programming and testing. For others, the programming is performed using existing specifications, including PAPI questionnaires, editing specifications, and other instrument documentation. Finally, for some applications, the programmer writes the questionnaire as he/she develops the CAI system.

House and Nicholls (1988) and Saris (1991) argue the merits of developing specifications that go beyond the initial questionnaire. This process forces developers to consider the needs of the CAI system prior to programming and thereby reduce the number of later content and design changes. Specifications encourage consistency when there are multiple question authors and programmers involved, and specifications provide testers with criteria against which to evaluate the accuracy of the CAI program.

6.3.2 Issues in Specifications Development

When an instrument is written without considering the needs of CAI administration, considerable effort may be required to develop the CAI program. PAPI questionnaires often do not provide routing instructions for "don't know" and "refused" responses. Skip patterns may be specified in a "go to" format while the CAI software requires an "if-then-else" approach. Similarly, PAPI questionnaires may not contain adequate specifications for customized wording or fills, on-line range and consistency checks, and help screen utilities (House and Nicholls, 1988; Kinsey, 1994). PAPI questionnaires also may have items that are left to the interviewer's discretion.

These problems complicate and extend the development process. Many designs decisions may not be made until CAI programming is initiated, creating a bottleneck in the process. Programmers may have to make design decisions on their own with limited knowledge of the objectives of specific questions. The quality of the data may suffer because of a lack of specified range, consistency, and data completeness checks. Some features may not be implemented as intended, resulting in time-consuming and costly program corrections. The resulting instrument may not perform as expected or collect the data needed for analysis.

Katz, Conrad, and Stinson (1996) describe two broad sources of error that create bottlenecks between designers and programmers. First, communication errors can occur when programmers misinterpret the specifications. Second, visualization errors can occur when designers misspecify the instrument; that is, although it meets the specifications, it does not meet their expectations. These flaws often lead to costly additional iterations between designers and programmers as the specifications are modified and additional versions of the instrument generated.

A separate specification process and elaborate specifications are not always warranted or feasible, especially for short, simple, one-time surveys or where the questionnaire is written directly by the programmer. Under such conditions the specifications may evolve as the questionnaire is programmed or be produced as a by-product of the system for documentation, rather than programming and testing purposes. Specifications, however, are beneficial for every instrument, regardless of size and complexity, especially when they are viewed as a communication medium between designers and programmers.

6.3.3 Specifications Development Methods and Tools

We can identify five general methods for specifying CAI instruments. These include both paper methods (written specifications and flowcharts) and computer methods (authoring systems, database dictionaries, and other input sources). The advantages and disadvantages of each of these methods depend on a number of factors, most important, the length and complexity of the instrument and the schedule for programming and testing. We discuss each of these methods in turn.

Paper Specifications. These may include editing specifications for internal consistency, range, other data completeness checks, expanded PAPI questionnaires to provide more detailed programming specifications, and various written instructions for implementing specific CAI features.

Editing specifications, which focus on data quality needs, can be developed before or during the CAI programming operation. In addition to cross-item consistency checks, these may include minimum standards established within the organization, such as the checking of all date fields against the interview date (see Bethlehem and van der Pol, Chapter 11). Editing specifications are typically used in combination with the question specifications to produce the CAI instrument. They document the edits that are to be performed during interviewing, as well as others for postprocessing. Provided that the paper questionnaire clearly specifies routing and other instrument features, this may give most of the instructions needed to program the CAI instrument.

Expanded paper questionnaires, which we call CAI reference questionnaires, or CRQs (see also Stratton and Hardy, 1996), are detailed programming specifications that provide question and answer choice wording and routing instuctions along with instructions for implementing all other CAI features. They are developed in a separate operation prior to programming and are expanded to provide instructions for programming features that are not clearly specified in the paper questionnaire (e.g., wording variations, routing for "don't know" and "refused" responses). They guide the programming and testing efforts and provide valuable documentation for analysts.

This separate specification task offers several advantages. Most important is that key instrument design and content issues are addressed prior to programming. This also entails the development of one central document that describes all instrument features. Because creating a CRQ is time-consuming and has the potential to delay the start of programming, CRQs are not consistently used.

Developers can assess the need for expanded specifications by considering the following questions:

1. Are there items that would be difficult or impossible to program without additional specifications?
2. Would the lack of programming specifications for some items hinder programming or testing?
3. Have routing instructions been provided for "don't know" and "refused" responses?
4. Is the intended respondent for specific wording variations of a question clear?
5. Can standardized blocks of CAI code be taken from existing CAI programs?
6. Have key variables been identified that may require consistency and range checks?

Concerns over the use of paper specifications or CRQs include the difficulty of maintaining the paper version as programming changes are made, the time needed to develop the specifications, and the level of detail required.

Flow Charts. Flowcharts are visual aids that show question sequencing and branching patterns. Jabine (1985) states that flowcharting is an important first step in programming because it guides the instrument development process. Flowcharts aid development by (1) determining the number of questions needed and how they will be sequenced or grouped within each instrument, (2) showing individual questions and branching patterns and helping reviewers understand complex questionnaires and spot faulty or awkward question sequences, (3) providing programmers with a visual representation of the question sequence and branching pattern, and (4) giving interviewers an overview of the instrument structure and flow during training. A fifth use, not mentioned by Jabine, is that flowcharts can enhance the testing of CAI programs.

Katz, Conrad, and Stinson (1996) report widespread use of flowcharts among more experienced CAI designers. On the other hand, CAI designers rarely intend their flowcharts to be used by others or to be used as a communication medium. Katz and Conrad (1996) have developed a software tool that uses flowcharts to facilitate communication between designers and programmers by helping designers more completely specify their instruments.

While flowcharts facilitate the design and specification process, their value may be limited for long and complex instruments. Flowcharts can be time-consuming to develop and the resulting diagram difficult to work with given its size and complexity. Still flowcharts are valuable for specific question series to provide a visual aid for programming and testing. And, as Jabine (1985) notes, there is merit in providing a broad overview of the instrument as a whole. In these applications the flowchart serves as a supporting, optional document accompanying the specifications.

Authoring Systems. Another specification method is the authoring system or front-end module. The systems vary widely, but in general, they provide designers with a structure within which to construct the questionnaire. At the Census Bureau, developers use a template in a word processor to provide programmers with the information for each question, including variable names, question text and answer choices, screening and routing instructions, decision rules, consistency and range checks, and screen display information (see also Hale and Dibbs, 1993). The template fosters consistency when constructing a questionnaire from scratch or when multiple question authors are involved.

At NORC, designers use a front-end module that works with the Survey-craft software for entry of question text, specification skips, and other features. The system uses this input to construct the basic screens and create first-stage code for simple routing patterns. The use of this system has removed all but the most difficult programming tasks from the hands of the programmers. The NORC module can then export the generated CAI instrument back into the module to generate survey documentation.

Several other organizations use similar authoring systems. They save time in the development process and encourage design consistency, while allowing programmers to focus on programming rather than on typing question text or implementing simple routing patterns. Yet, the systems are limited in what they can accomplish, and programmers must still add the code for complex features and intricate routing patterns.

Generally CAI authoring systems work well for more simple questionnaires. However, once the questionnaire becomes complex with intricate skip logic, consistency checks, and, particularly, the use of rosters and hierarchical structures, the general-purpose authoring systems have failed to perform well. Specialized programming efforts must be added to enhance the code produced by the system. At this point the authoring system no longer encompasses all the information about the CAI instrument, and all future changes to the instrument must be made outside of the authoring system. The authoring system is then reduced to a front-end application rather than the comprehensive CAI development engine originally intended.

Database Dictionaries. Database approaches are somewhat similar to authoring systems and are currently used by several organizations. A database serves as a central repository for all the CAI code — question specifications, routing, checks, fills, and so on. The system generates CAI code as well as a hard copy version of the instrument and supporting documentation. The RTI system was developed for a CASES environment, and it is used for instrument development, documentation, data delivery, review of the instrument in process, and documentation of question-by-question specifications. Because the system is created to represent the entire data collection instrument, all components are accessible through various user interfaces for specification, modification, and documentation (Simpson et al., 1995). These systems create more work during the design process but reduce the work required later when clients and analysts are anxiously awaiting their data (Simpson et al., 1995).

Other Input Sources. There are low-technology alternatives to authoring systems and database approaches that can expedite the programming process. These include importing word-processed versions of the instrument into the CAI system to eliminate keying question text and using expert typists to enter the question text directly into the CAI system.

Specifications Content. House and Nicholls (1988) offer a checklist of specification topics for CATI questionnaires, including items that are production oriented and others that concern the instrument content and design specifications. Table 6.1 presents an expanded list of instrument specifications to consider prior to programming, the suggested specification method, and the potential benefits. Basic instrument requirements, such as question numbers, question and answer choice wording, and routing instructions are not included.

Obviously these features are not present in every instrument, and there are additional system-specific features that are not reflected in this list. Nevertheless, question authors should consider each of these carefully when deciding what would be best attended to during or prior to programming.

Table 6.1. Instrument Features to Consider in Specifications Development Process

Feature	Suggested Specification Method	Benefit
Question or variable names	Use question numbers or question names that within the prescribed character limit, describe the intent of the question (e.g., MARSTAT for question about marital status)	Fosters consistency within and across instruments and provides data file documentation
Preloads	Specify input source (e.g., variable names) and where preloads are used (e.g., as fills, to drive routing, etc.)	Expedites programming and provides documentation for user
Wording variations and fill text	Specify "who" gets each version of the question and the preload variables or responses that drive the variants and fills	Eliminates programmer guesswork; facilitates testing
Routing instructions for "don't know" and "refused" responses	Indicate where these are allowable responses; indicate routing for each, or specify general rule that that these responses always follow the "no" response unless otherwise specified	Prepares instrument to handle the unexpected or less expected responses; prevents inappropriate questions from being asked in error
Format requirements	Specify input format for dates, telephone numbers, dollar amounts, etc.	Allows program to captures data in the desired format
Consistency checks	For each question specify any consistency or data completeness check that is required; indicate the specific variables or responses that are involved in the check	Identifies key items in the interview that require editing; allows interviewers to resolve problems with the respondent during the interview
Scripted probes	Specify the exact wording to be used to resolve detected errors; provide means to continue the interview when no resolution is possible	Assists the interviewer and respondent in resolving consistency problems detected during the interview
Range checks	Specify the allowable range for open-ended numeric fields	Checks for acceptable responses
Looping requirements	Indicate the number of times the program should loop through a series of questions (e.g., the number of jobs to collect in a job history)	Addresses data completeness needs by allowing adequate space in the data file
Field widths for open-end text responses	Specify the character limit for each field	Ensures that sufficient space is allowed for data that require coding during or after the interview; reduces truncation
Interviewer instructions or Q × Qs	Specify with CAPS to indicate they are not read to the respondent	Reduces the need for hard copy reference materials
Programmer instructions	Describe how rosters, tables, and other features are envisioned	Promotes open lines of communication with programmer
Date and time stamps	Indicate when the interview date is to be captured (e.g., at the start or at the end) and where timing data is needed (for each section, for a particular question block, etc.)	Captures interview data for analysis and timing data to identify sections that may need to be shortened in length

113

Specifications Maintenance. Regardless of the specification method, a critical issue is the maintenance of accurate, up-to-date specifications throughout the iterative design, programming, and testing process. As decisions evolve that affect instrument content and design, the specifications must be updated. Otherwise, specifications diverge from the program and lose much of their value as a programming and testing tool. This is especially problematic with multiple question authors and programmers. One solution is to appoint a "specification keeper," such as the primary question author, to maintain the specifications. All instrument changes are funneled through the specification keeper to ensure they are (1) appropriate, necessary, and accurate and (2) documented in an accurate and timely fashion. Updated versions of the specifications can then be dated and redistributed as needed.

6.4   PROGRAMMING

The CAI programming process is highly dependent on an organization's choice of CAI software and the capabilities and limits of that package. We do not attempt to evaluate software packages. Rather, we discuss the general issues that affect the programming process for most instruments and describe specific methods that expedite and improve the program.

6.4.1   Issues in CAI Programming

CAI programming is affected by many of the same issues that affect content and specifications development — communication problems between designers and programmers, missing or inadequate specifications, and late content and design changes with insufficient time to test them. Other problems include the maintenance of programming standards and version control.

Programming standards are essential to ensure that all components of an instrument are consistently implemented and that items similar in nature (e.g., code all that apply questions) employ the same administration method. Without standards, programmers may introduce inconsistent code for similar features within and across instruments. Version control problems are usually the direct result of multiple iterations of an instrument during the programming and testing processes.

Programming specifications, authoring systems, and other standardized processes and tools enhance instrument quality by reducing programming errors and discrepancies between the intended design and the resulting program. With clear, detailed specifications, regardless of the development method, producing the instrument should be no more difficult than programming the specifications.

6.4.2 Programming Methods and Tools

Organizations used several methods to address these problems, including programming standards, modular design, portability, and master file maintenance. We discuss each of these in turn.

Programming Standards. The primary goal of programming standards is to reduce inconsistencies in instruments. To this end, organizations have established standards for such things as question naming and formatting conventions, function key mappings, and conventions for implementing consistency checks. Standardized blocks of code for common or shared questions, such as front-end screening and interface modules, and standards for input and output files, have also been established.

Organizations continue to search for additional processes that would benefit from standardization, as well as methods for achieving a balance between standardized and tailored programming. One way to accomplish this is through formal staff training in which organizational standards are shared with all programmers.

Modular Design. Many authors (e.g., Nicholls and House, 1987; Saris, 1991) argue for the use of modular design, especially for complex CAI instruments. With modular design, blocks or modules of questions and utility routines are self-contained and more easily revised, tested, and documented, with less of an effect on the rest of the instrument. Modular design enhances portability, the re-use of modules for other surveys, and the ability to test in a modular fashion. A disadvantage is that temporary questions may have to be added to each module to drive wording and routing across modules.

Portability. Portability entails the re-use of questions and CAI code from other surveys. This offers several benefits. The questions have usually been field tested in a previous survey so that question authors do not have to write questions from scratch. This saves both development and testing time. For example, NASS has maximized its use of existing question blocks to quickly field surveys with relatively short data collection periods. Their "specifications database" allows programmers to re-use crop and livestock questions from survey to survey, thereby reducing the amount of time required to design, program, and test instruments for various state surveys (see Pierzchala and Manners, Chapter 7). This strategy must be considered in the design stage to achieve maximum benefit. Many organizations do not consistently re-use blocks of questions or CAI code. In part, this stems from the difficulty in building a flexible and comprehensive question bank.

Master File Maintenance. Master file maintenance is a version control technique. One master version of the program is developed and maintained and all modifications emanate from this version. This becomes especially important when more than one programmer is involved. With this technique, programmers (1) date all versions of the program and specifications, (2) designate one person to update the master or working version of the program, and (3) maintain a version number on the data file, especially when changes are made to the instrument after the start of data collection. Similar procedures are implemented to maintain multiple versions of an instrument for field offices and separate interview settings, when applicable.

6.4.3 Late Changes

One of the critical tasks for a programmer is determining the effect of late content and design changes on the program. This requires a high degree of familiarity with the program and the ability to foresee potential problems and adequately project the level of effort required for corrections. Late changes are required to fix bugs, or when question authors or clients alter the design after viewing the questionnaire on-screen. Organizations handle late changes in various ways depending on when the change is requested (e.g., during programming or after the onset of data collection), the severity of the problem and its potential effect on data quality, and whether the change can be adequately tested. Program modifications must be weighed against the likelihood of introducing other errors into the system.

Some organizations maintain lists of key variables to determine which items are affected by requested changes. When a change is requested, the list is checked to determine the sections or blocks of code that will be affected. SRC has developed an automated utility to assist programmers with this process. The utility tests the questionnaire logic and references items affected by the proposed change, and documents where specific variables involved in the change are referenced.

Both of these processes aid in the early detection of the effect of programming changes. For example, if a key question is being deleted, this process pinpoints the items or sections in the instrument that will be affected by the deletion. Thus, if the deleted question drives wording or routing elsewhere in the instrument, the program can be modified to accommodate its removal. Because problems arise with insufficient testing of late changes, design tools such as updated specifications and mock scripts can facilitate this important task. In retesting late changes, all paths into and out of the affected items should be examined. As one contributor stated, "if you can't test it, don't do it."

Some organizations also allow for expansion or changes within the program structure and output data sets during initial programming. This requires additional planning at both the design and programming stages. It may also require a more modular approach to programming so that self-contained question modules can be easily revised and retested with less of an impact on the rest of the instrument.

6.4.4 Data Structures

During programming, attention is mainly focused on producing an instrument that functions as intended and works well during administration. The structure of the data should be examined as well so that the resulting data records can be easily understood and analyzed. Furthermore programmers should determine whether computed variables used to make routing or wording decisions need to be stored in the data file.

6.5 TESTING AND DEBUGGING

Testing is the "process of executing a program with the intent of finding errors" (Meyers 1979). With today's long and complicated CAI instruments, however, it may be impossible to manually test all paths and features. As a result organizations are searching for tools and techniques that will foster more effective and timely instrument testing.

During testing, organizations face several challenges, especially with complex instruments. First, lack of adequate specifications make it difficult to gauge the accuracy of the program. While flowcharts and other tools may help identify routing problems, these do not necessarily help testers determine the accuracy of the question wording or other instrument features such as consistency checks. Second, tight schedules and late changes make it difficult to test sufficiently the entire program. As a result the beginning of the questionnaire tends to receive extensive testing, while the end may not. Third, a lack of sufficient testing tools slows the testing process and reduces the efficiency of the testers. Finally, the lack of a structured testing process and testing plan affects the effectiveness of the testers.

6.5.1 CAI Testing Methods and Tools

Several methods and tools are used to test and debug CAI instruments. In this section we discuss these testing methods and the types of errors that can be expected from each.

Functionality Testing. Dumas and Redish (1994) describe functionality testing as "making sure that the product works according to specifications." In a CAI application this means gauging the instrument against the programming specifications. This includes testing all question and answer choice wording, routing, wording variations and fills, consistency checks, and other specified instrument features. To test functionality, testers use the specifications to (1) identify and traverse as many paths as possible through the instrument and (2) trigger all programmed consistency checks. They note any discrepancies between the two and report problems to the programmer.

In addition to specifications, organizations use a number of additional tools to test functionality. These include flowcharts, checklists of instrument items, interview scenarios, and mock interview scripts. At RTI, for example, we use checklists and interview scenario grids to facilitate testing. The checklists contain all questions in the instrument and track the items and paths that have been tested. Interview scenario grids list the major sections and routing points in the instrument, and testers are assigned to test specific scenarios to ensure that all major points are reached under at least one scenario.

Testing against the specifications is time-consuming and tedious, requiring an eye for detail and an understanding of the instrument. Question authors are generally well-suited to this task because of their knowledge of the instrument. Programmers, however, are not the best functionality testers (Meyers, 1979; Marick, 1995). Meyers points out that the job of the tester is to find errors, and as such it may be hard for programmers to find fault with their own work. Also programmers tend to test whether the program does what they programmed it to do, and not necessarily what the specifications require. However, all organizations rely on their programmers to do some level of testing. For some, this may be extensive testing against the specifications, while for others, programmers may only be expected to identify and resolve glaring errors before turning the system over to other testers.

For long or complex instruments, multiple testers are often necessary to ensure coverage of as many paths and questions as possible. A testing plan and a structured testing approach, established prior to testing, can maximize the effectiveness of a team of testers. One such testing plan and other structured approaches are discussed later in this chapter.

Usability Testing. Simpson (1985) argues that usability is important when many people will use the program, when error consequences are serious, or when the program will be used repeatedly. All apply to most CAI instruments. Usability focuses on the human-computer interaction, whether it be the interviewer (in CATI or CAPI) or respondent (in CASI or CSAQ). Dumas and Redish (1994) note that usability is a product of the entire package, including hardware, software, manuals, and training. Usability has to be considered at each development stage to ensure success; user involvement throughout development is also an essential part of the process.

Because interviewers are often the end users, they have a vested interest in the usability of the instrument. Their experience with a broad spectrum of CAI applications enables them to evaluate specific requirements; for example, they might indicate that a roster is difficult to complete or that the process for coding "all that apply" responses is problematic. For CASI instruments, different usability standards may be required. Usability testing requires a structured approach, similar to those offered by cognitive or usability laboratory methods (see Hansen, Fuchs, and Couper, 1997).

Program Inspections. Program inspections, walkthroughs, and reviews are testing techniques employed by programmers to compare their source code to the programming specifications. While these examination techniques are not used widely in CAI programming, they offer several advantages, especially for instruments that require extensive customized wording. This permits programmers to review the wording of questions and their variants in the CAI code without actually having to enter test data through various simulated interviews. The programmer also can compare his/her logic to the specifications.

The obvious disadvantage of this technique is the inability to actually see the CAI screen. As a result it is impossible to determine what actually happens at run time, since the fills and other features are not visible.

Module Testing. Module testing involves testing one program unit at a time for functionality, usability, or both. The unit can be one or more sections of the instrument depending on the programming structure. Module testing is a way of managing a potentially cumbersome testing task, since attention is initially focused on smaller units of the program (Meyers, 1979). It eases the debugging process by narrowing the domain. It allows testing and pro- gramming to occur in parallel. Additional benefits of module testing are the relative ease in generating appropriate tests for the most recently added module, the ability to test the system structure early, and the ability to provide users with a preliminary version of the instrument to ensure usability needs are met (Rauscher and Ott, 1987).

Module testing can be used on any type of instrument, although its greatest merit is the ability to test and debug large and complex instruments in smaller pieces. Once the modules are combined for production, the entire instrument must be retested to ensure that the modules work together as intended and that data are passed forward from module to module as needed.

Crash Testing. "Crash" testing is a method that puts as many testers on the system as possible and has them randomly test various instrument features. Crash testing enables a greater number of testers to test the instrument — the greater the number of eyes, the greater the number of bugs detected. Experi- enced interviewers and supervisors are excellent crash testers, since they can use their previous interviewing experiences to challenge the system. While most bugs are uncovered through specification testing, crash testing may identify bugs that occur only rarely. For example, these include problems generated by the routing of "don't know" and "refused" responses if these have not been taken into consideration during development.

Interviewers and supervisors also provide valuable input on the usability of the interview program, pointing out design problems that may hinder the interview process. A number of organizations use their CATI staff for this purpose, assigning them to crash test both CATI and CAPI applications. This type of "volume" testing is especially useful for large or complex surveys that will involve large numbers of interviewers.

The disadvantage of crash testing is its haphazard nature. When there is no structure to the process, the type and number of discovered bugs are limited generally to more unusual errors that would not occur in most interviews. Generally, crash testing is a supplementary process that should be conducted in addition to specification tests.

Pretesting. Pretesting with the CAI program allows developers to simulate the CAI interview process and test the CAI program. Additional benefits of pretesting include the ability to (1) conduct a structured functionality test, (2) evaluate usability of the instrument and any supporting systems, (3) collect and examine test data (e.g., via frequencies and machine edits), (4) identify items that would benefit from on-line help, consistency checks, and the like, and (5) conduct split-ballot experiments on question order, wording, and so on.

Conventional pretests involve the use of field or telephone interviewers to administer the instrument to a selected sample. In addition to administering the questions, interviewers listen to the respondents' remarks and questions and enter any comments so that any issues or other problems can be addressed prior to production interviewing. Following the pretest, debriefing sessions provide interviewers with an opportunity to discuss their overall experience as well as their experience with specific items in the questionnaire. Presser and Blair (1994) stress the importance of specifying the goals of the pretest to maximize its benefits. For a CAI application these may include whether to test the instrument content and design, the CAI program, the data collection procedures, various methodological experiments, or some combination of these.

Identifying and resolving bugs and other instrument problems prior to production interviewing is a critical facet of pretesting with the CAI program. It is important that the pretest be conducted with the version of the instrument that is expected to be used for production interviewing so that costly and time-consuming changes are minimized at this stage of development.

Hardware Testing. Prior to a formal pretest or any production interviewing, the instrument should be tested in the production environment, with the equipment and systems that interviewers and respondents will use. This is called a load test in some organizations. This ensures that system constraints (e.g., size, memory) are not violated, the program loads and runs in a reasonable time, and the screen appearance is appropriate for the administration mode. Additionally the testing process should confirm that the capacity and speed of batteries, modems, hard disks, and other hardware meet the needs of the study.

Review of Test Data. Earlier we noted the importance of programming with the data structure in mind. The review of test data at this stage allows developers to verify that the data are stored as expected, and in a manner that makes them easily understood and analyzed. Test data are also an excellent tool for identifying bugs in the CAI instrument. Organizations report several such uses of test data, including reviewing frequencies to identify routing problems and ensure that the needed data are collected. Early examination of the data from production interviews is also encouraged. Machine edits on test, pretest, and production data are also useful in identifying data problems and bugs.

Establishing a Formal Testing Plan. In this chapter we have discussed the advantages and disadvantages of several testing techniques and demonstrated how these can ensure various requirements of instrument quality. Here we discuss the merits of developing a formal testing plan and the benefits of a more structured approach to testing.

Testing has traditionally been approached in an ad hoc manner, without formal planning, terminating when developers felt the testing was adequate or when the time or money allocated had been consumed. Few of the organizations we talked to reported any structured testing methods beyond assigning testers specific instrument features or scenarios to test.

To produce a higher-quality instrument at reduced cost, structured testing is critical. This involves determining (1) the testing methods and tools best suited to and available for the instrument, (2) how to implement the selected testing method, (3) who should test, (4) how to document, correct, and retest detected problems, and (5) a testing schedule. Selecting testers and defining their roles is a key element of a testing plan. Marick (1995) recommends a strategy that assigns parts of the testing to those who can do them most cost effectively. In addition to having programmers test their own programs, he suggests that independent testers take part because they will find more, and more serious, faults in the program. As we argued earlier, for example, question authors may be best suited to specification testing, and experienced interviewers are usually excellent usability testers, although usability experts such as survey staff and methodologists may also be needed.

In defining testers' roles, it is important to consider who will perform specific tests, how those tests will be done, and what training, if any, the testers need (e.g., specification or CAI system training). Regardless of scale, a standardized testing technique improves testing. For example, at RTI we often begin with one-way testing for each question, followed by the testing of complex routing, consistency and range checks, and other instrument features. One-way testing checks all allowable responses for a question and the immediate resulting path. As part of this test, wording, consistency and range checks, and other features are examined.

One-way testing covers most items in the instrument and identifies a large percentage of the bugs. Testers should document the items tested and the results, thus identifying questions or features that were missed in the one-way test and require further testing.

A number of other organizations are developing standardized procedures for testing CAI programs. A key element of these approaches is the involvement of multiple staff, which sets the groundwork for a team approach to testing.

6.6   TEAM APPROACH TO CAI DEVELOPMENT

In this chapter we have offered a number of methods and tools for developing CAI instruments. Organizationally a critical aspect is a collaborative, team approach to CAI design, programming, testing, and debugging. It combines the skills and perspectives of staff at each stage of development — question authors, usability experts, and programmers. It encourages teamwork to produce a higher-quality product and saves valuable development time. Finally, it promotes cross training among development staff.

When the roles of question author, programmer, and, perhaps, usability expert are merged, development teams may have fewer members, and there may be less of a need for formal methods and tools that enhance communication and interaction. This discussion, however, focuses on the skills and perspectives needed on the team, regardless of the division of labor.

Question authors bring their knowledge of the study objectives, instrument requirements, and the prerequisites of data analysis to the team. In this role they lead the content development, content testing, and specifications development tasks. Usability experts bring valuable insight into how interviewers will administer certain items and which methods are best suited for specific populations. Programmers determine what is possible within the constraints of the CAI system, and cost effective methods for implementing specific items. Programmers may also be aware of existing blocks of CAI code that can be re-used in the survey.

By involving all key development staff early in the process, team members have a vested interest in fostering efforts that will faciitate the design, programming, testing, and documentation phases. Improved communication should not be underestimated. In fact, when organizations describe some of their more significant CAI development hurdles, they note problems that are due, in part, to miscommunication about instrument requirements and misunderstanding about the CAI system's capabilities. Therefore team involvement in the decision-making process is critical.

Of particular importance is programmer involvement in the overall development process. In addition to defining the capabilities and limits of the system, programmers weigh the effect of requested design chances and determine the level of effort required to implement and test the changes. Thus, once programming is initiated, there should be no surprises as to what the system can and cannot do. Additionally, there should be consensus on how specific instrument features will be implemented and presented to the interviewers. A team effort promotes a consistent vision of the final instrument.

The team approach also facilitates testing by involving all team members in the process. Because of their collaboration throughout development, team members have an interest in both the quality of the program and its user friendliness. As well, members have a broad knowledge of the instrument contents and goals that increase testing quality.

To ensure the success of this development approach, it is important to establish (1) standards for implementing instrument-specific features, (2) an overall development schedule, including a firm final deadline for instrument changes, (3) a standardized means of distributing the system to testers, and (4) a communications protocol for reporting bugs and their resolutions to team members and other testers. These guidelines will help the team produce a quality CAI product.

6.7 CONCLUSION

In this chapter we have described the key steps in the CAI development process and the issues associated with each. In discussing the various CAI development methods and tools, it is clear that no one method is suited to every instrument or organizational structure. Moreover a number of CAI applications, depending on their length and complexity, the development schedule, and organizational preferences, are successfully developed without undergoing all of these steps. However, these steps are generally applicable to all surveys, even though there may be no need or time for an iterative process for some applications.

We have also suggested a team approach for developing CAI systems. Although roles are merged in some organizations, the lines of responsibility between designers and programmers are becoming more and more blurred in organizations where these roles are separate. To meet these challenges, organizations are considering cross training or merging the question author and programmer roles for some surveys. With the advancement of authoring systems and other design tools, it is apparent that designers will need to become more familiar with programming basics and that programmers will have greater visibility and responsibility for CAI processes so that they will need to become more familiar with general survey procedures.

For example, SRC is currently undertaking a formal training program to teach programming basics to nonprogrammers. Their goal is to reduce time spent by CAI programmers on basic screen construction tasks, such as entering question and answer choice wording and implementing basic routing patterns. This will also free programmers to implement more difficult CAI features and, more important, will focus their attention on developing better design and testing tools.

Regardless of the division of labor, organizations recognize the need for more and better CAI development tools. Although significant advances have been made with authoring systems, database dictionaries, and other design tools, these have generally been accomplished at an organizational level rather than by the firms who build CAI software development systems. As a result many survey research firms are creating proprietary, supplemental systems to facilitate instrument development. As organizations are forced to field test instruments on tighter schedules and with greater complexity, increasingly advanced design tools are required.

Additionally attention is needed in the area of testing, especially in the development of automated testing tools. While organizations are making strides in the area of usability testing, including usability inspection methods, end-user testing, and usability laboratory methods, developers continue to search for methods and tools that will reduce the manual effort currently required to test CAI systems.

Finally, as instruments become more complicated, and the development process more demanding, it will be critical that a systematic approach involving well-defined procedures, tools, and trained personnel be implemented so that instruments can be developed in a timely and cost efficient manner.

CHAPTER 7

Producing CAI Instruments for a Program of Surveys

Mark Pierzchala
Westat

Tony Manners
Office for National Statistics, U.K.

7.1 INTRODUCTION

As survey organizations adopt CAI as their primary data collection mode, issues arise concerning coordination of the production of instruments for the multiple surveys an organization conducts. These issues are especially salient to government statistical agencies that conduct large-scale continuous and longitudinal surveys. However, market research and private contract research firms must also deal with these issues. The high costs of all survey research, and especially of in-person surveys, combined with difficult budget environments have stimulated survey organizations to seek ways to combine survey tasks by relying heavily on computerized interviewing technology. To use the technology efficiently, survey organizations need to develop CAI tools and standards for use across the range of surveys they carry out.

This chapter examines the factors designers may need to take into account in producing CAI instruments for a program of surveys. It describes strategies for dealing with the additional complexity and the opportunities for greater efficiency.

A strategy for producing CAI instruments comprises all the systematic steps that an organization takes to meet the common needs and opportunities of a program of surveys. Successful strategies for CAI instrument design and production reduce costs and timetables and improve survey quality. Such

Computer Assisted Survey Information Collection, Edited by Mick P. Couper, Reginald P. Baker, Jelke Bethlehem, Cynthia Z. F. Clark, Jean Martin, William L. Nicholls II, and James M. O'Reilly.
ISBN 0-471-17848-9 © 1998 John Wiley & Sons, Inc.

strategies usually focus on common ways of doing things and implement them as standards. They also keep the standards under continual review for improvement in the light of experience and innovation.

The main type of programs of surveys, many of which overlap, include:

- surveys carried out by the same organization;
- surveys carried out by the same interviewers;
- the same survey with regional variations;
- the same survey over time — continuous, follow-up, repeated, and panel suveys;
- surveys linked by type of sampling unit — business, houshold, and individual;
- surveys linked by topic — smoking among adults/schoolchildren/elderly, and so on;
- surveys that share a common core of questions — omnibus and integrated surveys.

There can be different perspectives within the same survey organization. In some national statistical offices there are separate organizations for demographic (household) surveys and economic (business) surveys. These may have different instrumentation strategies (Nicholls and Matchett, 1992).

7.2   EFFECT OF TECHNOLOGICAL PROGRESS

Rapid advances in hardware and software continually open up new opportunities for producing CAI instruments. Faster processors, larger active memory and disk storage, database advances, and other innovations already give significant leeway to the software developer and instrument writer. Three examples are:

- Instrument code may increasingly approach English language. Code does not necessarily have to be used to maximize program performance. Researchers without programming expertise can write code for complex questionnaires that meet interviewing needs and produce well-structured outputs. The increasing simplicity and transparency of code will make it easier to re-use code in a program of surveys.
- Very large instruments — larger than could conceivably be used in a single interview — have become possible. As a result there are now alternative approaches to the implementation of a program of surveys. Some organizations are combining several survey instruments into one very much larger instrument for integrated houshold survey programs.
- Individual instruments in a program of surveys can call common classific-

ations and other information from common files and libraries. Such files and libraries help to reduce development time and to avoid unintended differences between surveys in a program.

7.3 STRATEGIES FOR PRODUCING RELATED INSTRUMENTS

There are four main strategies for producing CAI survey instruments. The choice of a particular strategy is determined by the nature of the survey program itself. In addition structural and cultural factors may affect the choice of instrumentation strategy for a program of surveys. If external survey clients are involved in key technical decisions, they may limit the choices.

7.3.1 Produce Instruments for Each Survey Individually in a Coordinated Manner

Producing instruments for each survey individually, but in a coordinated manner, is a very common strategy. A survey organization's range of work may vary depending on different clients and subject matter content. However, usually an organization will try to fit all its work into a coordinated program and use standard methods as far as possible.

In this strategy the instruments use common CAI software and operate within a standard infrastructure. The CAI infrastructure may include administrative elements within the instruments themselves. For example, a standard set of questions and precodes for administrative information will minimize the need to amend the case management system on a survey-by-survey basis. The instruments may contain standard blocks of code for common administrative requirements such as a common identification structure, outcome codes, name and address verification, appointment, and nonresponse modules.

Another type of standard block is possible where there are common questions for a group of surveys. Often there is a standard household roster. Other examples include the collection and coding of classifications such as occupation and industry and educational qualifications; still other examples include the handling of batteries of attitude questions.

The infrastructure includes interfaces with sampling frames, respondent management systems, survey management systems, telecommunications, data handling, in-office editing, documentation, and output of files for analysis. Instrument designers for individual surveys work within this framework. The efficiency of a CAI system can to some extent be measured by the ease with which survey specific data and the metadata can move around together and hook into any of the basic processes.

Organizations usually try to ease the training burden and improve interviewing quality by developing standard screen interfaces for interviewers. More generally, organizations will work toward standard procedures for all aspects of their work.

7.3.2 Produce Large Instruments Covering Several Surveys

Another strategy for a related set of surveys is to encompass them in a large matrix instrument. Two main types have been developed: multiple variant and integrated instruments.

Multipe Variant Instruments. A set of surveys may seek essentially the same type of information about different types of sampling units. Examples may include surveys in which each asks about outputs in a different industry and takes account of varying processes and measures, and also surveys in a single industry, such as agriculture, which then would take account of different crops and environmental conditions in different regions.

In these cases a basic design or template is used for the different versions produced. A system that runs the surveys automatically from electronically held specifications tends to be less error prone and more cost effective than reliance on manual intervention by programmers. Pierzchala (1995) describes a CAI system developed at the U.S. National Agricultural Statistics Service (NASS) for handling the June Area Frame Survey with 45 versions in one instrument all driven automatically.

Integrated Survey Instruments. Surveys of similar content may be grouped together in a single mega-survey instrument combining the core questions with the questions specific to each survey. The U.K. Office for National Statistics (ONS) has held trials of a system in which four of the country's major continuous household surveys were combined in a single sample design, and interviewers had workloads comprising the four surveys (Manners and Deacon, 1997). Common questions formed the basis of a number of discrete question blocks placed at certain points throughout the instrument.

This mega-instrument represented an aggregate mean interviewing time of five hours, though any household would receive only one of the four surveys. While this survey was developed and tested as a single instrument, it could equally well have been managed as four surveys built from common and survey-specific question blocks.

As Statistics Netherlands, an integrated survey program has taken this idea further by using a discrete core of common questions (Heuvelmans, Kerssemakers, and Winkels, 1997). This core provides a significant amount of thoroughly tested and reliable code in each instrument. NASS's Quarterly Agricultural Survey (QAS) is another example of an integrated instrument. The QAS currently has sections for crops, grain stocks, and livestock. NASS is considering incorporating a section for hogs in the same manner as the mega-versions introduced by Pierzchala (1993). For the most part hog farmers would be the same respondents as they are for the current sections of the QAS survey. In the mega-version approach the sampling scheme for hogs would be separate from that of the rest of the instrument. Different sections of the integrated survey instrument would apply according to which samples a particular farmer fits in.

7.3.3 Produce Instruments Tailored to Respondents to Multiple Surveys

Some respondents, especially those in establishment surveys, may be subjected to many inquires throughout the year. A tailor-made instrument may reduce the burden on them by collecting common information only once. There are various ways to tailor instruments to respondents. One way uses one instrument but runs it individually for each respondent with routing based on prior information held in an external file. Another way is to build a separate individualized instrument for each respondent. Sikkel (Chapter 8) discusses yet another method that allows the respondent to determine the instrument flow and content dynamically. The first two approaches require a respondent management system with a detailed database specifying for each respondent the information the survey is to collect.

Run a Distinctly Separate Instrument for Each Respondent. Statistical organizations that conduct business surveys often have advance knowledge of the enterprises that the firms engage in. Therefore, an external file of toggles can be used to guide a firm on its individual path through the instrument. For example, the U.S. Bureau of the Census is developing a system for the Census of Retail Trade to automate mailing electronic instruments to businesses for computer assisted self-reported (McDonald, 1997, personal communication). Each business receives a single instrument that contains only the appropriate question sections for that firm. This method customizes the instrument to each respondent, without having to generate separate instruments for each firm.

A panel survey that uses dependent interviewing might also consider developing a unique instrument for each respondent (see Brown, Hale, and Michaud, Chapter 10). In this way the instrument can access information about the respondent from previous interviews. Some examples are discussed in Section 7.3.4.

Individual Instrument for Each Respondent. A strategy that some economic surveys may take in the future is to individualize the production of instruments. Each year there are many surveys in which large businesses are asked to cooperate. Tailoring the data collection to these respondents may make them more willing to cooperate. Also survey organizations would only need to contact them once, and would avoid the duplicate questions that can arise in separate surveys of the same unit. The idea is to have an instrument that can customize data collection to whatever way an organization can provide information. This might require, for example, customized wording, collection dates, units of production, ordering of questions, and omission or inclusion of specific sections or questions. This strategy would require the generation of instruments from a respondent database. A bigger challenge here would be integrating all the data into a survey database in a coherent manner.

7.3.4   Modify Instruments in Successive Waves of Longitudinal Surveys

Successive surveys within a longitudinal sample design include NASS's Quarterly Agricultural Survey (QAS) series, the Survey of Income and Program Participation (SIPP) and National Longitudinal Survey of Women conducted by the U.S. Bureau of the Census, the Medical Expenditure Panel Survey (MEPS) conducted by Westat and the Labour Force Survey conducted by the U.K. Office for National Statistics. These organizations modify their instruments from period to period according to their needs.

Many of them have a common core of questions which stay the same except in such minor details as date references. They have recognized that it is effective to re-use well-tested code for core questions and to build instruments by adding supplements for each wave. Even when there is need for substantial change, designers re-use as much code as possible. The SIPP, for example, created a core questionnaire which it has used from wave 2 onward by modifying the wave 1 instrument. The core questions only need to be changed if there is a change in the general survey methodology or procedure.

Something like this is used for the QAS which take place every June, March, September, and December. For this series the December instrument for one year is modified to become the December survey for the succeeding year, and so on, for each quarter. For agriculture surveys the subject matter is more alike year to year than for quarter to quarter.

7.4   REVIEW OF TECHNIQUES

In this section, we review a variety of techniques for building on common factors and implementing instrument production strategies for a program of surveys.

7.4.1   Common Precoded Responses

An organization can define a set of common response codes, independent of particular questions, to use for all of its surveys, such as yes/no, agree/no opinion/disagree, survey management codes, and time/date stamps. Other response codes may be common to a class of surveys. Household surveys, for example, may include standard precodes for categories of age, sex, and income. Some CAI systems allow designers to attach names to precoded responses. Then in the survey code it is possible to refer to the names rather than retyping the response text. A pseudocode example is:

> PRECODELIBRARY GlobalPreCodes
>
> PRECODELIBRARY BusinessPreCodes

This code would make the libraries available to the instrument. Each of the libraries could include standard precoded responses. The *GlobalPreCodes* library might include precoded responses that could apply to any survey. The *BusinessPreCodes* library might include precoded responses that would apply to economic surveys. A few global precoded responses might be:

PRECODES GlobalPreCodes

Gender = (Male, Female)

YesNo = (Yes, No)

A standard precoded response could be attached to a question such as:

QUESTIONS

GenderOfAdult "Gender of adult": Gender

GenderOfChild "Gender of respondent's child": Gender

The existence of a standard set of precoded responses does not necessarily mean that the designer is tightly constrained. The set may include standard alternatives. In the absence of such standard sets of responses, surveys in a program may develop their own idiosyncratic versions. The differences may confuse interviewers and hinder cross-survey comparisons.

7.4.2   Common Data Structures

Data structures in social surveys can be complex. It is helpful to interviewers and analysts if they are treated in the same way across surveys. For example, hierarchical data structures are common in social surveys. Households contain members, members have attributes such as jobs and diseases, jobs and diseases have characteristics, and so on. Interviewers need to be clear about which job of which member of which household they are currently asking about, and how to get to the same question for another member's job if necessary. Hierarchically structured information is gathered by means of rosters. Designs for rostering and modules of code can be shared between survey instruments. Such standardization helps interviewers and also simplifies the extraction of data from complex structures.

Where a complex data structure is difficult to implement, it may be useful to store the solution as model code, which can be shared by the surveys in a program. An example will illustrate the problems for data structure that the practical requirements of interviewing may present and how surveys can share the solutions. Several surveys carried out by Social Survey Division of ONS require concurrent interviewing of many different household members (Manners, 1992). The aim is to collect individual information from each respondent but to do it efficiently conversationally by asking a question once and collecting each respondent's answer. This is a useful way of reducing the response burden in lengthy interviews when similar sets of questions are asked of all adults in a household. It can also improve data quality in situations involving complex financial questions on shared accounts, loans, benefits, and the like, where there is risk of double counting and omissions. Of course such concurrent interviewing is difficult to implement. The CAI instrument would have to follow each individual's interview, with its own particular routing, at the same as that of all others, and allow for interruption without stopping the interview for others.

The ONS method uses a table to set up the data on each household with each member assigned a row of the table. There are separate tables for different small groups of questions which individuals can answer, if applicable.

Tables can be survey specific. In the case of household composition and demographics data, they can be standard and set up in well-tested standard code modules. In all the tables, standard code can be used to translate sets of questions into a tabular form that provides indexing for individuals and their routing, and uses conditional text substitution to display appropriate names and similar items. Figure 7.1 gives an example of such standard code. To place a set of questions for concurrent interviewing in a table, the instrument designer would only change the table and block names and the questions in the block.

In using the table's model code, the designer can focus on the subject content in row *BILO*. In this row the designer concentrates on the questions, flow, and edits as they apply to one person. The code for the table determines how more than one respondent in a household is handled. The model code assures that the responses will appear to the interviewer on the screen in a standard way across surveys. This saves the designer much time and effort in design and testing.

7.4.3 Question Libraries and Block Libraries

Question libraries are a resource for constructing questionnaires, whether for paper or for electronic instruments (Baker, 1988). With an electronic question library, it is possible to construct ways of bringing the question text and related information (e.g., valid codes) directly into a CAI instrument. It is more difficult to do the same for other important associated elements, such as routing (if complicated), edits (i.e., "hard"/mandatory and "soft"/warning checks on range and consistency), and details about data storage (if required by the CAI system). This information may have to be added manually, which precludes construction of instruments by simple aggregation of questions.

A possible way around this is to store blocks of code that contain groups of questions and their associated routing, edits, and other elements. It is good practice to design questionnaires as a series of such blocks (Kinsey and Jewell, Chapter 6). This flow and edit information can be held in a library of CAI code. This is essentially what is used by NASS for the automated generation of 45 versions of the QAS.

```
TABLE ILO
Looper : INTEGER  {Looping variable}

{Above is model code}
{Below is code provided by the designer}

  ROW BILO  {Provided by the designer}
    QUESTIONS
      Working "Did you do any paid work in the 7 days ending last Sunday
      either as an employee or as self-employed?" : YesNo

    {Other questions, routes, and edits from the designer}
    ENDROW

{Above is code provided by the designer}
{Below is model code}

  QUESTIONS  {Overall instrument level}
    ILOarray : ALLOW N ROWS OF BILO
  FLOW  {overall instrument level}
    FOR Looper := 1 TO HhldSize DO  {Up to household size}
      ILOarray[Looper]
    ENDDO
ENDTABLE
```

Figure 7.1. Model pseudocode for a table for interviewing.

For blocks of code, a library is nothing more than a directory/subdirectory structure with appropriately named files. In the example below, the library contains a subdirectory (called ID in the example) with two standard blocks (Ident1.Inc and Ident2.Inc) for dealing with case identification. There are other subdirectories containing, for example, code that screens the sample, links with case management systems, concludes the interview, makes appointments, and records details of nonresponse.

```
\LIB\ID\Ident1.Blk
        \Ident2.Blk

    \SCREEN

    \MANAGE

    \CONCLUDE

    \APPOINT

    \NONRESP
```

A librarian can be appointed to safeguard the files and to ensure that updates are made appropriately. The library may contain options to cover the range of surveys in an organization.

7.4.4 Common or Similar Administrative Blocks of Questions

Sharing administrative blocks of code between survey instruments is particularly useful. Such blocks include those for appointments, survey management, nonresponse, name and address, rostering structures, characteristics of households, people, and firms. In the pseudocode example below, files of predefined code are incorporated into the instrument.

> INCORPORATE "Ident.Blk" {Identification structure}
>
> INCORPORATE "Manage.Blk" {Management variables}
>
> INCORPORATE "DateTime.Blk" {Date and time stamp block}
>
> INCORPORATE "Appoint.Blk" {Appointment module}
>
> INCORPORATE "NonResp.Blk" {Nonresponse module}

In just five lines of code, the designer has provided an identification structure, a place for survey management information to be stored, a way of keeping track of date and time information for each section, and standard ways of handling appointments and nonresponse. As noted above, each block includes all within-block data definition, relationship, routing, edits, and computations. The designer can state how the included blocks should interact with each other and with survey-specific blocks. For example, the appointment block would not be available if the nonresponse block has been filled in, and neither the nonresponse block nor the appointment block would be eligible if all subject matter sections are filled in. Whether a designer can include blocks of code in any instrument without modification depends on the CAI system, the unique requirements of each survey, and the foresight used in defining the blocks in the first place.

Interrelations between incorporated blocks of administrative code may be standard for a program of surveys. If so, the blocks may be nested inside an administrative shell as described next. The challenge of course is in designing, implementing, and maintaining these standard common blocks so that they can be used in different instruments.

7.4.5 Common Administrative Shells

It is possible to extend the idea of including predefined administrative blocks within an instrument by providing an administrative shell, or small set of shells, for the designer (Schou and Pierzchala, 1993). A standard shell includes the relevant standard blocks and all interactions between them. A program of surveys where administrative interfaces and management requirements are reasonably similar can use a single administrative shell. The designer cannot alter the shells or, if permitted, does so at his or her own hazard. For example, NASS provides two shells, one each for probability and nonprobability surveys.

In another example, the Bureau of the Census re-uses sections of administrative code across its surveys. The sections deal with control functions such as breaking off and restarting an interview, stopping an interview for one person in a household, and jumping to an interview for another person and then returning to complete the interview of the first person.

7.4.6 Standard Interfaces to Administrative Databases

Survey organizations usually have administrative databases for sampling frames and for a survey management system. When these are common to a program of surveys, it is possible to use standard interfaces for the transfer of administrative data between the sampling frame and the CAI software. The interface might be a common intermediate file format that allows administrative sampling data to be transferred back and forth between the sampling frame and the instrumentation. Alternatively, it might be a module that allows the instrument to read from the administrative databases and write directly to them. If every survey uses the same information about sample elements, it is possible to use the same blocks of code in the instruments for identification structure, name and address recording or verification, and survey screening.

If surveys make use of the same external survey management infrastructure, the instruments can use the same survey management blocks of code. A statistical office will often have the following kinds of reports: check-in (percent complete by classes), missing reports (which reports are not in), the progression and location of each case in the system, review of appointments and nonresponse, review of frequencies of questions, and a record of the progress of the interview (date/time stamps). The desire not to keep rewriting these reports also encourage use of common blocks of code.

7.4.7 Common or Similar Subject Matter Sections

Surveys in a program will often share some common subject matter sections, for which standard code can be provided. At a minimum many household surveys collect the same basic demographic information.

Integrated surveys have taken the idea furthest. The Integrated Survey project at Statistics Netherlands is built around three levels of integration: a core of questions that can be fully standardized for all the constituent surveys with one set of CAI question blocks, even to the extent of comprising a discrete part of the interview; a second level where some of the component surveys share blocks; and a third level that is specific for each survey (Heuvelmans, Kerssemakers, Winkels, 1997). The experimental Integrated Household Survey development at ONS was also built on the idea of harmonized blocks of questions, but it allowed greater freedom for component surveys to place the blocks as they felt appropriate within their own particular contexts (Manners and Deacon, 1997).

If the block of employment questions (ROW BILO) in Figure 7.1 were a standard block, the survey organization could store it in a file (e.g., BILO.Blk) and call it into the standard table code (TABLE ILO) by substituting:

INCORPORATE "BILO.Blk"

In this way the survey organization could assemble the code in Figure 7.1 automatically for any survey and be sure that it would conform to standards.

7.4.8 Database of Metadata Specifications

A program of repeated surveys may have a database of questions and related meta-information. These metadata can be used to generate, or to guide the execution of, CAI instruments (Pierzchala, 1992, 1995). For example, NASS faces a severe challenge to its CAI instrument design strategy. There are up to 45 versions of each of its Quarterly Agricultural Surveys and the June Area Frame Survey. The crop and grain stocks sections of the questionnaires vary from state to state. Each state must carry out its part of the national crop estimation program, asking only questions appropriate to its agriculture. Each state varies from the others in the selection and precise text of crop questions, the order in which the questions are asked, and the units in which data are collected.

There are item codes (for keypunch) and SAS variable names that predate CAI. For the purpose of automatically producing CAI instruments, NASS combines then into a database that also includes question text, question name, question description, valid range, order of asking, units of collection, and other information that defines an instrument's data structure. There is one record of metadata for each state for each survey. NASS also stores an indication of which block structure is valid for each set of questions. Stating that a particular question belongs to a particular generic block structure is equivalent to defining its flow and general edit information.

NASS uses this information to generate automatically the 45 versions of the instrument that it needs. The instruments are each used in multiple data collection modes — CATI, CAPI, and Interactive Editing. NASS uses a variation of this method for the June Area Frame survey which also has 45 versions. Here one instrument is guided 45 different ways during run time from the database of specifications for CAPI and Interactive Editing.

This method of generating (or guiding) instruments from a database of specifications has many advantages. The database's accessibility makes it easier to maintain the instrumentation and introduce changes accurately. The program code for the generic block structures is simpler than it would be if there had to be numerous hard coded references to differences between states. It is also easier to review a state's specification by browsing the database than by inspecting thousands of lines of code.

This technique was made easier to implement because approximately 20 generic block structures could be preprogrammed and applied as needed to each crop in each state for the QAS. Encompassing flow and edit information within generic block structures eliminated the need for entering such within-block relationship information into the database.

7.4.9   Question and Screen Formatting Conventions

Where interviewers work on more than one survey in a program, interviewing quality can be improved by a consistent implementation of interviewing conventions. Some organizations know these conventions as screen design standards. For example, it may be standard that on-screen auxiliary information, such as an instruction to the interviewer, appear in uppercase letters, a different font, or a particular color. If there is a single system of conventions, interviewers spend less time learning how to use the software and more time learning about the survey.

As graphical user interfaces develop, the number of potential screen layout options is almost limitless. Standard screen layouts help to avoid confusion from different styles and conventions. In some software an organization can store the range of screen layouts that it needs in a library file. The layouts can be called and changed within an instrument with a few key words.

CAI software may allow reassignment of the uses of function keys. As with screen layout, standards for function keys help interviewers avoid confusion on a program of surveys.

7.4.10   Common Edit Conventions

CAI instruments usually include checks (edits) that answers are within the permitted range and consistent with other answers (Kinsey and Jewell, Chapter 6). When a check fails during the interview, an error message appears in a pop-up window or as a question screen, depending on the CAI software. Such checks must be designed, presented, and processed following a standard model. Since the interviewer has to break the normal interviewing sequence to correct error messages, the need for standardization goes beyond screen design to cover such issues as the number and complexity of edits that are reasonably dealt with in an interview. The range of CAI edit writing conventions found in survey organizations include (Pierzchala, 1994):

- kinds, number, and severity of edits;
- order in which the questions involved in the check are to appear on screen for the interviewer or editor to jump back to;
- key words to advise the interviewer of the severity of the edit, for example,

[SOFT ERROR] to mean the interviewer can either fix the problem or suppress the edit and keep moving on, and [HARD ERROR] to mean the problem has to be fixed before proceeding;

- edit message standards, for example, an edit heading that summarizes the edit succinctly followed by a more detailed edit message;
- identification of the edits that are soft in the interview but hard in postcollection review;
- text formatting conventions; and
- identification number of edits.

7.4.11   Standard Procedures and Subroutines

Code libraries can store standard procedures or subroutines that perform specific tasks. CAI software usually provides the capability to write a generalized procedure within the system language or to call a subroutine external to the system. For example, the software will probably provide a RANDOM function that selects one random number at a time, but it may be necessary to write a procedure to extend it to select m out of n elements randomly and without repetition. Such a general procedure can be built one time and used by any survey. A well-conceived procedure will operate independently of the specific names of questions that are used to pass information into it.

7.4.12   Standard Coding Frames

A program of surveys will often need to use standard coding frames. A coding frame is a file of descriptions and related codes of jobs, automobiles, commodities, chemicals, and any other long list of items for coding. For example, in the U.K. Labour Force Survey (LFS), the shortest coding frame (a list of countries) has some 600 items, including alternative country names, and the longest (place of residence) has about 30,000 items. A CAI coding module helps interviewers and editors navigate a coding frame to find matches for answers. It then stores the corresponding codes. Coding frames are often suitable for re-use on other surveys. Other U.K. government surveys have used the LFS coding frames.

Where coding frames embody national standard classifications, such as for occupation and industry, their use on many surveys is likely. A sponsoring organization may specify particular coding software as well as the coding frames it wants the survey organization to use. The survey organization's CAI software has to be capable of calling the sponsor's coding frame at appropriate points in the interview or data entry.

7.4.13   Standard External File Formats

External files can hold standard information for use by the surveys in a program. For example, one use is to customize the behavior and appearance of an instrument by geographical region. In NASS, external files of edit limits are used to handle state differences in agriculture. Another common use is to hold responses from one or more previous waves of a survey, to prompt respondents, or to check against answers in the current wave. Standard external file formats make it easier to implement this technique for a program of surveys.

7.4.14 Standards for Instrument Writing

Standards are useful for writing the instrument code itself. A program of surveys will usually have several instrument designers. A standard way of laying out instrument programming code makes it easier to understand, maintain, and change each other's code quickly and accurately. There needs to be a careful balance between prescription and fussiness that impedes good questionnaire writing. For example, prescribed tab sizes for indentations to indicate nesting may sound fussy, but in complex questionnaires it can enable a reader to spot instantly how a variable relates to others, by the level of nesting. Naming conventions can aid in the creation of self-documenting instruments. In survey work, designers develop new instruments or change existing ones at great speed. There may even be a need for wholesale restructuring. Naming conventions can indicate what type of element is being named and therefore where to find its definition. There are many similar conventions for instrument writing that are less important in their details than in the simple fact that they are standard and convey a consistent meaning to readers.

Instrument Writing Interfaces. Several CAI organizations have developed instrument writing interfaces or producing systems with varying degrees of success. This is done to shield the designer from the intricacies of the system's programming code, to keep the designer organized, to make sure that details are not omitted, to reduce syntax errors during compilation of the instrument, to generate reports about the instrument that the CAI system itself may not be able to generate, and to reduce testing.

In the interface, the designer is led to fill in necessary pieces of information, for example, type of question, the question name, valid range, other identifier (data-entry item code, SAS/SPSS variable name), and question text. Flow and edit information may be entered in some systems. From this information, which is now held in an electronic database, it is possible to generate CAI application code which is then compiled into a working instrument (see Kinsey and Jewell, Chapter 6, for further details).

7.4.15 On-line Context-Specific Help

On-line context-specific help is a useful CAI feature for any survey. There are two main reasons why it is particularly valuable in the context of a program of surveys:

- On-line help can absorb a significant part of the learning burden where interviewers must be able to perform effectively on a number of new surveys concurrently or within a short period.
- Libraries of standard code for questions can include the associated detailed interviewer instructions and help.

ONS developed on-line context-specific help when it was faced with these two demands during its experimental work on an integrated survey. This project combined four complex surveys in each interviewer workload, with the result that new interviewers could not build up their knowledge one survey at a time. New training methods relied on providing interviewers with access at the right time to the right information through on-line help which was tailored to the current question. Early trials showed that trying to train interviewers by traditional methods on even two of the component surveys in combination was unsuccessful. There was too much information to absorb. In combination with carefully structured distance learning, and a relatively high level rather than detailed briefing, the use of on-line context sensitive help allowed a successful field trial of the integrated survey with newly recruited interviewers (Manners and Deacon, 1997). This experimental integrated survey project showed the advantages of on-line help in a dramatic form, but the benefits may apply to any program of surveys.

7.5 CHECKLIST OF STRATEGIES AGAINST MAIN TECHNIQUES

Table 7.1 summarizes the use of some main instrument producing techniques against the instrument producing strategies that have been identified. The techniques of question and screen formatting conventions, common edit conventions, standard procedures and subroutines, and standards for instrument writing would all have "yes" in their columns if they were included in the table.

7.6 NEW MANAGEMENT CHALLENGES IN A PROGRAM OF SURVEYS

Standardization, centralization of expertise, coordination, and exploiting economies of scale are mainstay management strategies. Achieving them, however, is often not straightforward. Examples abound of organizations in which the results of standardization and centralization initiatives have been mixed or even disastrous. Careful planning and close attention to existing processes, relationships, and staff capabilities are critical.

Table 7.1. Applicability of Instrument Producing Techniques by Survey Program Type

Technique Strategy	Coordinated Individual Surveys	Multiple Variant Instruments	Integrated Survey Instruments	Run One Instrument Distinctively for Each Respondent	Modified Instruments in Successive Longitudinal Waves
Common precoded responses	Many precodes shared between instruments	Many	Yes	Yes	Yes
Common data structures	Parts of instruments share common structures	Yes	N/A	N/A	Some
Question libraries and block libraries	Some questions and blocks shared	Yes	Yes	Yes	Yes
Common or similar administrative blocks of questions	Most instruments share most or all of such blocks	Yes	N/A	N/A	Yes
Common administrative shells	Most instruments share the shells	Yes	N/A	N/A	Yes
Standard interfaces to administrative databases	Most instruments share interfaces	Yes	N/A	N/A	Yes
Common or similar subject matter sections	No	Some	Some	Yes	Yes
Database of metadata specifications	No	Depends on program	Depends on program	Respondent management system	No
Standard coding frames	Depends on individual survey	Some	Maybe	Yes	Yes
Standard external file formats	Depends on individual survey	Yes	Yes	Yes	Yes
On-line context-specific help	No	Yes	Yes	Yes	Yes

7.6.1   Tools for Standardization and Harmonization

Providing good tools may be the best way to implement standardization and harmonization within an organization. Simple insistence on standards tends to require an overhead for inspection and runs counter to the ideas of total quality management. As described earlier, an effective CAI system will allow a designer to build questionnaires from standard programs, classifications, and blocks of questions through such means as direct insertion, external files, and libraries. If designers can incorporate standard blocks in their surveys as plug-in modules, they are likely to do so. Standards will be followed by default, if not by intent.

   If, however, designers need to tailor the standard modules, with painstaking referencing of variables and other dynamic items, there is a risk that they will think it easier to build their own versions from inside their own instruments. This risk means that it is worthwhile investing in a support service to implement standard items on particular surveys. It may also be useful to allow space for survey-specific needs in a standard module, such as a specific range of outcome codes, within the standard set, which are allowed to vary by survey to deal with nonstandard outcomes.

7.6.2   Increased Organizational Flexibility

Standardization may make it possible for an organization to involve re-searchers in producing CAI instruments rather than relying on a division of labor between researchers and programmers. It may also make it easier for staff to switch between surveys.

Involving Researchers in Producing Instruments.   Some survey organizations would like researchers (subject matter specialists) to be as involved in produc-ing CAI instruments as they were in producing paper ones (Manners, 1992). An instrumentation strategy based on standardization and re-use of code is particularly important if the instrument writers are not professional pro-grammers. Standardization allows these staff to "stand on the shoulders" of experts by using modules of well-tested code. They can plug the modules directly into their instruments or use the modules as templates for their own blocks of questions. Standardization and re-use of code also enable the organization to maintain control of programming quality.

Flexibility in Use of Staff.   Using one set of tools and standards makes it easier to shift staff between surveys and allows them to concentrate on learning the new subject matter. Surveys in a related program will almost certainly have more than one designer. The instruments will need to be understood and referred to by many people. While this is obvious for closely related contem-poraneous surveys, it is sometimes overlooked for surveys that are related over time, such as follow-up and repeated surveys.

Of course organizations vary in the types of staff that they employ to write instruments. In PAPI, survey instruments were designed by survey researchers. In many organizations this approach changed with the advent of CAI methods. Survey researchers now design specifications, which they pass to programmers to implement. Some organizations, like ONS, Statistics Netherlands, and NASS, continue to have survey researchers produce the instruments. Computing specialists concentrate on tasks that more obviously require their particular expertise, like developing systems for case management and telecommunications.

These differences in practice seem to be unrelated to any differences in the kind and complexity of tasks facing the organizations. Instead, the differences seem to depend on (1) the degree to which the survey researcher was responsible for all aspects of the project in the pre-CAI division of labor and (2) the characteristics of the CAI software chosen — itself influenced by (1).

One of the arguments used by those who continue to have researchers write instruments is the cost of doubling up on specification and programming: having two staff where there used to be one. However, a program of surveys offers the opportunity to organize researcher-programmer teams in a cost effective fashion. This argument is not decisive, however, for many organizations give great weight to the researcher's direct control and understanding of all aspects of the instrument. There seem to be three main ways that survey researchers may be involved in producing quality CAI instruments.

First, the producer may be a team of researchers and programmers. This approach avoids communication problems. The difficulty lies in making it cost effective and in finding the optimum balance of different skills. For example, Westat has instituted an instrument design group for the Medical Expenditures Panel Survey. The group consists of analytical staff, survey design staff, and programming staff who work together to ensure that draft instruments meet the analytical goals, are compatible with existing standards, and can be programmed in a cost effective way. When issues arise, the group meets to iron out solutions.

Second, survey researchers should be trained as CAI programmers, and these programming skills should be kept fresh. This approach also minimizes the potential for miscommunication. NASS uses this method. Subject matter specialists in headquarters produce almost all CAI instruments. Additionally, in the state offices, many statisticians have programmed CAI instruments for state level surveys with or without formal training in the system. This has been aided by headquarters staff providing administrative modules and preprogrammed links to the sample frame system that all states use. NASS headquarters is now providing CAI training to the states' statisticians.

Third, each survey researcher may program his/her own part of a CAI questionnaire, using standard modules and templates. Under PAPI, researchers in many organizations had "hands-on" control and knowledge of the design of their instruments. ONS has approximately 50 survey researchers who design and usually analyze their surveys. They carry the prime responsibility for writing the instruments for their surveys. For the more senior researchers this means knowing enough about the interaction of CAI design with other survey design requirements to supervise the junior researchers who will write the code.

A drawback of this approach is that a large number of staff have to be trained. Researchers exercise their skills in writing CAI instruments only intermittently, and these skills may tarnish over time. One method for overcoming these problems is for the researchers to use model instruments that embody standards and pay due attention to efficient design. At ONS, a small part-time team comprising researcher, methodologist, fieldwork, and computing specialists provides the models and quality assurance of CAI instrument designs.

The success of an instrumentation strategy like ONS's depends on an effective environment of case management and data handling, provided by computing specialists. It is likely that the culture of an organization will play a strong role in deciding between these and other instrumentation options, such as the division of labor between researcher and programmer.

7.6.3 Standard CAI Software

An organization is extremely unlikely to implement the strategies and techniques mentioned above unless it standardizes on one CAI system, or for large and diverse organizations, at most two. The reason is that CAI systems tend to be so different from each other in terms of programming language, functionality, capacity, data storage, and in other ways that it is virtually impossible for them to share code, or even to take the code of one and convert it to the code of the other. Where two or more systems are used, there will be duplication of effort in providing connections to sampling frames, survey management systems, and so on. The choice of a CAI system is not an easy one. Still there are ways of choosing a system rationally. A list of necessary and desired features may be constructed, in itself not an easy task (NASS, 1992), and a statement of organization goals and systems needs may also be necessary.

7.7 SUMMARY AND CONCLUSIONS

Organizations that engage in the ongoing production or maintenance of CAI instruments for a program of surveys should employ a strategy, or strategies, for producing them. The benefits of implementing a good strategy include improved productivity, lower costs, better comparison of results across surveys, consistent interfaces between instruments for interviewers and data editors, quicker instrument production, re-usable interfaces between the CAI system and other organization systems, and organizational flexibility.

Four major strategies identified include producing instruments for each survey individually in a coordinated manner, producing large instruments covering several surveys, producing instruments tailored to respondents to multiple surveys, and modifying instruments in successive waves of longitudinal surveys. Such strategies do not come naturally from using one CAI system, though this is a good first step. Implementers of a strategy should expect and work for these improved benefits. They may have to sell the idea in their organizations and then work to ensure that the strategy is implemented correctly.

An instrumentation strategy should allow for experimentation and innovation, but in a testable way. Fifteen techniques for realizing the strategies were discussed. An excellent way of getting a jump-start on implementing an instrumentation strategy is to use the work of other organizations that have already done so.

ACKNOWLEDGMENTS

Much of the work referenced in this chapter was done when Mark Pierzchala worked at the U.S. National Agricultural Statistics Service. The following people have provided information on practices in their organizations: Asa Manning of NASS, Bill Mockovak and Barry Fink of the U.S. Bureau of the Census, Jane Shepherd of Westat, and Evert Blom of Statistics Sweden. Jim O'Reilly of Research Triangle Institute provided valuable editorial guidance.

CHAPTER 8

The Individualized Interview

Dirk Sikkel
CentERdata

8.1. INTRODUCTION

From the earliest days of CASIC, managers and researchers have worked on the feasibility and applicability of the new technology. At the start the questions were: What kind of logic should be available to the questionnaire designers? Will interviewers be able to handle computers? Will respondents become suspicious, for example, about confidentiality issues? And, can we increase the timeliness and lower the costs by using CASIC? These concerns are understandable because for a statistical bureau it is a major decision to change from paper and pencil data (P&P) collection to computer assisted methods. At present, these issues have found equilibrium. Both for CATI and CAPI it is clear that (1) they operate efficiently, (2) they present no serious stumbling block for interviewers, and (3) they do not lead to higher nonresponse rates (see Couper and Nicholls, Chapter 1). With respect to data quality, the status of computer assisted interviewing is less clear. Theoretically data quality should be higher than in corresponding P&P questionnaires. The existing research suggests that CAI leads to some improvement of data quality, but that the improvement is not very significant (e.g., see Hox, de Leeuw, and Snijkers, 1993; Bergman et al., 1994; Nicholls, Baker, and Martin, 1997).

These small advances, as reported in the literature, are in sharp contrast with the high expectations raised by the early developers of CAI (e.g., Freeman, 1983; Sikkel, 1987; Saris, 1989). Freeman (1983, p. 144) described the nature and the pitfalls of the development of computer assisted interviewing by the following comparison:

> Some years ago, I leased a fancy word processor, only to find that it was not used much and my staff did not have a high regard for it. We finally broke the lease. Only

Computer Assisted Survey Information Collection, Edited by Mick P. Couper, Reginald P. Baker, Jelke Bethlehem, Cynthia Z. F. Clark, Jean Martin, William L. Nicholls II, and James M. O'Reilly. ISBN 0-471-17848-9 © 1998 John Wiley & Sons, Inc.

then did an executive of the word processor company provide the crucial insight. He told us that our error was in thinking that we had obtained a better typewriter, instead of thinking of the word processor and the typewriter as distinct entities. Rather than using CATI as we foolishly did our word processor, I think we should seize the opportunity to examine how we do things, and also exploit the new technology for new kinds of substantive research.

This, indeed, should be CASIC's most spectacular contribution. When research procedures are developed that have no equivalent in the paper and pencil mode, we can say that CASIC has fulfilled its promise, provided of course that the new research procedures are sound. In that case comparison with a paper and pencil questionnaire becomes pointless, at least from the measurement point of view, because a P&P version simply does not exist. This chapter highlights the development of new substantive research. Gradually we leave the traditional problems and step into the exciting field of research that was unthinkable in the precomputer era. Our main topic is the individualization of questionnaires. The challenge is to design questionnaires so that a wide variety of the respondent's individual characteristics can be entered and used in the analysis in a meaningful way. This chapter explores how such questionnaires can be designed, with less attention paid to comparison of different (competing) methods.

This chapter deals with two types of CASIC. The first is self-administered questionnaires (SAQ), specifically a telepanel. A telepanel is a household panel where the panel members have a PC at their disposal (see Saris, Chapter 21). The Dutch foundation CentER*data* has two such panels. One panel is representative of the Dutch population and consists of 2000 households who complete a 30-minute questionnaire every week. The second panel consists of 500 households from the highest income decile in The Netherlands. They complete an hour-long questionnaire each month. These panels collect data used in both scientific and market research. The chapter also examines some CAPI applications. Here we follow new market research applications. Some of them are already used with great frequency; others still have an experimental character.

8.2 FEEDBACK TO RESPONDENT IN SELF-ADMINISTERED QUESTIONNAIRES

8.2.1 Use of Summary Screens

An attractive subject for telepanel research is income studies. From the substantive point of view, detailed longitudinal data about income, savings, pensions, assets, mortgages, and the like, are of great interest for example, in studies of the effects of governmental income policy or the different income management styles within households. From the measurement point of view, a telepanel is a promising research instrument. Respondents do not have to disclose sensitive data to an interviewer, they can fill out the questionnaire at their own pace, and they have the time and the opportunity to find all the documents needed to provide correct answers. Still income reporting is complex, particularly for respondents. Therefore it is necessary to give respondents ample opportunity to check the information they give and to allow correction of previously given answers since interviewers are not present to assist. An effective tool is the summary screen or summary correction screen (Saris, 1991). The sequence in Figure 8.1 shows a (relatively simple) series of screens in which questions are asked about income and taxes.

In the first screen of the figure, a question on income is asked (based on the most commonly used definitions). The second screen shows the question about taxes and premiums. The third screen is a warning and appears only when income and taxes create an implausible ratio. The fourth screen is the summary screen, which appears for every respondent. If the respondent indicates that there is an error, he/she will be given the opportunity to correct it, after which the summary screen appears again.

```
- 1 -
What was the amount of the total
GROSS salary in 1996 you
received from Tilburg University?
Do not take into account: child
allowance and employer's pension
premium
Take into account: withheld
taxes and national insurance
premiums.

Amount in guilders: 80,000------
```

```
- 2 -
What was the amount of withheld
tax on salaries and premiums of
national insurance in 1996 from
Tilburg University?

If you don't know type 0 (zero).

Amount in guilders: 800------
```

```
- 3 -
The amount of withheld tax on
salaries and premiums of national
insurance of 800 guilders is
relatively low.

If your answer is not correct,
please modify your answers on the
next screens.

Press ENTER.
```

```
- 4 -
Employer: Tilburg University

period: 1 January to 31 December

Gross salary: fl. 80,000
Withheld taxes: fl. 800

Is all of this correct?
0. no
1. yes
```

Figure 8.1. Example of summary screen used in a telepanel survey.

The number of changes a respondent makes can be recorded by the software. Regular income, as would be reported in an income tax form, is not corrected in the vast majority of the cases. Pensions and benefits are more difficult to report, and here the summary screens prove very useful (assuming that the corrected figures are better than the original ones). There also appears to be an age effect: Older respondents make more corrections than younger respondents. A probable explanation is found in Table 8.1, as corrections made for pension are naturally made only by older people. This effect disappears when adjusted for pensions.

It is possible to include correction sheets in paper and pencil surveys, but these are likely to annoy the respondent and may lead to new errors if the respondent is required to perform the calculations him/herself.

8.2.2   Embedded Measurement

The idea of summary screens can be extended to another problematic area of research: retrospective questions. In many areas it is desirable to obtain information about the respondent's past. Examples of such areas are health care (medical consumption), victimization, banking, housing, and expenditures on durable consumption goods. Retrospective questions often are the only and always the least expensive way to collect such data. Such questions, however, are notoriously unreliable, due to memory and recall effects (e.g., see Sudman, Bradburn, and Schwarz, 1996). The literature indicates that in the case of P&P questionnaires, working with calendars is a good procedure (Friedman et al., 1988; Linton, 1975; Wagenaar, 1986). It also appears that memory can be stimulated with "anchor points" in time. A respondent may not know exactly when he/she moved from one house to another but does know whether that happened before or after a certain historic event (the anchor point) (Linton, 1975; Loftus and Marburger, 1983; Strube, 1987). Such methods are difficult to implement in a large-scale P&P survey. First, they require special skills from the interviewers who need a thorough understanding of the purpose of the questions. Second, such procedures are expensive and time-consuming. In a telepanel, however, there are ways to overcome such problems.

Table 8.1. Percentage of Cases with One or More Changes on Summary Screens in a Survey on 1995 Income

Subject	Percentage Changed	(N)
Income from employment	7.7	(2119)
Income as self-employed in 1995	2.5	(239)
Income as self-employed in 1994	13.4	(239)
Pension	79.2	(590)
Illness benefits	43.9	(458)
Social benefits	85.0	(283)
Other sources	4.7	(358)

We tested two techniques in a small qualitative study. The first was forward recall. The respondent was asked to reconstruct his/her past in a fixed order from birth to present. Questions were asked about the birth itself, the family residence at time of birth, kindergarten, the birth of a sibling, moving to another city, primary school, and so on. This experiment did not give satisfactory results, which can be explained by the fact that memory is not linearly ordered in time.

The second approach used embedded measurement in the survey proper. Here we constructed a list of the respondent's most salient memories and used them as anchor points in the rest of the interview. For example, a respondent can report moving from one residence to the next. This is illustrated in Figures 8.2 to 8.4.

The three screens in Figure 8.2 show how the method is introduced and the events reported. Once the events are recorded, the respondent is given the

```
         - 1 -
In your life there have probably
been events that you remember
very well.  Such events can be
anything, like a broken arm, a new
bike, a meeting which made a big
impression on you, losing a friend,
a special sports event, etc.

press ENTER
```

```
         - 2 -
We will ask you to tell us such
events that you remember well,
spread out over your life in order
from your birth up to now, up to a
maximum of 8 events.

For the remainder of this
questionnaire it is essential that you
enter at least 5 events.

press ENTER
```

```
         - 3 -
We ask you to enter at least 5 and at
most 8 events (one line per event).
Please spread the events as much as
possible from your birth until now.

1. new brother---------------
2. robbed in Palermo-------
3. found first job------------
4. trip to Sri Lanka---------
5. mother died--------------
6. got Ph.D. degree---------
7. ----------------------------
8. ----------------------------
```

Figure 8.2. Example of recall aids in telepanel survey: Collection of key events.

opportunity to explain the reason why events are so special for him/her. All this information is irrelevant for the subject of interest (moving), but it helps the respondent freshen up his/her memory. The next step is to link the respondent's residences to the events mentioned, as shown in Figure 8.3.

The information obtained in this way is displayed in a summary screen and can be modified by the respondent. Once the list is checked and corrected, the respondent is asked to complete his/her history of residences, from birth up to the interview. This summary screen is illustrated in Figure 8.4.

A similar procedure was followed for data about childhood events, education, relationships, and jobs. When all data were collected, the complete life history of the respondent was shown to him/her in the form of a calender. The respondent was again given the opportunity to make changes. This proved crucial, as shown in Table 8.2.

We see that those who have complex life histories, and hence have many events to report, make most of the changes. For example, 7 percent of respondent account for 20 percent of all the changes to date of move, with a maximum of eight changes by any one respondent. In general, we conclude that giving feedback to the respondent about aspects of his/her life history generates reactions and corrections to previously given answers.

Can this approach be used in a P&P survey? When an interviewer is present, this is certainly possible, but time-consuming, and the interviewer needs extensive training. Using a computer may structure the process and lead to an increase in efficiency. For self-completion questionnaires the com-

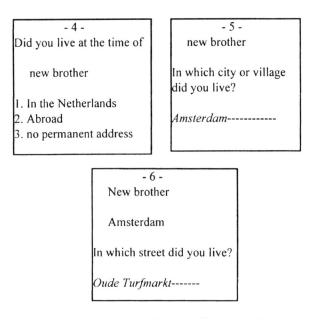

Figure 8.3. Example of recall aids: Use of key events for moves.

```
                    - 7 -
    1. new brother
       Amsterdam, Oude Turfmarkt
    2. robbed in Palermo
       no change
    3. found first job
       Leiden, Breestraat
    4. trip to Sri Lanka
       no change
    5. mother died
       Rotterdam, Hofplein
    6. got Ph.D. degree
       no change
```

Figure 8.4. Example of recall aids in telepanel survey: Summary screen of moves.

puterized approach is a necessary condition, since it is implausible that a respondent will carry out the procedure correctly when relying solely on written instructions.

8.3 NATURAL GROUPING

Natural grouping is an exploratory method to determine the structure of a market (Verhallen, 1988). It is a sorting method whereby every respondent can indicate his/her own perceived structure of the market. The respondent is

Table 8.2. Types and Numbers of Corrections in Moves and Education

Type of Correction	Maximum Number of Changes Made	Percentage of Respondents Making Changes	Percentage of Data Items Changed
Moves			
Date of move changed	6	7	20
Address changed	6	6	20
Address deleted	9	13	28
Address added	10	9	31
Education			
Date changed	6	12	46
Education deleted	8	7	36
Education added	5	4	18

presented with a set of stimuli, which may consist of products, commercials, brand names, and the like. He/she is asked to split the set into two groups, using his/her own criteria. When the respondent has completed the task, he/she is asked to give the reasons for the split. This process is repeated for the two groups that are the result of the split and continued until the respondent has exhausted the criteria to split the stimuli. We can illustrate the data collection method by the following example.

Let us assume that the set of stimuli consists of broad food categories:

$$S_0 = \{\text{cheese, potatoes, jam, meat, milk, fish, eggs, nuts and crisps, rice, peanut butter, meat products, fruit}\}$$

After the respondent is asked to split this set into two groups, he/she may form the following groups:

$$S_1 = \{\text{cheese, milk, eggs, meat products, peanut butter, jam}\}$$

and

$$S_2 = S_0 - S_1$$

The second group is the complement of the first. Then the respondent is asked why he/she split the total set of stimuli into these particular groups. His/her criterion may be that S_1 consists of products that are used with cold meals. Then he/she is asked to split S_1 again into two groups, and may form the following:

$$S_3 = \{\text{peanut butter, jam}\}$$

and

$$S_4 = S_1 - S_3$$

Again, he/she is asked to provide reasons for this split. The reason may be that the products are sweet and that they are not very healthy. It is essential that the interviewer presses the respondent to give extensive reasons for the split, thus generating as much information as possible. The grouping process stops when the respondent runs out of split criteria.

The data for one respondent have a tree structure as given in Figure 8.5. The vertices correspond to the sets; the horizontal edges are labeled by the reasons for splitting or attributes as given by the respondent. This example was simplified for the sake of clarity. Usually respondents give more voluble explanations during the grouping process. The tree may, hovever, realistically represent a natural grouping after recording of the data into a small number of categories. As Figure 8.5 shows, it is not necessary that the branches of the tree have the same length. In one branch the grouping may stop at an earlier

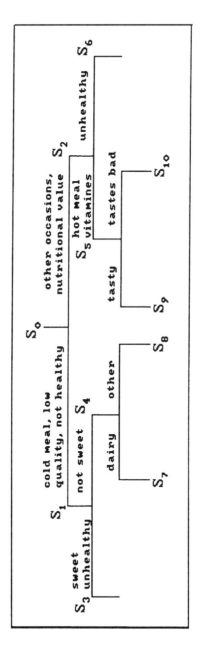

Figure 8.5. Example of a natural grouping tree: Sets and attributes.

level than in an other branch. Also the sizes of the groups may differ. And finally every respondent is allowed to construct his/her own tree.

The procedure sketched above can be applied to a wide variety of stimuli. These can be products, display cards with the names of the products, photographs, story boards of previously shown TV advertisements, and so on. The role of the interviewer is critical. He/she has to extract as much information as possible without leading the respondent. It is our experience that in a good interview the most interesting information emerges in the middle part of the grouping process. The first splits usually refer to broad general categories that contain only general information to the researcher (although the agreement among the individuals on these categories still may be of interest). The final splits often are idiosyncratic descriptions of the individual stimuli. It is in the middle where the respondents have to make nontrivial splits and have to express precisely their thoughts about the stimuli presented. Experience shows that the respondents find the natural grouping task pleasant and interesting. They are able to handle sets that contain over 60 stimuli.

Note that the resulting data set is of a highly individualized nature. Each respondent can build his/her own tree, which represents his/her personal perception of the market. The open answers on the "Why did you make this split?" questions are also highly individualized. Thus one of the advantages of natural grouping is that the respondent does not have to give opinions that he/she does not have, or to answer questions which he/she found irrelevant. Recoding (i.e., adding attributes to the open comments of the respondent) plays an important role in the data analysis stage. Good recoding software is vital. The data analysis consists of content analysis, scaling of stimuli and recodes, and tabulation, allowing various ways of aggregating the data. The technique has been used for market positioning, image research, fine tuning of product development, and exploratory analysis for markets whose structures are unknown.

A recent, rather surprising example of an application is the study of consumer values. The basic, most influential study into general values is that of Rokeach (1973) who identified 18 *end values* (which are a goal in itself) and 18 *instrumental values* (which serve to reach the end values). The goal of the present study is to determine the completeness of Rokeach's list and the applicability of that list to the Dutch society. One of the problems we encountered was the difficulty of conveying to respondents precisely what a value is. The Rokeach definition of a value is very abstract, and in general, it is very hard to avoid terms that have no meaning to the respondent. So natural grouping was used (1) to communicate what a value is through learning by example, (2) to identify which values have identical meanings to the consumer, and (3) to identify which values have ambiguous meanings. A list of 49 values was used.

After two or three splits, what was meant by values became clear to respondents. The interviews generated much intimate individual information, even to the extent that respondents felt embarrassed by having revealed so

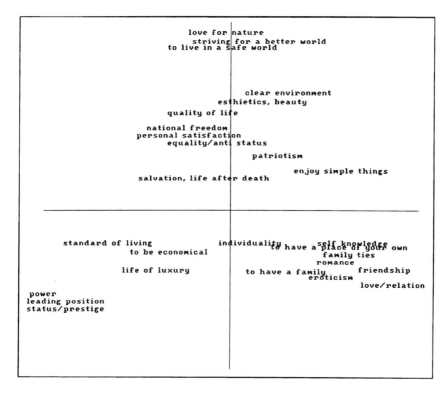

Figure 8.6. Correspondence analysis on values based on natural grouping (values in the middle of the figure are omitted).

much personal information. The structure of the values was analyzed using cluster analysis (a measure of similarity based on the number of times values appeared in the same group) and correspondence analysis (a measure of distance based on the number of times values appeared in different groups). The results are presented in graphs, such as in Figure 8.6. Here the end points of the axes show clusters of values that appeared to have a similar meaning to the respondents, such as, the cluster {power, leading position, status/prestige}. It is possible to find more clusters with similar meaning by plotting more axes. In this way the computer assisted procedure generates a large amount of relevant data that can be collected and analyzed efficiently using the grouping structure.

Can this be achieved in a data collection procedure with paper and pencil? In fact during the first projects with natural grouping, P&P procedures were used. This had, however, two impeding effects on the projects. First, the recording task for the interviewer was too heavy. He/she had to describe groups of 40 or more stimuli by numbers, without making errors, which led to long waiting times. After that the data entry and error checking proved difficult, since it was hard to repair an inconsistency in the data. As a result a specialized CAPI program was developed. This allowed for data entry at the time the respondents actually constructed the groups and for easy visual inspection for correctness.

8.4 PREFERENCES AND DECISIONS

One of the most important but also most complex problems in market research is the process of how consumers make purchase decisions. The complexity is caused by the individual differences between consumers. Not only do they have different values, standards, and evaluation criteria, but they also behave differently in the process of reaching a purchase decision. This makes it difficult to do quantitative market research whose results can be used for product development and commercial communication. It also may be a reason for the popularity of qualitative research in which idiosyncratic behavior is not a problem. CAI may provide ways to overcome the problem of individual differences between consumers. The fundamental concept is based on analysis at the individual level, where the outcomes are aggregated to make global predictions about the market. This puts great demands on the software or questionnaire designers. In this section we first review the most important technique in this field, conjoint analysis. Next we examine two tailor-made designs for the analysis of preferences and decisions.

8.4.1 Adaptive Conjoint Analysis

Conjoint analysis was introduced by Luce and Tukey (1964). It now is a widespread technique with thousands of commercial applications (see Cattin and Wittink, 1982; Wittink and Cattin, 1989). The basic idea is to describe a product as a bundle of attributes. For a car such attributes may be color, top speed, size, room for luggage, noise level, and price. Each individual consumer has preferences with respect to these attributes. One consumer may prefer an aggressive looking red car with a high top speed which may be expensive, whereas another consumer may prefer a big but solid car that can be used to transport a large family. In conjoint analysis the values u_{ijk} are calculated. They represent the utility for individual i of attribute j at level k (attribute j may be color, level k may be yellow). Then the total utility of a product p for an individual i is

$$U_{ip} = \sum_j u_{ijk_{jp}}$$

where k_{jp} is the level of attribute j in product p. The utilities are derived by presenting the respondent with different products and by asking which product he/she prefers. This can be done by rank ordering or pairwise comparison.

After showing the respondent a certain number of products, sufficient information has been obtained to calculate the utilities.

Although the approach described above is very attractive in many ways (the results are relevant to management because individual differences are taken into account), there are also some drawbacks. One controversial issue, which we will not discuss here, is the validity of the model. A second problem is that in a P&P interview (even with a moderate number of attributes and attribute levels), the respondent faces the task of comparing many different products to provide the basis on which the u_{ijk} are estimated. In a CAPI situation such problems can (at least partially) be solved by tailoring the interview to the situation of the respondent. This is implemented in ACA (adaptive conjoint analysis) (e.g., see Johnson, 1987). The basic idea is that certain attribute levels are unacceptable for an individual consumer so that they can be filtered out before a conjoint analysis starts, thus reducing the number of comparisons a respondent has to make. ACA consists of the following steps:

1. Elimination of unacceptable levels. When the color of a car may be red, blue, yellow, white, or black, the respondent indicates that under no circumstances does he/she want a red car.
2. *Preference of levels.* For some attributes (e.g., color) the respondent may give a preference order. For a variable like price, such a question is not asked because it is assumed that inexpensive is preferred over expensive.
3. *Selection of the most relevant levels.* ACA can start with nine attribute levels, but the conjoint analysis proper is restricted to five levels.
4. *Importance of attributes.* The respondent is asked to give an importance rating to each attribute between 1 and 4. This rating is used to calculate initial values for the utilities.
5. *Pairwise comparison.* Products are compared which differ with respect to two or three attributes. These attributes are selected so that in each step the standard error of the utilities decreases maximally. This requires a matrix inversion between each comparison.
6. *Calibration with purchase intention.* The respondent is asked to indicate the purchase likelihood for products with different utility. These results can be used as input for models of market shares.

Leaving validity issues aside, the program and its success in the market research world is a triumph for CAI. The program uses both substantive theory and complex mathematics, respects the idiosyncrasy of the consumer, and opens up research areas that were inaccessible before the advent of CAI.

Can this be achieved in a data collection procedure with paper and pencil? When the number of attributes is low and the number of levels per attribute is low, a P&P interview is possible because all products can be constructed and ranked by preference. From this ranking, utilities at the individual level can be calculated directly. Also partial rankings may be sufficient to calculate the utilities (Addelman, 1962; Green, 1974). This, however, allows only for problems of relatively low complexity. In ACA, from steps 2, 3, and 4, initial values of the utilities are calculated; they are updated in step 5 so that with each comparison the maximum amount of information is added. Between each question a matrix inversion is necessary — a task far beyond the capabilities of most interviewers. ACA can be used both for CAPI and CSAQ.

8.4.2 Reactions to New Technology

In survey research, as well as in many other fields, electronic technology creates new possibilities that may dazzle the ordinary consumer. Market research in this area is difficult because the consumers cannot make judgments about options that will not exist until some time in the future. This makes conjoint analysis unsuitable for products and services that are radically different from their present equivalents.

One possible new development is the introduction of a new type of bank office where, apart from banking products, many other products are sold electronically. Examples are theater tickets, trips abroad, florist services, and opening a savings account. A project was conducted to determine whether consumers would appreciate such an electronic shop or whether they would prefer obvious alternatives, like ordering by telephone, Internet, or going to a traditional shop or counter. The research was conducted using a telepanel.

As is typical in commercial market research, there are budget and time restrictions limiting the time for questionnaire completion. The questionnaire had the following structure: First it was determined which of the products were relevant to the respondent. For the relevant products the respondent had to choose the most important attribute from a list. If this most important attribute favored the new electronic shop, the respondent was presented with the disadvantages associated with the electronic shop. If the most important attribute favored another distribution channel, the respondent was confronted with the advantages of the electronic shop. In both cases the respondent had to indicate his/her preferences in a pairwise comparison of two situations.

The procedure is illustrated in Figure 8.7 for the product "theater tickets." The first two screens show the part where the importance of the different aspects is determined. When the most important attribute is "you get detailed information about the contents of the shows from the salesperson" a series of comparisons start, where the advantages of the new electronic shop are pointed out. The third screen shows one such a comparison. If the most important aspect is "you can buy a ticket seven days a week, 24 hours per day," the disadvantages of the electronic shop are pointed out in a series of comparisons. The fourth screen depicts a comparison where the item "personal contact with the salesperson of the theater" has been chosen as the most important aspect.

In contrast to conjoint analysis, the end result here is not a set of utility values but a series of individualized arguments (pros and cons), and an impression of how important these arguments are. This information is then

```
- 1 -
YOU WANT TO BUY
TICKETS FOR A CONCERT,
SHOW, SPORT EVENT, ETC.

On the next screen you will see a
number of aspects that may be
relevant for buying tickets.

You have 20 points that you can
divide over these aspects. An
aspect that you consider to be
more important you give more
points than an aspect that you
consider to be less important.
```

```
- 2 -
- personal contact with the
salesperson of the theatre
- the salesperson recognizes you
(knows who you are)
- you get basic information from
the salesperson
- you get detailed information
about the contents of the shows
from the salesperson
- you can buy the ticket 7 days a
week, 24 hours per day
- you don't have to stand in a
queue
- you can buy the tickets close to
where you work or live
```

```
- 3 -
SITUATION 1
You can get detailed information
about the shows from a salesperson,
but you run the risk that you have to
wait in a queue to buy tickets

SITUATION 2
You can get detailed information by
calling an information number, but
you don't have the risk of waiting in
a queue to buy your tickets.

Which situation do you prefer?
```

```
- 4 -
SITUATION 1
You have 24 hours per day, 7 days a
week to buy tickets, but you have no
personal contact with a salesperson
from the office that sells the tickets.

SITUATION 2
For buying tickets you are restricted
to the opening hours of the office.
You have personal contact with the
salesperson.

Which situation do you prefer?
```

Figure 8.7. Preferences for products: Theater tickets.

used to further guide product development and communication to potential clients.

An example is presented in Figure 8.8 — a simple tree showing the number of people who chose the (personal) sale by counter compared with the (electronic) sale by terminal for theater tickets, and the number of people who changed their minds when presented with a potential disadvantage. The first level separates those who subscribed to the argument for the counter or for the terminal. The second level shows the percentage who changed their minds when confronted with arguments against their previously chosen alternative.

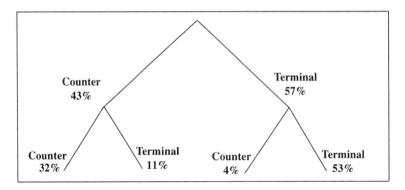

Figure 8.8. Decisions of respondents regarding arguments for counter versus terminal for theater tickets.

This approach allowed for detailed analysis of different arguments, which coincided with product attributes. The project appeared to be of great value in the design stage of the final concept.

Can this be achieved in a data collection procedure with paper and pencil? In a P&P interview it is possible to use display cards. It is then necessary to have a set of cards for each argument that emerged as most important from this first round. Given that there were, on average, eight arguments per service, and six services were treated in the interview, this requires approximately 6 * 8 *8 = 384 cards which have to be kept in order, a difficult task for an interviewer. For a self-administered survey the use of a computer is absolutely necessary.

8.4.3 Scenario Approach

One of the main criticisms of conjoint analysis and similar approaches is that consumers often do not reason in the same way as the computer assisted procedures do (Kuijlen, 1993). In a study on mortgages, an attempt was made to tailor the interview so that these individual decision paths could be captured and analyzed. A complex questionnaire was programmed in Blaise (Bethlehem et al., 1989). The project was conducted by the market research agency, Research International. The target group consisted of people who had bought a new house during the preceding year.

The questionnaire sought to divide the decision process into an number of steps which were relevant to the respondent (and hence also to the marketer). In each step a part of the market was eliminated until the respondent reached his/her final choice. The market consisted of 20 Dutch banks. A simplified version of the process is as follows:

1. The banks the respondent is not familiar with are eliminated. No further questions are asked.

2. The banks the respondent judges as unfit to do business with are eliminated. The respondent is then asked to motivate this judgment.
3. The banks the respondent judges as unfit to do mortgage business with are eliminated. The respondent is then asked to motivate this judgment.
4. The banks that the respondent did not ask for information are eliminated. The respondent is then asked to explain.
5. The banks that the respondent did not consider for an offer are eliminated. The respondent is then asked to explain and for a comparison of product attributes.
6. The banks that were not the respondent's final choice are eliminated. The respondent is then asked for a detailed comparison of product attributes.

This scheme describes in detail the decision process. Respondents who thoroughly analyze their options can answer the questions without problems, but so can the (surprisingly many) respondents who blindly follow the advice of an estate agent. The resulting data set is very rich and lends itself to a multitude of analyses.

To illustrate the potential of such a data set, we show a hypothetical analysis of the Post Bank and two of its competitors. In Figure 8.9 the "survival curves" of these banks are drawn. They indicate the percentage of respondents in each stage that have not eliminated the bank from their list of options. Compared

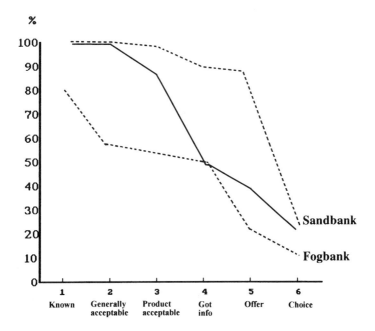

Figure 8.9. Survival curve of the postbank and two competitors.

to the Post Bank, the Sand Bank is very well known and accepted, but it loses many potential clients in the last stages of the process. The Fog Bank is less known and seems to have an image problem. This type of information shows the relative strengths and weaknesses in the marketing process. The analysis can be applied to many nonroutine consumer decisions. With statistical techniques that relate the survival curves to explanatory variables, including the reasons a given competitor is eliminated at a certain stage, the analysis can be refined.

Can this be achieved in a data collection procedure with paper and pencil? Again, in a P&P interview it is theoretically possible to use display cards showing company names. However, in practice, there are severe limits to the types and extent of manipulation and record keeping. In the qualitative stages before the main survey, we found that the computer was an absolute necessity in avoiding errors. Especially in stage 6, where product attributes were compared, it proved very difficult to ask questions about relevant product attributes (which depended on previous questions) of the remaining mortgage providers (which also depended on previous questions).

8.5 DISCUSSION

Freeman (1983) expressed the idea that the computer contained a large amount of knowledge to be able to ask detailed questions about specific subjects as these subjects appeared relevant. Examples were health surveys, where specific questions could be asked for every disease that was mentioned by the respondent, or victimization surveys, where every type of victimization could be treated in a specific manner. In other words, the computerized questionnaire turned into something of an expert system with detailed knowledge of every subject that could arise during the interview. In many cases the computer "knows" which answers are mutually exclusive and cannot be simultaneously true.

In the examples discussed above, however, the questionnaires allow for a lack of a priori knowledge. We do not know how the memory of a respondent is structured, so we allow him/her to build his/her own individual structure to aid the answering of retrospective questions. We do not know how the perception of a market is structured by an individual consumer. So in the natural grouping procedure, we allow him/her to divide a market in any manner for any reason. We do not know which attributes play a role in the trade-off for a market choice. In ACA all options are open at the start of the interview. We do not know what aspects of a new technology we should focus on, so the respondent makes this choice for us. We do not know how a decision process for a mortgage is structured, so the questionnaire is designed so that all possible structures can be entered and recognized. Instead of imposing structure and knowledge upon the interview, the computer behaves as if it has an "open mind."

Allowing for individual variation and different structures has many implications. From the philosophical and substantive point of view, we obtain new options in constructing questionnaires that limit a priori theoretical notions or, even more important, the researcher's prejudices. This, however, does not mean that questionnaires are disconnected from substantive theory. On the contrary, they allow for many more theoretical constructs and hypotheses a posteriori. In, for example, the mortgage questionnaire, many measures can be constructed for the effort a consumer makes to find the optimal bank; banks can be segmented into the type of consumers they attract but also into their performance at different stages of the purchase process.

From a technical point of view, the richness of the questionnaire may cause problems. The design stage of a project usually is more time-consuming than with conventional questionnaires. Most clients accept this as normal. Data analysis, however, is a stumbling block for such highly individualized questionnaires. Much work is required to extract the information from the large and sparsely filled data records. For CentER*data* this has led to frustration for the clients who failed to realize the problems presented by data analysis. Moreover the statistical techniques used for data analysis may be complicated and may even have to be tailored for a given questionnaire. Still in our opinion the gains from the individualized interviews both for market research and for fundamental scientific research make the extra effort worthwhile.

So far this discussion has been mainly research centered. We can also look at the new developments from the respondent's point of view. For the respondent it is important that being interviewed is a pleasurable experience. This can partly be achieved by conveying to the respondent that his/her answers are taken seriously as the next question is obviously based on the respondent's previously answer. When we look at the future of information technology, this argument can be developed even further.

In his book, *The Road Ahead*, Bill Gates (1995) describes the computer of the future as an agent. The computer knows (by experience) what a user wants and is programmed to obtain information from the Internet that will please the user. In other words, the computer becomes a friend and companion and develops a kind of relationship with the user. So one day someone may turn on its computer and find the message: "Hi Mary, I was just told that the people from the Census Bureau would like us to have a conversation about violence and law and order." Then a pleasant discussion starts between the computer and Mary which encompasses all relevant questions about Mary's experiences with violence and the police. Moreover audio-CASI (see Tourangeau and Smith, Chapter 22, and Turner et al., Chapter 23) can make this perspective even more vivid, since the computer actually speaks in a human voice. When voice recognition techniques become applicable on a large scale, typed answers can be replaced by spoken answers. Whatever technological advances loom in the future, the work described in this chapter is clearly a step toward the goal of the completely individualized interview.

CHAPTER 9

Incorporating Experiments into Computer Assisted Surveys

Thomas Piazza
University of California, Berkeley

Paul M. Sniderman
Stanford University

9.1 INTRODUCTION

The purpose of this chapter is to increase awareness of the interesting methodological possibilities offered by computer assisted interviewing (CAI). More specifically, we want to help researchers take advantage of the type of questions that can be addressed by incorporating experiments into CAI. We describe some of the different ways in which experiments can be incorporated into surveys, and for each of those ways we will provide specific examples. We also present some results that could not have been obtained by surveys that did not incorporate experiments.

Survey research has traditionally focused on the representation of some population, using probability sampling, to make inferences about the prevalence of certain characteristics of that population. Although surveys are usually conducted with good samples that allow generalization to a larger population, the validity of the results is often difficult to establish — that is, to show that the results actually demonstrate what we say they do.

The results of an experiment, on the other hand, are generally more persuasive in that regard. If some characteristic of a situation is varied at random, and an effect is found, it is not difficult to argue that the effect must have been due to the characteristic that was varied. Nevertheless, most experiments conducted by psychologists and social scientists are not done on

Computer Assisted Survey Information Collection, Edited by Mick P. Couper, Reginald P. Baker, Jelke Bethlehem, Cynthia Z. F. Clark, Jean Martin, William L. Nicholls II, and James M. O'Reilly. ISBN 0-471-17848-9 © 1998 John Wiley & Sons, Inc.

groups selected at random from a population, and it is therefore difficult to assert that the effect found in the experiment would also hold true in the population of interest. Experimenters have tried to compensate for poor samples by attempting to control explicitly for potentially confounding variables, but the experimental method would certainly benefit by being combined with good survey samples.

It turns out that CAI provides a solution to some of these problems. CAI can readily be used to combine the inferential power of experiments with the representational strength of surveys, and thus it can produce an enhanced version of both methods. A point we want to make in this chapter is that because of CAI, survey methodology has shifted to a new and more sophisticated plateau. The implication of this shift is that researchers will increasingly be expected to use computer assisted techniques not only for administrative efficiency but also for reasons integral to the substantive goals of the surveys themselves.

9.2   EVOLUTION OF DATA COLLECTION METHODS

The emergence of these new methods did not occur in a vacuum. They were made possible by the convergence of certain trends. In order to understand the current situation, it is useful to sketch at least briefly the shape of those trends. We do not intend to pass in review the history of survey research (for that, see Converse, 1987). Instead we will present a very schematic summary of developments in two areas of survey methods — the mode of interviewing, and the understanding of what constitutes a good questionnaire.

In regard to the mode of interviewing, the fundamental trend has been the shift to telephone interviewing and the eventual computerization of that process. Face-to-face interviewing has been replaced by telephone interviewing whenever possible. The driving force for this change is quite simple — it costs substantially less to call respondents on the telephone than go to their homes. Since telephone interviewing is often organized in a centralized manner, with telephone banks and supervisors, it was a prime candidate for computerization. The first systems for computer assisted telephone interviewing (CATI) appeared in the 1970s. (See Shanks, 1989, and Couper and Nicholls, Chapter 1, for a summary of these developments.)

The point we want to emphasize about the widespread diffusion of CATI systems is that the real impetus was administrative and not primarily methodological. Although the early market research advocates of CATI promoted its use because it could facilitate the randomization of the order of questions and response categories, the administrative aspect was the decisive one for most organizations, including academic survey research centers and government agencies. CATI was adopted because it helped survey organizations manage samples more efficiently; it provided up-to-date information on the progress of field work, and it also ensured that interviewers would follow skip patterns correctly. There was also the hope that increased efficiency would produce lower costs, but (fortunately) that was not an essential goal.

With the advent of portable computers, the new technology could be extended to surveys that, for reasons such as insufficient telephone coverage, remained as personal face-to-face interviews; hence the acronym CAPI for computer assisted personal interviewing. The same methodology is also extended to self-administered surveys. Currently we speak simply of a generic computer assisted interviewing (CAI) methodology.

Only within the last several years, however, has it gradually become clear to many researchers, especially in academic survey centers, that this new tool of computer assisted interviewing is a methodological breakthrough as well as an administrative one. CAI not only allows us to do more efficiently what we did before with paper and pencil, but it also allows us to do things that simply were not possible before. This is the essential point that we emphasize in the following and illustrate with examples, but first let us sketch briefly one other trend.

The second development we want to mention is the progression of the concept of the "good" questionnaire. Social research begins with the observation of behavior, and part of this basic observation is unstructured conversation with the people being studied. Participant observation, as a social science methodology, still relies on relatively unstructured conversation as a means of understanding the motivations, beliefs, and attitudes of people.

The problem with unstructured conversation is that it is difficult to make comparisons. If we want to know whether two or more people hold the same beliefs or attitudes, it makes sense to ask them all the same questions. This structuring of conversation, and the development of the standardized questionnaire, gave rise to survey research. The standardization both of questions and of the type of interaction with an interviewer is a basic premise of the method.

Standardization alone, however, is no guarantee that the questions really mean the same thing (or anything at all) to all respondents. This insight has led to a whole new field of study in which the lens of cognitive psychology has been focused on the survey questionnaire. (See Sudman, Bradburn, and Schwarz, 1996, for a discussion of these developments.) The result of this effort has certainly been an improvement in survey questions, at least for large-scale surveys with development budgets large enough to incorporate these refinements into questionnaire testing and design. But standardization itself is also at issue.

The gains of standardization are real, but we also pay a price. What we usually want to know is how people evaluate what they observe and how they will act in situations in which they are likely to find themselves. Is the restricted environment of the survey interview the best way to find this out? As we have argued elsewhere (Piazza, Sniderman, and Tetlock, 1989), the ideal is to evaluate the responses of a random sample of persons within a random sample of situations. The practical goal is to expand the range of situations in the typical survey to reflect more closely what people will encounter in everyday life. This has led us to use counterarguments and consistency checks, as well as experiments in our research. To some extent, therefore, we can now develop a questionnaire that is tailored specifically to each respondent, whether as a function of random numbers or of previous responses given by the respondent. This variation, however, must be carried out in a controlled manner so as not to relinquish the major benefits of standardization.

This controlled tailoring of the questionnaire could not take place on a large scale without the development of computer assisted interviewing. It is the convergence of these two trends — of the mode of interviewing and of the definition of the "good" questionnaire — that defines what we consider to be the current state-of-the-art in survey research methodology. And this evolution of methods provides the context for the remainder of our discussion.

9.3   CHARACTERISTICS OF THE NEW USE OF EXPERIMENTS IN SURVEYS

Experiments have been incorporated into surveys for many years, and we do not wish to imply otherwise. At the same time we want to make clear that there has been a shift in the focus and scope of the use of experiments. Before we look at individual examples, we would like to spell out the meaning of this shift.

The main characteristic of this new use of experiments is its substantive focus. The design and implementation of the experiments are guided by the theoretical goals of the research agenda. In contrast, the split-ballot experiments used previously in paper and pencil surveys tended to focus on measurement issues — the effects of alternate question wording and question order. The results of those experiments (summarized by Converse and Presser, 1986) greatly enhanced our understanding of the fragility, as well as the robustness, of the survey method. Yet their focus was primarily on issues of measurement and not on the substantive areas that may have been covered in the content of the questions.

The new experiments are focused on issues such as racism, discrimination, political values, and attitudes about AIDS. And the experiments are designed to shed light on those substantive areas by assessing the effect of the randomized factors embedded in the experiments.

The other distinguishing characteristic of the new use of experiments is their sheer number within a single survey. One survey can now contain dozens of independent experiments, and this is only possible using CAI. The practical limits imposed by the paper and pencil method on the number of experiments within a survey were really quite severe. You could have a "Form A" and a "Form B" and maybe even a "Form C," but the idea of having literally hundreds of distinct questionnaires for a single survey was simply unworkable.

The use of CAI, on the other hand, effectively removes that constraint. In a survey with many independent experiments (i.e., using random numbers generated independently of one another) there could potentially be as many distinct questionnaires as there are respondents. And this can happen in a manner completely transparent not only to the respondents but even to the interviewers.

This added complexity is a mixed blessing. It is not a trivial matter to design, test, and document such an instrument. Serious errors can arise if the random numbers are converted into code values that are not anticipated in the survey instrument. Furthermore the analysis of the resulting data requires care and ingenuity. These problems, however, are welcome ones, since they are the price of reaching a new methodological plateau and are symptoms of great potential opportunity.

From a technical point of view, all experiments in CAI have the same basic form: respondents are presented with a question that differs in one or more randomized aspects across respondents. This randomization is achieved in CAI by including random numbers in the case record when each case is set up for interviewing. Depending on the value of one or more of those random numbers for a particular case, a specific version of the question is selected or constructed by the CAI program and displayed on the interviewer's computer screen, ready to be administered to the respondent. The value of the random number is preserved on the output data file, and so indicates which version of the question was administered to this respondent. Although some CAI programs can generate random numbers during the interview itself, it is safer to prepare random numbers with discrete integer values ahead of time and to check that each one has the correct range (corresponding to the number of levels in that factor), the correct distribution (usually a uniform distribution across the values of the factor), and the desired correlation between random numbers (usually no significant correlation at all, unless some combinations of factors are to be excluded).

Now we will turn to a discussion of some examples of various types of experiments that have been incorporated into surveys. Some examples are taken from our own work on racial attitudes, but we also highlight some experiments carried out by others. We distinguish between the various types on the basis of the functional differences between the experimental factors. For example, we can modify at random the elements of a choice presented to the respondent. Alternatively, we can present respondents with the same choice, but we can modify at random certain elements that make up the context of the choice. There are many possibilities, and our intention is to provoke interest in this approach and invite others to develop additional variations on these themes.

9.4   EXPERIMENTS THAT MODIFY CHOICES

The simplest experimental items, from a conceptual point of view, present one random group of respondents with one version of a question and the other group(s) with the other version(s) of the question. If the response differs between groups to a degree that is statistically significant, one can argue that the difference in response is due to the difference in the question. We will give a few examples of this approach.

One very straightforward example is taken from the National Race and Politics Survey, a national random digit telephone survey (also called random digit dialing, or RDD) conducted in 1991. The survey included a question about university admissions for blacks, and there were two versions of the question, each to be administered to a random half of the sample. The main purpose of the experiment was to determine whether white opposition to affirmative action in university admissions hinged on explicit racial preferences or whether there was similar opposition to any extra effort at all. The wording of the questions, and the results, are shown in Figure 9.1.

As is clear from the percentages shown in the figure, the differences in the results for the two versions of the question are quite large. Relatively few whites, 26 percent, are in favor of explicit preferences, but 63 percent are in favor of "extra effort" to ensure that qualified blacks be considered for admission. (For other experiments with alternative formulations of affirmative action, see Kinder and Sanders, 1990.)

To analyze further these results, one could compare different groups of whites — men and women, for example — to see whether the effect of the experiment is as large in one group as in the other. This could be considered an analysis of variance problem. Alternatively, the analysis could be set up as a regression problem, in which the response is the dependent variable, the experimental factor is one independent variable, and various respondent characteristics are the other independent variables.

However the analysis is conducted, the main point we want to make is that the substantive information obtained from the experiment could not readily be

---

Version 1 (Racial preference):
> Some people say that because of past discrimination, qualified blacks should be given preference in university admissions. Others say that this is wrong because it discriminates against whites. How do you feel — are you in favor of or opposed to giving qualified blacks preference in admission to colleges and universities?

Version 2 (Extra effort):
> Some people say that because of past discrimination, an extra effort should be made to make sure that qualified blacks are considered for university admission. Others say that this extra effort is wrong because it discriminates against whites. How do you feel — are you in favor of or opposed to making an extra effort to make sure qualified blacks are considered for admission to colleges and universities?

Percent of white respondents in favor of the policy:
Version 1 (Racial preference):	26%	(n=889)
Version 2 (Extra effort):	63%	(n=911)

Figure 9.1. University admissions experiment. Source: National Race and Politics Survey.

obtained in another way. The nonexperimental survey approach would have us ask all respondents *both* versions of the question; we would then compare results for the two versions and conduct further analysis on both variables. The problem with asking both versions is that some individuals will feel pressured to answer them both in the same way, and we will be unsure of the magnitude of that correlation. The experiment removes that unknown correlation, and we can then concentrate on studying the effect that we are really interested in.

Let us turn now to a more elaborate example of an experiment that modified the choices presented to respondents. Here we wanted to know about support for providing government help to people who had lost jobs. Is support for, or opposition to, government help contingent on the characteristics of the person being helped? We expected that such characteristics would have an effect, but we were not sure which ones or how great an effect. The experiment we used to explore this question in the Bay Area Race and Politics Survey, a 1986 RDD survey of the San Francisco–Oakland Bay Area, is shown in Figure 9.2.

The person to receive government help is described in terms of five factors, each with two or more levels that varied at random. All together there were 96 different combinations of the five variables. From a technical point of view, what we did was program the CAI instrument to construct and display on the interviewer's screen a description of each person to be helped. The description was generated by using a different "fill" for each factor, based on the value of the random number set up for that particular factor.

We were able to use this experiment to learn something very interesting about the joint effects of the race and the work history of the person receiving government help. An analysis is presented in Sniderman et al. (1991), also in Sniderman and Piazza (1993). That analysis incorporated some of the personal characteristics of the respondent in order to show how the effects of the experimental factors interacted with certain respondent characteristics such as political ideology. To make the results intelligible, however, we had to simplify

---

The next questions are about three different people who were laid off because the company where they worked had to reduce its staff. Think for a moment about each person and then tell me how much government help, if any, that person should receive while looking for a new job. The first person is a [RACE] [SEX] in his/her [AGE]. He/she is [MARITAL-PARENTAL STATUS] and [WORK HISTORY]. How much help in finding a new job do you think the government should give to this person — a lot, some, or none at all?

Randomized characteristics:
  RACE(2): black, white
  SEX(2): male, female
  AGE(3): early twenties, mid-thirties, early forties
  MARITAL-PARENTAL(4): single, a single parent, married, married and has children
  WORK HISTORY(2): is a dependable worker, is not a dependable worker

Figure 9.2. Laid-off worker experiment. Source: Bay Area Race and Politics Survey.

the factorial design by averaging over the factors that did not have a significant effect on the response.

This problem of presenting results is something that should be considered in designing experiments for CAI. If many factors are included, perhaps for exploratory purposes, the resulting analysis can be very complex, particularly if there are interaction effects. If respondent characteristics are to be included in the analysis as well, it will become even more complex. On the other hand, the factor we drop might have the most interesting effects; that is always our fear. Nevertheless, it is not possible to include everything in a single question, and we have to accept that reality.

In retrospect we might say that the five-factor experiment shown in Figure 9.2 is probably too complicated. However, this example was one of our first attempts to use this method, and we were still operating under a "more is better" premise. We were influenced in this design by the vignette analysis approach used by Rossi and Nock (1982); the number and levels of the factors they used were even more numerous than ours. They also had each respondent make judgments about many different persons, each described in terms of the randomized factors. We simplified this latter aspect by asking respondents to consider only three persons. In addition, to further simplify our analysis, we used only the first description presented to a respondent; in that way we avoided having to take into account the correlations between the three judgments given by the same respondent.

There is a lesson in all this: It is sometimes easier to design complex experiments than to analyze the resulting data. Just as for survey data in general, the analysis of items with embedded experiments must be guided by theoretical considerations; it is not possible to examine the relationships between everything and everything else.

9.5   EXPERIMENTS THAT MODIFY THE CONTEXT OF A CHOICE

Another way to incorporate experiments into surveys is to modify at random the context or the frame of reference of a choice that the respondent is asked to make. Here the core question remains the same for all respondents, but some receive additional information that may (or may not) influence the choice they make. The purpose of the experiment is to study the effects of providing this additional information.

The importance of context, of course, is a recurring theme in methodological studies of questionnaire design. The context referred to in such studies is usually the set of questions asked before the particular question of interest. Schuman and Bobo (1988) examined the context (in that sense) of some racial attitudes, and they also conducted some experiments in order to draw substantive conclusions. The method described here has the same intent as those earlier experiments, but it supplements that multiquestion perspective of context, which certainly remains an issue, with an experimental manipulation of the contextual information provided as part of the question itself.

An application of this methodology to attitudes about job quotas based on race is described by Stoker (1998). She makes the distinction between context-free and context-specific choices and points out that surveys generally ask respondents to express their views without reference to a specific context. In a legal setting, on the other hand, judges deal with issues of discrimination in specific circumstances, and those circumstances often affect strongly the decision. She therefore decided to see whether a specification of circumstances would also affect survey responses on the issue of job quotas.

In the experiment Stoker designed, everyone was asked whether large companies should be required to give a certain number of jobs to blacks. Respondents were distributed on the basis of a random number prepared for each case across three conditions that differed in regard to the circumstances that might justify such a policy. There was a baseline condition that provided no justifying context at all; a second condition supplied the information that blacks are underrepresented at those companies; a third condition presented the information that the companies have employment policies that discriminate against blacks. The full text and the results are shown in Figure 9.3.

As can be seen in the figure, support among whites for job quotas is low in the context-free condition (23%) and also in the condition of underrepresentation (21%). Support is substantially higher (41%) in the discrimination condition. The decisive issue consequently is not simply whether or not a justifying context is provided; it matters what the context is. A history of discrimination matters in a way that simple underrepresentation does not.

From a methodological point of view we note that in this case a factor with only two conditions would have led to ambiguous results. If the experiment had included only the baseline condition and the discrimination condition, we might have concluded that the increase in support for job quotas was due to the presence of a justification—and that any justification might do. The presence of the underrepresentation condition has clarified the issue. Although

---

Version 1: (Blank)
Version 2: There are some large companies where blacks are underrepresented.
Version 3: There are some large companies with employment policies that discriminate against
      blacks.

Do you think (that/these) large companies should be required to give a certain number of jobs to blacks, or should the government stay out of this?

Percent of white respondents in favor of job quotas for each condition:
Version 1 (Baseline):	23%	(n=559)
Version 2 (Underrepresented):	21%	(n=627)
Version 3 (Discrimination):	41%	(n=606)

Figure 9.3. Job quotas experiment. Source: National Race and Politics Survey.

simplicity is a virtue in the design of experiments, we must take care not to oversimplify.

Another example of modifying the context of a choice is taken from the AIDS and Stigma Survey, a national RDD survey conducted in 1997. The purpose of the experiment was to measure the degree of sympathy toward people with AIDS. Does the level of sympathy vary if contextual information is provided about how a person contracted the disease? The experiment shown in Figure 9.4 was designed to answer that question. (See Herek and Capitanio, 1993, for a report on the research leading up to this experiment.)

The AIDS transmission experiment has three factors and a total of 32 variations. In addition to varying race and gender, the experiment varies sexual orientation and the transmission route. Preliminary results show that there is indeed a substantial effect of context.

A somewhat different approach to varying the context of a choice comes from providing (or not) the information that a prestigious figure supports or is responsible for a certain policy or action. There are many possible variants on this theme. A relatively straightforward example was included in the 1994 National Multi-Investigator Survey, a national RDD survey, and is shown in Figure 9.5. Respondents were asked to choose between two electoral methods for dealing with situations in which no candidate received more than half of the votes: the plurality method in which the candidate receiving the most votes wins, versus a run-off election between the top two candidates. Inserted into that question was an experiment to assess the effect of invoking a prestigious name as being in favor of the first choice (the candidate with the most votes wins).

This experiment had four randomized conditions. Three of the conditions mentioned a famous person (Jackson, Perot, or Kennedy); the fourth men-

---

The next questions are about a particular person with AIDS. This person is a [RACE] [GENDER-SEXUAL ORIENTATION] in his/her thirties. He/She definitely got AIDS from [TRANSMISSION ROUTE]. How sympathetic do you feel toward this person? Would you say you feel very sympathetic, somewhat, a little, or not at all sympathetic?

Randomized characteristics:
RACE(2):
    black, white
GENDER-SEXUAL ORIENTATION(4):
    heterosexual woman, heterosexual man, bisexual man, homosexual man
TRANSMISSION ROUTE(4):
    having sex with his/her partner — the only person he/she has had sex with in the last 15
        years.
    having sex, but he/she is not sure who he/she got it from because he/she has had sex with
        a large number of partners in the last 15 years.
    sharing needles for drugs during the last 15 years.
    receiving a blood transfusion about 15 years ago.

Figure 9.4. AIDS transmission experiment. Source: National Aids and Stigma Survey.

Now a question about two different ways of electing people to office.

Version 1: Some people such as Jesse Jackson
Version 2: Some people such as Ross Perot
Version 3: Some people such as (Senator) Ted Kennedy
Version 4: Some people

... think that the person who receives the most votes in an election should win, even if that person received fewer than half of the votes. Other people think that if the leading candidate does not receive more than half the votes in an election, there should be a run-off election between the two top candidates. What do you think? Should the person receiving the most votes win, even if the leading candidate does not receive more than half the votes, or should there be a run-off between the two leading candidates if no candidate gets more than half the votes?

Percent of respondents in favor of most votes winning (no run-off):
    Version 1 (Jackson):   41% (n=375)
    Version 2 (Perot):   47% (n=375)
    Version 3 (Kennedy):   46% (n=354)
    Version 4 (no name):   50% (n=339)

Figure 9.5. Prestigious name experiment. Source: National Multi-Investigator Survey.

tioned no such person and served as the baseline condition. As we see in Figure 9.5, there was no increase at all in the level of support for the "most votes win" position if a famous politician was mentioned as being in favor; in fact support in all three cases was less than the baseline condition with no name mentioned, significantly so when the politician mentioned was Jackson. Famous persons do not necessarily influence people to alter their choices, and when they do, it is not always in the expected direction.

In general, we have found that it is not easy to swing people's opinions by mentioning some authority figure or famous person. Of seven such experiments reported on elsewhere (Piazza, Sniderman, and Tetlock, 1989), only one had that effect. Nevertheless, there are reasons to expect such an effect in certain circumstances. Smith and Squire (1990) studied the effects of prestigious names on respondent support for state Supreme Court justices. They showed that if respondents are told the name of the governor who had appointed a specific justice, they were more likely to express an opinion on whether or not a justice should be confirmed; furthermore the direction of that opinion was also affected.

Survey respondents often must make choices about which they have insufficient information — they have an information shortfall. How severe that shortfall is depends on the amount of relevant information they already have on hand. In their study Smith and Squire (1990) went on to show that the strength of the effect of mentioning, or not, a prestigious name varied inversely with the respondents' level of formal education. This suggests that those who are less well-informed are more likely to take advantage of the extra information supplied by the mention of a prestigious name, in order to compensate for their information shortfall. (On the effect of political sophistication, see also Sniderman, Brody, and Tetlock, 1991.)

More generally, we could say that the choices people make depend not only on the nature of the choices they are presented with but also on the context of the choices and on the informational resources and personal characteristics they bring to the task. All of these can and should be analyzed together. The ability to combine experimental factors and respondent characteristics in the same analysis is one of the major benefits of integrating experiments into surveys based on good samples. Particularly when respondent characteristics can take on many possible values, as is the case for education and political sophistication, it is important to base the analysis on distributions of those variables that reflect the full range of values found in the population of interest.

We could add other examples of how the context of a choice can be modified experimentally (see Sniderman and Grob, 1996). Our purpose here, however, is to illustrate how the technically simple procedure of using one or more random numbers to alter the wording of a question in CAI can provide us with very interesting data. Both the elements of a choice and the context of that choice can be altered at random to allow us to gain valuable insights into the substantive questions that concern us.

9.6 UNOBTRUSIVE MEASUREMENT OF ANGER

It is possible to design experiments in surveys that depart somewhat from the pattern we have discussed and illustrated up to now. One very interesting type of experiment was designed by Kuklinski and his colleagues for the National Race and Politics Survey and has been used in various forms since then (see Kuklinski et al., 1997). The idea was to create an unobtrusive measure of the prevalence of people's anger over some situation, event, or policy. We will refer to this method as a "list experiment."

In a list experiment one randomly selected subgroup of respondents is presented with a list of a few situations, events, or policies and is asked how many of the situations make them angry or upset. Note that the respondents are not asked about the individual situations; they are only asked to give a number (from zero to the number of situations presented) indicating how many of the situations make them angry. Another randomly selected subgroup is presented with a second set of items that includes the first set plus one additional situation; that additional situation is the object of the experiment.

An example of this method is shown in Figure 9.6; it is taken from the National Race and Politics Survey. One randomly designated group of respondents was presented with version 1 of the list, containing the basic three items — gasoline tax increase, big salaries in sports, and corporate pollution. A second randomly designated group was presented with version 2 of the list, containing the basic three items plus an item on affirmative action. (There was

---

Now I'm going to read you (three/four) things that sometimes make people angry or upset. After I read all (three/four), just tell me HOW MANY of them upset you. (I don't want to know which ones, just how many.)

Version 1: (Basic three items)
(1) the federal government increasing the tax on gasoline;
(2) professional athletes getting million-dollar-plus salaries;
(3) large corporations polluting the environment.

Version 2: (Add affirmative action item)
(1-3) (Same as basic three items)
(4) black leaders asking the government for affirmative action.

How many, if any, of these things upset you?

Mean number of items reported by white respondents:
Version 1 (Basic three items):   2.20 (n=598)
Version 2 (Add affirmative action):   2.75 (n=596)

Figure 9.6. List experiment. Source: National Race and Politics Survey.

also a third random group receiving yet another treatment, but we can ignore that added complexity for our present purposes.)

The measurement of the effect of adding the extra situation to the list is very straightforward—it is the difference in the mean number of items that respondents say they are angry or upset about. In the example shown in Figure 9.6, white respondents who were presented with the basic three items (version 1) said, on average, that 2.20 out of the three items made them upset. On the other hand, white respondents who were presented with the list including the extra item (version 2) said, on average, that 2.75 of the four items made them upset. Assuming that the second subgroup of respondents (like the first) is upset about only 2.20 of the basic three items, we conclude that the increase of 0.55 (from 2.20 to 2.75) is due to the extra item. This increase of 0.55 implies that 55 percent of the respondents in the second group are upset about the extra item.

Although our focus in this chapter is on the design rather than the analysis of experiments, we should point out that the analysis of a list experiment requires a different approach from the other experiments described above. In a list experiment we do not have case-specific outcome data—that is, we do not know which respondents are upset about the extra item. All we know is that some estimated percentage of the respondents are upset. Although such information is valuable, we naturally want to know more. Is anger over affirmative action more likely to be found in some segments of society than in others? How would one carry out such an analysis?

One approach is to divide respondents into different groups based on some characteristic of interest and then analyze the list experiment separately within each group of respondents. The results for each group can then be compared.

Table 9.1. One Possible Analysis of a List Experiment

	Mean Angry Items		Estimated Percent Angry About Affirmative Action	N
	Basic 3 (Version 1)	Affirmative Action (Version 2)		
Age				
18–39	2.16	2.57	41	(568)
	(0.05)	(0.06)	(7)	
40–97	2.23	2.91	68	(617)
	(0.04)	(0.06)	(7)	
Total	2.20	2.75	55	(1194)
	(0.03)	(0.04)	(5)	

Source: National Race and Politics Survey.
Notes: Totals include nine cases with missing data on age. Estimated standard errors in parentheses.

A simple example of this type of analysis is shown in Table 9.1. In the row labeled "total" we have the analysis for the entire sample. As mentioned above, we estimate that 55 percent of whites are upset about the affirmative action item. That estimate is based on the mean number of items that upset respondents who were presented with only the basic three items (first column), compared with the mean for those presented with the baseline items plus the extra item (second column).

In the other rows of Table 9.1 we have replicated this analysis for two subgroups of respondents—those under age 40, and those aged 40 and over. As seen in the third column, although both groups are angry over affirmative action, older whites are more angry than younger whites (68% versus 41%), and we can therefore conclude that age has some bearing on opposition to affirmative action. More generally, this relatively simple form of analysis can be used for many different variables in order to assess their effect on anger over affirmative action.

Although this list experiment method can generate some very interesting results, as it did in this example, it can also have problems that we should mention at least briefly. There can be ceiling effects if some groups of respondents are angry or upset about all of the items; in such a case the measurement is no longer unobtrusive. There can be contrast effects if the test item is much more objectionable than the baseline items; in that case the baseline items may seem less upsetting when presented together with the test item than when presented alone. Also some respondents may report a higher number for the test group of items simply because there are more items to choose from, perhaps because they respond by merely sampling from the available alternatives. This last problem would suggest that the test item should be substituted for one of the baseline items rather than be added to that set of items. Some of these, and other, potential problems are currently being investigated, in part, by analyzing a more complex version of this list experiment that was included in the National Multi-Investigator Survey.

The point, however, is that the list experiment is one more example of how investigators can integrate experimental designs into a computer assisted survey. No new method is free of problems, and it will take some experience to figure out the best way to structure the lists. Nevertheless, it is possible to learn some things with this method that cannot readily be learned in other ways.

9.7 RANDOMIZED COUNTERARGUMENTS

Another computer assisted method we have described elsewhere (Piazza, Sniderman, and Tetlock, 1989; Sniderman and Piazza, 1993) is the use of counterarguments. Respondents first answer a standard question. They are then presented with a counterargument tailored to the specific answer that each respondent has given, and they are asked whether, in light of the counterargument, they would change their mind. The purpose of the procedure is to obtain a more realistic assessment of the strength of committment to a position than can usually be obtained by asking whether someone "strongly agrees" or only "somewhat agrees" with a statement.

Since the basic counterargument method does not use random numbers, we have not included it in this discussion of incorporating experiments into surveys. However, there is a further develoment of that method that we do wish to illustrate here. A criticism of the basic counterargument method is that some respondents will change their mind just because they are challenged — a type of acquiescence effect. An extension of the counterargument method randomizes the counterargument presented. For each initial position (in favor or opposed) two counterarguments are prepared — one of a substantive nature, and another that is basically content free but which should attract those respondents who tend to acquiesce to any challenge. Figure 9.7 shows the English translation of a series of questions in the Italian Prejudice Survey, a directory-based telephone survey of Italian adults conducted in 1994, which used this randomized counterargument technique.

In the Italian survey, respondents were first asked to state their position on the issue of increased autonomy for the major regions — an issue that is discussed extensively at the present time. For each initial position (in favor or opposed) one of two counterarguments was selected at random as a follow-up question. In this series of questions, notice that the nonsubstantive counterargument is actually the same both for those initially in favor and for those opposed.

There is a lot of discussion in Italy these days about autonomy for the regions and about the role that the central government should play. What do you think personally about the possibility of giving greater autonomy to the regions in Italy? Are you very favorable, somewhat favorable, somewhat opposed, or completely opposed?

(If very or somewhat favorable, ask:)
   Version 1 (substantive counterargument):
      Considering that there would be a risk of increasing government bureaucracy, ...
   Version 2 (nonsubstantive counterargument):
      Considering the complexity and the uncertainty of problems in Italy nowadays, ...

... are you still in favor of greater autonomy for the regions, or would this make you change your mind?

(If somewhat or completely opposed, ask:)
   Version 1 (substantive counterargument):
      Considering that with a centralized system public services could become even less efficient than they are now, ...
   Version 2 (nonsubstantive counterargument):
      Considering the complexity and the uncertainty of problems in Italy nowadays, ...

... are you still opposed to greater autonomy for the regions, or would this make you change your mind?

Initial positions		Change mind after counterargument
Favorable:	83%	Version 1: 34% (n=650)
		Version 2: 14% (n=684)
Opposed:	17%	Version 1: 35% (n=118)
		Version 2: 17% (n=120)
Total	100%	

Figure 9.7. Randomized counterarguments. Source: Italian Prejudice Survey.

The results at the bottom of Figure 9.7 show that the nonsubstantive counterargument attracted about 15 percent both of those initially in favor and of those opposed — this would be our estimate of the acquiescence effect. Both substantive counterarguments attracted about 35 percent of both groups. Net of acquiescence, we could conclude that about 20 percent (35–15) of each group was persuaded by the substance of the counterarguments. Notice that the majority of each group did not change their minds after hearing the counterarguments.

Further analysis of this series of questions would focus on trying to determine which segments of the population were more likely than others to hold a particular initial position and to maintain that position when challenged. Being able to identify that group of "firm believers," and to distinguish them from the pliable respondents, opens up the study of the dynamics of political persuasion.

9.8 CONCLUSIONS

Our own enthusiasm for the advances offered by this methodology has not overcome our awareness of substantial constraints. We therefore want to pull back now from the details of these examples of the use of experiments in surveys and to reflect on what this method has to offer, and what it does not.

Recall that we are still dealing with general population samples who respond to interviews that take place mostly in their own homes. Whereas experiments conducted in laboratory settings can manipulate and control many aspects of the environment presented to respondents, experiments within surveys are limited to rather narrow variations in stimuli presented verbally. All of the examples of experiments we have mentioned here are simple variations of the wording of questions.

Some of this narrowing of the scope of experimental variation may be self-imposed rather than necessary. Since our method is computer assisted, we tend to limit our thinking to whatever will fit on the interviewer's computer screen, with the consequent abandonment of any attempt to present to the respondent other stimuli such as visual objects or aids. Interviewing by telephone of course entails limits in what can be made available to respondents, but even computer assisted interviewing conducted in person may have become too reliant on what can fit on a computer screen. Standardization is easier to enforce if everything is focused on the computerized questionnaire, but as we have argued above, standardization is not the only goal and should not stifle other ends.

Exclusive reliance on verbal stimuli may lead to other problems that have an ethical cast. In the examples we have given, attempts were sometimes made to alter the context in which respondents made choices. What this meant in practice was that different respondents were given different items of information that might or might not influence their responses to a question. Should respondents be given information that is false to see whether their judgment is affected by that information? We think not, and we have attempted to construct alternative scenarios for respondents that are plausible and not patently false.

This constraint is due to the lack of an opportunity in the usual survey situation for a debriefing session. Such sessions are used by experimental psychologists to clarify for their subjects the purposes of the experiment and to correct any false information that was provided as part of the experiment. Absent a debriefing procedure, these experimental manipulations within a survey should probably be limited to those that will not deceive or misinform respondents.

All of these constraints and potential problems, however, should not keep us from seeing the advantages of incorporating experiments into computer assisted surveys. First and foremost, we want to emphasize that the results of experiments are more persuasive than results based on correlations between items in a survey. Correlations can be suggestive, and they may even lead us to correct conclusions, but they have nowhere near the inferential power of an experiment.

If this is so, why has it taken so long for survey researchers to incorporate experiments into surveys? After all, many of us have been doing computer assisted interviewing for almost 20 years. The answer is probably that we have learned to live with earlier, pre-CAI technology, and it has served us reasonably well. Our older models of questionnaire design and data analysis are continually improved and refined, and we have not seen any pressing reason to change. Although we have adopted CAI technology, most of our energy has gone into figuring out how to make it all run correctly. But those days are ending.

The challenge for us now is to rewrite the script of survey research. Experimentally driven techniques allow us to pose our questions in different ways and to incorporate meaningful experiments into almost every data collection effort. These new techniques should become routine and taken for granted, instead of being considered novel or cutting edge. Survey research as a whole seems to be approaching a new methodological plateau, and we are all being challenged to participate in one of the most interesting developments in the short history of this research method.

CHAPTER 10

Use of Computer Assisted Interviewing in Longitudinal Surveys

Ann Brown, Alison Hale, and Sylvie Michaud
Statistics Canada

10.1 INTRODUCTION

This chapter discusses longitudinal household surveys using computer assisted interviewing (CAI) in a decentralized environment where the interviews are conducted either by telephone from the interviewer's home or in-person at the respondent's home. We illustrate how CAI can be exploited for longitudinal household surveys as well as some of the pitfalls to be avoided. This use of CAI is consistent with the view noted in Dippo et al. (1992) that computer assisted instruments are "expert systems or software tools with built-in knowledge bases that aid users in performing complex cognitive tasks" and by Nicholls, Baker, and Martin (1997) who point out that CAI is "a platform for new approaches to data collection that could yield important future gains in the control and reduction of survey error." This chapter covers three themes of the use of CAI: maximizing response rates by tracing respondents who have moved, minimizing and preventing respondent recall error or seam effects by using dependent interviewing, and developing CAI instruments that handle the complexity of longitudinal surveys via some of the unique functionality of CAI.

Longitudinal household surveys contact the same respondents at regular intervals in order to measure change across time. The longitudinal respondent is the main unit of response and is followed in the subsequent waves of the survey. Additional members of the longitudinal respondent's family or house-

Computer Assisted Survey Information Collection, Edited by Mick P. Couper, Reginald P. Baker, Jelke Bethlehem, Cynthia Z. F. Clark, Jean Martin, William L. Nicholls II, and James M. O'Reilly. ISBN 0-471-17848-9 © 1998 John Wiley & Sons, Inc.

hold may be included in each wave while they are living with the longitudinal respondent. While the content for subsequent waves of the longitudinal survey are for the most part a repeat of that used in the initial wave, modules or components may be added or removed. Duncan (1992) assessed "the promise and problems of household panel surveys as sources of data for analyses" of change, gross flows across social class boundaries, distinctions between transitory and persistent characteristics such as poverty, and more sophisticated behavioral models. Also included was the discussion of quality concerns involving nonresponse and respondent error.

Nonresponse is a particular concern for longitudinal surveys, not only in terms of reducing the potential of bias from the initial nonrespondents but also in reducing the bias from accumulating attrition over subsequent waves. Since a sizable portion of attrition is from movers (other sources include no contacts, deaths and refusals), substantial efforts are invested in tracing or locating respondents for longitudinal surveys. We discuss this issue in Section 10.3.

Respondent recall error in longitudinal surveys has been researched in a number of studies. Recall error occurs when respondents forget or misplace events and either telescope or underreport activities occurring at the boundary between the two reference periods of adjacent waves of the longitudinal survey. This recall error results in a "seam" effect where there are more transitions or changes in status than at any other point in time (Hill, 1990; Martini, 1989). The research studies investigating this recall error often include techniques for preventing, reducing, or coping with this error source. Bassi, Torelli, and Trivellato (1996), Lemaître (1992), Young (1989), and Burkhead and Coder (1985) include analysis methods, while Dibbs et al. (1995), Dippo et al. (1992), and Egan et al. (1990) discuss using information from the earlier wave to reduce or prevent the seam effect, a technique referred to as dependent interviewing. Section 10.4 contains examples of how CAI can be exploited to reduce or prevent this recall error.

10.2   TRANSITION TO CAI

Given the importance of maintaining ongoing data series during the transition to CAI, longitudinal surveys have been the subject of a number of studies of the effect of CAI (e.g., see Olsen, 1992; Dippo et al., 1992; Edwards, Sperry, and Schaeffer, 1995; Kojetin, Kurlander, and Rope, 1994; Laurie and Moon, 1996). These have included exploration of the effects of the transition on the estimates, cost, timeliness, and data quality (validity and consistency edits, data flows, item nonresponse). Laurie and Moon (1996) weighed the advantages of moving the British Household Panel Survey (BHPS), an existing longitudinal survey, from paper and pencil interviewing to CAI against the possible effect on the existing survey processes. While the BHPS is delaying its conversion, many major households surveys around the world already use or are in transition to computer assisted interviewing.

Since 1993 Statistics Canada has implemented four new longitudinal surveys, the Survey of Labor and Income Dynamics (SLID), the National Population Health Survey (NPHS), the National Longitudinal Survey of Children and Youth (NLSCY), and the Self-Sufficiency Project (SSP). While the surveys differ in content, they are all conducted in a decentralized environment with the interviewers' working out of their homes. All the surveys consist of a mix of telephone and personal interviews, and they are all longitudinal household surveys that were designed from their inception to use CAI. The experiences from the implementation of CAI for these new surveys forms the basis for this chapter. We focus specifically on three areas: (1) tracing to improve response rates, (2) using dependent interviewing to prevent and reduce recall effect, and (3) using the improved functionality of CAI in the development of these complex surveys.

10.3 TRACING

10.3.1 Statistics Canada's Experience with Tracing in a Decentralized Environment

For the new longitudinal surveys at Statistics Canada that were implemented using decentralized CAI, tracing was developed within a case management system. It is available on all the interviewers' computers. Even though the surveys use different sources for tracing, there is a general capability applicable to all surveys.

Tracing at Statistics Canada is split into two steps. The initial trace is done by the interviewer. If this is unsuccessful, then the case is transferred to a central location, one of the six regional offices (ROs) where additional sources and different methods are used to try to trace the respondent. Currently the functionality of the tracing unit in the RO is almost identical to that of the interviewer. The tracing unit in the RO does have access to more tools to help them in tracing.

Tracing by the Original Interviewer

At the beginning of each collection cycle, interviewers are assigned their regular workload and there are no cases pre-assigned for tracing at the interviewer level. Every household has a database associated with it that contains some tracing information from previous collection cycles. This database may include, for example, the names and phone numbers of contact persons and the respondent's work telephone number. When a case is marked as requiring tracing, the case is automatically moved to a special "folder" where all the cases requiring tracing are kept and the tracing information is recorded. The case may involve the entire household or a subset of the original household if only some members have moved and require tracing.

Once in the trace folder, the tracing application begins with a "view and select" screen that allows a tracer to see all the cases that require tracing. After a case is selected from the view and select screen, a standard tracing screen is displayed (see Figure 10.1 for an example).

The top half of the screen displays case information such as the names and demographic information of the household members and whether each is an original (longitudinal) respondent in order to be prioritized for tracing. The bottom half of the screen gives a summary of the tracing actions to date and the date and time of last access for each source, along with the description of the source. These fields are blank when a source has not been used. The source indicates to the interviewer where the information came from. There is a standard list of possible sources that has been built for all surveys that include, for example, the long-distance operator, neighbors, and administrative records. Each survey chooses the sources that are appropriate. The last result of an attempt using the source and the details on the information collected from the source are called tracing notes. The number of times the source has been accessed is counted. In addition the last result of the tracing database that registers all the different calls that have been made is displayed on the screen.

The interviewer contacts one of the sources that are preloaded in the computer for the first attempt at tracing. Having the interviewer initiate the tracing is an efficient way of tracing for many cases. Interviewers conducting personal interviews can also ask neighbors whether they know where the respondent has moved, since close to 50 percent of people move within a fairly close neighborhood (based on information from Canada Post). If an interviewer is unable to trace the respondent, the case is sent to a tracer in the RO, since more tracing sources are available there.

Tracing in the Regional Office

The application used for RO tracing looks very similar to the tracing application used by the original interviewer. The system was designed this way because there is not yet a dedicated pool of interviewers whose job is only to trace respondents. Tracers are interviewers who are particularly skilled at tracing people. They trace individuals and usually interview them once they are found — though once traced the case could be transferred to the interviewer originally assigned the case.

The additional sources and tools that are available at the RO for tracing respondents include reverse directories (which list dwellings by street address) and electronic telephone books on CD. Since tracing is a time-consuming process, tracers at the RO are usually given some extra time (up to a month) beyond the normal interviewing period to do the tracing. If the RO tracers are not able to locate the respondent after that period, the case is sent back to Head Office for processing.

There is some added functionality for the RO's tracers. First, they can see which interviewer originally had the case and, if they have any questions or need clarification on a tracing note, can contact the interviewer. Cases can also

Interviewer Tracing Function
Case Management

CMINTTRCE1

Sample Id: T3333 10   **Tel:**(819)555-1234   **Note:** NO
Dwelling Address: 998 Graham Blvd
Mailing Address: 998 Graham Blvd.

Assignment # 09888

Given	Surname	Age	Sex	Ms	FID	Membership
Jane	Jones	32	F	1	A	0
Bob	Jones	34	M	1	A	0
‡ Carol	Jones	12	M	4	A	0
‡ Luc	Jones	8	M	4	A	0

Id	Date	Time	Source	No.	Result	Tel/Tracing notes
1.	/		Contact1	0	None	Marco Fioriano, 15 River St.
‡ 2.	21/03	11:15	Contact2	3	Busy	613-555-5555, Sue Jones, 102MainSt.

F1=Help F2=Add info F3=Update Trace Note F4=View Trace Note Tab=Switch screens F8=Record of trace calls
F11=Notes/Appt   F12=Continue ↵Start trace

Figure 10.1. Example of a tracing screen used in CAI.

189

be moved between ROs if a case requiring tracing was found in another province. This is particularly important for surveys done via personal interviewing. At the beginning of a collection period, cases can be sent directly to the tracers in the regional office so that cases not traced from a previous collection period can be tried again. After a predetermined number of collection cycles, which varies by survey, untraceable cases are dropped from the sample unless new information is received that may help to trace the respondent.

10.3.2   Effect of CAI on Tracing

The quality of the tracing process is one of the key elements in the success of a longitudinal survey for the reduction of unit nonresponse and its potential bias. The people who are being traced have one thing in common — they have moved — and therefore a change has occurred. To the extent that the move is related to other characteristics measured by the survey, the tracing operation will have an important effect on measures of change. The purpose of this section is not to discuss the process of tracing respondents (for a review of tracing methods, see Burgess, 1989) but rather to look at how CAI can be used in this activity.

The implementation of CAI can affect the tracing operation, especially in a decentralized environment. CAI can be used to enhance this operation in a number of ways. First, an increased amount of information can be provided to the interviewers with relative ease, since it was originally collected in an electronic format and can be returned systematically in either the current or future collection cycles.

Second, by using the communications system for the transfer of cases between interviewers, cases can be sent for tracing in different geographical areas in less time than it would take to send a paper questionnaire by mail or courier. Depending on the survey, if it was felt that it would be beneficial to use the same interviewer and if there was time, the case could then be transferred back to the original interviewer to finish the interview once the respondent had been traced.

Third, still related to the communication aspect, the "unable to trace" cases can be sent to a unit specialized in tracing. Statistics Canada has implemented the use of a centralized unit for tracing in its six regional offices since 1994.

Finally, more management tools can be developed in a CAI environment than for a paper survey. The system can easily store a record of every attempt that was made to trace a household, including the method (i.e., trace source), the date, time, and result of the attempt. This information can be used later by both managers and interviewers to determine the best strategy and tools for tracing.

However, the benefits of CAI tracing do not come without a price, and one has to be aware of the potential risks. First, tracing in a computer assisted interview can create a data management nightmare. Because it is easy to gather a lot of information, the resulting database can be very large and difficult to manipulate. For the accumulated household information one has to decide what information to keep and what to delete, using some sort of systematic approach.

Second, since the electronic transmission process for the transfers between persons and geographical areas is not perfect, there exists a risk of losing information. The more transfers occur, the higher is the risk of losing information somewhere along the "electronic highway."

Third, reports have to be designed carefully to show exact response rates and to accurately reflect the results of tracing. The issue of who the transferred case belongs to requires resolution for these reports, since it could belong to the original interviewer or to the interviewer who received the transfer.

10.3.3 Future Work toward Improving Tracing

Tracing with CAI is still relatively new, and work continues to improve the methods used in tracing. The following is a list of some of the outstanding issues:

Information management. Because there is the potential to accumulate a great deal of information, additional functionality is required to help the tracers to manage the information they have. This could include the ability to sort the information on different fields and to delete any information which is no longer needed.

Link between the tracing application and the survey instruments. Often the notes about tracing are recorded in an unstructured format and cannot be loaded directly into the interviewing environment. For example, if a person was traced by telephone to his/her new home address, and the telephone number was not entered at a fixed point in the tracing note, it may need to be re-asked once the survey interview has started. With more experience in CAI tracing, surveys should look at ways in which information collected in tracing could be easily transferred to the files used in interviewing.

Centralized tracing: tracers only or both tracers and interviewers? If a full-time permanent central tracing unit were created, the role of these tracers would likely change, since tracing could then occur outside of the collection window. This would affect the functionality required for the centralized tracing application. For example, the tracer could just collect the updated address and telephone for use in the next regular collection cycle and not conduct the interview, or the tracer could do the interview. Either way, the information collected during this separate tracing operation would have to be integrated with the main survey data before the next collection. Some analysis of the advantages and disadvantages is needed.

Linking of different tracing sources. There are a number of ways in which information useful for tracing can be collected — updates from respondents via change of address cards, use of sources on the Internet such as on-line telephone directories, or administrative files such as postal files. The issue is how to electronically link the information from various sources to a case in a specific survey. Also any link would need to be dynamic, since these external sources are revised regularly and it is the updated or new information that is required for a case.

10.4   ENSURING ACCURACY THROUGH TIME

As was mentioned earlier, one of the main concerns of longitudinal surveys is the accurate capture of events across time. Many longitudinal surveys use dependent interviewing to detect any changes in a respondent's situation and to ensure that any changes reported are in fact "real" changes.

Dependent interviewing involves using information from previous interviews to either remind the respondent or to probe for changes. CAI allows for greater use of dependent interviewing because it removes the logistical constraints that exist with paper and pencil methods. Also, with CAI, longitudinal surveys can tailor the use of dependent interviewing to the situation of the respondent and the subject matter being covered.

10.4.1   Seam Problems and Detection of Changes

A number of studies have looked at the "seam problem" in surveys with relatively long recall periods (up to two years depending on the survey). To aid respondents when they are asked to remember events in the recall reference period, the longitudinal surveys use some information from previous interviews in the course of the current interview (see Hale and Michaud, 1995). This dependent interviewing can be grouped into "proactive" and "reactive" approaches.

Proactive Approach to Feedback
The proactive approach to feeding back information involves reminding the respondents of their situation at the time of the last interview. For the Australian Survey of Employment and Unemployment Patterns (SEUP), the respondent is asked, at the start of the interview, to confirm their status of a year ago, that is, whether they were looking for work, working, or going to school. This initial reconciliation process precedes the collection of the current information where the respondent can confirm, deny, or update this information prior to collection of the data for the new reference period.

In other longitudinal surveys the information from previous interviews is used only when the relevant data are being collected for the current reference period. The following example from the Survey of Labor and Income Dynamics (SLID) illustrates the use of proactive feedback for employment information since the time of the last labor interview:

> Based on our interview of a year ago, [respondent's name] was working for [employer's name] around the beginning of January 1995. Is this correct?

After this information is confirmed, the information on his/her work with the employer in the current reference period is then collected.

In some surveys this method is used selectively due to the nature of the information collected. For example, in the National Population Health Survey (NPHS), information on a chronic medical condition is fed back only if the interview is being conducted directly with the person who has the condition (i.e., a nonproxy interview). In this way there is no disclosure of a medical condition to another member of the household who may not be aware of the respondent's condition. Only through the use of CAI can this be done with confidence. This also relieves the interviewer of the task of trying to keep track of various pieces of information in order to control the flow of the interview — with CAI they can now concentrate on using their interviewing skills.

Reactive Approach to Feedback
In some cases a reactive approach to feedback may be appropriate. In the reactive case the computer application compares the answer given in the current interview to that from a previous interview. If there is a conflict, the interviewer probes in an attempt to resolve the conflict and to determine whether the change is real or is an error in capture from the previous survey.

This approach may be preferred when the information fed back may bias the response given in the current interview. For example, to measure wage changes since the last interview, the previous year's wages would not be proactively fed back to the respondent because they may not report a change — thinking the dollar amount they gave last year was "close enough." Instead, the current wage would be first collected and then compared to the answer given in the previous interview. The interviewer would next probe to confirm that any change was real. In fact, with the use of CAI, the probing can be done selectively. For example, SLID used this approach for all wage increases or decreases of more than 10 percent.

In the NPHS the reactive approach to feeding back information is used when asking about a respondent's activity limitations or disabilities. For persons who reported a disability at the time of the last interview, two years earlier, and not in the current interview, the following question is asked:

> During our last interview in [date of last interview], there were activity restrictions or disabilities reported for [respondent's name] BUT this time there were not. Is this due to the disappearance or improvement of an old activity restriction or disability,

to use of special equipment (e.g., artificial limb), or to something else?
Possible Responses
Disappeared or improved
Currently uses special equipment
None at last interview
Never had
Currently has activity restriction or disability
Other (specify)

In this way the interviewer can determine whether there was actually a change in the respondent's condition since the previous interview.

Identifying Changes

Longitudinal surveys can use feedback to identify a change in circumstances, following which a series of questions about the change is asked. Complicated question flows based on responses in the current interview and previous interviews can only be done with the use of CAI. In a paper and pencil interview, the interviewers would be constantly referring back to previous responses causing lengthy, awkward, and error prone interviews to try to duplicate what a computer can do in a split second.

An example of this detection of change and flow through a series of questions about the change is from the SSP's module on marital history. If a respondent reported being married at the time of the last interview, 18 months earlier, and still reports being married, she is asked:

Are you married to the same partner that you were at the time of your interview on [date of last interview]?

If the respondent responds that she is not married to the same partner, she is then asked the date of the current marriage as well as when the previous marriage ended.

In some circumstances the change may in fact be a reversal of a previous change — such as the return of a former household member or return to work for a previous employer. These are the changes that longitudinal surveys are interested in identifying. CAI makes it possible to do this systematically.

If, in the course of an interview, it is discovered that one or more persons have joined the household since the last interview, these changes are systematically identified by SLID. The interviewer is presented with a list of the people who were household members in the past but had moved out. If the new household member is a former member who has returned, the interviewer highlights this person's name on the screen and the person is added to the list of household members. In this way background information that would normally be asked of new persons in the survey does not have to be re-asked, since it would have been collected previously. Also the return of a former household member is a substantially different situation from the addition of a new household member, and SLID wants to be able to distinguish between the two events. Once the data are ready to be processed, it is simple to link the information collected in past interviews with this new information and to add a variable that flags the household reformation.

This functionality is quite important for longitudinal surveys, since changes in household composition can have a major effect on an individual's well-being. However, it should be noted that, in practice, the file of former household members is quite complicated to create. Over the life of a longitudinal survey, a surprisingly long list of former household members can result. If this list is too long, it becomes difficult for the interviewer to quickly find the name of the person who has returned to the household, and the interviewer may disregard this function and add the person to the household roster as if he/she were someone who had never lived with any of the current household members. One idea being considered is to include only former members who are related to one of the current household members. While this is a good idea in principle, it will add more complexity to the creation of the file used to feed back this information.

A similar approach is taken to identify cases where an individual has returned to work for a former employer. If a new job is reported and the respondent has already reported working for other employers in past SLID interviews, the list of former employers (at least those known to the survey) is displayed and the interviewer can highlight the name of the employer on the list. This is important so that the survey can record the seasonal worker who keeps returning to the same job each year. It is also important to be able to identify those people who have gained seniority over time by working for the same employer each season, and with CAI this is possible.

Picking Up Missed Information

Information from previous surveys can be used to pinpoint information missed at the last collection either due to nonresponse or because the respondent was not eligible at that time. An example comes from the National Longitudinal Survey of Children and Youth (NLSCY). Two approaches are used to collect the number of times a respondent has moved since birth. If the respondent gave this information in the last interview, then he/she is asked only for the number of moves since the last interview. However, if this information was not collected previously, then the respondent is asked about the number of lifetime moves. Thus the survey ensures the same data for all respondents, regardless of what happened in the previous interview.

Respondent Reaction

There has been some concern about respondent reaction to the interviewer having some of the information from the previous interviews. Indeed various surveys have found that respondents are quite aware of the fact that their information is being entered directly into a computer and expect the interviewer to be able to access it in later interviews. Of course, when feeding back information from previous interviews, the respondent may deny the information that has been fed back. In the January 1996 SLID interview, the amount of denial varied slightly depending on the information being fed back, but overall there was denial for less than 1 percent of the cases. While this does leave longitudinal surveys with the dilemma of what to do about the inconsistent data collected in previous interviews (whether the previously collected data should be corrected or left as inconsistent), it is still better to systematically collect information about denials to monitor the respondent's reaction to feedback and to ensure that the most up-to-date information about the characteristics is being collected.

Variation in Amount of Feedback
The amount of feedback varies among the Canadian longitudinal surveys. For example, the NLSCY uses feedback to minimize keying errors and to reduce interviewer burden by displaying the previously collected address and employment information. The main reason further feedback is not used is that there was an unacceptable risk that changes would be missed with the two-year period between interviews. On the other hand, SLID uses feedback at the start of each module to remind respondents that they were working, looking for a job, attending school, or receiving government transfer payments at the end of the last reference period. This was adopted to reduce false transitions because previous surveys with similar content showed peaks in the number of transitions at the "seams" between reference periods. It was felt that the data quality benefits outweighed the possible suppression of changes that could occur with feedback. The amount of feedback used will vary by subject matter and only time will tell how much feedback is appropriate for any given survey.

10.4.2   Improving Recall with CAI

Dependent interviewing also promotes internal consistency of the data by prompting for related information in a subsequent interview. While SLID collects labor information in January for the previous calendar year, the income information for this reference period is collected later in the year to coincide wih the completion of income tax forms — since this is when most respondents would have the income information available.

In the January SLID labor interview, respondents are asked whether or not they had received any one of three types of government transfers in the previous calendar year — unemployment insurance, social assistance (welfare), or worker's compensation. A simple yes/no question is asked for each type of compensation — for example, "Did he/she receive any income from unemployment insurance in 1995?" and a flag is set for each of these income sources indicating what the respondent had reported. However, the actual amount of compensation received is not collected until the income interview a few months later when they are asked for their incomes from each of these sources.

At the time of the income interview, a reactive approach is used to feed back this information because receipt of these transfer payments is somewhat sensitive and is often underreported. If an expected income source is not reported, a probe is displayed for the interviewer to ask about the missing amount. The prompt supplied to the interviewer is designed to be nonthreatening and tactful, again because of the sensitivity of the topic. If a respondent did not report income for a government transfer payment but one was expected, the interviewer asks:

> Based on our January interview, we thought we would get an amount for [type of income]. Did we miss it?

This approach increased the reporting of amounts for government transfer payments by at least 20 percent (Dibbs et al., 1995). In this way the flow of the interview is only disrupted if necessary.

10.5 ONGOING DEVELOPMENT OF LONGITUDINAL SURVEYS WITH CAI

In this section we look at how the expanded functionality available with CAI can be utilized by longitudinal surveys.

10.5.1 CAI Functionality: Effect and Issues for Longitudinal Surveys

CAI survey instruments have unique functionality that can be exploited for longitudinal surveys. Longitudinal surveys tend to be very complex surveys collecting information on a variety of topics with complicated flow patterns. Automated question flows, on-line edits, time stamps of start and stop times of each module, and the capacity to enter interviewer comments for each question provide new opportunities for monitoring processes, for assessing data quality, and for continuous improvement of longitudinal survey instruments.

With the use of on-line edits, consistency of the data can be ensured over time — so that the CAI application must be able to compare the current information to previously collected information and display various alternatives to the interviewer. An example of this, mentioned in Section 10.4.1, occurs when the follow-up survey collects historical data, such as marital status, and the respondent provides event dates. In the subsequent contact, event dates may be reported inconsistently with those reported in the prior contact, although feedback helps to reduce these types of inconsistencies in reporting. The alternatives for the interviewer could consist of either accepting the current information or changing the entered value.

10.5.2 Development of CAI Survey Instruments for Longitudinal Surveys

In general, after the initial development of the questionnaire for a longitudinal survey is completed, changes in the variables being collected are kept to a minimum. This is to ensure consistent data over the life of the survey and to avoid redeveloping the survey prior to each collection. However, this raises the issue of how to best incorporate necessary changes and highly desirable enhancements.

Errors in the survey should be corrected as soon as possible but with caution. (See Walker, Brown, and Veevers, 1996, for an approach to validating longitudinal data collected using CAI.) All CAI surveys face the issue of what to do when a problem (e.g., an error in the question flow) is detected once a survey is in the field. If this error does not cause the instrument to freeze (for which all further collection would be terminated and the interviewer would have to recontact the respondent), the possible consequences would include sending out a "quick fix" during the collection phase, living without the data, or removing any unnecessary data resulting from the incorrect question flow. For longitudinal surveys one must also look at how any changes affect the subsequent waves of the survey.

For the subsequent waves the level of resource investment required for development should be substantially reduced when compared to that of the first wave, primarily due to the re-use of the survey instrument for the follow-up waves. The need for change can arise if shortcomings in the instrument are discovered in the initial wave—something that may occur depending on the experience of the persons involved in the survey and the amount of time allowed for development. Even if all goes well in the first wave, changes due to technological improvements may be required—something that most longitudinal surveys will not be able to avoid. A good example of this would be the move from a DOS-based surveying application to a Windows-based application. DOS was the operating system for most personal computers in the 1980s, but with the prevalence of Windows in the 1990s, there is a need to move to this newer technology.

Longitudinal CAI surveys must learn to adapt to changing technology to ensure that these sorts of changes do not adversely affect the data being collected. If the data are changed in some way because of the introduction of new technology, this would adversely affect the longitudinal use of the data, since there would be information for a respondent before and after the change. If systems are well documented and changes planned far in advance, there is a good chance that the implementation of any new technology can go ahead smoothly.

10.5.3   Effect on Interviewers (Field Collection Staff)

CAI allows for the collection of a great deal of data related to interviewer performance—not only response rates but also detailed information such as average interview time per section of the questionnaire and number of keystrokes per minute. One needs to investigate how all of the data related to performance should be analyzed and how to then get this information back to field staff in an effective way. In fact, while case-specific problems could be fed back to interviewers for correction in subsequent interviews, this would add an extra level of complexity to what is already a complex system. There is also the risk that if too many statistics on interviewer performance are produced, the interviewers will start to feel a threat from "big brother" watching over them at all times — which may then adversely affect performance.

Another issue affecting field staff is how to train interviewers throughout the life of the longitudinal survey. There will always be new interviewers who need more training than the majority of interviewers who are familiar with the survey and only need to be made aware of any changes. Few surveys can afford to develop multiple training packages depending on the experience of the interviewer. This may be an area where computer-based training could be developed that the interviewers could then tailor to their needs (see Wojcik and Hunt, Chapter 17).

10.6 SUMMARY AND CONCLUSION

The use of CAI technology has made it possible for household longitudinal surveys to be conducted effectively and to implement procedures that were too complicated for paper and pencil interviews. All procedures must be implemented carefully so that collection does not suffer. For example, the use of dependent interviewing increases the requirements from the group responsible for preparing files used in collection. This is often a complicated process involving the search of a number of large data files to find the relevant information.

Many of the examples in this chapter were taken from new longitudinal surveys that were developed for computer assisted interviewing. As discussed by Laurie and Moon (1996), the advantages of moving an existing longitudinal survey from paper and pencil interviewing to computer assisted interviewing must be weighed against the possible effect on the existing survey processes before the move to CAI can take place.

One of the key issues for longitudinal surveys is the ability to trace respondents as they move. While the use of CAI does allow surveys to transfer a case to the person who is best equipped to trace a respondent, still an issue is what sort of information is required for tracing and how the results from tracing should be stored and maintained.

It must be noted that in some ways technology has not quite caught up with the needs of longitudinal surveys — for example, one may want to move cases quickly between various interviewers yet this is not always technically feasible or easy to monitor. Technology advances can continue to be applied to longitudinal surveys to collect accurate data for persons while they are in the longitudinal survey — be that for one year or twenty. There is the need for flexibility in development of longitudinal surveys and also the need to control the survey process — the issue is one of striking a balance. We hope that the use of CAI will help us in this endeavor.

ACKNOWLEDGMENTS

We would like to thank the editors and the following individuals for their valuable suggestions on a previous draft of this chapter: Dave Dolson, Karen Roberts, Richard Veevers, Charlene Walker, and Maryanne Webber. Any errors or misrepresentations in the text are the responsibility of the authors.

CHAPTER 11

The Future of Data Editing

Jelke Bethlehem and Frank van de Pol
Statistics Netherlands

11.1 INTRODUCTION

The sample survey is a fallible instrument subject to many forms of bias and error. Data editing is one means of controlling and reducing survey errors, especially those arising from the interchange between respondents and interviewers or between respondents and self-administered forms during the data collection process. Data editing is the process of detecting errors in survey data and correcting those errors; correction can take place during the interview or in the survey office after data collection has been completed. Traditionally statistical organizations, especially in government, have invested great amounts of time and resources in data editing in the belief that it is crucial to the preparation of accurate statistics. Current data editing tools have become so powerful that question is now raised as to whether too much data editing occurs. A new objective is to minimize the amount of data editing performed while still guaranteeing a high level of data quality.

This chapter presents a conceptual and historical overview of survey data editing and the increasing role computing technology plays in these activities. Section 11.2 begins with the role of data editing in survey methodology, provides a formal definition of data editing, and considers the major approaches to editing. Section 11.3 discusses the history of data editing and the increasing role of technology in its development. Sections 11.4, 11.5, and 11.6 examine three alternative approaches to data editing: automated editing, selective editing, and macro-editing. The final section reviews stategic issues in the combined use of these three forms of editing.

Computer Assisted Survey Information Collection, Edited by Mick P. Couper, Reginald P. Baker, Jelke Bethlehem, Cynthia Z. F. Clark, Jean Martin, William L. Nicholls II, and James M. O'Reilly. ISBN 0-471-17848-9 © 1998 John Wiley & Sons, Inc.

11.2 FUNCTIONS AND DEFINITION OF DATA EDITING

The major function of data editing is to reduce errors in survey data and thereby in the population estimates derived from those data. Both the value and limitations of data editing are best understood in the broader context of survey errors. A variety of errors can occur, both before the collection of information from the respondent (e.g., noncoverage and nonresponse errors) and during the data collection process (measurement errors) (see Groves, 1989, for a typology of survey errors). Some errors can be reduced by taking preventive steps in survey design, but other errors will remain. It is therefore important to review collected data for residual error and, where possible, to correct any errors detected. This activity is called data editing.

11.2.1 A Definition of Data Editing

In this chapter we define *data editing* as the process of detecting errors in the survey data and then correcting any errors detected. Data editing may occur in many phases of the survey. Traditionally data editing takes place after data collection. With CAI methods, edits can be incorporated into the interviewing program, which means that data editing is conducted concurrently with data collection. In other methods, editing can be conducted using tables or graphs of the distribution of one or two variables. Such edits are possible only when a substantial part of the data collection is completed. Nor is data editing restricted to within-record editing. Between-record edits and edits on aggregated quantities are also useful methods that fall within the definition of data editing.

The literature is not entirely consistent on the meaning and scope of data editing, especially whether it encompasses both the correction and the detection of errors. Fellegi and Holt (1976) and Gosselin et al. (1978) restrict the phrase "data editing" to checking the data at the final stage of survey processing, just before tabulation. They do not include the correction process in the definition of data editing but see it as a separate activity called imputation. Others (e.g., Granquist, 1984; Federal Committee on Statistical Methodology, 1990a) define data editing to include both the detection and correction of errors. We adopt the latter view in this chapter.

11.2.2 Types of Data Editing and Data Errors

When data editing takes place at the level of individual forms or questionnaires, we will call it *micro-editing*. The forms are checked and corrected one at a time. The values of the variables in a form are checked without using the values in the other forms. Micro-editing typically is an activity that can take place during the interview or during data capture. When data editing takes place at the level of aggregated quantities obtained by using all available cases, we call it *macro-editing*. For macro-editing, a file of records is required. This is typically an activity that takes place after data collection, data entry, and possibly after micro-editing. Following Pierzchala (1990), data editing can be seen as addressing four principal types of data error:

Completeness Error

The first step in processing survey forms is to determine their degree of completeness. Forms that are blank or unreadable, or nearly so, are unusable. Incomplete forms can be treated as unit nonresponse, scheduled for callback, deleted from the completed sample, or imputed in some way, depending on the importance of the case.

Domain Error

Each question has a domain (or range) of valid answers. An answer outside the domain is considered an error. This is easily determined for numeric questions, since domain errors are defined as any answer that falls outside the allowable range. For questions that ask for values or quantities, it is sometimes possible to specify improbable as well as impossible values. For a closed question, the answer has to be chosen from a list (or range) of alternatives. The error may consist of not choosing an answer, choosing more answers than allowed, or choosing an answer outside the allowable range. For open questions, the domain imposes no restrictions; any text is accepted as an answer.

Consistency Error

Consistency errors occur when the answers to two or more questions contradict each other. Each question may have an answer in its valid domain, but the combination of answers may be impossible or unacceptable. A completed questionnaire may report a person as being an employee, or under five years of age, but the combination of these answers for the same person is probably an error. For instance, a firm known to have 10 employees should not report more than 10,000 person-days worked in the past year.

When a consistency error is detected, the answer causing the error is not always obvious. A correction may be necessary in one, two, or more questions, and resolving one inconsistency may produce another. So it is easier to detect consistency errors than to solve them.

Routing Errors (Skip Pattern Errors)

Many questionnaires contain routing instructions. These instructions specify conditions under which certain questions must be answered. In most cases closed and numeric questions are used in these instructions. In paper questionnaires routing instructions usually take the form of skip instructions attached to answers of questions, or of printed instructions to the interviewer. Routing instructions ensure that all applicable questions are asked, while inapplicable questions are omitted.

A routing error occurs when an interviewer or respondent fails to follow a routing instruction, and a wrong path is taken through the questionnaire. As a result the wrong questions are answered or applicable questions are left unanswered.

11.3   HISTORY OF DATA EDITING

In traditional survey processing, data editing was mainly a manual activity. Domain errors were identified by visually inspecting the answers one at the time. Consistency errors were typically caught only when they involved a small number of questions on the same page or on adjacent pages. Routing errors were found by following the route instructions and noting deviations. In general, manual editing could identify only a limited number of problems in the data.

The data editing process was greatly facilitated by the introduction of computers. Initially these were mainframe computers that only permitted batchwise editing. Tailor-made editing programs, usually written in cobol or fortran, were designed for each survey. Later, general-purpose batch editing programs were developed and extensively used in university survey research centers. These programs performed extensive checks on each record and generated printed lists of error reports by case ID. The error lists were then sent to subject matter experts or clerical staff, who attempted to manually reconcile these errors. This staff then prepared correction forms that were keyed to update the data file. The editing process consisted of the repetition of these steps.

Batch computer editing of data sets improved data editing because it allowed for both greater volume and greater complexity of error checks. This in turn led to the identification of even more data errors. However, the cycle of batchwise checking and manual correction proved labor intensive, time-consuming, and costly (see Bethlehem, 1997, for a more complete analysis of this process and its disadvantages).

With the emergence of microcomputers in the early 1980s, completely new methods of data editing became possible. One of these approaches has been called computer assisted data input (CADI) or computer assisted data entry (CADE). CADE and CADI provide an interactive and intelligent environment for combined entry and editing of paper forms by subject matter specialists or clerical staff. Data can be processed in two ways: either in combination with data entry or as a separate step. In the first approach, the subject matter staff process the survey forms one by one using a microcomputer. Data are entered "heads up," meaning that the staff tend to watch the computer screen as they make entries. After data entry is completed, staff run the checks that test for various kind of errors (omission, domain, consistency, and routing errors). Detected errors are displayed and explained on the screen. Staff can correct the errors by consulting the form or by contacting the supplier of the information. After elimination of all detectable errors, a "clean" record, that is, one that satisfies all check edit criteria, is written to the file. If the staff member does not succeed in producing a clean record, it can be written to a separate file of problem records. Specialists can later deal with these difficult cases using the same CADI system. This approach of combining capture and editing works best for surveys with relatively small samples but complex questionnaires.

In the second approach, clerical staff (data typists or entry specialists) key data through the CADI system "heads down," that is, without much error checking. When this entry step is complete, the CADI system checks all the records in a batch run and flags suspicious cases. Subject matter specialists then take over. They examine the flagged records and fields one by one on the computer screen and try to reconcile the detected errors. This approach works best for surveys with large samples and simple questionnaires.

A second advance in data editing occurred with the development of computer assisted interviewing (CAI). CAI offers three major advantages over traditional paper and pencil interviewing (PAPI):

1. CAI integrates three steps in the survey process: data collection, data entry, and data editing. Since interviewers use computers to record the answers to the questions, they are also entering the data at the same time. In many surveys the data editing is carried out during the interview, rendering separate editing steps superfluous. Once all of the interviewers' files have been combined into a single data file, the information is clean and ready for further processing.

2. The interview software determines proper question order and ensures that entries are within their domains. Hence routing and range errors are largely eliminated during data entry, which also reduces the burden on the interviewers as routing is handled by the program itself. Interviewer energy is freed to focus on getting accurate and complete responses.

3. With CAI it becomes possible to carry out consistency checks during the interview. Since both the interviewer and the respondent are still available when inconsistencies are detected, reconciliation can take place immediately. Experience has shown that data editing during the interview produces better quality data than editing after data collection.

Computer assisted interviewing moves data editing to the front of the statistical process, which allows the interviewer to assume many of the data editing tasks. This raises the question of whether all data editing should be carried out during the interview, thereby avoiding a separate data editing step. There is much to say in favor of this approach. As part of his quality vision, Deming (1986) admonishes against mass inspection of the final product. It is an ineffective and costly procedure. Rather, quality control should be built into the production process and be part of every production step. CAI with simultaneous data editing can be seen as a procedure that reflects the spirit of the Deming message.

On the other hand, data editing concurrent with interviewing entails some drawbacks. First, checks built into the interviewing program can be very complex, resulting in error messages that are difficult for interviewers and respondents to understand. Correction of some detected errors may prove a very difficult task. The developer of the interviewing program has to recognize that the interviewer is not a subject matter specialist. Editing programs should be written in such a way that the interviewer can easily handle on-the-spot error reconciliation.

Second, numerous checks in the interviewing program will increase the length of the interview as the interviewer stops to correct detected errors. In principle, interviews should be kept as short as possible. In longer interviews the respondent has a propensity to lose interest and data quality can suffer; such effects may offset any quality gains from additional editing.

Third, not all forms of data editing are possible during the interview. When comparisons of entered data are necessary with information from other sources, such as large databases, notebook computers used in CAPI may not have sufficient memory to perform the required tasks.

Fourth, improperly specified checks can virtually obstruct the completion of an interview. For instance, if the interviewing software refuses to accept entries that violate programmed edit checks and the respondent maintains that his/her answers are correct, an impasse can result. Fortunately most interviewing software can circumvent these deadlocks by permitting both hard checks and soft checks.

Hard Checks result from errors that must be corrected. The interviewer cannot continue the interview until the program is satisfied with the reconciled information. *Soft checks* alert the interviewer that the information is highly improbable, although possible. If the respondent insists that the answers are correct, the interviewer can accept the answers and continue. Soft checks, instead of hard checks, should be used wherever there is a risk of generating an impasse. It is also possible to combine soft and hard checks. A soft check with somewhat relaxed conditions is used to detect suspicious cases, whereas the same type of check with more strict conditions is specified as a hard check.

Despite the potential extra interviewing burden, and the limitations imposed by hardware, considerations of time, money, and quality, it is generally thought that as much data as possible should be edited during the interview. The only editing that should be conducted after data collection is that which is not feasible during the interview. This requires careful planning at the design stage of both the interview and the postinterview editing instruments.

Performing data editing during a computer assisted interview is greatly facilitated when the interviewing software allows specification of powerful checks in an easy and user friendly way. Although edit checks can be hard coded for each survey in standard programming languages, this is a costly, time-consuming, and error prone task. Many CAI software packages now offer very powerful tools for micro-editing that permit easy specification of a large number of checks, including those involving complex relationships among many questions. Editing during CAI is now extensively used both in government and private sector surveys.

Whether micro-editing is carried out during or after the interview, the entire process may have major disadvantages, especially when carried to extremes. Little and Smith (1987) have mentioned the risk of overediting. Powerful editing software offers ample means for almost any conceivable check, and it is sometimes assumed that the more checks one carries out, the more errors one will correct. But such assumptions entail both risks and costs.

First, the use of too many checks may cause problems in interviewing or postinterview data correction, especially when the checks are not carefully designed and thoroughly tested prior to use. Contradictory checks may cause virtually all records to be rejected, defeating the purpose of editing. Redundant checks may produce duplicate or superfluous error messages, which slow down the work. And checks for errors that have little effect on the quality of published estimates may generate work that does not contribute to the quality of the finished product.

Second, since data editing activities consume a large part of the total survey budget and many statistical agencies face budget reductions, cost effectiveness is an important consideration. Large numbers of micro-edits that require individual correction will increase the cost of a survey. Every attempt should be made to minimize data editing activities that do not improve the quality of the survey results.

Third, it must be recognized that not all data problems can be detected and repaired with micro-editing. One such problem is that of outliers. An *outlier* is a value of a variable that is within the domain of valid answers to a question but is highly unusual or improbable when compared to the distribution of all valid values. An outlier can be detected only when the distribution of all values is available; thus outliers require macro-editing.

The remaining sections of this chapter describe three alternatives to editing that address some of the limitations of traditional micro-editing. These alternatives could replace micro-editing in some situations. In other situations the alternatives could be carried out in combination with traditional micro-editing or with each other. They are:

1. *Automated editing* attempts to automate micro-editing. Since human intervention is eliminated, costs are reduced and timeliness is increased.

2. *Selective editing* attempts to minimize the number of edits in micro-editing. Only edits having an effect on the final survey results are performed.

3. *Macro-editing* offers a top-down approach. Edits are carried out on aggregate data rather than on individual records. Micro-editing of individual records is invoked only when problems are identified by the macro-edits.

11.4 AUTOMATED EDITING

Automated editing checks and corrects the records directed by a software package. Since no human participation is involved, this approach is fast and inexpensive. For automated editing the usual two stages of editing (error detection and correction) are expanded to three:

1. *Error detection.* As usual, the software detects errors or inconsistencies by reviewing each case using the prespecified edit rules.

2. *Identification of the variables causing the error.* If an edit detects an error that involves several variables, the system must determine which variable caused the error. Several strategies have been developed and implemented to solve this problem.

3. *Error correction.* Once the variable causing the error has been identified, its value must be changed so that the new value no longer generates an error message.

There is no straightforward way to determine which of several variables causes a consistency error. One obvious criterion is the number of inconsistencies that a variable is involved in. If variable A is related to three other variables B, C, and D, an erroneous value of A may generate three inconsistencies, with each of the other variables. If B, C, and D are involved in no other edit failures, A seems the likely culprit. However, it could be that no other edit rules have been specified for B, C, and D, in which case they could also be candidates for correction.

The Fellegi-Holt (F-H) method takes a more sophisticated approach (Fellegi and Holt, 1976; United Nations, 1994). To reduce dependence on the number of checks defined, the F-H method performs an analysis of the pertinent edit checks on each variable. Logically superfluous checks are removed, and all implied checks that can be logically derived from the checks in question are added. Records are then processed as a whole, not on a field-by-field basis, with all consistency checks in place to avoid the introduction of new errors as identified errors are resolved. The smallest possible set of imputable fields is located with which a record can be made consistent with all checks.

In the F-H method erroneous values are often corrected with hot deck imputation. Hot deck imputation employs values copied from a similar donor record (another case) not violating any edit checks. When the definition of "similar" is very strict or when the receptor record is unique, it may be impossible to find a similar donor record. In this situation a simple default imputation procedure is applied instead.

The F-H method has been programmed and put into practice by many government statistical agencies. Several editing packages exist. For editing categorical variables they include Discrete (USA), Aero (Hungary), DIA and

Lince (Spain), and DAISY (Italy). The U.S. Census Bureau program SPEER was designed for valid value checks on ratios of numerical variables. The Chernikova algorithm, which can only handle numerical variables, was further developed by Statistics Canada for the general editing program GEIS. All these programs identify fields that are likely to contain errors and impute estimates of their values. These programs run on diverse computer platforms and operating systems.

In practical applications, many ties occur; that is, several variables are equally likely to be in error. For one check and two inconsistent values, there is a 50 percent chance that the wrong variable will be changed, an undesirably high percentage of erroneous corrections. Ties are less frequent when a greater number of edit checks are specified, but the F-H method makes increased checking costly as greater computing resources are required (e.g., see Informatica Comunidad de Madrid, 1993). When checks are interrelated, there can be hundreds of thousands of implied checks, using a vast amount of computing time for their calculation. Nevertheless, a greater number of original checks can avoid ties later in the editing.

When there is a large number of checks, one approach to automated editing is to partition the checks into independent subsets. This approach was suggested by the designers of DAISY (Barcaroli and Venturi, 1995). It is nevertheless not always easy in practice, for edit checks do not always fall into naturally separate and independent subsets. Another approach is to use the Chernikova algorithm, which has been implemented in GEIS (Schiopu-Kratina and Kovar, 1989; Kovar and Whitridge, 1990). This algorithm is particularly useful for editing continuous variables and is applicable in many situations. For small data sets this algorithm is faster than the F-H algorithm, since its computing time depends primarily on the size of the data set. The F-H algorithm requires a large computational investment to analyze the edit checks before passing through the data; this computational investment pays off only when many records are processed. Nor is the Chernikova algorithm particularly fast, since a considerable computation effort is required for each record. If edit checks are restricted to ratios of two variables at a time, a simple and fast algorithm emerges. This algorithm has been implemented in SPEER (Winkler and Draper, 1994).

Bankier et al. (1995) have recently proposed a new approach that can handle both categorical and numerical data. This method uses a distance metric to select donor records that are similar to the original (erroneous) record. It also minimizes changes to the original erroneous values. In our experience the latter property can be a drawback when very large errors occur, such as a value that is 1000 times too large. Then donor records that have very large values on most variables may be selected. Extreme outliers should not be handled in this way. De Waal (1996) has proposed alternate distance functions to handle ties in the Chernikova algorithm.

There is an additional problem with the Chernikova algorithm, caused by the weakness of its user interface. Although the Chernikova algorithm will handle a string of checks connected by the logical *and* operator, it cannot handle several strings that are connected by an *or* operator.

Our impression is that the current state of automated editing allows for only limited applicability of these techniques. This is disappointing because powerful automated data editing tools can substantially reduce survey costs. Currently automated editing should be used only for detecting and correcting errors that have no substantial effect on the published statistics. Furthermore automated editing never should be the only data editing activity. To avoid imputation of values of wrong variables, it should be used in combination with other editing techniques.

11.5 SELECTIVE EDITING

The implicit assumption of micro-editing is that every record receives the same treatment and the same degree of effort. In business surveys this assumption may not be appropriate or cost effective, since not every record has the same effect on the computed estimates. For instance, some firms make substantially larger contributions to published estimates than do other firms. Instead of conserving editing resources by fully automating the process, greater efficiency and cost effectiveness may be achieved by focusing resources on the most necessary edits. Necessary edits are those that have a noticeable effect on published figures, including outliers. We call this approach selective editing.

To establish the effect of edits on population estimates, estimates based on unedited data can be compared with estimates based on edited data. Boucher (1991) and Lindell (1994) compared unedited data with edited data and found that for each variable studied, 50–80 percent of the edits had virtually no effect on the estimate of the grand total. Similar results were obtained in an investigation carried out by van de Pol and Molenaar (1995) on the effects of editing on the Dutch Annual Construction Survey. Figure 11.1 plots the value of the production estimate for three groups of firms against the number of edits carried out, showing that editing had little effect on the estimates. The level on the (unedited) left side of each graph is almost the same as the level on the (edited) right side for all three. A strong editing effect was found for only one of the 12 variables studied, namely, trade benefits (not shown here). Editing reduced this amount from about 2 percent of the production value to less than 1 percent.

From this investigation, and other research in this area (e.g., Boucher, 1991; Lindell, 1994), it becomes clear that only a few edits have a substantial effect on the final figures. Therefore data editing efforts can be reduced by identifying the edits that actually have the greatest effect. One implementation is to use a criterion to split records into critical and noncritical streams. The *critical stream* contains the records that have a high risk of containing influential errors and therefore require thorough micro-editing. Records in the *noncritical stream* could remain unedited or could be limited to automated editing.

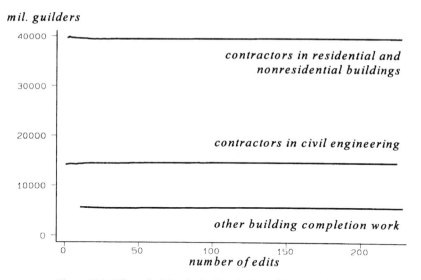

Figure 11.1. Effect of editing in the Dutch Annual Construction Survey.

The basic question of selective editing is whether a criterion can be identified that splits records into a critical and noncritical stream. One might think that only records of large firms will contain influential errors. However, van de Pol and Molenaar (1995) show that this is not the case. Both big firms and small firms generate influential errors. A more sophisticated criterion is needed than just the size of the firm. A more powerful and yet practical criterion would take account of inclusion probabilities, nonresponse adjustments, size of relevant subpopulations, relative importance of record, and, most important of all, a benchmark to determine whether an observed quantity may be in error. Examples of such benchmarks could be deviations from the sample mean (or median) for that quantity.

For business surveys Hidiroglou and Berthelot (1986) probably were the first to use a score function to select records containing influential errors. Their approach was followed by Lindell (1994) and Engström and Ängsved (1994). A somewhat modified approach was developed by van de Pol and Molenaar (1995). They concentrate on edits based on ratios. Let

$$R_{ijk} = \frac{Y_{ik}}{Y_{ij}}$$

be the ratio of the values of two variables j and k for firm i. This ratio is compared with the median value M_{jk} of all the ratios, by computing the distance

$$D_{ijk} = \text{Max} \left\{ \frac{R_{ijk}}{M_{jk}}, \frac{M_{jk}}{R_{ijk}} \right\}$$

or, equivalently,

$$D_{ijk} = e^{(|\log(R_{ijk}) - \log(M_{jk})|)}$$

A cutoff criterion may be used to set D_{ijk} to zero when it is not suspiciously high. Next a risk index is computed as a weighted sum of the distances for all edits in a record:

$$RI_i = \frac{I_i}{\pi_i}\left[\sum e^{(W_{jk}\log(D_{ijk}))} - Q\right]$$

The number of ratios involved is denoted Q. The quantity I_i denotes the relative importance of firm i. It is included to ensure that more important firms get a higher edit priority than less important firms. The inclusion probability π_i is determined by the sampling design. The weight W_{jk} is the reciprocal of the estimated standard deviation of the $\log(D_{ijk})$.

This risk index can be transformed into an OK *index* by carrying out the transformation

$$OK_i = 100 - \frac{100RI_i}{Med(RI_i) + RI_i} \qquad (11.5.5)$$

Low values of the OK index indicate a record is not OK and is in need of further treatment. The transformation causes the values of the OK index to be more or less uniformly distributed over the interval $[0, 100]$. This has the advantage of a simple relationship between the criterion value and the amount of work to be done: The decision to micro-edit records with an OK index value below a certain value z means that approximately z percent of the records are in the critical stream.

The OK index can be used to order the records from lowest to highest OK index value. If micro-editing is carried out in this sequence, the most influential errors will be edited first. The question arises of when to stop editing records. Latouche and Berthelot (1992), Engström and Ängsved (1994), and van de Pol and Molenaar (1995) discuss several stop criteria. The last suggest that editing records with an OK index value under 50 would have little effect on the quality of estimates. It is generally sufficient to edit only half of the records.

At present, there is no standard software available to implement the concepts of selective editing just described. In some situations selective editing can be applied using existing software, such as the Blaise System developed by Statistics Netherlands (1996). First the Data Entry Program (DEP) is used for heads down data entry. Then the data manipulation tool (Manipula) is used to compute the values of the OK index, and these values are added to the records of each case. One approach could be to split the data file into a file of critical records (i.e., records with OK index below a specified threshold) and a noncritical file. The critical file will then be subject to micro-editing. Another approach could be to sort the data file by the OK index value from low to high, and then allow the analyst to continue working with the file. The analyst can then decide on a case-by-case basis how much editing the records require.

Selective editing is a promising approach to data editing. However, the method is still in its infancy. This approach has been shown to work in specific cases, but a general framework is needed to provide more tools for deciding which records require micro-editing.

11.6 MACRO-EDITING

Macro-editing provides a solution to some of the data problems left unsolved by micro-editing. It also can address data problems at the aggregate, distribution, and higher levels. The types of edit checks employed by macro-editing are similar to those of micro-editing, but the difference is that macro-edit checks involve aggregated quantities. In this section we focus on two general methods of macro-editing.

The first method is sometimes called the aggregation method (Granquist, 1990; United Nations, 1994). It formalizes and systematizes what statistical agencies routinely do before publishing statistical tables. The current figures are compared with those of previous periods to determine their plausibility (e.g., see van de Pol and Diederen, 1996; Laflamme et al., 1996). When an unusual value is observed at the aggregate level, the individual records contributing to the anomaly are edited at the micro level. The advantage of this form of editing is that it concentrates editing activities to those points that have an effect on the final results of the survey. No superfluous micro-editing is carried out on records that do not produce unusual values at the aggregate level. A disadvantage of this approach is that results may be biased in the direction of one's expectations. There is also the risk that undetected errors may introduce bias.

A second form of macro-editing is the distribution method. The available data are formed into distributions of variables, and the individual values are compared with their distributions. Measures of location, spread, and covariation are computed. Records containing values that appear unusual or atypical in their distributions are candidates for further inspection and possible editing.

Many macro-editing techniques analyze the behavior of a single observation in the distribution of all observations. We discuss some of these techniques and restrict ourselves to the analysis of quantitative variables.

Exploratory data analysis (EDA) is a field of statistics for analyzing distributions of variables (Tukey, 1977). Some of the most powerful EDA methods employ graphical representations that provide greater insight into the behavior of variables than do numerical techniques. Many of these techniques can be applied directly to macro-editing and are capable of revealing unusual and unexpected properties that might not be revealed by numerical inspection and analysis.

Here we focus on two groups of techniques. The first group analyzes the distribution of a single variable and concentrates on detection of outliers. The second group analyzes the relationship between two variables and looks for records with unusual combinations of values. All techniques discussed can be seen as special cases of the distribution method of macro-editing.

A technique that displays the distribution of a single variable in its most elementary form is the one-way scatterplot. In such a plot each individual value of the variable is displayed on a horizontal scale, as shown in Figure 11.2.

The variable displayed in the plot is the yearly manure production by farms in 98 municipalities in the Dutch province of Zuid-Holland. Each small square represents the manure production within one municipality. A one-way scatterplot can be used to analyze the following properties of the distribution:

- *Outliers.* Outliers appear as single squares that are widely separated from the other observations. Outliers require further analysis, since they may represent incorrect values.

- *Grouping.* When values are grouped, they appear to be clustered in separate groups rather than being evenly spaced across the whole domain of possible answers. Grouping may indicate that the observations originate from separate subpopulations that behave differently. It may be more appropriate to analyze these groups separately.

- *Concentration.* Concentration denotes values that are clustered around a specific location. Means, medians, and modes are common measures of central location.

- *Symmetry.* Symmetry assumes that the distribution is symmetric and is centered around one location. Symmetry permits the use of numerical techniques that assume normality.

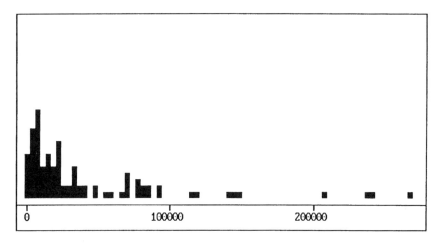

Figure 11.2. One-way scatterplot of manure production.

In the example of Figure 11.2, we see four values on the right side of the distribution that might be considered outliers. There is no grouping of observations. The distribution is far from symmetric, making it difficult to use numerical quantities like means and standard deviations to describe the distribution.

For asymmetric distributions one might consider a transformation. If the transformed distribution resembles the normal one, further techniques are available for analyzing the distribution. In the case of skewed distributions, like the one in the example, taking logarithms or square roots might help.

Another technique of displaying the distribution of a single variable is the box-and-whisker plot, which portrays the shape of the distribution in a schematic way without showing every individual value. Figure 11.3 contains an example of a box-and-whisker plot.

The rectangular box represents the central part of the distribution. It extends from the first quartile to the third quartile. The line within the box is the median (the second quartile). The whiskers connect the box to what are called adjacent values. The upper adjacent value is defined to be the largest value less than or equal to the third quartile plus 1.5 times the length of the box. Similarly the lower adjacent value is the smallest value greater than or equal to the first quartile minus 1.5 times the length of the box. Any value falling outside the interval between the two adjacent values is considered an outlier and therefore plotted as an individual point. The following properties of the box-and-whisker plot are important for analyzing the distribution:

- *Outliers.* Outliers appear as single points. These points require further analysis, since they indicate suspicious values.
- *Symmetry.* If the box-and-whisker plot is symmetric, the underlying distribution is symmetric. Only if the distribution is symmetric and concentrated around one location is it acceptable to use numerical techniques that assume normality.
- *Length of whiskers.* If the length of the whiskers is much shorter or much longer than the 1.5 times the length of the box, this indicates a deviation from normality. One should be careful with numerical techniques that assume normality of the underlying distribution.

A more traditional way of portraying the distribution is a histogram. The

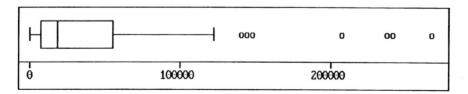

Figure 11.3. Box-and-whisker plot of manure production.

domain of possible values is divided into a number of intervals of equal width. The number of values in each interval is counted, and these counts are depicted as bars with heights proportional to the frequency counts, as Figure 11.4 shows.

Choosing the appropriate number of intervals is critical. Too few intervals will obscure the detail, and too many intervals will highlight irrelevant detail. One rule of thumb is to set the number of intervals roughly equal to the square root of the number of observations, provided that number is at least 5 and no greater than 20.

The primary use of the histogram is to analyze the symmetry and concentration of the distribution. In some cases a histogram also might suggest grouping of values. On the other hand, the histogram is not a useful device for detecting outliers.

There are other numerical ways to characterize the distribution and to search for outliers. One of the most frequently used techniques is based on the mean and variance of the observations. If the underlying distribution is normal, then approximately 95 percent of the values must lie within two standard deviations above and below the mean. Outliers can now be define as values outside these intervals. This simple technique has two important drawbacks:

1. The assumptions are satisfied only when the underlying distribution is approximately normal.

2. Traditional statistical analysis is very sensitive to outliers. A single outlier can have a large effect on the values of mean and variance and may therefore distort what is really going on in the data.

For these reasons this numeric technique should only be used after graphical methods have justified assumptions of normality.

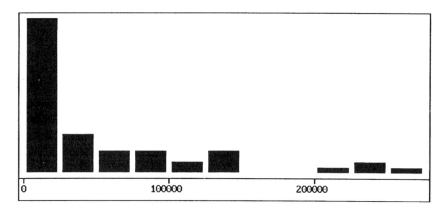

Figure 11.4. Histogram of manure production.

Numeric techniques based on the median and quartiles of the distribution are less vulnerable to extreme values. In line with this reasoning, the box-and-whisker plot can be applied in a numerical way. Values smaller than the lower adjacent value or larger than the upper adjacent value can be identified as outliers.

An obvious technique for displaying the relationship between two variables is the *two-dimensional scatterplot*. Suppose that we have two variables X and Y. For all N cases in the survey, assume that we have the pairs of values $(X_1, X_2), (X_2, Y_2), \ldots, (X_N, Y_N)$. The scatterplot, such as in Figure 11.5, displays each pair (X_i, Y_i) as a point with coordinates X_i and Y_i. The special properties of a scatterplot for macro-editing are twofold:

1. *Outliers.* Outliers appear as single points that are widely separated from other cases. Outliers require further examination, since they may represent incorrect values.
2. *Grouping.* Grouping describes points that appear to be clustered in separate entities and are thus not evenly distributed over the entire, two-dimensional domain of all possible answers. Grouping may indicate that the observations originate from separate subpopulations that behave differently. It may be more appropriate to analyze these groups separately.

If the points in a scatterplot form a clear pattern, this indicates a certain

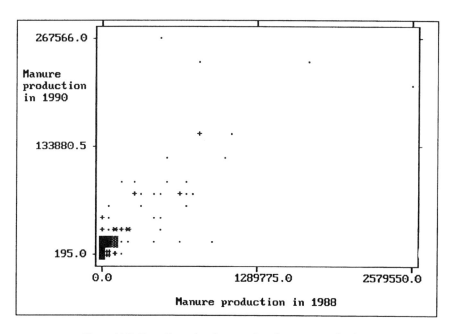

Figure 11.5. Two-dimensional scatterplot of manure production.

relationship between the variables. The simplest relationship is linear. In the linear case all points will lie approximately on a straight line. When such a relationship appears present, it is important to look for points that do not follow the pattern, since these may indicate error in the data.

The X-axis represents the yearly manure production in 1988 by farms in the municipalities in the Dutch province of Zuid-Holland. The Y-axis represents the same variable for 1990. The plot seems to indicate a linear relationship between the two variables, but there are many points that do not conform to the pattern.

If the assumption of linearity is not unreasonable, it can be summarized in the form of a regression line, in order to detect outliers. First, define the residual R_i for case i by

$$R_i = Y_i - a - bX_i.$$

The residual is equal to the difference between the value Y_i and the prediction $a + bX_i$ for Y_i based on the regression line. It is the (vertical) distance between the point and the regression line. Large residuals indicate outliers. It can be shown that the studentized residual T_i, defined by

$$T_i = \frac{R_i}{S(T_i)}$$

has an approximately normal distribution with mean 0 and standard error 1. So 95 percent of the values should lie in the interval $(-2, 2)$, and more than 99 percent should lie in the interval $(-3, 3)$. Hence, studentized residuals with values outide these intervals can be considered outliers. Such cases require further investigation. Since these standardized residuals measure the distance of a value from the mean, they can be used for constructing a record editing priority index like the one described in Section 11.5.

Computation of the regression coefficients a and b is sensitive to outliers. A few large outliers can have a major effect on the slope and intercept of a regression line. The scatterplot of unedited data should be carefully examined before computing the regression. A reasonable approximation also may be obtained with edited data from previous rounds of data collection.

There are methods of computing a regression line that are less sensitive to outliers. One method is an iteratively reweighted regression. The regression line is repeatedly calculated with adjusted weights assigned to each case, which are determined by the residuals of the previous regression. The larger the residuals, the smaller the weights (e.g., see Chambers et al., 1983). Another method of robust statistical editing, using the expectation maximization (EM) algorithm, has been proposed by Little and Smith (1987).

Software to implement the macro-editing methods described here should include the following two components: (1) graphical and numerical techniques, like those reviewed here, to describe and display the data and (2) tools to access and edit the individual records found to have unusual values. Standard software that implements this approach to macro-editing is limited. We now describe three systems mentioned in the literature.

The first is GRED, a microcomputer program developed by Statistics New Zealand (Houston and Bruce, 1993) that displays the individual values of a variable for different firms for consecutive survey years. In one plot it is possible to detect outliers and deviations from the trend. For unusual points the sampling weight of the record can be adjusted, or the record can be removed. Outliers can be highlighted in different colors. These outliers will also be highlighted in other graphic displays by linked plots. This permits easy identification of the influence of a specific observation on aggregate statistics.

The second is ARIES, which was designed for macro-editing the Current Employment Statistics Program of the U.S. Bureau of Labor Statistics (Esposito and Lin, 1993; Esposito et al., 1994). A session with ARIES starts with what is called an anomaly plot. This is a graphical overview of the important estimates; each node represents a specific industry. Related estimates are connected by lines. Estimates identified as unusual based on month-to-month changes are marked in a different color. Only suspicious estimates are analyzed in greater detail.

For industry groups ARIES can generate two types of plots: a scatterplot of the data values of the current month against the data values of the previous month, and a plot of the distribution of the month-to-month changes. By selecting points using a mouse, the data values can be displayed in tabular form on the screen. The adjustment weight of detected outliers can be modified interactively. Future versions of ARIES will produce simultaneous scatterplots of multiple variables. Linked plots will help the analyst study outlier cases from one plot in other plots for the same or additional variables.

A third example of a macro-editing system was developed for use in the Swedish Short Periodic Employment Survey (Engström and Ängsved, 1994). First, suspect estimates are selected, based on time series analysis and sample variance. This may be seen as a form of aggregate macro-editing. Next, a scatterplot is made of the data contributing to each suspect estimate, with its data values plotted against the corresponding values of the previous quarter. Outliers are displayed in different colors. Single-clicking on an observation displays information about it, including the weight assigned to the observation and a measure of its contribution to the estimate. Double-clicking an observation provides the analyst with access to the corresponding data record. The analyst can make changes in the data record. The system then rechecks the record for inconsistencies and updates the scatterplot and all corresponding parameters.

Current macro-editing systems have been designed for application to specific surveys. Thus they cannot be directly applied to surveys with different variables and data structure. There is clearly a need for general macro-editing software that can be employed for a wide variety of surveys.

A simple alternative is to load the data into a spreadsheet or data analysis program, sort the cases for each variable in turn, and identify outliers from the smallest to the largest values. Sorting criteria may include original values, regression residuals, and ratios of variables.

This simple nongraphical approach to macro-editing can be combined with data entry and micro-editing software in the following ways:

1. Enter the data with a standard data entry or data editing package and perform only limited micro-data editing on the data set.
2. Compute the statistics needed to detect outliers using a standard analysis package. For individual variables these may include means, variances, quantiles, and the like. For bivariate analysis including change from the last survey period, these could be regression residuals.
3. Identify and flag outliers through batch runs in which individual values are compared with distribution statistics.
4. Edit interactively the flagged records using a data editing package.

This simple approach lacks all the advantages provided by graphics. Its multistep, batch activities make it less efficient, even though batch processing also makes it less time-consuming. When possible, it is clearly preferable to build an interactive software module that reads the data file, generates diagnostic plots, makes it easy to flag outliers, and accesses individual data records to edit values.

Bethlehem and Hofman (1995) describe an application of the simple, nongraphical approach for the Blaise System. They also propose a way of extending this system to incorporate the preferred graphical, interactive approach. There still is a gap between the powerful nongraphical editing systems like Blaise and graphical, noninteractive statistical analysis systems. The survey community would benefit from a combination of the two approaches.

11.7 CONCLUSION

Selective editing, automated editing, and macro-editing are attractive alternatives to traditional micro-editing at a time when editing resources must be minimized in many survey organizations while still maintaining a high level of data quality. These alternatives are not equally appropriate to all editing situations.

Selective editing, which sets priorities among the cases to be edited at the microlevel, is especially applicable when (1) data editing is performed in the survey office rather than with computer assisted interviewing during data collection, (2) some cases are clearly more important than others because of their size or contribution to the final estimate, and (3) important data errors can be detected at the microlevel through inconsistencies or relational checks rather than by comparison with data from other cases.

Automated editing which employs both automated error detection and automated correction can be viewed as complementary to selective data editing. Automated editing can be employed to clean up remaining small inconsistencies or those in low-priority cases after selective editing. Automated editing also may be the only feasible alternative to edit data files that are too large for clerical review and correction, such as population censuses. For high-quality automated editing using the Fellegi-Holt method, large numbers of edit checks are necessary making the process slow and demanding on computing resources.

Macro-editing which identifies errors by examining aggregates, such as tables to be published, can be used when (1) the published estimates matter and the microdata are not released in any form to outside analysts, (2) historical data and other information sources are available to provide good predictions of published figures, and (3) an index, graph, or other means is available to trace problems in publication cells to individual cases.

Some researchers and survey managers wonder whether application of these approaches will result in lower data quality than would be obtained with exhaustive micro-editing. Managers who prepare data sets at the microlevel for release to the public or other analysts often must be convinced that these approaches are not harmful, especially for multivariate analysis.

The first reassuring point is that what is called microlevel analysis does not involve the inspection of individual records. Multivariate analysis brings along the estimation of parameters that always are some sort of an aggregate. The estimation of output elasticities for energy, labor, and material from, for instance, annual construction survey data turned out to differ less than one standard deviation when compared to no data editing, selective data editing, and exhaustive data editing. Only data files that are used for administrative purposes, such as tax files and social security files, need to be perfect. Nevertheless, even "perfect" files will contain some error.

Additionally all methods of data editing, including traditional exhaustive micro-editing, will leave some error unnoticed and uncorrected because not all errors are apparent. Other data fields will be changed without reason (overediting). Data editors are human, which means they can both introduce errors and not detect errors from time to time. However, data editors will make fewer errors if they have good tools to navigate through the data (macro-editing) and to distinguish between important and unimportant errors (selective editing). Because multivariate methods are often sensitive to outliers, methods that trace these outliers are a welcome contribution. Moreover correcting for measurement error has been an issue in micro-econometric and sociometric model building for several decades because survey data are known to contain error.

It is our opinion that efficient data editing is particularly valuable when editing staff are under time pressure or when resources need to be funneled to other areas, like reducing nonresponse.

In this chapter we have described various approaches to data editing. Our knowledge on these approaches is largely based on experience with specific cases. We wish to emphasize that the field lacks a general theoretical framework that incorporates these and other new ideas and that provides guidelines for practice. We hope that the rapid development of information technology will provide the means to better survey data collection and processing, including general editing software of the type envisioned here. Continuous research and development will be necessary for many years to come.

CHAPTER 12

Automated Coding of Survey Data

Howard Speizer
National Opinion Research Center

Paul Buckley
Abt Associates, Inc.

The traditional method of coding survey data, which requires coding experts to assign categories to survey response text in a review process separate from the data collection, is often expensive, slow, and subjective, and it can produce results that are difficult to validate. Consequently the survey research industry has identified the need, and made many attempts, to automate the coding process.

Coding automation has evolved over time. Systems have been developed to assist the clerical coding task (computer assisted coding), to provide coding functionality to interviewing staff (on-line coding), and to automate the coding process by assigning codes in a batch processing mode (automated coding). However, common problems such as developing adequate reference databases, text recognition techniques, and code assignment algorithms have arisen in the development of all the aforementioned systems. In this chapter we discuss the problems related to automating the coding process and review the current progress toward solutions.

12.1 SURVEY RESPONSE CODING PROBLEM

The process by which survey text data are translated into numeric or coded format is complex, yet often taken for granted. In the simple scenario the response to a survey question is anticipated and respondents, or interviewers,

Computer Assisted Survey Information Collection, Edited by Mick P. Couper, Reginald P. Baker, Jelke Bethlehem, Cynthia Z. F. Clark, Jean Martin, William L. Nicholls II, and James M. O'Reilly. ISBN 0-471-17848-9 © 1998 John Wiley & Sons, Inc.

translate a text response directly to an appropriate code by marking the correct entry.

Often a list of anticipated responses is established for a survey question (the question's "code frame"), and respondents are asked to "fit" their responses into the set of anticipated answers. When necessary, respondents are asked to modify initial answers so that the response can be classified into the code frame. This is accomplished by providing instructions to the respondent in the questionnaire, or guidance by the interviewer, to clarify the meaning of the question. The assumption is that respondent understanding will produce an anticipated answer.

There is a fair amount of cognitive processing that occurs as interviewers or respondents use auxiliary information to "fit" answers into the established code frame. However, for the most part, survey practitioners provide little if any computer support for this process. The translation of the survey text into the appropriate code is left completely to the interviewer and respondent.

Survey text translation is more complicated when an exact match is anticipated, but the code frame is very large, as when respondents are asked to provide the brand name for their cars or the name and location of their high schools. For this type of data, the survey designer is able to prepare a full list of the expected responses, but recording the correct response requires that an interviewer, respondent, or data coder identify the correct category in a long list of possible responses. There have been efforts to automate coding variables with extensive and exhaustive code frames. One such system was designed by Statistics Sweden in the 1970s (Andersson and Lyberg, 1983). This system matched survey text to a reference database of standard text phrases augmented with common misspellings and alternative phrases. A code was assigned if a direct match was identified to a record in the database; otherwise, the system passed the survey text to a manual review process for coding.

Partial phrases, or word fragments, can be used for more extensive matching capability (see Perloff et al., 1996). Most current commercial computer assisted interviewing (CAI) software packages include the functionality to directly match survey text to a database of examples for coding purposes.

The task of classifying survey data is more difficult when the verbatim text response offered by the respondent is not expected to match the established code frame. In these situations the text must be analyzed and a code assigned without the benefit of a one-to-one match existing between the text and an anticipated response. The survey question will, for example, ask the respondent to provide an answer in their own words. The challenge is to extract the meaning of the response and to classify the text into a predefined code set. Often adding to the complexity is that multiple data items might need to be considered to assign the correct code.

One common survey question that often requires code assignment is the classification of a respondent's job. Typically respondents are asked to describe in their own words the industry in which they work and the duties typically performed in their jobs. Based on these data, the respondent's job is categorized into a predefined list of industry and occupation codes.

In order to appreciate the challenge of industry and occupation coding, it is instructive to examine the question series used in the 1990 U.S. census long form. The following six open-ended questions were used to collect verbatim text data about respondents' industry and occupation (Creecy et al., 1992).

Industry

1. For whom does this person work?
2. What kind of business or industry is this?
3. Is it mainly manufacturing, wholesale trade, retail trade, or other?

Occupation

1. What kind of work is this person doing?
2. What are this person's most important activities or duties?
3. Is this person an employee of a private company, government employee, self-employed, or working without pay?

In the 1990 census, the response data to the three industry questions were used to classify the industry in which the respondent works into one of the 243 possible codes. Responses to the three questions related to occupation were used in conjunction with the industry code to classify the occupation into one of 504 possible occupation responses. Similar coding problems are faced by survey organizations around the world.

The task of correctly applying codes to verbatim industry and occupation data is not simple or trivial. Traditionally the coding task requires specially trained, expert staff and incurs significant expense in the overall survey data processing budget. The Census Bureau, for example, trained 700 expert codes and spent $7.7 million in 1980 to review and manually code the 17 million responses to the industry and occupation series asked in the 1980 census long-form questionnaire (Knaus, 1987; Appel and Hellerman, 1983).

12.1.1 Steps Involved in Coding Survey Data

Survey data translation from text to coded values involves four main processes. These processes are required in both the manual and automated environments. The steps are text recording, text translation, categorization, and process improvement learning.

As a first step toward coding text data, the verbatim text response must be recorded exactly as it is received from the respondent. This preserves the text for coding or for use with a different code frame or a different set of coding instructions at a later time.

In the second step the response text must be translated. To translate, the coder (or coding system) must recognize the words and understand the sentences in the response. Misspellings are common, and phrases that are understandable to a human are often quite difficult for a computer to comprehend. The context in which the response is given is often critical to understanding the meaning of the response, since words can have different meanings depending on the context in which the words were used. Occupations, in particular, depend very much on the industry in which the person works and education or certifications attained (e.g., without additional contextual information, "painter" applies equally to a person who paints houses or to a person who creates fine art). Finally the translation task requires the coder (or coding machine) to use the meaning of both the response text and the associated items in the code frame. We must consider both the data source and the reference database to understand the translation challenge.

In the categorization phase a match is made between the response and the appropriate category. The coder follows a set of logical rules to establish the link between the incoming text and the reference database text. The categorization process is the heart of most coding systems. There are a number of interesting components of this process, and the logical steps followed can be difficult to describe. Different coders following the same set of rules can make dissimilar assignments. It is also not always clear what the "right" answer is, and different schemes have been developed to adjudicate competing claims to this "right" answer. Perhaps the most common of these is to have a senior coder resolve the toughest cases (e.g., see Biemer and Caspar, 1994).

The last phase in the coding process is a learning phase. In the learning phase the links established between the incoming text and the codes are examined to improve the performance of the system (or the process followed by clerical coding staff). In the learning phase errors are isolated and root causes of discrepancies are determined. This information is used to "tune" the coding system (or process) and to improve the consistency (or accuracy) of the code assignments.

12.1.2   Traditional Clerical Coding Process

The process of coding survey data typically employs coding "experts" to translate and code survey responses. The coding expert analyzes the verbatim response and ancillary data and determines the best category to assign the response. The translation and categorization phases, described above, are often combined into one cognitive processing step for the coder. The text is read and translated, and based on a set of underlying rules of logic, the coder arrives at a classification for the response.

Often the underlying assignment rules are documented in a coding task manual. Typically the manual will have multiple representations and explanations for each code in the code frame. Exclusions and dependencies are listed in the manual and examples are usually provided to support code definitions.

Expert coding staff often concentrate their efforts in particular specialties. With experience, coding staff will memorize all or most of the code frame and will be able to categorize survey text without referring to a coding manual. The manual is used only for ambiguous or difficult situations.

Coding response data following a traditional manual review and assignment process is expensive and subject to error. The batch orientation of the process increases costs, for data must be extracted and then examined before the final value of the data item can be determined. The cost to examine and code each entry is expensive. In addition the costs of finding and training competent staff are significant. Training and retaining competent staff is problematic, since the manual coding process requires skilled operators for a very monotonous task, often for a long period of time.

The survey research industry has pursued automation options to reduce the cost, and impove the quality, of coding. Controlling the quality of a manual coding operation is difficult. Achieving consistency across staff is particularly challenging. Extensive efforts are required to review and compare coding results to correct errors and to ensure consistency. One of the benefits of automating the coding effort is the level of control, and therefore consistency, that an automated system provides.

12.1.3 Automated Coding and Its Place in the Survey Process

Survey response coding is a required component of data preparation activities in many survey applications. An open-end question is often the only way to collect appropriate data. This is especially true with questions for which research staff have developed a code frame to explain a complicated set of responses that have little relevance to the respondent. Industry and occupation descriptions, medical procedures, and field-of-study or degree concentration are examples of data items for which open-end questions are typically used.

A second example of data commonly collected and then coded are "other–specify" questions. An "other–specify" option is included when the response set for a question is not necessarily complete and it is anticipated that responses not fitting in the coding frame might be offered. These data must then be examined and either coded back into the predefined response categories or into newly created categories.

Open-end or other–specify coding can occur at different points in the data capture and preparation process. Computer assisted systems can be used to support interviewers to code "on-line" (as the data are collected). These computer assisted systems are also used by clerical staff to code self-administered surveys in a post data-collection process. Alternatively, and using slightly different techniques, automated coding systems are developed to code survey data in "batch" mode. Coding systems that analyze data for coding in "batch" mode are, for the purpose of this chapter, described as automated coding systems.

Computer assisted systems are typically implemented to support interviewers when the survey data are relatively simple to code. These applications include long-list and one-to-one matches — code frames that can be taught to an interviewer and are not too complicated to be administered in the course of an interview. By coding on-line, the interviewer can benefit from having the respondent available during the code assignment process.

Computer assisted systems are used to support clerical coding staff when the survey data are not in electronic form and for more complicated coding applications. It is possible, and still common, to code verbatim text without first converting the data to electronic form. Computer assisted systems assist clerical staff working from hard copy documents in searching the necessary databases and identifying the appropriate code. Computer assisted applications are also used to support clerical staff for the most difficult coding applications — those for which multiple data items must be considered to accurately code the text response. Human judgment is extremely difficult to automate, and many survey data coding applications simply provide the data items to expert coding staff who make decisions rather than incorporating the logic and rationale into the data capture system.

Automated systems attempt to assign a code to survey data without the assistance of clerical or interviewing staff. The automated system assigns a code based on the information and processing rules established by the system designer. Automated systems are designed to code a certain percentage of the data entries, examples that are problematic for the system are referred to clerical coding staff for resolution.

12.2 AUTOMATED CODING SYSTEMS

There are quite a number of coding systems currently being used in survey and market research. In this section we highlight a few of the major systems and provide a summary of their approaches and capabilities. The systems selected are unique in various aspects and taken together represent a survey of the techniques and approaches that have been used to solve the most difficult coding problems.

The U.S. Bureau of the Census is credited with the first production level experiments with automated coding systems. The initial effort was an attempt to code data collected for the 1967 Economic Census. The objective of this effort was to improve the quality and consistency of coding for industry and principal product/activity responses received from small business establishments. The original automated coding routine, developed by census staff, was called the O'Reagan algorithm (O'Reagan, 1972). This algorithm was developed and tested with data from the 1967 census and refined and tested with data from the 1970 decennial census.

The Census Bureau experimented with other automated coding routines in the 1970s (see Corbett, 1972). The algorithms that were tested relied on a set of previously coded responses as a base from which to develop a set of rules for coding new responses. The O'Reagan algorithm and the Corbett approach, although following different methods, developed association maps between phrases and codes by analyzing the probabilities that certain words inferred certain categories. The association map defined the rules for processing incoming phrases. The words in each phrase were analyzed according to the association map to determine the most likely code. The words analyzed in the survey text had to match exactly a record in the reference database in order for the system to use the word in the decision process. There was no attempt to standardize either the survey text or the database entries. (See O'Reagan, 1972; Corbett, 1972; Appel and Hellerman, 1983; Lyberg and Dean, 1990 for a more detailed description of the O'Reagan and Corbett algorithms.)

These early experiments convinced the Census Bureau of the potential advantages provided by a successful automated coding system. The O'Reagan designed system was able to achieve coding production and accuracy rates similar to clerical coders for industry classification. However, Census Bureau staff were unable to develop a production system for the decennial census using the research of the early 1970s. In late 1976 a decision was made to manually code industry and occupation responses from the 1980 census long form. This decision was based on the belief that the costs of the automated system were not appropriate for the benefits gained. This was true for many reasons, including the fact that all text entries from the 1980 census had to be keyed before they could be coded with an automated system.

12.2.1 Automated Industry and Occupation Coding System (AIOCS)

In the late 1970s research staff at the Census Bureau began developing an automated system for industry and occupation coding that could be used for the U.S. decennial census. Instead of concentrating efforts at defining a word-to-code association map, the new system attempted to replicate the steps taken by a clerical coder. The reference database for the new system was the coding manual as used by clerical coding staff. The processing strategy for the system was to translate the response text, match it to the coding manual, and make coding decisions based on the result of the match. Significant advances were made through the 1980s and the resulting system, AIOCS, was used successfully for the 1990 census.

Instead of using a sample of questionnaire responses and validated codes as the basis for the reference database, AIOCS incorporated the contents of the coding manual used by the clerical coding staff. In order to improve the system's ability to match to this database, AIOCS processed the response data through a word replacement and standardization routine. This program, for matching purposes, replaced words used in the response with synonyms and abbreviations found in the coding manual. Misspellings were corrected, the suffixes for each word were parsed, and the text was formatted to match the words used in the coding manual database.

The AIOCS system incorporated a technique developed by Knaus (1981, 1987) for identifying critical words in matching to the reference database. In the Knaus methodology a heuristic weight was calculated for each of the words in the coding manual. The weight measured the descriptive power of each word and was calculated such that the fewer number of times a word was used in the manual the higher the heuristic weight that was assigned.

After words used in the response text were "standardized," AIOCS processed each word based on its heuristic weight; processing the word from the response text with the highest weight first. The system extracted all of the entries in the database containing the highest weighted word. Each potential match was then scored on its closeness-of-fit following a phrase-scoring technique developed by Hellerman and Appell (1983).

The scoring algorithm scored each phrase based on the heuristic weight of the words in the phrase that matched the response text. The more words that matched, and the higher the heuristic weight of the words that matched, the higher was the overall score assigned to the phrase. When a database record scored above preset thresholds and significantly above other alternatives, it was selected for assignment. Otherwise, if no best-fit database records were identified, the system cycled through the other words in the response text that had lower heuristic weights until all words had been considered. Response text entries for which no acceptable database record could be matched were referred to clerical coding.

The AIOCS was developed to code industry and occupation survey responses; it was not generalizable to other applications. Expert system programming code was included in the system to assist in the industry and occupation coding task. For example, the industry code established a domain for defining possible occupation codes. This allowed context to be considered before an occupation code value was assigned. The system was also programmed to treat certain industry, employer, and occupation values separately and to directly assign a code based on specific values in these fields without going through the process of matching the incoming text to the reference database.

Assigning closeness-of-fit scores and establishing thresholds for assigning codes to response text requires trading productivity for accuracy. The higher the productivity, or the more matches recommended by the system and accepted, the greater are the chances that an assignment is inaccurate. Initially the Census Bureau developed a technique, called the "certified method," to control error rates and therefore determine the overall productivity of the automated AIOCS system. The certified method was an all-or-nothing approach. If the system was able to code a particular code category within an acceptable rate of error, the system was certified for that category. Otherwise, all response texts for which the system assigned a noncertified code were referred for clerical coding.

A slightly more effective approach was developed by census research staff and used in the 1990 census. In the revised approach, termed the "cutoff method," the closeness-of-fit score was used to determine whether or not to accept a code assignment within each category. A cutoff value was calculated for each code category based on the results of coding a known set of responses. The cutoff value was the score above which the code had a predetermined chance of being correct. Using a set of responses for which validated codes had been assigned, census staff showed that this certification scheme yielded a 10 percent productivity gain without adding appreciably to the error rates (Chen, Creecy, and Appel, 1993).

The AIOCS system, using the cutoff method for error control, was employed in the 1990 census which collected industry and occupation data from slightly more than 22 million respondents. Data were entered into electronic format and coded first by the AIOCS system. Responses for which the AIOCS system could not assign both the industry and occupation code were referred to an expert coding staff who used an interactive coding system to determine the final classification. The AIOCS system was able to code about 50 percent of the data with error rates averaging about 10 percent. (Productivity for industry codes was significantly higher than for occupation codes and error rates were lower; see Mersch, Gbur, and Russell, 1992.)

12.2.2 Parallel Automated Coding Expert (PACE)

A very different approach to the industry and occupation classification problem was published by census research staff in Creecy et al. (1992). The authors developed a coding system that significantly outperformed the AIOCS system. The PACE system used "data parallel" computation techniques and was implemented on a massively parallel supercomputer. PACE eschewed the expert system approach taken by AIOCS for a "brute-force" type strategy which took advantage of the extensive computation resources to identify text matches in a very large database.

The PACE system matched text data to a database of prior-coded and adjudicated response data. For testing purposes PACE relied on a sample of 132,247 industry and occupation responses from the 1980 census which were triply coded and adjudicated by coding experts. These data were coded by census staff in 1986, originally to test the AIOCS system, and they provided a "truth" set of text response and code pairs. This database became the training database for the PACE system. The PACE system matched new response text to the entire training database and assigned a code based on the codes (already validated) of the "nearest neighbors" in the database.

The PACE system created "features" which consisted of all two-word combinations of the words used in each text entry of the training database. The words were linked to the question type so that a word used in the response to an industry question was considered separately from the same word used in the occupation series. Using the training database as the truth set, the system calculated conditional probabilities that the two-word combinations (features) in each of the text entries predicted a code. The effect of the conditional probabilities is similar to the heuristic weights developed by Knaus (1987); features with low relative occurrence received higher conditional probabilities.

The PACE system used an application of Memory-Based Reasoning (MBR) to identify nearest neighbors in the training database. PACE considered, in parallel, all of the words provided in the response to the set of six industry and occupation questions. The features in the input stream were compared to all of the features in the training database and by considering conditional probabilities, the system produced a set of possible near neighbors. The system then sorted through the possible matches and based on a scoring scheme designed to maximize productivity and minimize the number of erroneous assignments, determined whether or not a suitable match had been found.

The conjunction of words into two-word features for matching proved to be significantly more predictive of the correct code than the use of single-word features. Using this scheme, the system factored the interaction between the industry and occupation entries when calculating the conditional probabilities of code assignment. The PACE system was therefore able to incorporate this interaction into the knowledge base and code response text in context. It is important to emphasize that the PACE system had very strenuous computation requirements. Sixty-five thousand training examples produced over 4.5 million features. Calculating their conditional probabilities and matching to each example was a complex and computationally intense task.

The PACE system proved to be a significantly better coding engine than the AIOCS system. Using a similar test set, and coding at an accuracy rate equivalent to the AIOCS system (and to the accuracy level of clerical coders), the PACE system coded 63 percent of the industry classifications and 57 percent of the occupation codes. This was a significant improvement over the AIOCS system which coded 57 percent of the industry codes but only 37 percent of the occupation entries.

12.2.3   Automated Coding by Text Recognition (ACTR)

Statistics Canada began developing the ACTR (Automated Coding by Text Recognition) system in the mid-1980s. The system is not application dependent, and has been used successfully by the agency to process and code many different types of text survey variables (see Hale, 1988; Rowland and Kinack, 1994; Tourigny and Moloney, 1992).

ACTR is a text coder using near-matches identity techniques similar to those used by the AIOCS system. Using the ACTR system, survey staff define a database for matching purposes which can include precoded and verified survey responses as well as coding manual entries. A sophisticated parsing algorithm, whose many features can be controlled by the user, is applied to standardize the entries in the database. The same parsing routines are applied to the incoming text before being matched to the reference file.

The ACTR system attempts to directly match incoming response data, which have been standardized to the records in the database of examples. A Compressed Key (CPK) is calculated and used for each database entry to reduce the overall storage requirements and quicken the direct matching procedure. The CPK is calculated for the response text and the system rules out a direct match before moving to the indirect match phase.

The indirect matching scheme used in ACTR is a simplified version of the routine developed for the AIOCS system. "Heuristic" weights are calculated for each unique word in the reference database and scale the predictive power of the word as compared to other entries in the database. A measure of closeness-of-fit is generated for each database entry determined by the number of matches to the input text and the weight of the words that match. Thresholds are established that provide guidance to the system for when to accept an entry, when to present multiple options to the user, and when to refer the case for clerical coding.

The ACTR system allows the user to open different databases for coding within the same application. These are described as different "contexts" and allow the user to code multiple variables at the same time. However, the system currently does not support coding fields that depend on ancillary information. The user must supply the programming logic to code-related fields such as industry and occupation.

The ACTR system was redesigned and updated in 1996 (Wenzowski, 1996). The updated system is more portable across computing platforms than its predecessor; it provides upgraded tools to support the development and maintenance of its reference databases, as well as an application programming interface (API) to call the system from within another program. Statistics Canada has plans to market ACTR for general distribution.

12.2.4 QUID/SICORE

The French National Statistics and Economic Studies Institute (INSEE) developed a powerful survey data coding system in the mid-1980s. The system was upgraded in the mid-1990s and is currently being used by the French agency and other, mostly European, clients. The original system was called QUID (see Lorigny, 1988), and the similar updated system was renamed SICORE (see Schuhl, 1996).

The coding database for the French system is developed from prior-coded data. System performance is dependent on the quantity and quality of the examples in the database. The SICORE coding system is designed to code any survey text variable, and it is not optimized for a particular application. The QUID system runs on an IBM mainframe; the SICORE system is written in C and will run on a variety of platforms.

The SICORE system prepares the reference database by normalizing the data and then splitting each text record into multiple letter combinations. Most applications use two-letter combinations — or bigrams. When bigrams are used, each reference database example is represented as the set of two-letter groupings; this makes up the text of the phrase. The record is represented in the database as a subset of the bigrams which uniquely define the phrase. This technique takes advantage of a binary-tree representation for each record and results in a significant reduction in storage space for the reference file while increasing the speed by which the software can match incoming text to the reference database.

The French system handles dependent variable coding by incorporating a logical rule processing component in the software. The system considers additional survey variables and applies a rule-based decision depending on the results of the coding process and the value of the auxiliary variables. The rules are established and maintained in a decision table structured as a series of if–then statements.

The redesign and development of SICORE was undertaken to provide a more generalizable, and robust, system for coding. The updated system can run on multiple computing platforms, and it has improved ability to code in context. The new system provides more structure to the reference database and more tools for establishing and maintaining the different inputs to the system.

12.2.5   NCES System

The U.S. Department of Education's, National Center for Education Statistics (NCES) developed for use in agency-sponsored surveys a coding system that has some unique capabilities. The NCES system is a computer assisted on-line coding system designed strictly for CAI applications. In this environment the overriding objective is to expeditiously provide a reasonable set of options for an interviewer to use in making a code selection. It is less important for the software to select and assign a code (see Bobbitt and Carroll, 1993).

The NCES on-line coding system processes the input text, as typed by the interviewer, and provides a set of best code options to the interviewer. The system attempts to match each of the words in the response text to a reference database. The interviewer is asked to select alternative words, for matching purposes, when the word supplied by the respondent and typed by the interviewer cannot be located in the database. The system chooses possible alternative words by identifying words in the database with a similar root structure, or with similar initial syllables. In this way the coding system forces the interviewer to correct abbreviated or misspelled words and uses only interviewer sanctioned words in the matching process.

The coding system uses a very simple matching approach for determining the classification codes. All codes that contain the words in their definition, or database entry, are considered. The codes are ordered by the number of matches contained in their definition and presented to the interviewer in that order for selection. The interviewer chooses the best code, and the response text and selected code are stored in the database. The advantage of the on-line system is that the interviewer can clarify ambiguous situations directly with the respondent when determining the best code. The algorithm for identifying possible matches is kept simple, and other design steps are taken to ensure that the system can quickly provide a selection of codes to the interviewers to facilitate the flow of the interview.

The on-line coding system tracks the words processed from the response text that could not be matched to the reference database. These words are added to the database to supply new definitions and provide more productive matching capability for later examples.

12.3 AUTOMATED CODING SYSTEM FEATURES AND COMMON ISSUES

12.3.1 Expert Systems

Software developers have taken two distinct approaches to developing automated coding systems. In expert systems, as defined here, software developers attempt to create a system that emulates an expert clerical coder. These automated coding systems attempt to be natural-language interpreters, and they arrive at categorization decisions by translating and understanding the input stream. Expert systems are typically focused on a particular coding application; they manipulate and standardize the input and reference database text, and they rely on decision rules for coding variables that are context sensitive. Improvements in expert systems are dependent on review by expert coding staff of uncodeable response text and false positive assignments. Expert systems are very common, and they are represented by the AIOCS, ACTR, QUID/SICORE, among other examples (see Dalton and Keogh, 1996; Eilas, 1996; Stoudt, 1996).

The reference database is a critical component of any coding system. Designers of expert systems typically rely on expert coders to build the reference database, often using the coding manual as a base. Significant time is committed to examining each entry and continually modifying examples in the database to improve the system's performance. The task of matching to the reference database in expert systems is complicated because respondents do not use terms and phrases commonly represented in the manual. Even when the manual has a good set of example phrases that can be included in the database, problems such as mismatched tense, misspelled words, and slang or jargon-filled phrases cause difficult translation and matching problems. The solution most often used by expert systems is to parse and standardize both the incoming text and the data in the reference database to increase the likelihood of matching the two sources.

The ACTR system employs a powerful parsing algorithm that is illustrative of the approach taken by many of the expert systems. In a character-oriented parsing phase, the system trims extraneous characters, replaces abbreviations, deletes unproductive characters, and breaks the character string into a series of words. In the word-oriented parsing phase, the system applies a number of different rules to standardize the incoming words and prepare the phrase for matching. These routines include, among other steps, standardizing the treatment of hyphenated words and multiple word occurrences, removing trivial words, replacing words with more common or consistent synonyms, removing prefixes, and removing suffixes. In a final step duplicate words are deleted and the remaining words sorted (see Wenzowski, 1996). These steps are applied to the records in the reference database and improve the chance of a direct match between the input text and the database.

Parsing and standardizing text for matching purposes can introduce problems and thus must be applied carefully. The risk of parsing English text is that subtle changes to the meaning of the text can be affected with seemingly inconsequential changes to the input stream. The effect of reducing words to their roots, for example, is that more words are considered relevant in matching to the input response text. For example, reducing the word "heating" to the root "heat" effectively adds phrases which include the words: "heats," "heated," and "heater" to the number of database examples that must be considered.

In tests measuring effective retrieval of data from library on-line catalogs, "strong" parsers (parsers that reduce all suffixes) are shown to produce significantly more false-positive matches than "weak" parsers (parsers that reduce only a limited number of suffixes) (see Harmon, 1991; Walker and Jones, 1987). An example is supplied by the authors of the PACE system (Creecy et al., 1992) who demonstrate that the word "attorneys" implies a different occupation than when the root word "attorney" is used. By stripping the "s" from the input text of this word, a coding system would inadvertently change the meaning of the phrase. This example illustrates the parsing and word standardization issues that are important for consideration in any text-matching system.

Coding in context is perhaps the most difficult challenge for automated coding systems. To be successful, the system must take into account multiple inputs to assign accurate codes. In expert systems the rules that govern the relationships among related variables are developed as additional decision rules considered in conjunction with the variable being coded. The AIOCS system, for example, uses the assigned industry code to control the domain of possible occupation codes. The SICORE system builds a table of if–then decision rules and executes it during the coding process in order to factor all inputs in the decision process. In each case, for expert systems, the designers must employ the knowledge of coding experts to build and maintain the decision mechanisms, and each system must be tailored to the coding application.

12.3.2 Nonexpert System Approaches

The PACE system, and others not described here (see, e.g., Lawrence, 1992), approach the coding problem from a different perspective. These nonexpert coding systems do not rely on specific application knowledge in the development of their underlying databases or in the processing of the input text. Instead, the objective is simply to find the closest example in the reference database that can be used as a proxy to determine the best code. These systems are "taught" the rules for coding by a training database of examples that includes validated links between the text and the codes. To achieve this objective, the training set must include all of the rules that the system will need to code the data. In other words, the system requires a complete and accurate database of possible outcomes. More comprehensive training sets impart more knowledge, raising the productivity of these systems.

Generating and validating a representative training database requires substantial time and resources. The data for this must come from a comparable population of respondents who are asked exactly the same survey questions. Examples from a different population or data from even slightly different questions may not prepare the system properly. This requires either coding a subset of the data from the same cohort to develop a training database or using survey text and classification codes that have been validated from prior rounds of the survey. Calculating the appropriate database size is an issue of concern. Ultimately size depends on the number of codes and the dispersion of answers among these codes. New database examples must be added as they are discovered in the production data.

Using a validated set of observations generated from the 1980 census, the PACE system was able to outperform the AIOCS system, coding significantly more observations at the same error level. The PACE system required 4 person-months to develop, compared to a period of 192 person-months required by the AIOCS system (Creecy et al., 1992). The intelligence of the PACE system, and others like it, is in its ability to pattern-match text data and to distinguish "good" matches from false-positive matches. Once the kernel of this methodology has been developed, these systems can support different coding applications without the setup and maintenance costs associated with expert systems. However, it is important not to trivialize the work involved in developing and implementing the pattern-matching algorithms inherent in these systems. Also, in the case of the PACE system, the authors did not factor into their cost comparison the significant cost of the hardware that was required to run the system.

12.3.3   Data Storage and Access Methodology

The reference databases used for automated coding systems are large, and the methods used for storage and access of these data have a significant effect on performance. Each of the coding systems we reviewed uses data compression and storage techniques designed to reduce computer resources and accelerate access times.

The most common approach is to use compression and data reduction methods to reduce the size of the database. The AIOCS and ACTR systems, for example, compress database entries by reducing the number of characters required to represent each word or phrase. Much of this savings stems from the parsing strategy followed by each system. Space is further saved in the ACTR system by defining an expanded alphabet that can be represented in a byte (eight bits) of data. Common double- or triple-letter combinations are defined and added to the set of possible characters represented in the eight-bit code. Using this technique, the ACTR system can reduce the entries by an estimated 50 percent (Wenzowski, 1996).

The INSEE systems define a new word representation model with its use of bigrams and trigrams. Each bigram, for example, is assigned a unique binary number that reduces by half the number of bytes that are needed to represent a word. The size of the reference database is significantly reduced beyond the smaller size needed to represent the bigrams by the use of a binary tree to represent each phrase. Since well-defined algorithms are used to determine the smallest number of bigrams needed to represent each word or phrase, only a subset of the bigrams in a phrase are needed and stored for each entry (see Schuhl, 1996; Lorigny, 1988 for a discussion of the use of the Shannon Entropy to define the binary tree). In thus reducing each entry to its shortest representation, the INSEE systems can dramatically improve the access time to each database example.

12.3.4 Management Systems

Automated coding systems are becoming integrated, and further efforts are being made to develop more complete systems. Initially automated and computer assisted coding systems were developed and implemented in pieces. The software for matching text entries would be implemented with one set of programs and tools, while other components, like database development routines, were implemented and managed separately.

As the systems have matured, software developers have made significant advances in producing complete coding-system packages. Tools are being incorporated into coding systems for managing the entire process from developing the reference database through the measuring of production and accuracy to improve performance. As the interfaces for systems such as ACTR and SICORE are improved, the automated coding process can be managed by survey professionals rather than data processing staff. Many of the coding systems developed in the late 1980s and early 1990s have been or are being rewritten in portable software so that they can run on multiple computing platforms.

At the time this chapter is being written, very few of the automated systems have been commercially marketed. Developers continue, however, to express an interest in marketing their systems and we anticipate that the trend toward complete coding system packages will progress rapidly.

12.3.5   Interactive On-line Coding

The interactive CAI system developed by the Department of Education demonstrates that on-line coding is technically feasible. The NCES system has been employed in multiple CAI applications and is used to code industry/ occupation, field-of-study, and other long-list variables such as school name. Although we have highlighted the Department of Education's system, there are other examples of on-line CAI coding systems in practice (see Perloff et al., 1996; Legum, 1996).

On-line CAI coding moves the coding process to the interview. The on-line systems that we have reviewed, including the NCES system, change the task from determining a best code to providing the interviewer with options based on the information supplied by the respondent. The interviewer can review these options with the respondent and together they determine the best code. These systems follow the traditional automated coding system approach. The survey data in free-form supplied by the respondent are matched and analyzed to determine the coding options.

There is a tendency for interviewers to learn the limitations and methods of an on-line coding system and adapt their behavior accordingly. One concern is that on-line CAI coding systems mask errors by delegating the text-recording operation to the interviewers. In the case of the NCES system, for example, where certain words are ignored in the matching process and other words are replaced, it is natural for interviewers to want to modify the text supplied by the respondent in order to find the fastest route to a successful code and to reduce the burden on the respondent. This reduces quality, since the complete response is not considered in the decision process and the actual text supplied by the respondent is not saved.

If the interviewer records the text exactly as provided by the respondent, then a postprocess review can be conducted to determine error rates and to improve interviewer performance. For error rates that are high, as they might be for specialty coding performed by interviewers with limited experience and training as coders, a recoding effort might be required and the advantages of on-line coding would be reduced. The concern is that all of the relevant data are not recorded, reducing the observed error rate and eliminating the opportunity to check and correct entries.

12.3.6   Error Measurement and Standards Development

Coding quality is traditionally measured in terms of reliability (the probability that two independent coders will assign the same code) and validity (the accuracy of the code assigned). Coder reliability depends on a number of factors, including the amount and quality of coder training, the complexity of the code frame, and the quality of the input data. Computer assisted coding has the potential to increase reliability by providing a more focused and consistent set of information for coders to consider when assigning codes (Campanelli et al., 1997).

The consistency provided by computer asssisted coding systems can have the unintended effect of increasing correlated coder variability (Bushnell, 1996). Introducing "system bias" into the estimates is a risk inherent in using a system that forces a standard set of behaviors. For example, coders who blindly select a code suggested (perhaps because it is displayed at the top of a list) by a computer assisted system can introduce an overall bias into the results. This is especially the case in industry and occupation coding systems that, because of the complexity of the task, might suggest three or four very different codes to a coder.

In order to measure the validity (or accuracy) of an automated or computer assisted coding system, a precoded data set that establishes a correct code for each entry in the evaluation is needed. Establishing this baseline usually involves clerically coding a set of response text examples with multiple expert coders and following a discrepancy resolution process to identify a "true" code for each example. It is important also to measure the accuracy of the clerical coding effort for use as a point of comparison with the automated system.

The fundamental performance measures used to compare automated coding systems are production and accuracy. The production rate is the percent of incoming examples to which the system is able to assign a code. The accuracy rate is the percent of these entries that are coded correctly. Most automated coding systems contain decision parameters that can be manipulated to trade production for accuracy. For determining the benefit of an automated solution, it is important to determine the production rate that can be achieved by a coding system while maintaining an accuracy rate equivalent to that of clerical coders coding the same data.

One of the most difficult aspects of reviewing automated coding systems is comparing production and accuracy rates. Each of the performance measures quoted by system developers is subject to data, application, and environmental effects which are difficult to interpret and control. The critical factors that affect production and accuracy rates are the complexity of the code frame, the quality and quantity of response data, and the makeup of the sample members from whom the data are collected. It is impossible and irresponsible to compare rates between systems unless all of these factors are distinctly similar. As an example, the language (words, phrases, etc.) used by a younger population is not equivalent to that used by a more elderly sample. As such it is likely that production and accuracy rates for these groups will be different even if the survey question(s) and code frame are alike.

Standardizing performance measures has not been achieved as of the writing of this chapter. Different measures are provided by the developers of each system and very little data that can be used for comparing systems are available.

12.4   NEXT-GENERATION TECHNIQUES

Since the dawn of the computer age, one of the standard tests of artificial intelligence has been the ability of a computer to play chess. In a fundamental way the cognitive work involved in playing chess is very much akin to the cognitive work involved in coding. The chess grandmaster (whether human or electronic) has to assimilate information and decide on the right next move — analyzing the board and choosing one of a closed set of possible moves. The coder similarly assimilates a complex pattern of information and chooses one of a closed set of possible codes.

In chess, all possible information is immediately available — the moves allowed each player are well defined at any point in time. The player can construct a decision tree consisting of each combination of possible future moves and countermoves and evaluate each node in that tree. Players — both human and electronic — construct and traverse this tree, avoiding paths that weaken their position or lead directly to a loss and seeking out those that lead to a win.

For human players the ultimate size of the resulting tree is daunting. The number of potential combinations become so huge that "pruning" is necessary. Some paths through the tree are identified as promising and are examined in detail; others are discarded as unlikely to lead to a win. Various pruning methods were developed for early computer chess-playing programs.

While elegant programs became increasingly sophisticated, the continued development of computer hardware steadily undermined their *raison d'être*. As the number of moves that could be evaluated per second increased, the need to prune the decision tree decreased. As a result chess-playing programs became less elegant. Today nearly all the advantages in a chess-playing program comes not from its ability to "look deeper into the board" along promising paths but to look broadly in the decision tree, exploring more paths than previous programs and certainly more than any human player could manage.

The resulting "brute force" computer chess players follow the pattern of previous efforts to automate human tasks. Machine looms (the hallmark of the Industrial Revolution) do not resemble weavers. Typewriters do not look or behave like human scribes. Automobiles no longer look like horseless carriages. Mechanical dishwashers bear no resemblance to human dishwashers. In each case the evolution of the automated process depended on the characteristics of the machine — optimizing the work to take advantage of those characteristics — rather than mimicking the human process that it replaced. It is no surprise that the resemblance between computer and human chess players has diminished. We should expect the same evolution in computer-based coding programs.

Compared to chess players, coders start with a poorly defined problem. Nevertheless, the strategy followed is similar — successively narrowing the list of possible codes by examining the information at hand and evaluating which of a number of alternatives best fit. Frequently the available information is inadequate to identify one code as "best." Even when coding is done on-line and the respondent can immediately provide additional information, neither the interviewer nor the respondent may know exactly what additional data are critical to determining the correct code.

For many code frames there is an underlying structure, which can be described as a decision tree. Jobs, for example, can be classified first into relatively large categories (e.g., service) and then into successively smaller categories until a unique occupation code is selected. A skilled coder uses this structure to quickly eliminate all but a handful of codes (pruning the tree) and choose from among the remaining alternatives.

To date, most coding program developers have acted like the chess program developers of 20 years ago. Except for PACE, coding programs have also been designed as if the computer storage and processing demands for a complete solution are beyond reach. Moreover the problem has been defined in terms of duplicating the operations of a skilled human coder. Simply stated, coding programs attempted to use whatever information the respondent provided, with limited prompting, to identify the best code. As in chess, however, the ways in which computers differ from humans allows for a totally different approach to the problem.

If the respondent is available to participate in the task, coding can be conceived as no more than proper questionnaire design. Rather than asking for a description and then attempting to "decode" that text and encode it into the coding frame, the respondent can be asked for each of the pieces of relevant information.

The underlying decision tree for a coding frame can be expressed as a hierarchically ordered series of questions with the answers to each question determining which question to ask next — coding can be done as a game of "20 Questions." Using this approach, a set of simple questions can be asked to uniquely identify each code. If only yes/no questions are used, the total number needed to construct the decision tree will be equal to the number of codes minus one. Thus, for a code frame of 1000 codes, 999 yes/no questions would be needed. In use, however, only eight need to be answered by a respondent to reach any code. If questions with more than two possible responses are used (e.g., when an industry code is used in determining an occupation code), the total number of questions needed will be smaller and the length of each respondent's path will be shorter.

Using current CAI programs to accomplish this task is possible but unreasonable. Questionnaires are assumed to be composed of questions that are asked of all or most respondents. They are not designed to handle a very large number of questions, each of which is asked of only a few respondents. It is of course obvious that the work involved in writing, reviewing, and accurately generating the questions needed is substantial, but the benefits are similarly substantial. Being a good coder comes down to knowing which questions to ask and when to ask them. With a well-designed set of questions, and the approach to automation outlined above, an ordinary, well-trained interviewer can do the job.

ACKNOWLEDGMENTS

The authors are grateful to system developers who shared many "secrets" with the authors. Special thanks to Dan Gillman who was a patient and helpful advisor through the entire process and to Bob Bailey who helped code the information we collected.

CHAPTER 13

Automated Management of Survey Data: An Overview

William E. Connett
University of Michigan

13.1 INTRODUCTION

Two primary functions characterize automated interviewing systems. The first is the collection of interview data, and the second the management of the interviewing system and the collected data. At the risk of oversimplifying, the management system can be thought of as the glue between the system user and the survey software. Typical functions of case management software are providing the interviewer with sample cases, feeding the appropriate information to the interviewing system to set up a particular interview, and accepting case-related information back from the survey software upon completion of an interview. There are many variations on this theme that add astonishing complexity to the complete CASIC case management system. As the definition of the data collection system expands to include postcollection processing, development of metadata, and automated dissemination of survey results over the Internet, the boundaries of the case management system grow fuzzier by the moment.

The next few chapters explore many facets of CASIC case management but generally limit the definition of case management to those functions closely tied to the data collection process. In particular, this chapter provides an overview of case managment design issues and discusses some specifics of case management from the central office or "server side" of the case management system. Other chapters discuss specific aspects of case management in more detail. Hofman and Gray (Chapter 14) discuss features and issues of case management systems from the perspective of the client or interviewing side of the system.

Computer Assisted Survey Information Collection, Edited by Mick P. Couper, Reginald P. Baker, Jelke Bethlehem, Cynthia Z. F. Clark, Jean Martin, William L. Nicholls II, and James M. O'Reilly.
ISBN 0-471-17848-9 © 1998 John Wiley & Sons, Inc.

Smith, Rhoads, and Sherpherd (Chapter 16) detail the intricacies of telecommunications that are necessary to connect distributed elements of the case management system, and they provide thoughts on future possibilities for remote data communications. Edwards, Suresh, and Weeks (Chapter 15) share the results of an extensive survey of survey organizations regarding the use of CATI call scheduling and discuss possible reasons for, and implications of, current automated call scheduling practices.

13.2   GENERAL FUNCTIONS OF THE CASE MANAGEMENT SYSTEM

Case management systems originally grew out of enhancements to the early computer assisted telephone interviewing (CATI) systems. Nicholls and Groves (1986) provided a list of capabilities of CATI systems. These included sample management, on-line call scheduling and case management, on-line interviewing, on-line monitoring, automatic record keeping, and preparation of data sets. Edwards, Suresh, and Weeks (Chapter 15) provide a list of the "more important features" of typical CATI call scheduling systems. They differentiate between call scheduling and case management systems, indicating that call scheduling functions are a component of the broader CATI case management system.

In addition to call scheduling, modern CATI case management systems typically include:

- sampling routines providing for the use of random digit dialing (RDD) or alternatively the use of a list sample;
- call note features that allow interviewers to make notes about the case and a particular call;
- automated call history that keeps track of times and events related to the case;
- maintenance and reporting of call outcome data;
- management of interviewing staff log-ins, assignment to studies and teams;
- monitoring routines that allow display of interviewer screens on supervisor's machines and in some cases provide the ability for the supervisor to intervene;
- case transfer mechanisms to move partial cases from one interviewer to another; and
- ability to manage groups of interviewers and categories of calls.

The first CATI systems were developed on dedicated minicomputers or mainframes. More recent systems are based on networked microcomputers.

Regardless of the host architecture, CATI systems generally reside in one building or are linked by a network to a central database or set of files containing all of the case information for one or more studies. This architecture makes it relatively simple to manipulate CATI case information for case delivery, management, scheduling, and reporting purposes. The inherently centralized design of CATI systems allows the implementation of management utilities using common database approaches. As a result most CATI systems provide either a suite of turnkey management facilities or a set of tools that allow users to build their own case management utilities.

The advent of CAPI in the late 1980s created new demands on the case management system. Nicholls and Kindel (1993) listed some of the functions of CAPI case management. Hofman and Gray (Chapter 14) discuss these functions in detail. Features not mentioned by Nicholls and Kindel (1993) include the collection of ancillary data such as interviewer time and expense reporting, respondent address data, and other noninterview case-related data. In some situations the case management system is required to initiate special routines such as random sampling within a subset of sample households or respondents.

Case management is a critical function for efficient operation of an interviewing system. The lack of a good case management system can seriously affect the cost of a study, as can a case management system gone awry or misused. One of the problems with modern case management systems is that the high level of complexity may lead to the situation where only one or two people in an organization fully understand the details of system setup and operation. If the case management system fails, the result can vary from inconvenient and expensive to catastrophic (Connett, 1996). It is likely that every survey organization doing CAI surveying has experienced failure, to some degree, in their case management system and in a few instances the failures have resulted in significant data loss. Organizations are generally not particularly open about such mishaps.

In summary, the CASIC case management system is the engine that runs the automated interviewing process and all of its affiliated activities. Important measures of quality of a CASIC interviewing system should be the breadth of function and the ease of use of its management features. Modern systems development tools allow the development of CASIC management systems that can be operated by nonprogrammers to meet the management requirements of the great majority of surveys. There is no longer an excuse for case management systems that are difficult to use and require constant programmer intervention for setup and maintenance.

Figure 13.1 depicts typical data management and data flow functions of a modern CAPI case management system. As is evident from the diagram, there is a division of function between case management on the remote machine (laptop or palmtop computer) and the host machine (server). See Hofman and Gray (Chapter 14) for more detail on the client side of case management.

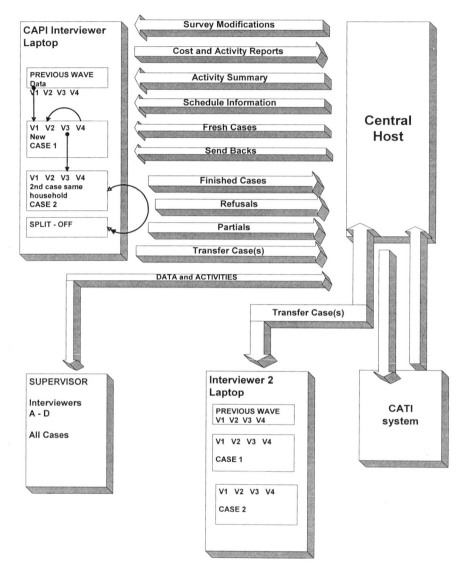

Figure 13.1. CAI data management.

13.3   COMPLICATIONS OF CASE MANAGEMENT

13.3.1   Mixed Mode Studies

Many survey organizations face the requirement of conducting "mixed mode" studies. These studies employ at least two different types of data collection from a list including mail, CATI, CAPI, distributed CATI, FAX, and other modes.

In these situations it is advantageous to have a single case management system operating across all of the data collection modes. The failure to include linkages between modes in the case management system can result in the inability to effectively manage mixed mode studies and can present serious cost impediments for the conduct of such studies. A basic mixed mode case management system might include the following additional capabilities:

- move CAPI cases to the CATI system,
- move CATI cases to the CAPI system,
- move direct data entry from paper into either CATI or CAPI modes,
- report sample statistics on each mode and on the combined modes, and
- allow the input of case disposition data for paper and pencil surveys.

The need to move cases between modes can occur in a variety of ways: Fresh cases may be initially misassigned to the wrong mode and need to be moved before interviewing begins; partially completed cases may be transferred from one mode to the other for completion, for example, a difficult CATI case may need to be transferred to a CAPI interviewer for completion; or it may be considered more cost effective to transfer end-of-study cases to CATI for long-term follow-up.

By 1997 a few major survey organizations had developed case management systems capable of handling some degree of mixed mode case transfers. In the future the ability to incorporate multiple survey modes and to move interview data seamlessly between modes will differentiate the best case management systems from those of limited usefulness. The increased mix of data collection modes including self-interviewing responses via the Internet, CAPI, distributed CATI, centralized CATI, paper and pencil, TDE, and other modes will demand that large survey organizations have case management systems capable of distributing surveys across all modes, moving partially completed surveys between modes and collecting data to a central point from all modes. The expected increase in the use of graphical user interfaces (GUI), audio, video, and other multimedia objects as part of the interviewing system will increase the complexity of managing the mixed mode studies of the future.

13.3.2 Multiple CATI Sites

Some large government and market research organizations have multiple telephone facilities in widespread geographic locations. In some instances these "regional" centers do self-contained regional surveys. However, in other instances a single survey is spread among multiple facilities. Obviously this complicates CATI case management. This complication is handled in a variety of ways including splitting the sample among the centers, on-line access to a central database of cases, or retrieval of cases and daily schedules from a central host with return of the information to the host at the end of the day.

13.3.3 Distributed CATI

The advent of microcomputers has led to the development of distributed CATI in which interviewers use telephones in their homes and enter the respondent's data on microcomputers. Case management issues related to distributed CATI are essentially the same as with CAPI, assuming that the interviewers are working off-line and not constantly connected to a central host computer.

The issue of call scheduling for distributed CATI or for CAPI has generally not been addressed by survey organizations or by vendors. With distributed CATI or with CAPI, call scheduling is related to a specific interviewer rather than a group of interviewers. With CAPI, scheduling is further complicated by the interaction of scheduling with geographic location, traffic patterns, and typically more flexible schedules of remote interviewers. Optimization of scheduling under these complicated conditions may be beyond the realm of current scheduling tools. It is likely that organizations will soon provide electronic tools to assist individual interviewers with their scheduling, logging, and reporting needs. Some organizations already provide individual time and expense logging and reporting tools as part of their case management systems.

13.4 CASE MANAGEMENT GENERAL DESIGN ISSUES

13.4.1 Users and User Requirements

The complexity of case management systems stems from the diversity of users who come to the survey process with different needs and requirements. Figure 13.2 depicts some of the users of a typical case management system and briefly lists some user requirements.

Connett, Mockovak, and Uglow (1994) have described interviewing and case management requirements from the user's viewpoint. Users range from the study's principal investigator(s) to the interviewers, and each is an important user of the system. While difficult to do, a well-designed case management system will meet the needs of each type of user in a way that reinforces the feeling that the system was designed specifically for them.

The development of modern computing systems requires continual input from the users during the development process. The use of a method for collecting user input, such as the USE CASES method described by Jacobson (1992) is highly recommended. The USE CASES method allows users to describe their requirements and specifications in normal conversational terms, focusing on what they wish to accomplish with the system. A USE CASES can be developed for each type of user of a system. This collection of USE CASES then guides the designer in the development of the system to ensure that all users' needs are met. Failure to consider the entire set of users' needs during the design stage can result in weakly developed software. Subsequent patches to the software enhance the weaknesses until, over time, the entire system fails prematurely. The use of a formal system for gathering user needs is important

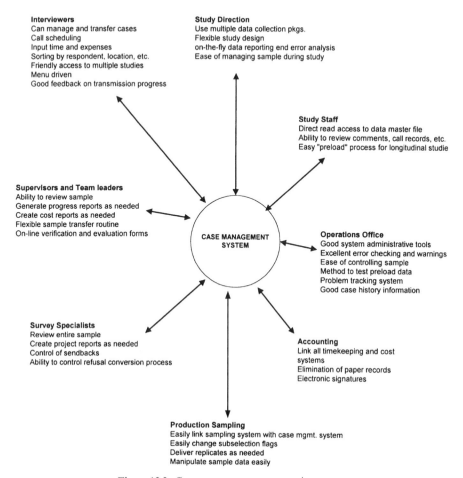

Interviewers
Can manage and transfer cases
Call scheduling
Input time and expenses
Sorting by respondent, location, etc.
Friendly access to multiple studies
Menu driven
Good feedback on transmission progress

Study Direction
Use multiple data collection pkgs.
Flexible study design
on-the-fly data reporting end error analysis
Ease of managing sample during study

Study Staff
Direct read access to data master file
Ability to review comments, call records, etc.
Easy "preload" process for longitudinal studie

Supervisors and Team leaders
Ability to review sample
Generate progress reports as needed
Create cost reports as needed
Flexible sample transfer routine
On-line verification and evaluation forms

CASE MANAGEMENT SYSTEM

Operations Office
Good system administrative tools
Excellent error checking and warnings
Ease of controlling sample
Method to test preload data
Problem tracking system
Good case history information

Survey Specialists
Review entire sample
Create project reports as needed
Control of sendbacks
Ability to control refusal conversion process

Accounting
Link all timekeeping and cost systems
Elimination of paper records
Electronic signatures

Production Sampling
Easily link sampling system with case mgmt. system
Easily change subselection flags
Deliver replicates as needed
Manipulate sample data easily

Figure 13.2. Case management user requirements.

in the development of software and particularly so in the development of case management software because of the diversity of users.

13.4.2 Organizational Control Issues

The following control issues are of great importance to users' needs: (1) issues relating to the need to control data loss or confusion including duplicate interviews, overwriting of cases, or the inadvertent changing of valid data; (2) control issues of a different type, regarding who controls what aspects of the case management system. This second category may represent valid system concerns important to the safeguarding of data, or it may have little to do with data and instead be based on past practices. This is common where one or more groups have thought of themselves as the gatekeepers of survey data and practices and may wish to perpetuate that position through the control of the case management system.

With changes in modern organizational structure tending toward empowerment and increased responsibility at lower levels of the organization, the modern case management system is more likely to be developed with the vision that each user group is of equal value to the organization with equally valid needs and priorities. The control to prevent accidents and manage studies is more likely to belong to the system itself than to one particular user group. Another way of expressing this is to say that the modern case management system is an open system with each user group having access to all of the information needed by that group without having to obtain permission from another user group.

For instance, where survey analysts may once have gone through an operations office to get current reports of case disposition, the modern system will allow the analyst direct access to the data. This in turn gives the analyst the ability to create his or her own progress reports, while continuing to protect the integrity of the data from accidental or intentional modification. Other groups, such as the operations office, can use the same data to create official standardized reports for study management and distribution, while a third user, the study manager, can conduct critical checks on data reasonableness employing user friendly reporting tools as soon as the data arrive on the central system.

13.4.3 Idiosyncratic Design

Over the last few years a belief has emerged that CASIC case management systems tend to be idiosyncratic to the organizations in which they reside, and that case management systems tend to closely reflect the structure and work methods of their users. Unlike interviewing software (the part of the system that paints screens with questions, controls routing in a survey, etc.) which appears fairly similar in function across organizations, CASIC case management may look quite different when viewed across organizations. It was the result of this observation which led the Panel of Experts on Computer Software for CASIC, convened by the U.S. Bureau of the Census in 1990–1991, to recommend that the bureau cease development of its own interviewing software and concentrate instead on the development of case management software specific to the bureau's needs. This view is further espoused by Hofman and Gray (Chapter 14).

That a case management system mirrors the organizational characteristics of the agency or company in which it resides may, in fact, be an anomaly based on a restricted view of the role and function of such a system. As the concept of the case management system is broadened to encompass the needs of all users, rather than just selected users within the organization, it is possible to envision the development of systems that have sufficiently broad features to function for all surveys within an organization and even between organizations.

13.4.4 Field Supervisory Issues

Most survey organizations have field supervisors. These individuals go by a variety of names depending on the organization; however, their functions are similar. They generally supervise a group of interviewers in a geographical area and provide guidance, motivation, problem assistance, and other support functions for their sets of interviewers. In addition, while not ubiquitous, it is common for supervisors to provide some level of quality control for the interviewer's work including reviewing a subselection of completed interviews and doing some callbacks of respondents to prevent falsification.

Supervisory functions are critical to the conduct of surveys and therefore are critical in the creation and operation of a case management system. Depending on the design of the system and the organizational expectations, the highly interactive nature of supervisory functions can be a major challenge. In organizations like the Census Bureau where supervisors reside in regional offices and are directly on-line to the main host computers, meeting supervisory requirements is not the challenge that it is in servicing supervisors located remotely working from laptop computers without the support of direct connections. At least one organization tracks all interviewers that belonged to a particular supervisor and transfers all copies of completed cases for those interviewers to the appropriate supervisor's laptop computer. In the early days of CAPI this worked well. In more recent times, where CAPI is the mode for a much higher proportion of survey work and with multiple studies conducted concurrently, supervisors have had thousands of cases residing on a laptop. This resulted in a high proportion of system problems stemming from the supervisory functions.

The availability of better and less expensive telecommunications (see Smith, Rhoads, and Shepherd, Chapter 16) makes it feasible, in the United States, to consider having supervisors work on-line using a host system and the Internet or similar wide-area connections. The ability for on-line supervisory access to the host computer is an appealing aspect of any new case management system. More than any other part of the system the supervisory function should maximally exercise the client-server capabilities allowing the selecting and subsetting of data, the selective downloading of data, and the production of reports run on the server and transmitted to the supervisor's screen in real time. While some of these features can be provided in a temporarily connected queue-based system, the advantages and declining cost of providing them on-line must be considered in any state-of-the-art case management system. Factors militating against supervisors working on-line are their need for frequent travel and the cost of more than one phone line, which would be necessary were they to spend significant amounts of time on-line. Given these considerations it seems that the most cost-efficient CAPI case management systems will provide the ability to conduct some supervisory functions either on-line or off-line.

13.4.5   Technical Design Issues

The distributed nature of CAPI systems—that is, interviewers with laptop computers spread far and wide, and the need for some type of bidirectional communications system between the central facility and each interviewer—requires much more of the systems designer. Software engineers building CAPI case management systems must have a grasp of the intricacies of distributed computer systems. It is no longer the simple situation of being able to create a new table or report from a database residing on the central computer. Building distributed systems requires that developers understand both telecommunications and client-server concepts as these elements must function in tandem on the remote and host computers. It essentially requires the development of three or more software systems that work perfectly together.

In the early 1990s the development of distributed systems was in its infancy, and there were few tools designed to make it easy. For personal computer-based systems, only simple database languages existed such as Clipper and Turbo Pascal. These tools were designed for accessing a resident database or an in-house network. The weakness of the development tools and vendor concentration on the interviewing aspects of CAPI resulted in weak or nonexistent CAPI management systems. Those organizations (vendors primarily) that did adapt their CATI interviewing systems to CAPI simply abandoned much of their previously developed CATI control systems, while those who developed CAPI software from scratch tended to concentrate on the interviewing aspects of the software rather than on case management issues. The result has been a general lack of case management systems for CAPI leading to the necessity for survey organizations to develop their own.

Some vendors have recognized the need for CAPI case management tools as a part of their interviewing software. However, this recognition has resulted in the provision of basic tools that allow the user to develop his or her own case management system or in attempts to include a turnkey case management system. The latter usually entails a way to transfer completed case information to a central facility, largely ignoring the other issues and functions required to make a system usable and complete.

13.4.6   Client-Server Nature of Case Management Systems

Case management systems are, by nature, client-server systems. A client-server system, in simplest terms, consists of two or more logical entities that work together over a network to accomplish a task (Orfali, Hankey, and Edwards, 1994). Client-server systems should not be confused with mainframe-terminal systems where the terminal was often referred to as a "dumb terminal." In client-server systems both the client and the server are smart and involved in the operation of the system.

In most case management systems the remote computers are the clients and the central host is the server. Processing is split into two or more parts with some processing local to the remote computer and some processing specific to the host server. One of the fundamental issues in developing client-server systems is the decision regarding which processes to execute on the client and which processes to execute on the server. These decisions are affected by functional decisions regarding the way interviewers and other users work.

For instance, interviewers who are expected to connect to the host system on a daily basis may have fewer functions on their computers than interviewers in an organization where interviewers are expected to connect to the host system on a weekly basis. One can imagine a continuum going from always working on-line to working mostly off-line and connecting to the host on, perhaps, a weekly basis. Remote users working on-line would require little processing ability on their computers resulting in a "thin client." A remote user communicating to the host only infrequently would require many functions on their computer resulting in what is referred to as a "fat client." In case management systems it is likely that there will be a fairly logical division between the function of the remote client and the host server and that this division will be dictated by the practical needs of the interviewers and by the degree of control that is afforded remote users versus central users.

The simplest client-server system is two-tiered and involves only clients and a server working together over a network of some sort. It is very common, however, to have multitiered systems with additional processing steps placed between the client and the server. These processing steps are often related to the telecommunications functions required to move data between client and server, or to reporting services that abstract or summarize data in a way that allows more efficient access. Another way to think about multitiered client-server systems is that some of the intermediate functions of the host or the client are "off-loaded" to additional processors. The result is thinner and more efficient clients and servers, but in numbers greater than what otherwise would be required.

13.4.7   Fundamental Decisions

There are several issues that must be taken into consideration when developing a client-server case management system. The first is the communication between the client and the server, and the second is the degree of association required between the case management software and the interviewing software.

On-line versus Queue Processing
Most of the existing case management systems are based on a messaging and queuing or "store and forward" model. In these systems the data received from an interviewer are stored in a file and processed by a host processor as time permits. Naming conventions are used to separate different types of data and to sort and merge the appropriate directories and files on the host system. Conversely, data that are to be picked up by interviewers are stored on the host in a directory specific to that interviewer. These data are usually in

compressed or packed format while awaiting transmission to the interviewer when she or he connects to the host. These data may consist of many files, diverse in nature, including fresh cases, cases transferred from another interviewer, and, perhaps, a new version of the questionnaire. All of these files are preprocessed by packing (zipping) them into one package on the host, transferring them to the remote computer when the connection is made, and then unpacking them and sending them to their appropriate places on the remote machine. Client-server systems employing this approach are said to be message oriented, and the software connecting the client and server is called message-oriented middleware or MOM.

The advantage of the MOM-based system is a lessened dependence on the telecommunications system and on the speed of the host processor. Interviewers connecting by telephone and modem automatically deposit files into a "receiving" host directory and pick up files waiting for them in a unique directory specified for each interviewer. After the connection is terminated, the server processes the received information before it is availale for use. It is evident that given several hundred or possibly thousands of interviewers, this system requires an equal number of directories and many times that number of files. If something goes wrong in the process of updating directories or files, the solution can be very confusing and time-consuming. Because of the level of work required to implement this solution it may be desirable to use several computers on the host side to handle the postprocessing and storage of the data.

Recent improvements in the speed of microprocessors and file systems, the advent of microcomputer based client-server databases and the reduced costs of on-line connections are creating affordable alternatives to the file-based store-and-forward approach. The ability to scan and modify databases with tens of thousands of records in a few seconds allows for synchronized databases on the server and the client. It is now possible to combine case information for all interviewers and for multiple surveys in a set of related databases on the host, portions of which can be replicated between client and host automatically. This approach eliminates much of the physical and logical complexity of previous systems and greatly simplifies host management. At least one major database vendor currently offers a software package that automatically handles the MOM aspects of synchronizing databases between host and client. The University of Michigan Survey Research Center is using this product to develop a new third-generation case management system.

An emerging possibility for managing remote clients is to simply have them log into the host system in order to synchronize databases directly. As the Internet and other long-distance telecommunications systems become more affordable and reliable, this may become a more common choice. Taking this approach eliminates the need for the message-oriented middleware, since there is no requirement to store data as messages.

A third possibility is simply to have remote clients log into the host system and work on-line the entire time directly accessing the databases on the host machine. This approach eliminates many of the problems associated with having remote clients and brings most of the issues back to those of the familiar centralized system. Speed requirements, telecommunications reliability and costs, and simultaneous availability of the client and the host systems are considerations in deciding which of the above approaches to employ. It is likely in the near future that we will see single organizations employing all three methods.

Loose Coupling with the Interviewing System

Social survey organizations are currently being called on to employ more than one interviewing system. There are a variety of reasons for this, including using interviewing systems best suited to the characteristics of particular surveys, cooperative arrangements between survey organizations where one organization uses the interviewing system of the other to simplify shared data collection, and government mandates requiring the use of a particular interviewing system for a government funded study. While there may be some advantages to using more than one interviewing system, there are many advantages to using only one case management system within an organization or in some situations between organizations.

Loose coupling is a computing term that implies that two systems can be connected without changing fundamental aspects of either system. Implementing loose coupling between interviewing and case management systems simply requires the development, as part of the case management system, of an interface module for each interviewing system. The interface module allows the case management system to view information from each interviewing system in a common format and provides information to the specific interviewing system in its particular format. The concept of loose coupling is very common in everyday software. Every user of a microcomputer word processor is familiar with the ability of one word processor to read and write the files of a different word processor. The concept can be applied to communications between a case management system and different interviewing systems as well.

The ability to use a single case management system with multiple interviewing systems is a very cost-effective approach for the survey organization. The interface modules required between the case management system and the interviewing system are reasonably simple and only need to be written once for each interviewing system.

13.4.8 Functions of the Case Management Server

A few of the functions of the host side or server side of a case management system include (1) data importation, (2) sample delivery, (3) call scheduling, (4) survey updating and modification, (5) case transfers, (6) data receipt and storage, (7) detection of system and data errors, (8) data consolidation and reporting, and (9) security. Call scheduling is dealt with in detail by Edwards, Suresh, and Weeks (Chapter 15) and will not be discussed further here.

Data Importation

Data importation is required for a variety of data that must be used by the case management system. These data include sample information such as respondent names and addresses, ancillary data for use in lookup systems such as the names of each congressional candidate in state districts, and data from a previous wave of a longitudinal study so that the interviewer has relevant information from previous interviews conducted with the respondent. In most cases data to be imported into the system are prepared by programs external to the case management system. The case management system handles data delivery to the interviewer regardless of whether a central CATI or a remote system is used.

Sample Delivery

Sample delivery is a function of all automated case management systems. With centralized systems the server simply accesses the centralized database and delivers the case to the interviewer when the interviewing software requests a new case. With remote clients the sample is generally apportioned among the remote clients by geographic region or some other method of apportionment and held on the server until requested by the remote client. The primary difference between central and client-server systems is that the sample is usually delivered in multiples to a remote client and one at a time in centralized systems. The number of cases delivered simultaneously to the remote client varies between organizations and is affected by the sampling design for the survey, the frequency with which an interviewer is expected to make contact with the central server, and size and duration of the survey.

Survey Updating and Modification

It is not uncommon, particularly in social surveys because of their size and complexity, for modifications to be made to a survey during the field period. The case management system must be able to convey any changes in survey format, question wording, or even structure to the interviewing system. It is also necessary to receive interview data that may be modified due to the changes in the survey. It is also desirable for the case management system to maintain some sort of version control for tracking and matching of different survey versions and their coresponding data.

Case Transfers between Interviewers

The transfer of sample information and partially completed survey results between one client and a different client is perhaps the most complex and important function of case control systems for either centralized systems or those with remote clients. Different organizations allow transfers to be controlled from the client or the host or both places. In the latter instance, rules are required to mediate conflict when transfers are initiated from both host and client. Another requirement could be the need to notify the initiator of the successful completion of the case transfer.

Data Receipt and Storage

All case management systems have the function of accepting data from the client system, running rudimentary checks on the completeness of the data, and incorporating this information into the central database with the appropriate time stamps and other identifying information. In addition it is necessary for the host system to maintain logs and histories of data sent and received, of transfers that passed through the system, and of all other relevant system activity.

Detection of System and Data Errors

Errors occur in all complex systems. The host case management system should be programmed to anticipate and detect as many different anomalies in operation and data as possible and to take appropriate action either to correct the problem or to notify the system users before errors compound. An example might be the duplication that can occur when the same case had necessarily been assigned to two interviewers. It is possible for the host system to receive two cases from interviewers for the same case ID. If this happens, the host system must detect the error, save the data, and resolve the situation or alert the appropriate individuals to take action.

Data Consolidation and Reporting

The system should provide data in formats that can be read by users with software they are familiar with such as SAS or Microsoft Access, and it should also make the data open, easily exported, and fully readable from the reception database.

Security

Security must be provided against intrusion on the host database server and against accidental changes due to careless data access. Most security problems occur due to carelessness rather than malice. The host system should both protect users from themselves and also institute protection controls to allow only authorized users access to the data. Access control should allow separate control over the ability to read the data and to write or modify the data. In dealing with remote clients, it may also be necessary for the host to compress and encrypt the data before transmitting it to the client, and to decompress and decrypt data received from the clients.

13.4.9   System Management—Host Business Rules

There are many rules that must be established in the development of a case management system. These automated decisions are commonly referred to as "business rules." The definition of business rules requires a complete conceptualization of how the system will actually work. Each of the functions mentioned in the previous section requires business rules that define the set of possible behaviors for that function. There are of course other functions on the *client* side that also require the definition of business rules. It should be obvious that the host and client rules must complement each other. These rules then become the controlling logic of the case management system. Some examples of business rules are as follows:

Rule 1. Supervisors cannot change a sample address.
Rule 2. A transfer cannot be initiated on a "completed" case.
Rule 3. A case with a transfer flag cannot be moved from the operational to the archival database.

Each business rule gives rise to one or more procedures in which appropriate actions are taken on the data. There may be hundreds of business rules built into sophisticated case management systems. Making these rules explicit and providing the ability to switch them on and off allow the system to be customized to the needs of different surveys without rewriting the program. This approach also makes many of the system's assumptions explicit and accessible to the users. These assumptions are often hidden in first- and second-generation systems and result in user confusion and added system maintenance costs.

In the future it may be possible to describe the business rules for a system and have a software generator automatically generate a new version of the software that incorporates all of the standard functions but is customized according to the selected rules. As we become better at making the case management requirements and assumptions clear and explicit, we will come closer to realizing this possibility. By having a switch for each software object (function) of a case management system, the software generation program could read all of the switches and build the customized system for a particular survey. This is the obverse of building a generalized case management system that includes all possible functions and then turning those functions on or off by the business rules.

13.4.10 Data Management

Every CAPI case management system must provide for the merging of interview data and for the production of a master data set that can be accessed by the study analysts. In addition the case management system should provide automatic backup of the data on a frequent basis and convenient archiving of the raw data and associated files at the end of a study. The system should provide for a standard set of reports produced "on demand," including operational real time statistics on cases contacted, case outcomes, current response rates, interviewer hours expended, and current cost per case.

Experience tells us that every study manager desires something unique to his or her study. The survey organization needs to be sensitive to this desire by balancing the time and cost required in providing unique information for a given study with the true need for unique data. One approach to satisfying the "need" without breaking the bank is to develop a standard case management system for all studies, with flexibility added by the incorporation of business rules, as discussed earlier, and through user-defined data in the case management databases. A system that employs user-specified data definitions allows a nonprogrammer or a novice programmer to add, within limits, multiple unique data elements to the case management system for a particular survey without altering the underlying case management system.

The use of on-line analytic processing (OLAP) should also be considered in the development of new case management systems. Coupled with the concepts of metadata and data warehousing, it is a logical step to extend the case management system such that data are automatically exported to a data warehouse database package that includes OLAP capabilities. The advantage of combining case management and data warehousing systems is the elimination of one of the "seams" that analytic users often find to be the most troublesome—getting the data from the CAI system into the analytic system. The enormous power and immediacy of this approach will allow analysts to deal with data far more efficiently and creatively. While a few CASIC systems incorporate analytic tools, they are generally far less sophisticated than general-purpose analytic tools such as SAS or SPSS. On the other hand, these tools are less integrated than data warehouse software systems that combine database storage and powerful analytic functions. Being able to combine the data collection software and the data warehouse storage and analytic functions in a single integrated system would be a major advance in the evolution of CASIC systems.

13.4.11 Other Server-Related Issues

Cost Accounting and Management are critical to the conduct of surveys. There is often a need for daily or even more frequent cost data in order to make intelligent decisions regarding the management of a survey sample. Just as the case management system should interface easily with the statistical programs, there is complementary need to interface well with financial systems to provide rapid and comprehensive cost reports.

E-mail among interviewers, field supervisors, and central staff has been, in some organizations, considered a function of the case management system. Most survey organizations have built some form of messaging into their existing case management systems, although it is most commonly included as a one-way messaging giving the central staff a way to send notices to the field staff. With the development and continued improvement of Internet-based e-mail systems, one must ask whether it makes sense to try to duplicate the features of e-mail in the case management system.

System backup is a necessity in all case management systems. There is a variety of alternatives, and it is probably wise to consider redundant backup not only of data but also of hardware systems. Most system administrators are aware of the difficulties of building new hardware configurations in the heat of battle. There are multiple levels of data backup, beginning with a redundant disk drive arrangement called a RAID array. With this type of disk system, a disk drive on the host system can fail without loss of data. A new disk can be plugged in to replace the failed drive, often without "downing" the system, and the data on the drive will be automatically rebuilt. In addition to providing backup safety for the disk drives, some type of tape backup is also recommended. Tape backup systems range from small tape drives for backing up a single server to robotic tape drives that back up multiple servers on a routine, usually nightly, basis.

13.5 SUMMARY

The case management system is an integral and critical element of the overall data collection system. It must answer to many users, and a failure in case management functions presents a major threat to the survey organization. Modern systems must meet the demands of a diverse collection of users. Organizational case management systems must provide a significant measure of flexibility to satisfy the demands of different study designs while adhering to a set of general standards for all studies. To date, CASIC case management systems have been designed and customized to meet the needs of particular survey organizations and often to particular surveys within the organization. Little attention, beyond the most basic features, has been given to CASIC case management by the major software developers. The use of modern client-server database tools and modern software development techniques provides the potential to develop case management software that can be used effectively for multiple surveys and even by multiple organizations. These approaches combined with the extension of case management though the concepts of data warehousing and OLAP can create powerful and innovative new approaches to the collection, management, and analysis of CAI data.

CHAPTER 14

CAPI Survey Management Systems: Case Management on Laptops

Lon Hofman
Statistics Netherlands

James Gray
Office for National Statistics, U.K

14.1 INTRODUCTION

This chapter focuses on the factors that lead to different CAPI survey management systems designs. As Connett (1996) has pointed out, "All of the organizations who do serious CAPI studies have developed their own remote case management systems." It is apparent that different organizations build different systems. Why is this, when most use third-party CAPI software?

We examine the reasons for differing systems. We focus particularly on large multisurvey organizations such as government statistical agencies, since they typically demand the most from management systems. These organizations more than any other have tended to build their own case management systems rather than buy an existing product. This is largely due to the fact that independent case management software is not yet commercially available. Management tools that are provided as part of the interviewing software are not always appropriate because of the widely varying nature of the work and structure of the organizations. In addition there is often the need to run several different varieties of interviewing software, or to supplement the main interviewing software with specialized third-party utilities. Management software that is specifically designed to integrate closely with a particular brand of CAPI software is often not capable of such operation. It is certainly also true in some cases that "developers like to develop," and management systems are some-

Computer Assisted Survey Information Collection, Edited by Mick P. Couper, Reginald P. Baker, Jelke Bethlehem, Cynthia Z. F. Clark, Jean Martin, William L. Nicholls II, and James M. O'Reilly. ISBN 0-471-17848-9 © 1998 John Wiley & Sons, Inc.

times built when they could have been adapted from existing software. In actuality, the nature of systems will reflect the organization's development style and the individuals who decide the development path. We examine some of these factors.

Connett (Chapter 13) gives an overview of the entire case management process. Here we are addressing solely the set of programs that runs on portable computers, supporting CAPI work (but excluding the actual questionnaire software)—the "client" side of the distributed case management task. Connett concludes that a single case management system for all organizations is possible, driven by a set of business rules to adapt it to different organizational requirements. We argue that the organizational differences are greater than can reasonably be accommodated in this manner. It is not impossible, but such a system would have to be highly extensible and capable of operating in a number of different ways. We think that in the longer term, groupware products from the major software houses will become sufficiently open and adaptable to take over the main case management functions. In the meantime organizations are still refining their needs and expectations and are building or adapting their own systems.

We start with a review of the purposes and functions of laptop case management systems. These are surprisingly difficult to define authoritatively; different organizations have very different backgrounds and expectations. Some of these differences are explored in Section 14.3, where we examine the factors that lead to different systems. Organizations are different. They have different cultures and internal political structures, they have disparate tool sets, and they regard their interviewers in various lights. These factors have led to a variety of differing systems and may, arguably, mean that there will always be a need for tailor-made case management software, or at least for a high degree of customization. We conclude with a brief case study of three different approaches to the problem from Statistics Netherlands and the Office for National Statistics (ONS) in the U.K.

14.2   PURPOSE OF LAPTOP CASE MANAGEMENT SYSTEMS

Systems for case management on laptop computers are caught in a tension between two main customers. The interviewer needs to be supported in his or her work with systems that are easy to understand and operate and that help them to work efficiently. On the other hand, the organization needs to control the work, and has requirements in terms of reporting on progress, and in managing and securing the data.

14.2.1   Needs of the Interviewers

The interviewers want a system that is easy to understand and operate. Consideration to the human-computer interaction aspects of the system is important to achieving usability. The system must be easy to use in itself, and it must also have an interface that is consistent with the interviewing software. Even systems that appear to work well may be quite difficult to use. Methods such as the examination of keystroke files for some CAPI software can reveal where errors are being made because of poor interface design (Couper, Hansen, and Sadosky, 1997).

Defining the specific interviewer requirements of a management system is not always easy. One critical question is the degree of automation that should be offered. Should the system manage just the interviewing process, or attempt to automate all the interviewer's tasks? There may be good reasons to leave some interviewer tasks on paper. Paper diaries, maps, and address lists, for example, are flexible, easy to use under a variety of difficult circumstances, and have no start-up delay.

The needs of the interviewers change over time. Interviewers without laptop experience will not have well-formulated expectations. Interviewers with greater CAPI experience are better able to define their requirements, but they also have greater expectations of the system. Sometimes the first version of a laptop system will help the interviewers (and the organization) to focus on what they *really* want from the system; the second system developed will be more appropriate to the newly recognized requirements. (See Brooks, 1982, for a more general discussion of this commonly recurring theme in software engineering.)

14.2.2 Needs of the Organization

Aside from fulfilling the basic functions of survey management on laptop computers (discussed in Section 14.2.3 below), organizations will have some underlying requirements about the way the system works and interacts with their data.

First, the organization wants a system that is reliable and controllable. The system should not fail, and data should not be irrecoverably lost. System failure is expensive and, if persistent, can also cause motivation problems for the interviewers. This is consistent with the interviewer's view of the system. Realistically, however, organizations will know that hardware and software will occasionally fail (or that laptop computers will be lost, stolen, or badly damaged). Organizations will build error managment systems that prevent, detect, and recover from the errors and failures that inevitably will occur. Systems that never fail under any circumstances are prohibitively expensive; what is needed is a balance between the interviewer's ideal (never to have to go back and re-interview because the system lost the data) and the organization's aims (to collect good data, inexpensively). Decisions will need to be made (either explicitly in the form of contingency plans, or implicitly in the form of the organization's development policies and culture) on what degree of data loss is acceptable under various circumstances.

For many organizations there is another potentially conflicting tension—that of confidentiality of data (Nicholls et al., 1994). This varies according to the nature of the survey data and also with the attitude (and regulatory regime) of the organization. Fully securing data against unauthorized disclosure is expensive, not only in direct financial terms but also in opportunity costs. Effective security measures usually include password-based access control and data encryption. These can impede the work of the interviewer (especially if multiple passwords need to be remembered for different parts of the system, or if passwords are changed frequently) and can make data recovery more difficult in the event of hardware or system failure. Appropriate steps may be measured against the security of paper forms (where data are easily readable together with their definitions) or against other systems in the organization (which may be used to handling very confidential medical records or census returns).

The organization will also want to be able to control the fieldwork. Because the fieldwork is decentralized, it is hard to control, but for the same reason there is a great need for control. To achieve this, survey managers will want to be able to generate a variety of reports on progress and quality of the fieldwork. The specific needs will vary between organizations according to the size and management structure of the interviewer force: The larger the force, the greater is the need for better and faster information.

14.2.3   Functions of CAPI Management Systems

The functionality expected of CAPI management systems can be categorized in various ways. Rather than try to place another structure on the functionality, we use the list of eight core functions proposed by Nicholls and Kindel (1993) as a starting point for further discussion. They are described as "core" functions because they are central to what the interviewer must do, and because they represented the current state of the art.

We regard these as functions that must be available in one way or another to the interviewer, but different organizations will have different views on whether they are implemented in the programs running on the laptop computer, or whether they are better provided by the broader work environment of the interviewer.

Accept and Store Interviewing Assignments
Each interviewer receives a list with assignments from the central office. In early systems these were more often sent by diskette; modem transmission is now more normal. In fact a few early systems used modems, and for some organizations transfer by diskette is still more appropriate. The system must be capable of reading these assignments and storing them together with any existing assignments.

The nature of exactly what is sent as an assignment will vary. The laptop software will need information to uniquely identify the sample points either by receiving a pre-assigned serial number as part of the assignment or by receiving enough information to generate identifiers during the course of the work. The interviewer will need information to plan the work and enough detail to correctly identify the sample point. In many cases it may be felt that paper address lists offer more flexibility than computerized records, particularly for work planning and for ease of access in the field. Conversely, there will be occasions when the speed of delivery by modem and the lower cost of reducing paper handling become more important. Organizations will make their own decisions, often varying with different surveys according to their requirements.

If the survey is based on a panel, the assignments will not only contain all necessary information to contact the member of the panel but may also contain information gathered in previous panel waves. The degree to which contact information and previously gathered substantive data can be standardized across surveys will vary. Different methodologies of dealing with the panel element will result in different solutions depending on the nature and degree of dependent interviewing (see Brown, Hale, and Michaud, Chapter 10).

In some cases the assignment itself will not identify the individual cases for interview but will identify a higher stratum of the sample. An example would be a museum visitor survey where the assignment is an interviewer shift that will generate a varying number of people to interview depending on visitor flow and on the chosen sampling interval. Here there will be a need to sample the target population in the field. In many circumstances it may be felt that this is most efficiently done with manual, paper-based methods. However, it may be desirable to build some sampling capability (e.g., quota sampling control) into the management system.

Display a List of These Assignments

The interviewer must be able to inspect the assignments available, for instance, in the form of a scrollable list. In many early systems the available assignments were not visible in this way on the laptop screen but were selected by the interviewer from the paper list; the appropriate serial number was typed into the laptop. Most organizations have moved away from this assignment method because it can be cumbersome and error prone.

The assignment list is one of the most important interfaces of the laptop case management system. Different approaches have been adopted for how the list is presented, dictated partly by the underlying data storage mechanisms and partly by the way the task is perceived by the developers. In some systems (e.g., Statistics Netherlands LIPS system) a menu-based approach allows the interviewer to select a survey and period and then view a list of all assignments within the selection. In others (e.g., the ONS Casebook) the hierarchy is flattened, and all outstanding assignments are viewed on the same screen regardless of survey. There are advantages to both approaches, and perhaps the ideal system would allow both views of the list. The menu approach works well for a highly structured field program, where the interviewer will be working on one survey at a time and where the assignments are amenable to being structured in a hierarchical way. Menus are good at restricting the list, to focus the interviewer, but less good where the interviewer needs to see the whole of the outstanding work at a glance. The flattened list approach gives an immediate view of all outstanding work and ensures that none is "lost" deep in an unexplored branch of a menu system. But, if the list is too long, it can overwhelm the interviewer. The key to a successful flat list is the ability to order it (e.g., to group by survey) and particularly to exclude completed work, thus reducing redundant information.

Some systems will display all assignments, including ones that are completed and have been transmitted back to the office; others only display those that require further work by the interviewer.

As Nicholls and Kindel (1993) suggest, interviewers may want to re-order, sort, prioritize, search, or annotate the list. By ordering on status, completed assignments and new assignments can be grouped. Thus interviewers can determine their remaining workload. In a multiple-survey environment, it should be possible to order assignments by survey so that remaining workload per survey can be inspected.

A geographic ordering, for instance, by postal (zip) code, can be useful to help the interviewer plan the workload efficiently. With the increasing use of graphical interfaces, one can envisage that one view of the assignment list could be in the form of a map on the screen, with selected samples points shown on the map.

Select a Case to Interview
The interviewer must be able to select a case for interviewing. If the original assignment is at a case level, then one case will be selected from the list, either by "point and click" or by keying in a serial number.

Where the assigned sample may contain more than one interview case, the system needs to take that fact into account. For example, the sample point may be a postal address that could contain several households. Depending on the survey a selection of, or all, available households will need to be interviewed. In this case there is not a one-to-one correspondence between the assignment and the cases the interviewer has to complete. One assignment can result in more than one case and therefore also in more than one interview. The system must be able to cope with such a situation. It could mean that extra cases have to be generated ("spawned") on the fly, so the system must guarantee the uniqueness of the identification of each case. This could be done by selecting identifiers from a list, creating them by an algorithm that guarantees unique-ness, or using random generation techniques that are extremely unlikely to produce duplicates (Daata and Wojcik, 1994).

One can envisage situations where case selection would not be an explicit interviewer action. For geographically based samples, laptop computers equip-ped with Global Positioning System (GPS) receivers and software could identify the interviewer's location and select the appropriate data records. It is unclear whether the benefits of this (fewer serial number errors, better geo-graphic positioning information, better auditing agaijt bogus interviews) would outweigh the investment. Where postinterview work needs to be undertaken (e.g., coding of occupations), interviewer-driven case selection may still be required, though variations on CATI scheduling could present outstanding work to the interviewer. This is one area where the degree of autonomy expected of the interviewer will have an effect on the system requirements (see Section 14.3.1).

Store the Interview Data
The system must control the storage of interview data. Here the system may share responsibility with the operating system on the portable computer. Storage implies maintaining the integrity of the data and controlling access to it. The system will need to strike a balance between ensuring that the interviewer and the head office can easily access the data and preventing unauthorized access if the machine falls into the wrong hands. As was discussed earlier, individual organizations will want to make their own decisions on how best to achieve the right compromise.

There is sometimes a tendency to disregard the physical storage of the data and to concentrate instead on the conceptual framework of the wider system. However, the physical storage can make a difference to the ease with which various actions are performed (as well as the difficulty of achieving the required functionality). For example, the ONS Casebook system stores each assignment as a separate file; this means that many processes (copying, transferring, deleting, etc., of individual assignments) can be largely delegated to the operating system, making the job easy and reliable, while others (e.g., coding job details, or recording call outcomes, across many cases) are less efficient.

Record the Status
Each assignment has a status that gives an indication of the current processing state. The status of a case can be set by the management system (e.g., "new assignment," "mailed") or by the interviewer (e.g., "complete," "nonresponse"). Effective status keeping is essential for CAPI survey management systems for two separate reasons: to enable the control systems to make appropriate decisions about case movement (e.g., is a case "clean" for transmission back to headquarters), and to provide information to help management of the surveys and fieldforce.

Interviewer accounting is essential for effective survey management. Status recording and accounting is conducted at various logical levels—for example, sample, visit, or travel. Depending on the management information needs of the survey organization, the case management system needs to be able to handle and transmit this information back to the head office, either with the data or separately from it (e.g., there may be a need to gather information about sample status before the data are ready for transmission).

Initiate Telecommunications
It is possible to rely completely on telecommunications to exchange all kinds

of data between the central office and the field. The central office sends new assignments to the field, and the field sends completed assignments, composed of accounting information and interview data, to the central office. Telecommunications can now also be used to send the questionnaire from the central office to the field. In the past, using telecommunications for the distribution of questionnaires was problematic, since the transmission speed was simply not high enough to efficiently handle the distribution of large questionnaires. Nowadays high-speed modems have solved this problem for many organizations.

Telecommunications are also used for several other purposes, for instance, to install software updates, for remote access to the laptop computer to detect, and to diagnose and solve system errors, software auditing, and the like.

Even in 1998 not all organizations feel that modem transmission is the appropriate solution to their data transfer requirements. For organizations with a fieldforce that is not widely distributed, or in a region with very good postal services (e.g., Northern Ireland), transfer of data on floppy disk can be cheap, reliable, and acceptably fast. In other cases (e.g., Statistics Iceland), direct visit by the interviewer to the central office can be appropriate (Helgadóttir, 1995). Telecommunications systems are treated more fully by Smith, Rhoads, and Shepherd (Chapter 16).

Perform Various System Functions

A variety of system functions will usually be available in a CAPI case management system. Examples are the ability to make backups, questionnaire management, and access to the operating system.

Making backups should protect the interviewers against possible data loss in case of malfunction of their computers. If data can be transmitted back to the office very quickly after the interview, there may be little need for data backup, since files are only at risk for a short period of time. Organizations may decide that backups are only worthwhile for occasions when partially completed cases are expected to reside on the laptop computer for more than a day or two. The frequency of any backup will depend partly on the consequences of data loss (the organization's culture and attitude will have an effect on the perception of this) and partly on the physical means of backup. For example, it may be possible to copy the relevant data files to another part of the hard disk every time any data are amended (in fact keystroke files can sometimes also fulfill this function)—this ensures absolute currency of data but does not protect against loss or major failure of the laptop computer. It is more usual to take copy data to floppy disk for greater physical security; this is more likely to be a daily activity. With faster modems and better telecommunications, it becomes possible for backup copies of all data to be made directly to the head office, without the need for removable storage media in the field. If handled appropriately, the same system can deal with collection of management data for incomplete cases by interrogation of the backup data set at head office.

Backing up data is an overhead on interviewers' time that has a cost associated with it. If data are backed up, there will need to be recovery systems. This introduces the issue of whether the systems should be field-based or restricted to HQ staff. If field-based, there is a danger of inadvertently overwriting good data with bad, by inappropriate use of the recovery tools; if HQ-based, timeliness may be threatened. The standards of acceptable loss will vary from one organization to another and can be measured either against other computerized systems within the organization or against previous practice on paper-based surveys (where there are no backup facilities but less chance of data loss).

What has to be done with the questionnaire program after all cases have been dealt with? One option is to automatically delete the program. This is typically the solution for a one-time survey. For a continuous survey the initiative to remove the questionnaire has to be taken by the central office. The instruction to erase a questionnaire has to be given using telecommunications. Another aspect to be considered is version management of the questionnaire and the questionnaire software. There will often be occasions when it is discovered that a questionnaire program contains an error; the management system will need to be able to accept a new version of the program and manage any compatibility issues that arise.

Should it be possible for the interviewer to go to the operating system or to start other programs outside of the case management system? There may be a need for this to be an option for diagnostic or recovery purposes where an unexpected error occurs, in which case the ability may be protected by password. For both LIPS and Casebook (discussed below), the option is protected by a password that changes on a daily basis. Some organizations prefer to offer the option to their interviewers as a matter of course (see Section 14.3.1).

Provide E-mail Communication

The interviewers need to communicate with their supervisors or the central office. The nature, direction, and mode of the communication will vary. The telecommunications function for transfer of assignments and completed work is outlined above. It is, however, noted that there is a further need for more loosely structured communications. These could cover a number of different topics, from queries on specific issues, through hardware and software problems, management issues, pay and expenses claims, to "social" contacts that maintain morale and build a community of interviewers.

The provision of e-mail may be an appropriate response to the communications needs of the interviewer in some cases, but for others (particularly smaller organizations, in terms of geographical coverage or number of interviewers) the function may be more effectively carried out by telephone or even face to face.

Depending on the organization's culture, any e-mail provision may be within very tight control with a few specified channels of communication (e.g., interviewers able to communicate with supervisor and survey officer only), or there may be wider provision to communicate directly with anyone in the organization and by external Internet e-mail. This is considered further in Section 14.3.

14.2.4 Other Functions

There is a simple model for CAPI work whereby addresses for interview are allocated and sent to individual interviewers; they then complete the work and transmit the completed interviews (or nonresponse returns) to the office. In practice, the problem is often more complex than that—interviewers suddenly become unable to do the work, and the work needs to be re-allocated, or nonresponding cases need to be re-assigned to a different interviewer for a second attempt. The whole system must be able to deal effectively with these issues without the risk of duplicate data (or worse, responding data being overwritten by the original refusal data). Some organizations have found that their first CAPI systems were actually slower at these tasks than the old paper-based systems that they were replacing (Martin et al., 1994), although in some cases 20 to 30 percent of interviews were re-allocated in this way (Nicholls et al., 1994).

The systems must not only support the interviewer but must also support the interviewer's manager in ensuring the accuracy of the information collected. On paper-based surveys, one important technique was "prechecking," where the supervisor peruses the paper forms looking for routing and other errors, and checking the quality of coding of open answers. Some of these checks are no longer necessary (e.g., routing is taken care of by the laptop computer), but quality assurance of coding is still needed. Not all systems support the supervisor in this in a reasonably efficient way (Martin et al., 1994).

Another technique for developing interviewers skills is occasional observation of the interviewing process, especially for trainee interviewers. This is a more demanding task for the supervisor under CAPI because the questionnaire cannot be followed without keying in the appropriate responses; this makes it more difficult to keep notes of how well problems are handled by the interviewer. Some organizations extend their laptop systems to allow linking of a supervisor's screen to the interviewer's laptop. This is done with the aid of a cable between the interviewer's and the supervisor's computers, and software that sends screen images between the two (see Goodger, 1995).

One test of accuracy is to check that the interviewer has actually called at the addresses and carried out the interviews claimed. Sometimes effective management information flows and audit trails can identify potential problem areas (e.g., if personal interviews at addresses 60 kilometers apart are completed within five minutes of each other). Most organizations will also conduct periodic checks on their interviewers. These may well create additional demands on the laptop systems, depending on the way they are organized.

It is possible to gather statistics on how the system is used. Based on this information improvements can be made in future versions; for example, use of keystroke files can identify poorly designed parts of the system (e.g., see Couper, Hansen, and Sadosky, 1997). It is also possible to gather detailed information on items such as call patterns. There may be an ethical problem here: Do you tell the interviewer what extra information is being gathered and for what purpose? The answer will depend on the regulatory regime of the country or state, and on the organization's attitude toward its interviewers.

Some surveys still require a paper element. Often this involves some form of respondent-completed questionnaire or diary. Sometimes the organization may need written permission from the respondent for special procedures (e.g., invasive medical tests). The paper may have to be moved around the organization in a physical way (though document imaging can be used at headquarters); the case management system then needs to keep track of the paper and ensure that all data are accounted for.

14.3 FACTORS IN LAPTOP CASE MANAGEMENT SYSTEM DESIGN

14.3.1 Role of the Interviewer

The laptop system has to support the role of the interviewer in his or her work, as well as maintain the data. The specific tasks undertaken by the interviewer will vary between and within organizations; this will have an effect on the specific functionality to be provided by the system. There are also more fundamental questions about the way the interviewer is perceived within the agency that may have a deeper effect on the types of systems they are given and may suggest basically different designs. The system that is developed within an organization will reflect that organization's culture and attitude toward their interviewers. That attitude is not constant across organizations. There is a spectrum of ways in which organizations view their interviewers— we highlight three models:

Interviewer as an Interviewing Machine
In many CATI units the interviewers have very little freedom to organize their own work and are expected to sit and interview.

Developments such as automated call scheduling and predictive dialing (see Edwards, Suresh, and Weeks, Chapter 15) enable the computer to control the whole process apart from gaining respondent cooperation and interviewing. This calls for a highly prescriptive, closed system, with an emphasis on control and monitoring. Few organizations extend this model entirely to their face-to-face interviewers (though some might like to) because the nature of the work calls for a degree of autonomy. Replacing the interviewers' independence with systems that do everything for them would call for a very highly specified, very detailed system. This is coming within greater reach as organizations learn more about interviewers' needs.

Interviewer as a Skilled Practitioner at the Job of Interviewing

Many organizations regard their face-to-face interviewers in this light. Here the system is likely to be a closed system (one cannot go outside the interviewing and case management environment) but will not be prescriptive, since the interviewer needs to organize the work rather than the other way round (though this may change in the more distant future even for face-to-face interviewers). The system provided may be as a series of menus or a set of tools to perform the tasks required. The emphasis will be on providing good access to the main functions identified above. There will be a tension between giving flexibility to the interviewer and retaining control (or at least strong monitoring) at the center. Both the Statistics Netherlands LIPS system and the ONS Casebook system are examples of systems that try to match this organizational view.

Interviewer as a Knowledge Worker with an Interviewing Specialism

Some organizations regard their interviewers in this light, and many regard field-based supervisors like this. Systems are characterized by their openness, with the interviewers having access to a variety of different software tools, not all of which are directly connected with the interviewing process. Typically there may be an emphasis on communications tools; for example, Statistics Norway provides full e-mail facilities, and encourages networking among interviewers (Degerdal, Hoel, and Thirud, 1995). Free exchange of communication may be frowned upon by organizations at the more restricted end of the spectrum, who would worry about the spread of bad practice unless all communications were funneled through the center. Very open systems indicate strong faith and confidence in the interviewers.

Each of these models represents a valid approach. The view adopted by an organization will interact strongly with such issues as recruitment and retention strategy, training, and remuneration policies. To a large extent the way an agency regards its interviewers is part of what defines that organization's position in the survey marketplace. Thus an organization with a large field-force, a highly variable amount of work, and working in a price-sensitive environment will tend toward a high turnover of interviewers working in a tightly controlled environment. An organization with a more stable level of varied and complex projects, but with an emphasis on high quality, will be investing more time and money to gain the best of their interviewers and is more likely to regard them as skilled practitioners or knowledge workers.

14.3.2 System Developers

Ultimately the types of systems developed by an organization to support its interviewers will reflect the development environment. The move from paper and pencil interviewing to CAPI is a major one, and it calls for different development strategies for a number of reasons.

First, obviously different tools are needed. In the early 1990s when many systems were being developed, survey IT specialists who were used to mainframe-based databases for storing, editing, and manipulating survey data suddenly found themselves faced with PCs, and DOS-based systems that were often less sophisticated than the systems the programmers were used to.

Second, the customers are different (Martin, 1993). The interviewer is suddenly an important user of IT systems. Not all organizations respond positively to this, and one still finds IT developers who will not talk to interviewers, or field managers who refuse access to interviewers. Certainly a lot of larger statistical bureaus suffer from "large organization" effects and find that IT development is controlled by a central IT department that is seen by the business areas as being unresponsive to their needs.

Different organizations manage the integration of IT developments with the business areas in different ways. Management philosophies can have an effect on the style and type of systems that emerge. The balance of power between the IT development area and the business area often reflects the organization's size and the type of other activities that are taking place within the organization.

There is a wide spectrum of practices in IT development. One end of the spectrum is the large organization that has major, high-profile, high-risk systems. Such organizations will tend to develop highly designed, fully structured, tested, and documented systems. The systems will often emerge as fully featured (sometimes overfeatured) and well-integrated after a fairly lengthy development path. At the other extreme will be the small groups working in a less structured environment, and they will rapidly produce systems that are just good enough for the current purpose and have a short production life. Between these two extremes will be organizations using rapid, iterative development techniques. Early systems are typical of the small organization in their speed of implementation; later systems build on these in a cumulative fashion as experience is gained and requirements become better understood by the users.

14.3.3   Size and Experience of the Organization

Training and support costs tend to rise with the number of interviewers. The cost of developing a system depends more on the functionality and design than on the number of users. Thus for very large organizations one would expect to see more fully developed laptop systems that are usable without much training, resilient enough to require little support from HQ, and functional enough to support a wide range of interviewer tasks, thus reducing field and administrative costs.

Conversely, a smaller organization which does not have the development resources nor a sizable fieldforce, will not have the financial imperative for such a system. Such organizations are faced with either buying in a system from outside or developing a smaller system that may require more training and support. An example is Statistics Iceland which has very few interviewers who are all within easy reach of the head office in case of problems. The systems to support them are fairly minimal and do not attempt to deal with every possible eventuality—exceptions are dealt with by direct support.

A survey organization with a large number of survey managers and developers will have different expectations than one with only a few. The survey manager and the IT specialist for a survey are the major customers of the laptop system. If there are many such customers, there is always a danger of conflicting interests. This can occur if the management system needs any adaptations for new surveys, and if coordination is poor, the build up of changes may destabilize the system (e.g., Gray, 1995).

An organization with only one CAPI survey in the field will have simpler, more definable needs from their laptop systems. The degree of complexity required depends on both the number and the variety of surveys. An organization with many surveys built around the same basic model will find it relatively easy to devise a management system to deal effectively with that model (likely they will have difficulty if they later wish to diverge from it). Conversely, if there are many different types of surveys, the organization must learn to deal effectively with the diversity.

Typically, a highly standardized organization will have a consistent sample design, all CAPI surveys will be implemented in the same software with similar structures, and they will have similar data formats and the same coding frames for management information. In such organizations it is usually beneficial to develop a system that is tightly integrated with the survey implementation. The system will know about and understand a great deal of the data, it will not be easily transportable to another organization, and it may even act as an agent against change within the organization—be it change in the survey structures or in the software used to conduct interviews. An example of a tightly integrated laptop case management system is the LIPS system which we describe in Section 14.4.

However, some organizations have more diverse and loosely coupled projects that use several different survey models, and may need to be able to react to new and quite different types of work. These organizations may accommodate different kinds of samples, use different interviewing software (or supplement the CAPI software with different tools to extend its capabilities), or hold the data in different formats. This very different type of system is one that is very loosely linked to the detailed implementation of each survey (see Gray and Anderson, 1996). There may be loss in efficiency (though this can be overcome), but there will be a big gain in flexibility. While maintenance costs may be slightly higher if the organization is highly standardized, for a more diverse organization, maintenance will be cheaper (since the closely coupled system is not needed). One example of such a system is the ONS Casebook system which we describe in Section 14.4.

Laptop systems developed by organizations tend not only to reflect the current situation of the organization but also the history of use of CAI in that organization (Gray, 1994). Systems tend not to just spring into existence but to develop over time, often undergoing major revisions along the way.

There seems to be a definite split between organizations that have extensive CATI experience and those that have not. Agencies with a history of CATI work are more likely to have case management systems already available that can be fitted to the laptop and the CAPI work environment, and to have a number of the skills and understanding of the requirements necessary to build a good system. Conversely, organizations that are moving directly from paper and pencil to CAPI with no CATI experience will be approaching the problem with fresh eyes. Agencies undergoing that first leap from paper to CAI often suffer a squeeze on development resource—frequently the first operational laptop case management system is a quickly developed tactical solution, built with simple tools, that is far from "state of the art" but fully adequate to get the surveys working. See Kuusela (1995) for a description of the Statistics Finland system, which is strikingly similar to the early ONS system (Gray, 1994).

The first generation of a laptop system will often only last one or two years before the needs change (e.g., more surveys move to CAPI) and the organization gains greater understanding of what its needs are. Developers can then build on the first system to implement features that were omitted due to time constraints. The first system is often never displayed outside the organization, or if it is, it is acknowledged, in a diffident manner, to be a tactical system.

14.4 TWO APPROACHES TO DEVELOPING A COMMON SYSTEM

Up to now we have discussed the design issues for an organization creating a CAPI management system for its own use. Most large agencies with many internal customers will try to create a "standard" system that works well across a wide range of products and can in theory be exported to other organizations. We now examine more closely two such systems. Then we discuss an alternative approach developed by Statistics Netherlands, one that is designed to enable organizations to build their own integrated CAPI survey management system. This tool is being used by several large organizations to develop tailor-made laptop case management systems; among them are the Australian Bureau of Statistics (ABS) (Wensing, 1995), the Institut National de Statistiques et des Etudes Economiques (INSEE) in France (Dussert and Luciani, 1995), Statistisches Bundesambt in Germany, and Statistics Netherlands.

14.4.1 LIPS and Casebook

In the early 1990s both Statistics Netherlands and ONS found themselves faced with the task of redeveloping their laptop systems. Each was a "second system," in that earlier, tactical, systems were already running successfully. In neither case was there a commerical system available that could be used in the organization.

Statistics Netherlands led with their LIPS (Laptop Information system for Personal Surveys) system in 1992 (Hofman and Keller, 1993). ONS followed in 1994 with their Casebook system. Each organization was committed entirely to surveys using Blaise, and as European government statistical agencies, they might be expected to develop similar systems. Although there are similarities, there are also differences in design; these partly reflect the tools and styles of the designers and partly the differences in the business culture of the two organizations. It is also true that the time difference was important—ONS was able to learn a great deal from the Dutch experience. LIPS and Casebook were not developed in parallel; many of the design decisions made in Casebook were directly influenced by LIPS, and ONS acknowledges the valuable insights gained from examination of the LIPS system.

Organization Differences

One fundamental difference between the two organizations was the degree of standardization between surveys. All social surveys carried out by Statistics Netherlands are standardized to fit within the definitions required by the LIPS system. They all use the same sample frame, the same accounting scheme, and so on. In the ONS, by contrast, there is a wide degree of variability between surveys, and a stronger culture of survey independence (though there has been a greater push toward standardization in many respects since Casebook was developed). So LIPS was developed as a tightly designed system that worked with a well-structured, standardized set of surveys, which it was able to exploit with a rich functionality that was built upon known data structures. The approach adopted by Statistics Netherlands was not appropriate to the different organizational structure of ONS; many diverse projects with differing needs would be very difficult to regiment into such a tightly integrated system as LIPS. A system was needed that acknowledged the variety of projects.

Both systems are somewhat "loosely coupled" with the CAPI software in that they could, in theory, operate with software other than Blaise. LIPS can be best described as a tailored shell around the questionnaire programs. It has the same look and feel as Blaise (version 2), but in practice it does not know that it is running Blaise questionnaires.

In the case of Casebook, loose coupling with the interviewing software was an essential design priority—it was designed to deal with Blaise III and Blaise 2 at the same time, although at the point of development, all that was known about the Blaise III data file structures and calling methods was that it would be different from Blaise 2.

The Casebook system uses techniques from object-oriented methodologies to create a case management system that is fully independent of the details of the CAPI software. The case management role is seen as handling a number of fully independent address-level objects, each with a common public interface for the items necessary for their management, and also internal representations of survey data (which are irrelevant to the handling systems) fully encapsulated and hidden from the management system.

Role of the Interviewer

Both organizations regard their interviewers as skilled practitioners at the interviewing job (the second model identified in Section 14.3.1 above). Both systems were designed with the aim of making it easier for the interviewer to concentrate on the main task of interviewing, and to control their own work. For example, both systems give the interviewer the freedom to fill in accounting information whenever it suits him/her best, either directly after the visit or in the evening at home. Both systems were designed explicitly with the aims of "visibility and control." Visibility, in that the interviewer can always see what is happening (e.g., when transmitting, Casebook displays the list of cases about to be sent back to the office); control, in that all actions are initiated by the interviewer and not by the system and in that all decisions are made by the interviewer (though LIPS and Casebook may disallow some).

14.4.2   Implementation of Specific Case Management Functions

Accept and Store Interviewing Assignments

In Casebook, interview assignments are accepted and stored in the form of sample point level objects. The contents of these vary; they can contain any of the types of assignment identified in Section 14.2.3. Spawning of extra cases in the field is usually done within the object (extra households are generated and stored together in the same file), though it is also possible to create new objects in the field. In this instance some of the unique serial numbering problems identified by Daata and Wojcik (1994) become important.

In LIPS, interview assignments are accepted and stored on a address level or on a person level. Each assignment has a standardized part and a nonstandardized part. The information contained in the standardized part is used by LIPS (e.g., in the scrollable lists). The information contained in the nonstandardized part is only stored by LIPS and given to the questionnaire program. It can be used, for instance, to store data of a previous panel wave. For an address based survey, LIPS takes care of the spawning of household cases.

Display a List of Assignments

A difference between the two systems is in the depth of menu involved. LIPS is completely menu-driven. By using the cursor keys, the interviewer scrolls through menus and lists. LIPS uses the survey as a top level of hierarchy, with addresses at a lower level. All the available surveys on the laptop are presented to the interviewer in a menu. Each survey in the menu is identified by the short description and the survey period, for instance, "Labor Force Survey, 01-11-1997/30-11-1997." While performing actions on the address level, the survey is always identified in that manner on the screen.

The sample addresses are presented to the interviewer as a scrollable list. The list can be sorted according to a number of parameters if required, such as on postal code, alphabetically on street name, and on status value. All assigned addresses are shown, including those that are completed and have been transmitted back to the office, though completed assignments are clearly marked and cannot be selected for interview. LIPS can present a list of all the addresses for all surveys available on the laptop. This is useful for determining the total workload and an efficient route to visit the addresses.

In Casebook, display of assignments is by displaying a list of all stored objects. Here the Casebook system differs from LIPS and many other systems in that a flat list of all outstanding assignments is displayed. This is possible because completed assignments are deleted from the laptop on successful transmission. It is desirable because it means that there is one less level of menus to navigate to find cases. In fact the main screen in Casebook is the list of outstanding assignments—there is no menu option to interview (it is the default action) or to choose the survey or period (all outstanding cases are displayed).

In Casebook there is only one order of viewing of cases because the system knows very little about the contents of each sample point. The sort order is determined by an "information line" attribute that typically refers to the survey, sampling period, and serial number. Thus the default sort order is serial number within period within survey. However, the system can also deal with cases that do not fit into this model, for example, nonsurvey objects for laptop housekeeping, audit, and the like. The need for interviewers to control the sort order is of reduced importance, since only outstanding assignments are displayed.

Select a Case to Interview

In LIPS the interviewer follows a step scheme to carry out an interview. First, the survey to work on is selected from the survey menu. LIPS presents a list of cases for the selected survey. The interviewer selects a case from the list. The interviewer provides sample account information for the case, for example, the number of households living at the address. By pressing a function key, the interview program can be started. After the interview, the account information can be provided.

In Casebook, by contrast, selecting a case to interview consists of highlighting the case and pressing the ⟨ENTER⟩ key. Control then passes to the interview method of the case object; this will control aspects such as subsample selection or rostering between multiple households at an address. When the computer is switched on, Casebook will automatically highlight the previously selected case.

Record the Status of Each Case

The manner in which case status is recorded and used is different between the two systems. LIPS uses the status value to determine the current point in the life cycle of the address in the survey. The status value is used by LIPS in many places, for instance, to determine whether a certain action can or should be performed with the address. Examples of status values are "approached," "completed," and "sent." The status value "approached" indicates that the interviewer provided information to LIPS about the address. If the status value is "completed," LIPS will send all the available information provided by the interviewer to the office when data communication is performed. If the status value is "sent," the interviewer cannot select the address anymore. LIPS will use the status information to order the addresses in the list. Sent cases will continue to appear on the list, but by default they will be sorted to the end.

With Casebook the system uses less information about address status. Since the functionality between the common systems and the survey systems is separated, the common management layer (Casebook itself) needs only to know whether the case can be sent back to HQ. Extra information is available to the interviewer and displayed as part of the information line for each case (but not interpreted by Casebook). Every address object is responsible for deciding whether it is clean or not, according to its own rules.

Initiate Telecommunications
In both organizations data communication between the office and the interviewer takes place via modem directly to a headquarters system. Diskettes are available in the event of difficulty, but they are seldom used. Data transmission is initiated by the interviewer in both organizations and controlled by password access control. With LIPS the interviewer can choose to delay the transmission to nighttime, but more often the interviewer will transmit immediately in order to observe the transmission.

Performing System Functions
With Casebook the contents of an object do not have to constitute a survey interview. The same case handling system now deals with many of the system and other functions identified in Sections 14.2.3 and 14.2.4. Although most of the expected systems functions are built into the core of Casebook, ONS can easily extend its performance of ad hoc system functions by sending new object files to the interviewers. The flexibility of the objects has been demonstrated many times—ONS can send programs and nonstandard data to the interviewers and handle them entirely within the standard case management system as if they were interviewing assignments.

14.4.3 Tool for Developing Integrated Survey Management Systems

As was described above, different organizations have been developing their own case management systems for use on portable computers, often using database packages such as Clipper and FoxPro, with high-level general-purpose languages such as Pascal and Basic or just DOS batch files. This means that the user interface may differ from the interview software. There may also be different data representations within the system and thus less control in case of error, since the management system is not fully integrated with the interview software. Writing tailor-made management systems is expensive and time-consuming, so naturally these organizations look to the CAPI software developers to provide a solution. Unfortunately, the variability between different organizations and their requirements means that the provision of a single case management system is not readily achievable.

When Statistics Netherlands, the developers of Blaise CAI software, found themselves in this dilemma, rather than attempting a universal case management system as part of Blaise, they opted instead to develop a tool set designed to make in-house development of case management systems much easier and less expensive for survey organizations.

The result is a tool called Maniplus. It was designed from the start as a high-level language for case management programming with the minimum of effort. There is an emphasis on providing good interface components—menus, dialog boxes, context-sensitive help, and so on. These are all easily defined in the language. With Maniplus it is possible to implement the core functions as described above; however, it also provides enough flexibility to add other desired or required functionality. There will always be specialized tasks, for instance, data communication, that cannot be directly executed within Maniplus, so it supports both access to the operating system and Dynamic Link Libraries (DLLs).

Maniplus can be used as a case management shell around most CAPI software. However, it has been designed to integrate very tightly with Blaise. It has, for instance, extensive alternatives for maintaining Blaise data files and can also carry out complex calculations, generate user defined reports, and so on. It gives direct access to two important components of the Blaise system, the interview engine and the Data Viewer.

This interactive aspect of Maniplus is a very important feature. Without too much effort, nice looking interactive systems can be built that, because of the integration with the interviewing software, offer a consistent user interface for all parts of the system the interviewer has to interact with.

Maniplus has already been used to rebuild the redesigned LIPS, but it could also be used to implement the object approach outlined in Section 14.4.2. What is important for a successful laptop management system is both a good philosophy and a good tool to implement that philosophy. A tool like Maniplus is of great help here.

14.5 CONCLUSION

Connett (Chapter 13) suggests that it is possible to develop a generalized case management system capable of functioning across multiple interviewing systems and organizations. We believe that although it may be possible, a generalized method is unlikely to come about in the near future. The core interviewing task is relatively easy to define and relates to fairly standard interviewing methodology. However, the organizations that run CAPI surveys are very different, and have different ways of working and managing their fieldforces. We have seen a variety of factors that lead to different solutions to the case management problem (or even ideas on what the problem is).

There is certainly scope for better provision of case management tools and software by the CAPI software houses. However, organizations that expect to run more than one variety of CAPI software will need to look elsewhere for their solutions.

Tailor-made solutions are expensive, but for now they offer the only real solution for large survey houses whose complex range of projects depend on a variety of software. The short- to medium-term solution is to use tools more extensively to "bolt together" big systems from small parts. In the longer term we can regard CAPI case management as a special case of electronic messaging and groupware. We can expect the functionality and adaptability of main-stream products from Microsoft, Lotus, and the rest, to become sufficiently broad to enable the CAPI case management problem to be merely a case of tuning an existing product.

We predict that the CAPI management system of the future will be an industry standard groupware product, tuned for better HQ control, transport-ing survey data, metadata, and programs as encapsulated objects.

CHAPTER 15

Automated Call Scheduling: Current Systems and Practices

Teresa Parsley Edwards, R. Suresh, and Michael F. Weeks
Research Triangle Institute

15.1 INTRODUCTION

The process of making the telephone calls to contact and interview eligible respondents and identify and close out ineligible and nonresponding cases is a necessary and resource-intensive component of every telephone survey. To structure, monitor, and optimize the call-making effort, survey researchers use some variant of a call scheduling system.

In the pre-CATI era, call scheduling was implemented manually by the telephone survey staff. While there were usually some standard guidelines about when to call for various types of cases (e.g., call back a "busy" number in 15 minutes), decisions about when to call were left to the discretion of the interviewer. With the advent of CATI, automated call scheduling systems have replaced the manual systems. These systems enhance the efficiency of the calling effort by reducing or eliminating the time-consuming manual processes and by employing sophisticated computerized calling algorithms to optimize the scheduling of calls instead of relying solely on interviewer judgment.

15.1.1 Goals of This Chapter

This chapter describes and discusses the features currently available in call scheduling systems, with an emphasis on how survey organizations use and value these features. We present an overview of current systems and practices in automated call scheduling and review and summarize the documentation for several commercially available CATI packages on the market in 1996 that

Computer Assisted Survey Information Collection, Edited by Mick P. Couper, Reginald P. Baker, Jelke Bethlehem, Cynthia Z. F. Clark, Jean Martin, William L. Nicholls II, and James M. O'Reilly. ISBN 0-471-17848-9 © 1998 John Wiley & Sons, Inc.

include an automated call scheduling component. We conclude with a summary of our findings on the prevalence and importance of call scheduling features and suggest improvements and future developments for call scheduling systems.

15.1.2 Survey Design

To summarize current call scheduling practices, we surveyed U.S. and Canadian survey organizations. To ensure a mix of academic and commercial organizations, we selected all U.S. and Canadian survey organizations in the Blue Book of the American Association for Public Opinion Research (AAPOR). We then added all the federal government agencies who conduct CATI surveys and academic survey organizations who subscribe to *Survey Research* published quarterly by the University of Illinois. Questionnaires were mailed to a contact person in mid-April 1996. A cover letter asked the recipient to give the questionnaire to the most knowledgeable person in his/her organization. Thank you postcards, follow-up mailings, and telephone contacts were used to maximize response rate. When mail nonrespondents were willing, an interview was conducted by telephone or by FAX. Overall, a 91 percent response rate was achieved.

Omitting cases where the organization had gone out of business since the mailing, the results for the 236 organizations in the sample are as follows: 157 responded by mail or FAX; 58 responded by telephone; 5 refused by telephone; 3 said they returned the questionnaire, but it was not received; 7 respondents were unable to be reached by mail or telephone; and 6 agreed to respond by FAX but did not. While there was a total of 215 respondents, only 110 had conducted any CATI surveys in the past year, and only 76 used automated call scheduling for any of their projects. The discussion presented in this chapter is based on the information provided by the 76 responding survey organizations.

Of the 34 organizations who did not use automated call scheduling, 13 did not have networked interviewing stations or use software that did not provide the capability. The remaining 21 did not use automated call scheduling because of insufficient programming resources or because they "just haven't developed that part of the system yet." Many indicated they planned to do so or were in the development stages. A few organizations said they had "no need for it" or thought the cost of implementing autoscheduling prohibitive.

We succeeded in surveying a wide range of organizations. The 76 responding organizations included 5 government agencies (2 CATI shops within a single organization are counted separately), 37 university organizations, 31 private companies, and 2 public companies. Each organization's annual number of surveys ranged from 2 to 1,120. The size of their largest survey ranged from 300 to 400,000 completed interviews. The organizations had anywhere from 10 to 900 CATI workstations. They used a wide array of CATI software packages for call scheduling, as shown in Table 15.1.

Table 15.1. Call Scheduling Software Used by Responding Organizations

CATI Software	Number of Users in Sample
Access	1
ACS Query	6
Blaise	2
Ci2 (Sawtooth)	1
Ci3 (Sawtooth)	14
Computer Assisted Survey Execution System (CASES)	18
Computer Assisted Software System (CASS)	2
DASHcati	1
EIS	1
Info 01 Interviewer	3
MacCATI	4
PC Survent/C Survent (Computers for Marketing Corporation)	8
Pulse Train (Bellview)	2
Quancept	3
QuizWhiz	1
Surveycraft	2
Survey System	1
Own proprietary software	6

15.2   ISSUES IN CALL SCHEDULING

15.2.1   Features of Call Scheduling

CATI call scheduling has been broadly defined as the system used to implement and control the call-making effort in a CATI survey (Weeks, 1988). Call scheduling systems are often confused with case management systems. We view call scheduling as a component of case management. While a call scheduling system performs functions like scheduling, monitoring, and controlling, the larger case management system offers additional functions, such as managing the selection of sample numbers and coordinating mailing of advance letters. (See Connett, Chapter 13, for more detail on case management systems.)

Typical CATI call scheduling systems offer a variety of features designed to improve the chances of a call resulting in a successful outcome while at the same time improving the general efficiency of the call-making effort. Some of the more important features include:

- keeping track of appointments and scheduling callbacks at the appointed time;
- controlling the scheduling of appointments based on the availability of telephone interviewers to keep the appointments;

- scheduling "cold calls" (i.e., first calls to cases and callbacks to cases where no prior contact has been made) at times when the chances of contacting an eligible respondent are best;
- implementing prespecified calling algorithms for various other types of cases (e.g., "busy" numbers, broken appointments);
- prioritizing so that the most important cases are called first;
- assigning special cases to appropriate types of interviewers (e.g., refusal converters, bilingual interviewers);
- adjusting for time zone differences so that cases are not called at inappropriate times;
- closing out cases automatically, or referring them to a supervisor for review, after a prescribed outcome or level of effort has been reached;
- projecting staffing needs for future work shifts, based on the number and types of cases available;
- producing a variety of status reports.

A CATI call scheduling system minimizes the total number of calls required to complete the survey. For example, the scheduling of appointments improves the success rate of subsequent callbacks, and using complex algorithms to schedule cold calls at optimal times helps improve the contact rate thus reducing the total number of calls required to reach an eligible respondent. The reduced number of calls means less interviewer labor is required to complete the survey. If the call scheduling system can predict the workload on future shifts, this information can be used by the telephone survey manager to optimize the staffing of these shifts.

CATI call scheduling systems are also thought to improve response rates. For example, the ability to make and keep appointments, control the timing of calls so that calls are not made at inappropriate times (e.g., after 9:00 PM at night, respondent time), assignment of calls to certain types of interviewers in special cases, and automatic referral of problem cases to a supervisor are features that can have beneficial effects on survey participation.

15.2.2   Previous Research

The early literature on call scheduling focused almost exclusively on the timing of calls to sample households and the effects of a variety of factors on call outcomes, including time of day, day of the week, seasonality, and type of area surveyed (e.g., Falthzik, 1972; Rogers, 1976; Groves and Kahn, 1979; Fitti, 1979; Wiseman and McDonald, 1979; Vigdehous, 1981; Kerin and Peterson, 1983; Warde, 1986). Weeks, Kulka, and Pierson (1987) reported that the chances of finding a respondent at home and conducting an interview were much better on weekday evenings and on weekends than during weekday daytime hours, and Kulka and Weeks (1988) experienced a reduction of almost

25 percent in the average number of calls required to contact a household when they used a three-call scheduling algorithm based on the probabilities of success for the various time periods in Weeks, Kulka, and Pierson (1987). Kulka and Weeks (1988) also examined the call outcome data from another random digit dialed (RDD) survey and found that the probability of contacting someone on a cold call is conditional on the timing of previous, unanswered calls to the sample number.

The first detailed discussion of computerized call scheduling appears in Weeks (1988). Based on a survey of leading survey research organizations, he outlined several advantages of an automated call scheduling system over a traditional manual system, including reduced potential for human error in making call scheduling decisions; reduced labor associated with managing the call-making effort; use of calling protocols complex beyond the capacity of a manual system; and use of numerous utility programs and system-produced reports that can forecast interviewer workload and optimize staffing, evaluate interviewer performance, and more effectively manage the overall call-making effort (Weeks, 1988).

Brick et al. (1996) reiterate the findings of higher contact rates on evenings and weekends and extend this result to second calls as well. Furthermore this work also demonstrates that households in census blocks with higher median income are more likely to be contacted on the first call, and those in areas with higher median years of education and higher percentages of black or Hispanic residents are less likely to be contacted on the first call—regardless of when the first call is made. Brick et al. (1996) also look at the timing of calls that resulted in completed interviews and of those resulting in refusals, and find the same patterns as in contact rates. Weeks, Kulka, and Pierson (1987) find no difference in patterns for completed interviews.

In 1990 Stokes and Greenberg proposed a conditional probability model for optimizing cold calls. Modeling call history data from a completed U.S. Census Bureau RDD survey, they proposed that prioritizing cases by their probability of contact on the upcoming call would increase calling efficiency by decreasing noncontacts. In their data, probability of contact was best predicted by a logistic regression model including the number of calls previously made, timing of past unanswered calls, time since last call, and outcome of previous attempt. The Stokes and Greenberg model was never implemented in an actual survey.

In a different approach Greenberg and Stokes (1990) used a Markov decision process on the same data to generate optimal calling procedures. The findings, however, are largely affected by the short survey duration (two weeks) and the Mitofsky-Waksberg sample design which required cluster screening, a procedure no longer used by most survey organizations.

15.2.3   Autodialing

A relatively recent innovation in call scheduling is autodialing. We classify autodialing into three types: simple autodialing, autodetecting, and predictive autodialing. In simple autodialing, a dialing device (e.g., a modem) at the local PC dials the telephone number upon command from the interviewer. Autodialing eliminates misdials and is faster than manual dialing. In autodetecting, telephone signal processing hardware detects certain outcomes such as busy signals, modems, FAX machines, and "tri-tone" nonworking number recordings. The interviewer hears the dialing but is not required to take any action unless a voice answers. The autodetection device reports no-answer outcomes directly to the call scheduling system.

Predictive autodialing extends autodetecting to another level. The system dials numbers automatically and transfers calls to interviewers the moment the call has been answered. No-answer, busy, nonworking numbers, and other such outcomes are dealt with automatically. Some systems also attempt to distinguish a live voice from an answering machine, thus requiring human involvement only on calls with a respondent. The accuracy of the answer versus no-answer distinction is typically greater than the respondent voice versus recorded voice distinction. Neither is perfect, and some cases are invariably handled incorrectly.

Predictive systems rely on algorithms that use variables such as the number of waiting interviewers, the average length of the interview, and the total call attempts to "predict" the availability of an interviewer to take the call. If an interviewer is not available, the call is "abandoned," that is, the system "hangs up on" the person who has answered the telephone. The researcher can set the abandonment rate so that the system adjusts its dialing pace to stay within the prescribed rate of abandoned calls. In general, a large number of potential interviewers can maintain a fast dialing rate while still minimizing the abandonment rate. A large pool of interviewers increases the probability that at least one interviewer will be available when the autodialer detects a live voice.

Predictive autodialers are used primarily in North America, especially by organizations that conduct large and frequent RDD surveys that typically involve large numbers of cold calls. Predictive autodialers may be stand-alone systems (some of which can be linked to a CATI system) or an integrated component of a standard CATI system. Stand-alone predictive autodialers have some call scheduling capability, as well as the capacity to conduct relatively straightforward interviews. Predictive autodialers that are already integrated into a CATI system permit use of the call scheduling and complex instrument capabilities of the CATI system.

Despite the potential advantages of predictive autodialers, some telephone survey organizations express concern about their cost, the need to abandon some calls, and the lack of sophisticated call scheduling capabilities. The price of a sophisticated system usually costs three to ten thousand dollars per interviewing station. In cases where the organization is already using a CATI system that does not include an integrated predictive autodialer component, there is the added complication of needing to integrate a stand-alone system into the existing CATI system. In addition there is typically a slight pause between the time a person answers the telephone and the time the interviewer takes control of the call. As predictive autodialers proliferate and public awareness of this technology increases, response rates to surveys using this technology may be affected. That is, potential respondents will correctly identify the lag in response as signaling a survey request, or (more likely) a telemarketing call, and hang up during the delay rather than wait for the arrival of the interviewer.

While the potential advantages of predictive autodialing are tremendous, there is no published report of actual cost savings in a telephone survey, and to our knowledge, no controlled experiments have been conducted.

15.3   FEATURES OF CURRENT SYSTEMS

The features we consider important in distinguishing one call scheduling system from another are shown in Table 15.2. The list was developed based on (1) our own experience with CATI call scheduling, (2) call scheduling features listed in various published checklists (e.g., Carpenter, 1988; deBie et al., 1989; Connett et al., 1990; O'Muircheartaigh and Murphy, 1991), and (3) documentation from the vendors of various CATI software packages. We asked responding survey organizations to indicate the importance of the features in Table 15.2 to the surveys they conduct. The scale ranged from 1 to 5, with 1 denoting "not at all important" and 5 denoting "very important." The mean score for each feature is shown in Table 15.2. Scores range from 3.4 to 4.7. Below we discuss each of these features in detail. In addition we contacted the vendors of nine commercially available CATI software packages that include call scheduling capabilities and noted the presence or absence of several of the call scheduling features in their software (see Table 15.3).

Before reviewing the features, we define two commonly used terms. "Time slots" refer to divisions of the calling week which divide the week into meaningful sections that are then used to schedule a case for its next call. For example, the time slots used by RTI are as follows:

Slot 1: Monday–Friday, 8 AM–5 PM
Slot 2: Monday–Friday, 5 PM–9 PM
Slot 3: Saturday, 9 AM–5 PM
Slot 4: Saturday, 5 PM–9 PM
Slot 5: Sunday, 1 PM–5 PM
Slot 6: Sunday, 5 PM–9 PM

Time slots are specified in the respondent's local time, and when combined, must cover all calling hours for the week.

A second commonly used term is a scheduler "queue." Some software systems use queues and some do not. The purpose of a queue is to assure that certain actions are performed on all cases in a given queue and that these actions are different from the actions performed on cases in other queues. For

Table 15.2. Perceived Importance of Call Scheduler Features

Feature	Importance
Minimizing call attempts and maximizing likelihood of contact	
Ability to customize number and definition of time slots	4.1
Built-in feature to scatter calls	3.8
Ability to specify time slot for next call to case	4.4
Ability to use priority scheme to prioritize cases within queues	4.2
Ability to customize priority scheme	4.1
Handling special situations	
Ability to put case "on hold" for specified period of time (e.g., a cooling-off period for refusals)	4.0
Ability to assign cases to certain types of interviewers (refusal converters, Spanish speakers, etc.)	3.9
Ability to customize number and types of calling queues	4.1
Ability for supervisor to assign a particular case to a particular interviewer	3.6
Ability for interviewer to request a particular case by entering ID number	4.3
Ability to access a case again quickly after being worked	4.1
Time zone and daylight savings time adjustments	
Automatic adjustments for time zone differences and daylight savings time	4.4
Appointments	
Ability to prioritize different types of appointments	4.2
Interaction with facility/project staffing	
Consideration of staff availability in allowing new appointments to be set	3.7
Ability to project staffing needs for future shifts	3.9
Availability of daily status reports	4.7
Availability of hourly status reports	3.4
Availability of status report at any time upon request	4.4
Built-in reports on interviewer productivity and performance	4.5
Ability to customize reports on interviewer productivity and performance	4.4

Sample size = approximately 72.
Scale: 1 = Not at all important, 5 = Very important.

example, initial refusal cases are often assigned to a refusal queue. As will be discussed later, many organizations use special interviewers to re-call these cases as part of refusal conversion. Furthermore queuing allows easy allocation to those interviewers qualified for such cases.

15.3.1   Minimizing Call Attempts and Maximizing Likelihood of Contact

A primary objective of automated call scheduling is to contact the sample unit as quickly and efficiently as possible—that is, to make as few unproductive calls as possible before achieving contact. When calling a new case or one in which there has been no previous contact, the most common strategy is to spread call attempts across varying days of the week and times of day. This is often referred to as "scattering" the calls. Weeks (1988) describes two basic ways of scattering calls. One is to use a preset algorithm specifying exactly which day and/or time slot will be assigned for the next call when the current attempt results in a noncontact outcome. This algorithm can be as simple as "try again next shift" or as complicated as specific patterns of day/time combinations to be tried in a specific order. Nevertheless, the algorithm always determines a general or specific date/time, which is stored by the system. The other basic approach is to calculate and store a priority score for each case. At designated intervals the scheduler searches through all cases and assigns a priority score based on specified criteria. Cases are then dealt to interviewers "off the top" of the prioritized list until the next scheduler run. Scattering of calls is accomplished by assigning a higher priority to cases that have not yet been attempted during the current time slot or that have been tried fewer times relative to other available cases.

The priority score approach is usually very flexible and can serve additional functions besides scattering cold calls. For example, if a subset of cases is designated for inclusion in preliminary data releases, the priority scores of these cases may be "boosted" so that they move toward completion as quickly as possible.

The preset algorithm and priority score approaches have advantages and disadvantages. Using prespecified algorithms is less flexible and requires good precision in specifying the algorithms. Since there are cases that should not be called during certain times, there can be some interviewing shifts that suffer from a severe shortage of caseload. While the predetermined optimal time for the next call may be at some point in the future, it might still be a relatively productive time to call many cases, and for staffing reasons, flexibility in the number of cases available to be worked is desirable. On the other hand, if there are no hard limits (as is the case with most priority schemes) and too many staff are available for the caseload, cases near or at the very bottom of the priority list will be called. This often results in wasted resources, since the probability of a contact for these cases is very low. Thus the priority scheme requires a very close integration of the scheduler and interviewer staffing. While we do not know of any system that provides such a feature, this issue might be addressed by a priority score approach that prohibits cases below a specified priority level from being called.

Our survey showed that 35 of the 76 responding organizations using automated call scheduling (46%) apply call scattering to noncontact cases. Our survey questionnaire included five items designed to measure the importance of features related to scattering calls. The first was the *ability to customize number and definition of time slots*. Depending on the population surveyed, the critical days/hours of calling may differ. The time slots shown earlier are used at RTI for household surveys. They are based on the assumption that for most

Table 15.3. Availability of Call Scheduling Features in Commercial CATI Packages (Version number in parentheses)

	ACS Query (4.4)	Blaise III (1.15)	Ci3 (3.1)	CASES (4.1)	MacCATI (2.25)	PC-Survent (7.0)	Pulse Train (6.96)	Quancept (7.0)	Survey-craft (7.2)
Ability to customize number and definition of time slots	Yes	Does not use time slots	Does not use time slots	Yes	Does not use time slots	No	Does not use time slots	Yes	Yes
Built-in feature to scatter calls[1]	Parameters	Fixed	None	Parameters	Fixed	By inst	By inst	Umbrella	Parameters
Ability to specify time slot for next call to case	Yes	Does not use time slots	Does not use time slots	Yes	Does not use time slots	Yes	Does not use time slots	Yes	Yes
Ability to use priority scheme to prioritize cases within queues	Yes	No	No	Yes	No	Yes	Yes	Yes	No
Ability to customize priority scheme[2]	Yes	N/A	N/A	Yes	N/A	Yes	Limited	Yes	N/A
Ability to put case "on hold," for specified period of time[3]	Yes	By sup	By sup	By sup	Yes	Yes	Yes	Umbrella	Yes
Ability to assign cases to certain types of interviewers[4]	By ID	By ID	By ID	By type	By type	By type	By type	By type	By type
Ability to customize number and type of calling queues[5]	Simulate	Does not use queues	Does not use queues	Yes	Umbrella	Does not use queues	Yes	Yes	Simulate
Ability for supervisor to assign a particular case to a particular interviewer	Yes	Yes	Yes	Yes	No	Yes	Yes	Yes	Yes
Ability for an interviewer to request a particular case by entering ID number	Yes	Yes	Yes	Yes	No	Yes	Yes	Yes	Yes
Ability to access a case quickly again after being worked[6]	Yes	Yes	Busies	No	No	Yes	Yes	Yes	Yes
Automatic adjustments for time zone differences and daylight savings time[7]	Yes	Offset	Time zone only	Yes	No	Yes	Offset	Yes	Yes
Ability to prioritize different types of appointments	No	Yes	Yes	Yes	No	Yes	Yes	Yes	Yes
Consideration of staff availability in allowing new appointments to be set	No	Yes	Yes	No	No	No	No	No	Yes

Ability to project staffing needs for future shifts[8]	No	No	No	No	No	No	No	Umbrella	No
Availability of status report at any time upon request[9]	Yes	Yes	Yes	Yes	Queue report	Yes	Yes	Yes	Yes
Built-in reports on interviewer productivity and performance	Yes	Yes	Yes	Yes	Yes	Yes	Yes	Yes	Yes
Ability to customize reports on interviewer productivity and performance[10]	No	No	No	No	N/A	No	No	Umbrella	No
Handling of missed appointments[11]	Next req	Next req	Next req	Next req	Reched	Next req	Next req	Reched	Pool
Ability to customize status report	No	Yes	No	No	No	Yes	Yes	Yes	Yes

[1] Parameters: User specifies parameters such as delaying calls, limiting calls per time slot, sorting calls within queues, etc., to accomplish scattering.
By inst: Logic programmed into the questionnaire instrument is responsible for scattering.
Fixed: Preprogrammed rules delay callbacks to noncontact cases for a predetermined amount of time.
Umbrella: An overarching program that controls sample management has the capability to provide this feature. Some programming may be required.

[2] Limited: Limited to status as well as number of calls and other selected variables.

[3] Yes: Cases can be put on hold either through automated (preprogrammed) means or by action by the supervisor on a single case at a time.
By sup: Cases can be put on hold by action of a supervisor on a single case, but no automated facilities are possible.
Umbrella: An overarching program that controls sample management has the capability to provide this feature. Some programming may be possible.

[4] By ID: Must be done by specifying an ID number or set of ID numbers of interviewers who may work the case.
By type: Cases are put into a special "bucket"; supervisors or others later assign particular persons to work the "bucket."
Simulate: Does not use queues, but has a feature which simulates queues and is modifiable by the user.
Umbrella: An overarching program that controls sample management has the capability to provide this feature. Some programming may be required.

[6] Yes: A case can be accessed again immediately after being worked.
No: After being worked, the case cannot be accessed again until the next "run" of scheduler.
Busies: Cases with busy signals will be delivered again soon. Other cases cannot be accessed again quickly.

[7] Offset: Time zones are defined as offsets from the facility time. These can include adjustments for daylight savings time.

[8] Umbrella: An overarching program that controls sample management has the capability to provide this feature. Some programming may be required.

[9] Queue report: Scheduler report shows number of cases in different queues.

[10] Umbrella: An overarching program that controls sample management has the capability to provide this feature. Some programming may be required.

[11] Pool: Missed appointments are dumped into a single pool. A supervisor must review each and determine the next action to be taken.
Next req: Missed appointments can be delivered to the next interviewer who requests a case.
Reched: Missed appointments can be automatically rescheduled for the same time next day.

households weekdays are fundamentally different from weekends and daytime hours are fundamentally different than evening hours. Weeks, Kulka, and Pierson (1987) and Brick et al. (1996) found, however, that there is no fundamental difference between weekday mornings and weekday afternoons. On the other hand, the strength of such a statement depends on the target population. In calling small businesses, for example, which may have restricted hours, different time slots for morning and afternoon might be wise. Separate time slots place some cold calls in the morning and others in the afternoon, rather than allowing all the calls (the same number of calls) to occur in the afternoon. Similarly for an elderly, or other population less likely to be confined by the traditional 40-hour workweek, differentiation between mornings and afternoons might be wise. Other customization of the number of time slots or their hours might be needed for other specialized populations.

The second scattering feature was a *built-in feature to scatter calls.* The call scheduling software packages we examined offered the full range of the continuum for this feature (see Table 15.3). One package predesignates the rules related to cold calls; others leave the decision and development of the algorithm entirely to the user. Some use parameters within the scheduling software itself: for instance, the program evaluates potential cases and assigns them to interviewers when a case is requested. The parameters include, for example, the minimum number of minutes (hours) between calls, maximum number of calls in a time slot, and specifications for prioritizing and sorting cases. In other systems the user specifies scattering logic in the questionnaire program. That is, the program takes into account the days and times of the previous calls and sets an appointment or determines the time slot for the next call before the interviewer exits the current case. Clearly, having a built-in feature that manages all call scattering is convenient if it accomplishes its purpose satisfactorily across all of the organization's studies. However, in our experience no single set of algorithms or parameters is capable of doing this. Therefore it is more desirable to have parameters that are easily modified by the user. This preference was prevalent among our survey respondents too. The mean importance rating for a *built-in feature to scatter calls* was 3.8 on the 5 point scale—one of the lowest ratings among all features.

As discussed above, there are two basic ways to scatter cold calls—the algorithm approach and the priority scheme approach. Within the algorithm approach there are at least three variations. An organization can use appointments that specify a date and time for the next call. We deem this method undesirable, particularly in a large survey with scarce interviewing resources. During peak periods or staffing shortages, interviewer labor should be devoted to actual respondents rather than cold calls. (Often this results in missed appointments—a situation handled more elegantly by some software packages than others.) A second and more flexible approach is to specify the time slot for the next call, but not the exact time or exact day. Thus, if interviewers are busy with appointments on Tuesday afternoon, the cold call can wait until Wednesday afternoon without generating a missed appointment. Nevertheless, both of these approaches can result in staff on hand with no cases to work. A third variation of the algorithm approach provides an "immediate override" feature that allows supervisors the discretionary option of releasing cases originally assigned to a future time.

We also asked respondents to rate the *ability to specify time slot for next call to a case*. All software packages should be able to scatter cold calls using appointments, since the logic and programming code to set the appointment can be built into the questionnaire program—with varying degrees of difficulty. We included this feature to determine how important organizations consider the ability to specify the time slot for the next call without bogging down the system and staff with appointments.

Because of the potential problems with the algorithm approach, we gradually developed a preference for the priority score approach to scattering calls. The primary advantage of this scheme is its flexibility. It ensures that within any given interval between scheduler runs, the highest priority cases are the ones actually worked. For instance, our organization uses a priority score in which one of the sorting variables is the ratio of the number of calls in the current time slot to the total number of calls. Thus a case with one out of seven cold calls in a time slot would be called before one with two out of four calls in the current slot. Because the scheduler evaluates all available cases at periodic intervals, a case's priority score changes as the time slot changes. Thus, in times that are less optimal (i.e., the likelihood of contact is lower), the case has lower priority and is less likely to be called. And yet, the case is not inaccessible should the available staffing level make it possible to attempt the case. Another advantage of the priority score approach is that the parameters and algorithms for calculating the scores can be modified easily over the life of a study. At RTI, changes in prioritizing parameters take effect on the next scheduler run— usually within the hour. Thus, if a client unexpectedly requests that cases from a certain region be given higher priority and worked sooner than the remainder of the sample, this change can be implemented immediately.

We also asked about the importance placed on the *ability to use priority schemes to prioritize cases within queues*. The designation "within queues" was intended to convey that we are not talking about prioritizing one queue over another (e.g., appointments enjoy the highest priority in every organization and all software systems reflect this fact), but rather we were referring to prioritizing cases within a "cold call" or "refusal" queue in which no appointments were involved. This item received a mean score of 4.2 on the 5 point scale. Apparently other organizations lack RTI's enthusiasm for the priority score approach.

An additional feature to be rated was the *ability to customize the priority scheme*. Among the software packages we examined, there were none that allowed for a priority score approach that did not also provide capabilities to design one's own set of parameters or modify the default parameters. In our survey this item had a rating of 4.1, which we interpret as further evidence of lukewarm sentiment for the priority score approach.

15.3.2 Handling Special Situations

During data collection, situations invariably arise that need special handling. Perhaps the most common example is attempts to convert initial refusals. In our survey of organizations, 76 percent of organizations (58 of 76) report giving special treatment to refusal cases. The most common approaches are to allow a "cooling-off period" before the next attempt and/or to assign refusals to special interviewers. For organizations using automated scheduling, it would seem desirable to use the software rather than manual procedures to manage refusal conversion. Only 37 of the 58 organizations, however, indicate that refusal conversion procedures are handled either wholly or partly by the scheduling software. Others rely on paper and pencil.

Another special situation occurs in studies using bilingual interviewers. Cases requiring the special language need to be identified and assigned to the bilinguals for calling. There are at least three ways a call scheduling system can assign particular types of cases to particular interviewers. All assume that each interviewer has an ID number or other unique identifier that the CATI system recognizes when the interviewer logs on to the system. In the first method, a given case is designated to be worked next by a designated interviewer who is identified by his or her ID number. The scheduler then delivers this case if and only if the designated interviewer is logged on and requests a case. Difficulties arise when the designated interviewer is not working at a time the case needs to be called. In the second method, the manager designates groups of interviewers, for instance, refusal converters, by their ID numbers. The case can then be worked by any interviewer belonging to the designated group. In the third method, cases are assigned to interviewers by grouping the cases into special queues (refusal conversion queue, Spanish language queue, etc.). The cases in a given queue are sent to certain interviewers only. The system recognizes these interviewers by their ID numbers which are used for logging on. These two approaches accomplish the same end but through different operational means.

One would think it efficient to handle special assignments through the call scheduling system, but our survey respondents gave the *ability to assign cases to certain types of interviewers* a mean score of only 3.9. Indeed only 48 percent of our respondents (36 of 75) indicated that they used a scheduling system that can deliver certain cases to designated interviewers. (For those that do, however, refusal converters, bilingual interviewers, locating experts, and supervisors were the most common categories reported.) Following the third approach, the *ability to customize number and types of calling queues* is a component of assigning certain cases to designated interviewers. This feature received a slightly higher mean rating of 4.1.

There are other situations that may not require attention by designated interviewers but require specification of the day and time for the next call. An example is when a business answering machine is reached during evening hours. The next call should be scheduled during the case's local business hours to maximize the likelihood of contact. Similarly, when language barriers or refusal cases are encountered, these should be called again during a time slot other than when the refusal or barrier occurred. When a modem or FAX is reached on an RDD survey, some organizations make one or more additional calls to the number before finalizing the case. It may prove beneficial to make the second call at a time of day other than when the first call was made, in case use of the modem/FAX follows a daily routine. By calling at a different time, the odds of a person answering may be improved. However, survey managers may not want to devote precious interviewing resources to appointments made for such calls.

Another use of this feature is to specify the time slot for the first call to the case. Based on data from an RDD survey, Kulka and Weeks (1988) recommended limiting the first three cold calls to evening or weekend hours. The feature for specifying the time slot for the next call can also be used to specify time slot for first call. Among the organizations we surveyed, 46 percent (35 of 76) indicated that they place restrictions on the time of the first call.

In addition to special categories of cases, a supervisor may want to assign an individual case to a particular interviewer. The *ability for an interviewer to request a particular case by entering the ID number* allows the supervisor to direct the interviewer to work on a particular case. The feature also allows a supervisor to review a particular case and determine the course of action in an unusual situation (e.g., a hostile or threatening refusal). Finally special situations might require an interviewer or supervisor to *access a case quickly again after being worked*—as in the case of the hostile refusal.

15.3.3 Time Zone and Daylight Savings Time Adjustments

One of the primary functions of an automated scheduling system is to keep track of appointments and ensure that interviewers receive cases at the appropriate time. Organizations conducting surveys across time zones must translate respondent time into local facility time. In addition, in the United States, whether or not the respondent's area and the telephone facility observe daylight savings time (DST) affects this translation. The translation is important not only in the case of specified callbacks but in determining whether the current time is within an accepted calling range (i.e., 9 AM to 9 PM respondent time).

15.3.4 Appointments

When the telephone facility is short staffed, choices may have to be made about which cases get called during a desired time and which do not. Appointments usually take precedence over cold calls, refusal conversions, and the like. Even within the appointments, however, some may be more pressing or have greater likelihood of resulting in a completed interview than others. For this reason most organizations (71%, or 50 of 70) use multiple categories of appointments.

The categories most commonly used include "hard appointment," "soft appointment," and "estimated" or "guessed appointment." While the labels are shared, the definitions are not. In some organizations, hard appointments are those set by the respondent himself/herself, versus a soft appointment suggested by someone else in the household, or an estimated appointment selected by the interviewer based on the call history. In other organizations, hard appointments are those for a specific date and time, while soft appointments refer to a time span, such as morning, afternoon, or Tuesdays. In still other organizations, a hard appointment is one with a short window of time in which the respondent may be reached (e.g., 7:00 PM to 7:30 PM), while a soft one refers to a respondent who will be available "after 7 in the evening." Some organizations also have categories for busy appointments (to be retried in the near future—such as 15 minutes) and for refusal appointments (initial refusals to be converted).

Once they have developed rapport with a potential respondent, interviewers often feel that they themselves should call back and complete the interview. We asked survey organizations whether interviewers in their facilities are allowed to "reserve" or "flag" callbacks for themselves. Thirty-seven percent (28 of 75) of the responding organizations allow such a policy, and of the 28, 16 say that these situations are handled by the scheduling software (as opposed to manual procedures).

A final issue regarding appointments is what happens when the appointment is not kept, either by the survey organization or by the respondent. We refer to the former situation as a *missed* appointment and the latter as a *broken* appointment. We did not ask our survey respondents how they deal with missed appointments, but we did evaluate the options provided by the nine software packages. In most cases a missed appointment is delivered to the next interviewer who requests a case, even though the appointed time has already passed. This can potentially lead to problems if the hour is now too late to call and the interviewer does not notice this before dialing the number. In two software packages, missed appointments of all types (hard, soft, etc.) are compiled into a single pool which must be reviewed by a supervisor to determine the next course of action. Two packages provide the facility for automatic rescheduling of missed appointments for the same time on the next day, if desired.

We asked the survey organizations how they dealt with broken appointments—situations in which the respondent was not at home or not available when called at the appointed time. Unfortunately, from the comments provided it is clear that many organizations interpreted the question to refer to missed appointments rather than broken ones. Of the 57 respondents who appear to have interpreted the question correctly, 33 (58%) said they are handled by rescheduling the appointment for the same time on the next day. Eleven (19%) said the case is returned to the cold calls queue, and 13 (23%) answered "other." In most cases "other" refers to a supervisor reviewing the case to determine the next course of action. While the automatic rescheduling of appointments for the same time next day may be an effective strategy, it can have limits. Rescheduling the appointment again for the same time next day when the respondent has shown repeatedly not to have been home may not be sensible (though it may provide an easy way of avoidance for the respondent!). Rescheduling strategies should be restricted to a limited number of days after which the case should be opened up for calling at other times or should be reviewed by a supervisor to determine timing for the next call.

15.3.5 Interaction with Facility and Project Staffing

Staffing both the telephone facility and the projects is a tricky business. Interaction with the automated scheduler—meeting all appointments, providing adequate labor to meet scattering goals—adds another level of complexity. To the extent that staffing situations can be anticipated when setting callbacks with respondents, staff shortages can sometimes be avoided. We asked our respondents whether their call scheduling systems consider projected staffing when setting appointments. We know of no software commercially available that does this. Indeed there were only four affirmative responses to this query, and of these four, only two provided follow-up information about how this is achieved. Both have written their own programs which keep track of future appointments and make this information available to interviewers when scheduling appointments.

Another feature we have long thought useful is the *ability to project staffing needs for future shifts.* While the staff needed to call appointments is known at least a short time in advance, the staff needed to adequately work nonappointment calls during a given shift may be harder to estimate. In a priority score approach, one ideally wants to schedule enough staff to work cases down to a certain priority score but not below that level, since the likelihood of contact is small. The algorithm method of scattering may be more adept in predicting staff needed in future shifts. In either case we envision a complex model based on historical data which predicts the staffing needs of each shift of a data collection period. We assume that the sample will be worked to completion by the closing date, and adjustments based on each day's productivity will be made. Again we know of no commercial software package that offers this feature. When asked about the importance of this feature, respondents gave it a mean score of 3.9. We found this surprising as this type of feature has great potential value.

15.3.6 Reports

Some of the commercial CATI software packages include preprogrammed reports. Others leave the development of such reports entirely to the user. Among our respondents, 83 percent (62 of 75) indicated that their software automatically produces reports—the most frequent types were sample disposition reports and interviewer productivity reports.

All facilities need information on interviewer productivity to operate efficiently. Large calling centers and those with high interviewer-to-supervisor ratios may rely more heavily on data from such reports than centers with frequent supervisor-interviewer interaction. Some software packages provide built-in reports on interviewer performance. Nevertheless, some organizations or projects will need additional information or differently formatted information than that provided as default by the program. At RTI, for example, interviewer productivity across cases is linked with information from the telephone and time reporting systems to provide statistics on labor time per completed interview, percentage of charged hours spent on the telephone, and so on. At present such reports have to be tailored by the individual organizations.

15.4 ORGANIZATIONAL DIFFERENCES

15.4.1 Importance of Features

The means shown in Table 15.2 show only small variation, ranging only from 3.4 to 4.7. We thought it possible, however, that the mean scores masked differences among organizations of various types. To test this, we fit linear regression models to determine whether the number of CATI workstations, size of "typical" survey, or number of years experience with CATI surveys are related to the importance scores assigned to the features. We used an analysis of variance model to evaluate the relationship between type of organization (government, university, private) and importance score. Analysis of covariance was used to assess the combined effect. For the analyses the features were grouped by the categories discussed in the preceding section and shown in Table 15.2. Within each category an organization's scores on the features were averaged, and the mean was used as the dependent variable.

In the one-way analysis of variance, private organizations rated the importance of automatic time zone and daylight savings time adjustments significantly higher than university organizations. This likely reflects the makeup of the organization's workload, with universities conducting more local or regional surveys within a single time zone. The difference remains significant in the combined model including the organization's size and years of CATI experience.

When controlling for number of workstations and type of organization, an organization's years of CATI experience is significantly and positively correlated with the importance assigned to features for minimizing call attempts and maximizing likelihood of contact, and with those for handling special situations. A plausible reason is that organizations with long CATI experience have invested in developing and using these features and appreciate their advantages and value. The notion that organizations develop their call scheduling systems gradually over time is supported by the fact that organizations that use automated call scheduling have a significantly higher mean number of years of CATI experience than those who do not (8.6 years vs. 6.2 years). Furthermore open-ended questions on the reason organizations do not use automated scheduling elicit responses like: "just haven't developed that part yet." There were no other significant effects in the regression and ANOVA analyses.

15.4.2 Call Scheduling Practices

We explored whether some organizations are more likely than others to implement the call scheduling practices described above. Government and university organizations are more likely to give special treatment for refusals— such as assigning them to special interviewers or allowing a "cooling off period" before the next attempt. The longer an organization has conducted CATI surveys, the more likely it is to restrict the days and times of first calls and to scatter cold calls across days and times. This supports the notion that organizations develop their call scheduling gradually over time, perhaps starting with simple aspects like automating appointments and moving on to more complicated features that minimize call attempts. There were no other statistically significant differences according to size of organization (number of CATI workstations), affiliation (government, university, private), or years experience with CATI.

15.4.3 Use of Autodialers

Of the 76 organizations surveyed, 33 (43%) reported using an autodialer. Only six, however, have a centralized autodialing system (and therefore the capability for predictive dialing). The others have a modem at each interviewing station that dials the number for the interviewer upon command. For most CATI facilities the latter setup is probably more cost effective. While predictive dialers hold great potential for cold calling, it is difficult to see their utility once contact has been made. Usually there are notes or comments left by the previous interviewer that should be reviewed by the current interviewer before making the call. Predictive autodialing precludes much of the time necessary for such review. For extremely large RDD surveys that screen out nonworking numbers, using a predictive autodialer for the screening phase can entail significant cost savings. Nevertheless, these savings must be balanced against the high purchase price of the system. A solution used by RTI is to send RDD sample numbers to a subcontractor who screens the sample using an autodialer before the data collection starts. It is likely that the use of single modems as autodialers will increase, since the cost of this technology is steadily decreasing. On the other hand, it is only in the largest survey organizations where predictive autodialers are likely to proliferate.

15.5   DISCUSSION

Having ascertained from our survey of organizations the importance of various call scheduling features, we wanted to determine the extent to which commercial CATI packages provide these features. Table 15.4 shows the features in rank order from most to least important, along with the number of packages that do or do not provide the feature.

All but one of the packages provide the two top-ranked features: daily status reports and built-in reports on interviewer productivity and performance. In

Table 15.4. Summary of CATI Packages Evaluated

Feature	Mean Score	Have Feature	Do not Have Feature	N/A
Availability of daily status reports	4.7	8	1	0
Built-in reports on interviewer productivity and performance	4.5	8	1	0
Ability to specify time slot for next call to case	4.4	5	0	4
Automatic adjustments for time zone differences and daylight savings time	4.4	7	2	0
Availability of status report at any time upon request	4.4	8	1	0
Ability to customize reports on interviewer productivity and performance	4.4	0	8	1
Ability for interviewer to request a particular case by entering ID number	4.3	8	1	0
Ability to use priority scheme to prioritize cases within queues	4.2	5	4	0
Ability to prioritize different types of appointments	4.2	7	2	0
Ability to customize priority scheme	4.1	4	1	4
Ability to customize number and types of calling queues	4.1	5	1	3
Ability to access a case again quickly after being worked	4.1	6	3	0
Ability to put case "on hold" for specified period of time	4.0	5	4	0
Ability to assign cases to certain types of interviewers	3.9	6	3	0
Ability to project staffing needs for future shifts	3.9	0	9	0
Built-in feature to scatter calls	3.8	3	6	0
Consideration of staff availability in allowing new appointments to be set	3.7	3	6	0
Ability for supervisor to assign a particular case to a particular interviewer	3.6	8	1	0
Availability of hourly status reports	3.4	8	1	0

fact status reports are provided not only daily but at any time upon demand from the user—a feature that ranked fifth. The ability to specify the time slot for the next call was provided by all packages which use time slots. (Not all packages use time slots, but some have other ways of accomplishing some of the same goals.) Automatic adjustments for time zone and daylight savings time are provided by seven out of the nine packages. Surprisingly none of the packages is capable of customizing reports on interviewer productivity and performance.

Looking down the third column of Table 15.4, as importance decreases, fewer packages provide the feature, indicating either that the software vendors apparently have a good idea of what is important to their users or that the users fail to realize the importance of certain features if they themselves do not have or use the feature. Exceptions are the last two items in the table which are rated low by the organizations but are provided by all but one of the packages. The lowest ranked item (availability of hourly status reports) is a feature provided by the packages by virtue of the fact that reports are available any time requested. The other feature (ability for supervisor to assign a particular case to a particular interviewer) is perhaps an easy feature for the software to provide and therefore offered though not considered important by many users.

We asked the organizations whether there were any other features of a call scheduling system that would improve the quality or cost efficiency of their CATI surveys. The most common response was a desire for more or better reports on survey management and forecasting staffing needs. A few cited the general cost of getting the scheduling system up and running, and a few others wanted better control of and reporting on scattering calls across various times and days of the week.

15.6 CONCLUSIONS

We were surprised by the small variation in the mean importance ratings for the features. For all features the ratings are concentrated in the high range of the scale. Typically the distribution for a given feature includes less than 10 percent of responses below 3, about 15 percent in the middle category (3), and the remainder in the high (4 and 5) categories. Exceptions include *the ability to specify time slot for next call to case, built-in reports on interviewer productivity and performance,* and *availability of daily and on-demand status reports,* which are more highly skewed toward the higher categories. Items with slightly more scores (19% and 25%) in the 1 and 2 range were the *ability for supervisor to assign a particular case to a particular interviewer* and *availability of hourly status reports.* As discussed earlier, there are few differences by organizational size, type, or length of CATI experience, though the sample size is too small for detecting any but the strongest differences. It is possible that the lack of variation reflects the inability of respondents to distinguish one feature from another or to understand the question's intention. However, missing data rates for the items are small (about 5% in most cases). The cover letter asked recipients to give the questionnaire to the most knowledgeable person in their organizations. Manual review of open- and close-ended questions reveals few responses that indicate respondent misunderstanding. Thus we believe that the high ratings indicate real importance assigned to the scheduling features by the responding organizations.

Even the most sophisticated automated call scheduling system can achieve its potential only when it operates in concert with interviewer staffing. If the appropriate number and types of interviewers are not available to work cases at the times designated by the scheduling system, the features will never yield their maximum benefit. For this reason we believe the area most sorely in need of future work is the interaction between the autoscheduling system and interviewer staffing. This includes modeling all aspects of the data collection process to predict staffing needs days or even weeks in advance. Such models would need to take into account various characteristics of the survey, such as desired length of data collection, the population surveyed, desired distribution of calling over weekdays, week nights, weekends, and the like. With each day of data collection, the models should adjust for differences in predicted as opposed to actual production on that day and revise the estimates for future labor requirements accordingly. Certainly such a system is no small order, but once developed, it would be of infinite value to the survey research community.

While the CATI software packages appear to offer the features most desired by users, we encourage vendors to make reports on interviewer performance and production modifiable by the user. Understandably the data needed to evaluate interviewer performance may vary across survey organizations or across projects within a single organization.

Finally we would welcome research comparing various methods of scattering cold calls, as well as an assessment of the level of improvement gained from more complex scattering schemes compared to simpler ones. While several researchers have studied the timing of first calls to a case and a few studied various schemes with the algorithm approach, we know of no research empirically assessing the algorithm approach versus the priority score approach. Such information would help researchers determine the optimal cold calling schedule and help software vendors focus on providing the required features.

ACKNOWLEDGMENTS

The authors thank Beth Berry for her assistance in contacting software vendors and obtaining product documentation and Helen Ray for telephone assistance with nonresponse follow-up.

CHAPTER 16

Getting from There to Here: Electronic Data Communications in Field Surveys

James Smith, Michael Rhoads, and Jane Shepherd
Westat

16.1   INTRODUCTION

The basic purpose of a field survey is to bring information from respondents "out there" to a place "in here" where the data can be assembled and analyzed. With the microcomputer revolution of the 1980s and the appearance of powerful lightweight portable computers in the 1990s, field surveys entered the age of computer assisted survey information collection (CASIC). Today interviewers equipped with notebook computers regularly perform computer assisted personal interviewing (CAPI) using electronic questionnaires with sophisticated data checks, complex question routing, and other useful features.

While computerized questionnaires have been a dominant topic in discussions of CASIC methods, there is much more to a CASIC survey than its questionnaire(s) alone. From the early days of CAPI it was recognized that "a CAPI system includes more than the software for a questionnaire" (Kovar, 1990), and within a few years it became apparent that "the development of CAPI communications and case management has been among the most important additions to personal interviewing technology in recent years" (Nicholls and Kindel, 1993, p. 637). Without effective methods for managing and moving data in CAPI or other types of CASIC field surveys, the other benefits of these survey methods can be reduced or lost.

In this chapter we address the topic of electronic communications in field surveys, with an emphasis on CAPI surveys. When planning electronic com-

Computer Assisted Survey Information Collection, Edited by Mick P. Couper, Reginald P. Baker, Jelke Bethlehem, Cynthia Z. F. Clark, Jean Martin, William L. Nicholls II, and James M. O'Reilly. ISBN 0-471-17848-9   © 1998 John Wiley & Sons, Inc.

munications, survey organizations and practitioners often find themselves facing a bewildering array of communications technologies, strange terminologies, little understood technology options, and unknown risks. Our purpose is to help survey planners address three key aspects of field survey data communications. First, we present a framework for identifying the communications requirements of a survey. Next, we discuss the general architecture of a communications system suitable for a field survey. Finally, we summarize some key communications services that are most likely to be used in survey situations, followed by a brief discussion of communications security. Although our discussion focuses mostly on CAPI surveys, much of the material can be generalized to other types of field surveys as well.

16.2   SURVEY COMMUNICATIONS REQUIREMENTS

A simple model can be a useful tool for identifying the communications requirements of a survey. Such a model helps to create a shared understanding between survey operations staff who will use the communications system and those technical staff who develop and support its operation. From this foundation, survey processes and communications paths can be specified with increasing detail, ultimately yielding a functional survey communications system.

A first step in formulating the model is to list the survey processes, players, and items to be communicated. For a very simple field survey the list includes *survey processes* such as sending initial case assignments, collecting completed work, and regular case status reporting; *players* such as home office, supervisors, and interviewers; and *items* to be communicated such as cases, case status data, and messages. As we will see later (Figure 16.4), these lists can become quite elaborate for a real survey, and completely identifying these elements for a complex survey is an important planning exercise in itself.

Using these lists, a process-player matrix is created as shown in Figure 16.1. The purpose of this matrix is to show the communications actions taken by each player in fulfilling each survey process. For example, for this survey's case assignment process, the supervisor receives cases from home office, assigns them to interviewers, and transmits these assignments back to home office. Interviewers then receive their case assignments from home office. Whatever the particular processes in the survey, a matrix of this kind forms a catalog of the communications actions that are required. This or some equivalent explicit specification with reference to players and processes is a useful bridge between the person-oriented view of survey operations typically held by operations staff and the more process-oriented, and often more mechanical, view of the same activities held by technical staff. Furthermore a communications flow diagram is created from the process-player matrix as illustrated in Figure 16.2. This is the foundation for elaborating more detailed flow specifications that ultimately show all of the communications paths required by the survey.

	PLAYER		
Process	**Home Office**	**Supervisor**	**Interviewer**
Case Assignment	• Originates cases • Assigns cases to supervisor • Relays supervisor case assignments to interviewers	• Receives cases for assignment • Makes case assignments to interviewers • Transmits case assignments to home office	• Receives cases for work from home office
Completed Work	• Receives finalized cases from interviewers	• May receive completed cases for QC review from home office	• Transmits finalized cases to home office
Case Status Reporting	• Relays interviewer case status information to supervisors	• Receives case status information from home office	• Transmits case status information to home office

Figure 16.1. A simple process-player matrix.

So far the model is devoid of metrics, that is, the various size, volume, and frequency measures that are so vital in planning any communications system. How many supervisors and interviewers will there be? Is there a single home office or multiple offices? What is the size and frequency of communication of each item across each path? A complete list of these metrics is needed and can be laid out conveniently in a table like Figure 16.3 for each of the processes, players, and items.

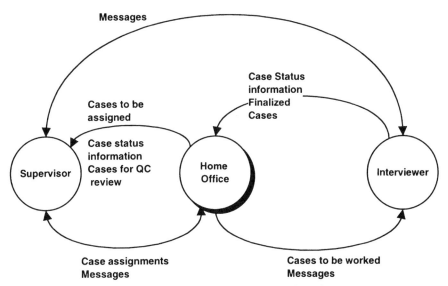

Figure 16.2. A simple field communications flow diagram.

MODEL ELEMENT		TYPES OF BASIC METRICS	EXAMPLES
Players	Home Office	Number / Distribution	1 home office
	Supervisors	Number / Distribution	5 supervisors in sampled PSU areas
	Interviewers	Number / Distribution	75 interviewers, approximately 15 per supervisor
Processes	Case Assignment	Sample Size / Plan for Field Release	Sample of 2,500 cases released in three waves of 1500, 500, 500 respectively
	Completed Work	Number of Cases / Projected Response Rate / Length of Field Period	Planned response rate of 90% 2,250 cases over a 5-week field period
	Status Reporting	Number of Reports / Periodicity	Status reports produced daily
Items	Case Data	Size / Length of interview	30 question, 30 minute administration time, approx. 1Kb per complete case
	Status Data	Size / Complexity	1 status record per case
	Messages	Type / Frequency	Verbal and brief written messages as needed

Figure 16.3. Example of communication metrics.

The process-player matrix and its corresponding path diagram and table of metrics give an orderly summary of a survey's communications requirements. Arriving at a clear and complete specification of these requirements is an essential planning step. Without this, there is substantial risk of confusion, extra costs, and reduced survey yield due to omissions and lapses in communications during the course of the survey. These risks are of particular concern in field surveys where the often wide geographic distribution of the users (i.e., interviewers) means that communications problems may be difficult to diagnose and fix during the data collection process. Carefully specified requirements not only reduce the chance of such problems but also provide a solid foundation for training field and home office staff in all the necessary aspects of communications prior to the start of data collection.

16.2.1 A CAPI Survey Example

Using as an example the Medical Expenditure Panel Survey (MEPS) Household Component sponsored by the U.S. Agency for Health Care Policy and Research and the National Center for Health Statistics, several elements of the communications model are shown for this CAPI survey in Figure 16.4. The first panel of this multiround, multipanel survey collected data from more than 10,000 households using over three hundred field interviewers. The players in this survey included home office, field managers, supervisors, interviewers, and quality control assistants who reported to supervisors. The survey processes were numerous and sometimes complex as were the items of information, so only a few are listed in the figure as illustration of the concepts we are discussing here.

The process-player matrix for this survey is shown in Figure 16.5 with a corresponding flow diagram in Figure 16.6. From the flow diagram it is apparent that home office served as a central communications hub in this survey, even for case transfer between field staff. One of the most challenging requirements of this survey was the need to make case transfers rapidly between interviewers while still maintaining stringent field supervisor control over all case transfers. In the planning stage of the survey, several alternative procedures for doing this were evaluated in terms of the number of communications required and the resulting elapsed time necessary to perform and confirm a case transfer. While the anticipated number of case transfers was not large, these transfers were important beyond what their numbers would suggest because some geographic areas were covered entirely by traveling interviewers who received their cases through the case transfer process.

16.3 CHOOSING A SURVEY COMMUNICATIONS SYSTEM

Although it must satisfy challenging requirements, a field survey communications system, like the familiar telephone system, needs to be extremely reliable

PLAYERS	SURVEY PROCESSES	ITEMS
Home Office	Initial Case Assignment	Case Data
Field Managers	Case Re-Assignment/Transfer	Case Status Data
Supervisors	Completed work	Messages
Interviewers	Case Status Reporting	Permission Forms
Quality Control Assistants	Quality Control (QC)	Quality Control Data
	Time and Expense (T&E) Reporting	Record of Calls Data
		Time and Expense Data

Figure 16.4. Selected communications model elements for the MEPS CAPI survey example.

PLAYER

Process	Home Office	Field Manager	Supervisor	Interviewer	Quality Control Assistant
Initial Case Assignment	Originates and supervisor makes initial assignment at start of field period to an interviewer	Reviews initial assignments with supervisor	Receives initial case assignments from home office and makes interviewer assignment; sends initial assignments to home office	Receives from home office at start of field period	No Action
Case Assignment	Originates and sends to supervisor on a flow basis throughout field period to assign to an interviewer	Reviews case assignments with supervisor	Receives and assigns to an interviewer	Receives from home office as new work on a flow basis throughout the field period	No Action
Case Re-Assignment (Case Transfer)	Receives re-assignment from supervisor; removes from source interviewer and passes to target interviewer; informs supervisor when transfer is completed	Reviews re-assignments with supervisor	Originates and informs home office and source and target interviewers	Source interviewer "returns" case to home office; target interviewer receives case from home office	No Action
Completed Work	Receives from interviewer	No Action	No Action	Originates and sends to home office	No Action
Case Status Reporting	Originates case status reports based on information received from interviewers	Receives reports from home office	Receives reports from home office	No Action	No Action
Quality Control	Receives information from supervisor and QC assistant; generates reports	Receives reports from home office	Sends information to home office; receives reports from home office	No Action	Originates and sends information to home office; receives reports from home office
Time and Expense Reporting	Receives from interviewer; generates reports	Receives reports from home office	Receives reports from home office	Originates and sends to home office	Originates and sends to home office

Figure 16.5. Process–player matrix for the MEPS CAPI survey example.

313

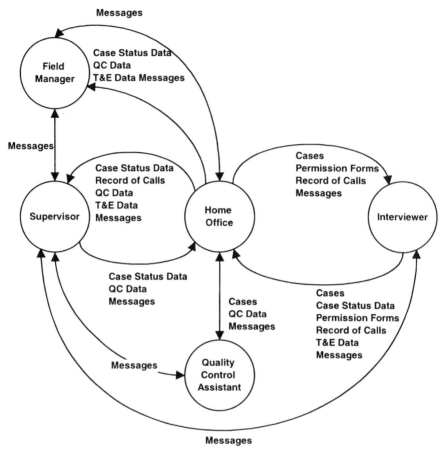

Figure 16.6. Communications flow diagram for the MEPS CAPI survey example.

and simple to use. Providing such a custom-built system from scratch for a single survey can be a major undertaking involving software development, the integration of multiple hardware and software components, and extensive testing. Fortunately most surveys do not require such a fully customized communications system and can therefore benefit from using existing experience and systems. Moreover, in order to control costs, it is usually in the interest of a survey research organization to re-use as much prior experience and technology as possible. In cases where surveys do have genuinely new communications requirements, these can usually be added incrementally to existing system capabilities. When carefully managed, this strategy of re-using existing capabilities along with incremental development of new capabilities can produce a snowball effect that accumulates operational experience and technical capabilities over time within the survey organization while, at the same time, distributing development costs and risks across many surveys.

A major challenge to this strategy of system re-use is rapid technology change in the marketplace. For example, yesterday's fastest laptop computers and modems are today's older and slower equipment, making them seem out of date only a year or two after their purchase. Unfortunately, it seems that today's rapidly changing technology marketplace pays much less attention to maximizing the use of existing computer and data communications assets than to promoting tomorrow's newest technologies. Nevertheless, by proceeding carefully with a strategy of system re-use and incremental development, a survey organization can make effective use of its tried and tested technologies and know-how and by this strategy position itself to capture important new technologies.

Whether a survey communications system is newly developed or re-used with incremental enhancements, it almost always involves integrating multiple technologies into a working whole. However, vendors and developers of the individual technologies do not guarantee that this integration will be entirely successful. Thus so-called "standards," such as modem protocol standards, give only partial assurance because the standards themselves are always evolving and there are often different interpretations of the same standard. Only by planning, testing, and tuning can it be proved that the modems from different manufacturers will in fact communicate at acceptable speed with each other; or that laptop computer CAPI or field management software will import and export the data correctly for transfer to home office; or that the communications "shell" or menu choice item on the laptop computer will use the correct disk directories and file names when passing files to and from the communications software. In order to address the many parts and interactions in a communications system without getting lost in details, an overall architectural framework is needed, to which we now turn.

16.4 GENERAL ARCHITECTURE OF A COMMUNICATIONS SYSTEM

Because of their complexity, computer-based communications systems are typically broken down into layers, ranging from a lowest layer that implements physical connectivity and circuit connections to a highest layer that includes the user's application software. Intervening layers handle tasks such as creating and addressing data packets, encryption and other aspects of data security, and the coordination of the sequence and flow of communications. A widely used industry standard for describing the layers of a communications system is the Open Systems Interconnections (OSI) model (Stallings and Van Slyke, 1994).

For simplicity, most survey communications systems can be adequately described with a simplified three-layer OSI model as illustrated in Figure 16.7. This example system is designed to hold information on either end—laptop computer or home office—until a dial-up connection is made. Thus it is an example what is called a "store and forward" communications system.

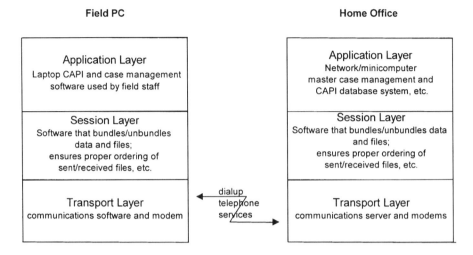

Figure 16.7. Architecture of a typical CAPI survey communications system.

From the field staff users' standpoint, communications is usually a menu item on the laptop computer, called "link to home office" or something similar. Beneath this menu item lies a series of processing steps that prepares and receives files, processes these files, and performs the actual data communications with the home office system. Ideally all of this happens automatically and invisibly without recourse to the user until the whole communications process is successfully completed. Thus the user interacts with the system through software in the application layer that makes up the survey or case management system (see Connett, Chapter 13; Hofman and Gray, Chapter 14).

16.4.1   Session Layer

No survey wants to require its field staff to understand the details of computer files, disk directories, and other technical details underlying CAPI or the other applications they use. Hence the need for session layer software that embodies the rules about what data are to be transmitted or received, and how to prepare and process these data before transmission or upon receipt. Other tasks typically performed by session level software are encrypting outgoing data and decrypting incoming data, compressing and decompressing of files, posting received data to application files or databases, extracting data to be sent from the laptop computer, managing the order in which these various operations are performed, and handling error recovery in ways that maintain data integrity throughout the communications process. In addition the session layer software invokes the transport layer software that performs physical communications.

There are two general approaches as to how the session layer may operate. One is as a transaction processing system that sends only those data representing new information or changes to previous information to home office. This requires the session layer software to extract and send only those data that have changed since the last communications session, such as case status records or newly completed interview data. Likewise, under this approach the session layer must be able to receive selected pieces of information and insert them into appropriate files in the laptop.

Alternatively, the session layer can be somewhat "dumber" and always transmit complete files that contain unchanged as well as changed information. In the latter case the burden falls on the home office processing system to decide which information is new and which is repeated from before, and can therefore be ignored. While at first glance this may seem wasteful, there are reasons for doing so. One reason is that the home office more easily serves as a data backup for laptop computers when it has a recent complete copy of data from field interviewers. Another reason is that transferring all data means the laptop software embodies fewer rules and processing steps than if it must extract only new data. Complex processing rules always imply a higher probability of incompleteness, inconsistency, or other errors, and it is desirable to minimize these risks on geographically distributed laptop computers, leaving complex processing to the home office. Of course this reduction in risk must be weighted against the extra data transmission burden implied by sending more data with every communications session.

Some brief examples of session layers in actual surveys will illustrate some of the approaches that can be taken. In early years of the U.S. Medicare Current Beneficiary Survey (MCBS) sponsored by the U.S. Health Care Financing Administration, a commercial modem communications package with a built-in script language was used for data communications. Each laptop computer contained a script program that performed the various tasks of the session layer. The home office communications server acted as an electronic bulletin board customized for the CAPI survey's communications needs. The script on the laptop simply sent and received files to and from the home office server according to the rules programmed in the script. More recent rounds of MCBS converted the session layer to a commercial file transfer system. In this case all of the session layer rules resided in a script on the home office communications server. Upon connecting to the server, the laptop became a slave machine to the server which then directed the laptop to gather, send, receive, and process files. This system had the advantage of localizing all session layer rules at home office, making it easy to maintain, fix, and modify the rules during the survey.

Finally the MEPS CAPI survey introduced earlier used the approach of sending data packages to laptop computers as "executable transactions." When received by the laptop computer, these packages contained their own executable program along with any relevant data. Thus the session layer rules were embodied in these programs which were created at home office and executed by the laptop computer when they were received. This approach gave very high flexibility for home office to change and modify rules and to customize the processing, including ad-hoc needs, on laptop computers.

16.4.2 Transport Layer

The transport layer of the communications system conducts the actual transmission of data bytes from point to point. The programming of this layer is an onerous task involving numerous technical details of hardware and external communications services. It is hard to imagine a survey where developing this software would be chosen over using a commercial product. However, while the use of commercial products is almost essential, there is the drawback that these products usually come with their own user interfaces. This is not desirable in most survey situations where the goal is to give field staff a simple menu option for performing communications without any additional software for users to learn.

Fortunately it is often possible to embed commercial communications products beneath other software, hiding their user interfaces from the user. For example, an e-mail program or a modem communications package may provide options or an internal script language that can be used to circumvent or suppress the use of the usual user interface. Better yet, e-mail or modem software may provide a separate version without a user interface that can be used for this purpose. In either case the goal is to use the commercial software for the technical tasks of establishing connections and sending and receiving data packets at appropriate speeds, along with error checking and correction, as invisibly to the user as possible.

Whatever software product is used, the transport layer is the interface between the user's computer and the circuits over which electronic communication occurs. Having found a suitable commercial product for this purpose, the actual communications circuits used in a survey will be supplied by a communications service provider. We now review several of the major types of services that are most relevant to field surveys.

16.5 COMMUNICATIONS SERVICES FOR SURVEYS

In traditional paper and pencil surveys, the national telephone and mail systems provided easy to use communications services. In today's surveys, managers and organizations face several choices for the electronic communications services they need, and these choices will become more varied and complex in the future. We consider these services in three broad categories: telephone services, network services, and emerging services including the Internet. A fundamental feature of any communications service is its data transmission speed. Table 16.1 lists several services we will be discussing below with a general indication of their speeds. In general, telephone services (using

Table 16.1. Typical Communication Service Speeds

Type of Communication Services	Typical Data TransmissionSpeeds[1]
POTS (plain old telephone service)	9.6 to 56 kbps
ISDN (integrated services digital network)	64 to 128 kbps
DSL (digital subscriber line)	1.5 mbps (HDSL) 6 mbps/224 kbps (ADSL)
T1/T3 (T-carrier services)	1.5 mbps (T1) 45 mbps (T3)
Frame relay	155 mbps to 1200 mbps
ATM (asynchronous transfer mode)	4 mbps
Cable	25–40 mbps expected

[1] Kbps = kilobits per second, Mbps = megabits per second.

modems) have been the basis for CAPI survey electronics communications and certain types of computer assisted self-administered questionnaire (CSAQ). Site-based surveys are often amenable to more sophisticated and higher bandwidth network services. In recent years the emergence of the Internet has opened the prospects for enhanced widespread electronic communications involving a mix of network and telephone-based services.

16.5.1   Telephone Services for CAPI and CSAQ

Many data collection efforts require easily available and highly reliable data communications systems. As we have seen earlier, CAPI studies require a wide variety of information to be sent back and forth between interviewers, supervisors, and the home office, such as completed cases, new cases to be worked, and case status information. Factors such as timeliness and accuracy, moreover, make it highly desirable that such communications be electronic. Given the decentralized nature of CAPI, interviewers' electronic communications need to be from their homes, or possibly from hotels if they are traveling, so easily available and highly reliable data communications systems are essential.

Widely available data communications systems are also critical to the success of many CSAQ methods, where respondents provide data directly from their homes or businesses. These include touchtone data entry (TDE), voice recognition entry (VRE), FAX, optical character recognition (OCR), and electronic data interchange (EDI). For these data collection modes, where wide availability and reliability are critical, it is important to remember that voice telephone services are still the most prevalent and proven communications services throughout the world. Over 90 percent of the households in the United

States have one or more telephones (Lebow, 1995), and similar high percentages prevail in other developed countries. Thus the telephone system is an existing, fully functional infrastructure that is readily available for survey voice and electronic communications at a moment's notice in field staff and respondent households. The remainder of this subsection will briefly discuss the major technical issues relating to telephone-based data communications systems: bandwidth, compatibility, and reliability.

Originally designed only for voice traffic using analog (nondigital) signals across simple copper wiring, telephone systems allow convenient point-to-point communications with telephone numbers as unique addresses. Digital communications between computers is also easily accomplished over telephone circuits using inexpensive modems that can perform telephone dialing and answering and that convert a computer's digital data into analog "sound" pulses for transmission in the telephone system. The size of modems has decreased over time, from early external boxes, to internal cards for desktops, to the current PC Card (formerly PCMCIA) standard for laptops which is no bigger than a credit card.

Through the years improvements in algorithms and circuitry have greatly increased the speeds with which modems can transmit data. In the early 1990s, when reasonably sized laptop computers finally started making CAPI surveys feasible, modems featuring speeds of 2400 bits per second (bps) were the "workhorse" of the industry. By the mid-1990s, most modems on the market claimed throughput rates of 28,800 or 33,600 bps, and 56,000 bps modems began to appear in the late 1990s.

Low modem speeds, up to 1200 bps, tend to be highly reliable over most dial-up telephone lines. Higher speeds, however, are more susceptible to line noise and other problems. Therefore the standards for faster modems typically include algorithms for error correction. In addition, to squeeze out faster speeds, manufacturers began adding data compression features to their modems that minimize the number of bits actually transmitted when the data tend to be repetitive. Even with these techniques, 28,800 and 33,600 are close to the theoretical limits for analog data transmission over telephone lines, which is about 35,000 bps. (56,000 bps modems achieve the rated transmission speed only in one direction and have other special requirements for the communications circuits, thus appearing to exceed this maximum.)

An important issue facing the survey practitioner is compatibility among modems. There are international standards for modulation method, error correction, and data compression so that, at least in theory, modems made by different manufacturers can communicate successfully as long as they use the same protocol. However, there may be detailed and subtle differences in the implementation of high-speed modem protocols that, in practice, can make a difference in reliability and efficiency when modems of unlike manufacturers connect to each other. These compatibility issues can make it difficult to establish or maintain connections, thus causing delays in such critical functions as getting new assignments or software to interviewers.

In a field survey the quality of local telephone lines is often an even more important factor than modem speed and reliability. In actual practice the highest possible modem transmission speeds are achieved only over relatively noise free, high-quality telephone circuits with data that are highly compressible. Thus surveys that rely on data transmission from interviewer's homes or other uncontrolled locations are unlikely to reach these maximums on a consistent basis, and allowance must be made for serious problems with modem communications in some cases, particularly outlying locations with older telephone systems. Rather than relying upon the technical speed capacity ratings of modems alone, survey practitioners should experiment and observe the actual throughput and reliability of modem communications, particularly where large numbers of modems are to be purchased and placed in field use.

Although data communications via modem over telephone lines may not always achieve the maximum claimed transmission rates, it is extremely reliable in terms of data quality. Modern modems use checking techniques that perform a mathematical calculation on each block of data to be transmitted. This calculation yields a specific pattern of additional bits that is transmitted along with the actual data, something like a checksum. The receiving modem uses this information to detect and to direct the sender to retransmit the entire block if an error has occurred. Thus line noise and other problems will reduce throughput by causing data to be retransmitted, but it is virtually impossible for erroneous data to find their way into the central survey database. Such error detection techniques have a reliability rate of 99.99995 percent (Bates and Gregory, 1995), and a communications session will simply terminate with a status of "unrecoverable error" before it will deliver erroneous data to the user.

Thus the ubiquity and reliability of what sometimes is affectionately referred to as POTS (plain old telephone service) continues to make it the technology of choice for data communications on most field surveys. Later we will discuss some expansions of and alternatives to this technology and how these technologies might affect data communications strategies for CAPI studies and other field surveys in the future.

16.5.2 Network Services for Site-Based Surveys

The survey types discussed so far all involve situations where the number of players in the data communications system, whether interviewers or the respondents themselves, may be quite large. However, not all projects fit this pattern; some involve data collection in a single central site, or a relatively small number of locations. This includes studies where the data are collected at treatment centers or other community service agencies, as well as those where dedicated sites may be established specifically for the project.

Site-based studies of this type may be able to use a basic communications strategy relying on telephone connections and modems, similar to what is typically used for CAPI surveys. On the other hand, the relatively small number of players in the communications system can make it cost effective to implement more advanced communications technologies than are practical in studies with hundreds of interviewers. Moreover such studies may have more demanding data communications requirements in terms of data volume or timeliness.

Several alternatives exist for these situations. When there is a long-term need for high speed communications between two sites, leasing a dedicated communications line between them may be the best approach. So-called T-1 digital lines provide 1.544 mbps of bandwidth which, if desired, can be shared among voice and data communications. The T-3 lines are similar in concept but offer much higher bandwidths (44.736 mbps). Digital line technologies such as E-1 are also available in Europe, although the data transmission rates and technical standards are somewhat different from those in the United States. The maturity and reliability of this technology makes it a good solution for situations requiring high-speed communications and where the frequency of communications is great enough to justify the cost.

Frame relay technology is an alternative solution for situations where persistent, high-bandwidth connections are required among multiple sites. Frame relay offers bandwidths up to 2.048 mbps, and as a "packet-switching" technology it allows a communications infrastructure to be shared among a multitude of users, thus eliminating one of the drawbacks of dedicated T-1 circuits. Despite this sharing of bandwidth, however, frame relay allows so-called private virtual connections between sites which offer guaranteed throughput. Frame relay also improves on the cost effectiveness of T-1 by offering highly scalable connections, so the bandwidth of the connections can be set at the optimum level. Furthermore the connections do not have to have the same bandwidth in both directions, which can often be an advantage in survey communications settings where the flow of data is much greater in one direction than in the other. Frame relay service is available from a variety of vendors, including major long distance carriers in the United States.

Designers of more elaborate survey communications systems should also keep an eye on Asynchronous Transfer Mode, or ATM. The ATM, like X.25 and frame relay, is a packet-switching technology, but it is designed to carry voice, video, and computer data at extremely high rates of speed. Currently bandwidths of 155 mbps can be obtained with rates of up to 1200 mbps expected by the turn of the century (Bates and Gregory, 1995). The ATM could well become the technology of choice in the future for site-based survey data communications applications.

The U.S. National Health and Nutrition Examination Survey IV (NHANES IV), which is conducted by Westat for the National Center for Health Statistics, is an example of a site-based study that takes advantage of some of these technologies. NHANES IV uses traveling teams of interviewers, field office staff, and medical examiners, along with Mobile Examination Centers (MECs). Data are collected for approximately two months in each location; at any given time two MECs and three field offices are in operation. The project requires constant coordination between the field offices, MECs, medical laboratories, and home office staff. The transmission requirements include high volume items such as X-ray images, bone scan files, and other document images, as well as more traditional survey data items. The data communications strategy centers on a national frame relay network, with T-1 lines used to link the MECs and the home office into this network.

16.5.3   New and Emerging Communications Services

The overall scenario of communications services in the 1990s may change rapidly in the early twenty-first century. This is largely due to the explosive growth in popularity of the Internet, which has two major implications for survey research. First, as access to the Internet becomes more widespread, it may become a useful alternative to telephone services for data collection modes that require high availability. Second, the popularity of the Internet is driving the development and deployment of higher bandwidth data communications systems that are technically and economically viable for residential as well as institutional use. The remainder of this subsection discusses these two trends further.

Internet use has increased dramatically over the last few years, although usage levels remain much lower in Europe and Canada than in the United States, and even lower in developing countries (Christie, Illingworth, and Lange, 1996). Some have estimated that the number of interactive Internet user accounts worldwide will exceed 152 million in the year 2000 (Meeker and DePuy, 1996). Like TDE, VRE, and FAX, on-line data collection via the Internet will not require particularly high bandwidths or sophisticated communications infrastructure. The critical factor for the success or failure of the Internet for data collection is likely to be the penetration of Internet access in terms of the target population of each survey.

The increasing residential popularity of the Internet, however, has served to expose the relative slowness of data transmission over standard telephone lines and modems that connect homes and many business users to the Internet through Internet service providers. Traditional analog modems are generally adequate for transmitting text, but unless telephone line quality is extremely high, they can be somewhat slow when dealing with the colorful and complex graphics that are featured on the Internet's World Wide Web. The increasing availability of real-time audio and video on Web sites exposes the shortcomings of these relatively slow transmission speeds to an even greater extent. Since it is not possible to squeeze much more in the way of speed from "traditional" modem technology, a number of newer systems and standards are emerging to compete for the residential data communications market. The increasing availability of higher bandwidth solutions could expand the horizons of what is possible in surveys. Large files such as segment maps or other graphical images could be easily sent back and forth between interviewers and the home office. Case management systems may also benefit from higher bandwidth (e.g., see Connett, Chapter 13).

At this point the best established of these newer technologies is Integrated Services Digital Network, or ISDN. The ISDN works over the copper wiring that constitutes most of the "local loop" into individual homes, but it differs from earlier systems in that the data remain in digital form throughout the communications path, rather than being translated into analog form at the point of entry into the telephone system. The type of ISDN service used in homes is generally what is called Basic Rate Interface (BRI), which provides two primary 64 kbps channels of service. The existence of these two channels allows simultaneous voice and data communication. Alternatively, the two channels can be combined to provide data throughput of 128 kbps, or roughly twice as fast as the fastest nondigital modem allows. While ISDN has been available for several years, its use is not widespread, largely because of high equipment costs, installation difficulties, and high monthly line charges.

The next great leap forward in data communications over the existing telephone infrastructure is a set of technologies known as Digital Subscriber Lines (DSL). Like ISDN, DSL takes advantage of the ability of existing copper wiring to carry high-bandwidth signals over relatively short distances. DSL technology, however, offers transmission speeds that are several orders of magnitude greater than what is possible with current ISDN services. Asymmetric Digital Subscriber Line (ADSL) is the form of DSL technology that is most often mentioned for potential use in residential environments. The ADSL takes advantage of the fact that most "home" communications applications, such as video-on-demand or browsing the Web, require much less bandwidth in the "sending" direction (home to host) than in the "receiving" direction (host to home). Therefore ADSL is optimized for high bandwidth in the "downstream" direction, where it can achieve speeds up to 6 mbps (Baines, 1995; Briere and Heckart, 1996). Bandwidth in the other direction is limited to 224 kbps (which is still faster than BRI/ISDN service). In terms of equipment ADSL requires two special modems, one in the subscriber's home and one in the telephone company central office. Another variant is High-bit-rate Digital Subscriber Line (HDSL) which provides transmission bandwidth of 1.5 mbps in each direction, equivalent to a T-1 leased line. The electronic equipment needed to make DSL work is quite complex and thus is likely to be relatively expensive initially, although prices can be expected to drop as economies of scale take effect.

Cable technology can provide much greater speed than ISDN or DSL can. Transmission rates over cable currently run around 4 mbps, as opposed to 64 kbps or 128 kbps for ISDN. Ultimately cable modems could provide speeds of 25–40 mbps. As with ADSL it is envisioned that these ultimate speeds would be achieved only when sending data such as Web pages or video images to the customer. Data sent by the customer would travel at much slower rates, although these would probably still match the speeds achievable by standard ISDN. The primary obstacle in the way of widespread deployment of cable technology as an Internet access mechanism is that cable was developed as a one-way system for carrying signals from the cable provider into subscribers' homes. Extensive system upgrades are required to allow two-way data communication.

If, fueled by technologies such as ISDN, ADSL, and cable modems, general Internet availability ever begins to approach that of telephone service, survey practitioners will have a wider range of options open to them for designing data collection and communications systems. Self-administered surveys could go beyond the limits of current methodologies such as TDE and VRE. Data communications systems for CAPI studies could also benefit. While the volumes of data involved in most CAPI data transmission sessions are modest, the higher speeds made possible by newer communications technologies could significantly shorten certain tasks, such as downloading new versions of software.

16.6 COMMUNICATIONS SECURITY AND CONFIDENTIALITY

Given the critical importance of data confidentiality in survey research, practitioners must give attention to technologies and procedures that will guarantee the integrity and confidentiality of survey data. Unlike the national mail systems used to communicate data in traditional hardcopy surveys, the electronic communications services used by CASIC surveys are seldom under the auspices of a single official organization that can assure the same levels of security, and the same penalties for violations, as national postal systems. This and the inherent diversity and complexity of electronic communications methods make it mandatory for survey research organizations to address key security issues and safeguards in the surveys they perform. Since the field of electronic communications security is vast, we will focus on only a few of the fundamental issues and procedures that are particularly relevant to CASIC field surveys.

Survey data security means protecting data at all stages of the survey process from corruption, loss, or inappropriate disclosure. This includes data at home office, on field computers, or in communications channels. These protections require a mix of technologies, procedures, and policies that are designed to prevent both accidental and intentional events that would compromise data security, along with methods for detecting real or attempted violations. In addition there must be procedures for recovering as completely and rapidly as possible from problems that may arise. To be maximally effective, data security measures require attention during the earliest planning stages of a survey, and then they require continual attention throughout survey operations.

Some of the types of survey events that can cause data loss or data corruption are accidental file deletion by field staff, programming errors in CAPI or other computer programs, computer viruses, and the theft of field laptop computers. Although data loss may be repaired in theory by reinterviewing respondents, this is not always possible, and it always raises difficult issues of respondent burden, reinterview biases in the data, and, of course, additional costs. Therefore it is worth the effort to invest in prevention through effective data security measures rather than expecting a cure through reinterviewing.

The most common security measure for dealing with data loss is data backup. Data backups may take the form of copies of laptop data files kept on diskettes in secure places by field staff, home office copies of all data transmissions from the field, or extra copies of files on each field computer's hard disk (in case of accidental erasure), or similar measures. Of course backups are only useful if they are actually kept current. This often requires continual reminding and reinforcement along with whatever automated procedures can be put in place. Furthermore backups are only effective if they can be retrieved and used to restore data in a reliable and timely way. Therefore, as part of a data backup system, data restoration needs to be tested from time to time to ensure that the backup data files are actually retrievable and usable.

The most prevalent threat to data security in today's computing environments is computer viruses — small programs that surreptitiously attach themselves to other programs or documents in order to corrupt files and disks. Virus protection is an essential data security measure, particularly where distributed equipment, such as laptop computers and electronic communications systems, is involved. Viruses may be placed inadvertently on field computers by field staff using personal software such as games or other commercial products, or through infected e-mail or other electronic communications, either as part of the survey process or through other authorized or unauthorized use of the computer.

Automatic and frequent virus checking can be built into programs and scripts used on field computers to create as many automatic virus checkpoints as feasible. For example, when a CAPI laptop computer is powered on, or when a communications session starts or ends, or when some other program such as a field management system is used, the virus detection and correction software can be invoked. However, care is needed to avoid having checkpoints that interfere with time critical tasks, such as when an interviewer refers to the field management system with a respondent present. Some operating systems, such as Microsoft Windows, may allow virus checking to be performed more or less continuously, but this requires careful testing to ensure that this will not negatively affect CAPI instrument performance or have negative side efforts, such as system lockups or crashes when used with communications or other software.

Beyond these fundamentals of backup systems, virus protection, and other basic security measures such as password controls in CAPI and other software, a comprehensive data communications security plan involves policies and procedures in several categories, such as the following: authentication and authorization controls like password schemes and biometric identification devices; confidentiality preservation through encryption; data integrity assurances such as digital signatures to ensure that a message is unaltered; and nonrepudiation, meaning "proof" of who sent a message and when (e.g., Sheldon, 1994). For survey researchers, encryption is the most universally applicable of these more advanced measures.

Data encryption schemes are designed to ensure that data are not understandable in any meaningful form except by those who hold the key. The data are encrypted through the use of a key that "scrambles" the data, and then decrypted by the use of a key that can unscramble the data into their original form. Encryption schemes have been devised that are extremely difficult to break without knowing the key. However, using any encryption scheme has its costs in terms of convenience, processing time, additional procedures for distributing and protecting keys, training, and so on. But for CASIC surveys some form of encryption is basic and essential for confidentiality protection because the data will be communicated over a telephone or network system that is not totally dedicated and protected for that survey alone. Thus there will always be the risk of unintentional (e.g., wrong addressing of a message) or intentional (e.g., hackers) exposure of the data while it is in communications channels, making encryption the surest method of data security.

While encryption schemes and tools can be varied and complex, their overall operation is conceptually simple (e.g., see Stallings and Van Slyke, 1994; Muller, 1996). In a simple one-key encryption scheme, such as provided with common file compression programs, the key in the form of a character string is used by the sender to encrypt the data and also by the receiver to decrypt the data. This is convenient when encryption is done as part of compressing the file for data transmission, making it easy to build encryption into the session layer software operating on field computers and at home office. A weakness of this one-key approach is that the key must be known to all senders (e.g., field staff), or perhaps built into their software, and therefore has relatively high exposure. Knowing this key, any sender could decrypt messages sent by any other sender, as could an unauthorized user who might obtain the key.

More sophisticated two-key schemes, commonly called public key schemes, allow for the sender to encrypt data using a "public" key that need not be kept secret. This public key will not work to decrypt the data. This requires a second key — the "private key" — known only to the recipient, for instance, home office in a CAPI survey. Today there is wide use of public key encryption schemes like this, and there are a number of procedures and approaches to key definition and distribution that make public key encryption desirable for ensuring a high degree of data confidentiality in a field survey communications system.

In most cases survey practitioners will rely on commercially available solutions for data security such as backup tools, virus protection software, and encryption schemes. In practice, planning a survey's data security system involves weighing the costs of the security scheme against the perceived and real risks of data loss, corruption, or inappropriate disclosure. Fortunately there are strong governmental and commercial interests in developing better and better security systems and tools for electronic communications. There are also numerous off-the-shelf products and procedures available at affordable prices for use in survey systems. Thus the cost of providing very high assurances of data confidentiality and integrity should remain minimal considering the essential need for data security in survey research.

16.7   CONCLUSION

The twentieth century has seen the development and pervasive use of telephone, radio, and television communications technologies that reach into virtually all households and establishments. More than halfway through the century, electronic communications came into regular use in CATI surveys. In the last decade of the century, as portable computing came into its own, field surveys also began to benefit from advances in electronic data communications. By the close of the twentieth century, a new electronic communications revolution had clearly taken hold, yielding a variety of technologies for cost-effective high-speed communications both within and outside of the telephone system. One key force for change has been the Internet, and particularly the World Wide Web, which has created pressures for advanced communications technologies to be made more widely available.

What does the twenty-first century hold? Predicting technology trends is a perilous business, and knowledgeable industry pundits have more than once been wrong about projected trends as unpredicted developments came to the fore. But despite the many twists and turns of various technology and service trends, the story seems always a happy one for survey researchers because of the ever-expanding availability of fast and reliable data communications. This trend is not just one of magnitude but rather of orders of magnitude. For example, a one-megabyte file transmitted by "fast" modem (28.8 kbps) over analog voice telephone circuits may take five or six minutes, while the same transfer takes about 80 seconds on digital (ISDN) lines, a few seconds over cable circuits, and under two seconds with ADSL.

Communications speed improvements like these will likely encourage more than just efficiency gains in survey operations. Such phenomenal increases in communications speed suggest whole new approaches to survey data collection. As households become connected to networks through cable, digital telephony, and the Internet, these technologies may be used in more and more versatile and creative ways for all types of electronic communications in truly multimedia ways. In most developed countries of the world, the trend has already started, and the early years of the twenty-first century will see relatively cheap and very fast data communications services as commonplace within national communications infrastructures. One implication is that field interviewers will be able to maintain more continuous contact with home office, supervisors, or other interviewers, from any household, and perhaps even while on the road or in the air with wireless technologies. Furthermore visual images, voice, or video, as well as data records, will be readily communicated without delays. Ironically all of this may motivate a swing back to a more centralized approach to computer assisted field interviewing, starting perhaps with interviewers connected by wireless network to a central host computer during a CAPI interview. Under this scenario interview data would be available instantaneously at home office without the need for special separate communications sessions, and there could also be CATI-like monitoring of field interviewing from home office using the wireless network. Extending this scenario even further, why have the interviewer in the field at all? Perhaps remote "in person" interviewing over advanced interactive television devices will someday become feasible as a new interviewing method that is a cross between today's CAPI and CATI methods.

These future scenarios are indeed risky to predict in any detail. Nevertheless, it is striking how quickly communications technologies that could make such scenarios possible will be upon us. In light of this, a key task facing survey methodologists and practitioners in the early twenty-first century may be to pave the way for changes in survey methods that will far exceed developments in CASIC during the late twentieth century.

CHAPTER 17

Training Field Interviewers to Use Computers: Past, Present, and Future Trends

Mark S. Wojcik
Abt Associates, Inc.

Edwin Hunt
National Opinion Research Center

17.1 INTRODUCTION

Over the past decade there has been tremendous growth in the use of computer assisted interviewing by field interviewers. While the majority of this work has been done using computer assisted personal interviewing (CAPI), other methodologies which take advantage of increasing hardware capabilities are evolving. Among these are computer assisted self-interviews (CASI) and audio computer assisted self-interviews (ACASI).

As each of these methodologies emerges, there is usually a good deal of written material on its implementation. Most of the literature focuses on data quality, acceptance of the new process by both interviewers and respondents, and limits that are encountered by the developers. However, discussions rarely address the training of interviewers to use the emerging technologies. The little that has been written about interviewer training is older and does not reflect recent trends and innovations. There is also no "recorded" history of the changes that have occurred, and continue to occur, in the training procedures for CAPI interviewers. It is our purpose to describe the historical trends in CAPI interviewer training as well as identify future trends that can vastly improve the manner in which interviewers are trained.

Computer Assisted Survey Information Collection, Edited by Mick P. Couper, Reginald P. Baker,
Jelke Bethlehem, Cynthia Z. F. Clark, Jean Martin, William L. Nicholls II, and James M. O'Reilly.
ISBN 0-471-17848-9 © 1998 John Wiley & Sons, Inc.

The little written about the training of interviewers suggests that interviewer training does affect the quality of the data collected (see Fowler and Mangione, 1990; Billiet and Loosveldt, 1988). Training costs represent a big expenditure of any project, often accounting for a figure in excess of 10 percent of the total budget.

This chapter reviews a number of issues relevant to CAPI training and the approaches that have been taken by various organizations in responding to these issues.

17.2   TRAINING CAPI VERSUS CATI INTERVIEWERS

While it is true that there are many issues involved in interviewer training, it is also true that the data collection mode effects the way interviewers are trained. While they share many similarities, the job of a CATI interviewer and that of a CAPI interviewer vary in significant ways. Thus the training that each receives must also vary. Table 17.1 identifies some of the issues that must be addressed differently depending on the mode of data collection.

Clearly the training issues are different between the two modes. Because of the added complexity in training CAPI interviewers, we will focus our discussion on the training of these interviewers.

17.3   BACKGROUND

In general, the type of interviewer recruited tends to be nontechnically oriented and possess limited computer skills (see Couper and Burt, 1994). Mostly this stems from the fact that the most desirable characteristic of an interviewer is his/her ability to find respondents, convince them to participate in the study, and accurately complete interviews. It is considered a bonus if an interviewer happens to have computer skills. Although computer skills of interviewers have improved over time, due to more selective recruiting and more experience with CAPI, on the whole there are still relatively unsophisticated computer users in the field. Of course, anxiety about using the computer can negatively affect performance (Kelley and Charness, 1995). However, with the rapid spread of personal computers, we can expect more interviewers to have some computer skills when they are hired. Even today in the field, interviewers are more likely to have used a computer on a previous study than they were at the beginning of the decade. Therefore computer anxiety may be less of an issue in the near future.

On the other hand, as CAPI has matured, it has also become more complex. Recent CAPI applications require interviewers to use systems with increased functionality. With each development come new challenges for the training

Table 17.1. Comparison of CAPI and CATI Training Issues

	Computer Assisted Personal Interviewing (CAPI)	Computer Assisted Telephone Interviewing (CATI)
Training/work environment	Interviewers are trained in an artificial site that does not resemble their actual work environment.	Interviewers are trained in the environment in which they will work.
Case management	Interviewers must play an active role in the management of their cases. They determine the scheduling of interviews.	Interviewers often play a passive role in the management of cases. Scheduling programs determine the timing of interviews.
Survey and computer support	Interviewers who have questions or problems must consult an off-site source for support.	Support for both survey questions and computer problems is readily available on-site.
System problems	Because interviewers are operating a laptop, problems experienced by one interviewer may not be experienced by others.	Because interviewers are working from a network, problems are more likely to be shared by all interviewers working on a specific problem.
Logistics of using the computer	By operating in a remote environment, CAPI interviewers must be trained to conduct tasks such as data transmission and case-transferring protocols.	By operating in a central environment, CATI interviewers do not need to be trained as extensively in the logistics of using the computer.

staff. Some examples of added functionality that are now common are:

- *More complex questionnaires.* Instruments are now designed specifically for CAPI, and designers take advantage of the technology to use more complex sampling techniques, skip patterns, inter-item consistency checks, and other conventions. The end result is that it is often difficult for interviewers to navigate through an instrument and to determine exactly where they are in the administration process (see Sperry et al., Chapter 18).
- *Addition of computerized sampling tasks.* While this represents one of the major breakthrough uses of CAPI, it is often difficult to train interviewers to understand both the steps in the sampling process and the logic behind who is selected for participation.
- *More (and more complex) administrative and record-keeping functions.* In addition to interviewing tasks, interviewers are being asked to use the computer to record and report a variety of administrative information. The functionality of these administrative tasks often differs significantly from that of the interview, creating the need for additional training.
- *New operating systems.* As CAPI software moves from a DOS to a Windows environment, the interviewers will require retraining. Already improvements in screen design and procedures in Windows may mean that there will be a decrease in the amount of training required.
- *New survey methodologies.* As new survey methodologies (e.g., audio-CASI) are developed, interviewers must also be trained in their use.

Therefore it is quite evident that the approach to training CAPI interviewers must be an ongoing one in order to develop the appropriate level of skills. This issue is not confined to the survey industry. John Hurley, president of the American Society for Training and Development (as cited by McKenna, 1993), notes that "we are still state-of-the-art in technology development, but we are far from state-of-the-art in training workers to reap the benefits of that technology."

Furthermore any discussion of the move to CAPI inevitably turns to the increased costs associated with training interviewers in its use. It is not surprising then that much attention is devoted to the development of more cost efficient ways of conducting this task. As was mentioned earlier, training costs often account for 10 to 15 percent of the overall project budget. Therefore the implementation of more cost-effective training strategies can have a big effect on the costs of conducting a CAPI project.

The current trend is a shift in training from formal classroom training to an increased use of self-study. Therefore the main challenge facing the designers of CAPI training is to develop programs that adequately use self-paced materials to prepare interviewers to complete more complex tasks with less reliance on in-person, classroom style training.

17.4 IN THE BEGINNING

In the early 1990s, when CAPI feasibility studies were prevalent, the survey research industry was faced with the need to convert experienced paper and pencil interviewers to CAPI. Needless to say, the early training models concentrated heavily on providing a great deal of personal attention to the trainees. Because these early efforts were feasibility studies, they often involved a very small number of interviewers (ranging from as few as 2 or 3 to about 25). As a result, the number of trainers (and interested observers who could also help provide personal attention) was often greater than the number of trainees. In effect, each interviewer received what amounted to personal one-on-one training.

At this time interviewers were usually also trained in the use of the paper questionnaire. This training served two main purposes. First, it provided an understanding of the structure and flow of the instrument, and second, it allowed interviewers to use a hard copy instrument in the event that the CAPI program crashed during an interview (Sebestik et al., 1988; Iverson, 1992).

While these early training sessions proved that field interviewers could learn to use computers, the associated costs were extremely high. In a small early study, training costs for CAPI interviewing were found to be 18 percent higher than for paper and pencil (Weeks, 1992). So work began on identifying ways of providing the same level of personal attention that was found in these early training sessions while cutting costs.

17.5 DEVELOPMENT OF IN-PERSON TRAINING MODELS

As CAPI became a more feasible methodology, considerable effort was given to the development of a training model that would provide interviewers with all of the information and support that they required to learn how to conduct computer assisted interviews. While they accomplished them in different ways, all CAPI training models sought to meet the following goals:

- *Help interviewers get over anxiety of using computers.* One of the main goals of any CAPI training model is providing the trainees with the chance to overcome their anxiety about the use of the computer and begin to feel comfortable with its use. While this anxiety is diminishing, it remains an issue that requires attention.

- *Teach interviewers to use the computer to complete interviews.* Since interviewers were not responsible for the entry of a great deal of case management data, the main focus of early CAPI training was to provide interviewers with sufficient computer knowledge to access and complete interviews.

- *Maintain the same level of training on the administration of the question-*

naire that was present when using a paper and pencil questionnaire. When interviewers were trained to use a paper and pencil questionnaire, a great deal of time was spent acquainting them with the content and flow of the instrument. While CAPI controls the flow, it also makes it more difficult for the interviewers to gauge where they are in the questionnaire (e.g., see Groves and Mathiowetz, 1984). So training interviewers on the content and flow of the instrument remains an important goal of CAPI training.

While the issues involved in the use of CAPI are easily identified, the manner in which they are addressed can vary tremendously. As a result many varied approaches have developed. Furthermore, as time progresses and better technologies emerge, these approaches continue to change. In the following section we identify a number of general issues that should be considered in CAPI training.

17.6 LOGISTICS

Although easily overlooked, all interviewer training sessions involve a number of logistical issues. The introduction of the laptop computer increases the number and importance of these issues, including (1) the physical requirements of a training site, (2) the general approach to training logistics (regional training vs. centralized training), (3) group size, (4) length of training, and (5) the types of staff that are used as trainers.

There are also many factors that influence these logistical concerns. The main influence seems to be the structure of the organization conducting the training. For instance, organizations with regional offices use their regional offices and staff to conduct regional training, whereas organizations with a central location train centrally.

17.6.1 Physical Requirements for a Training Site

The proper choice of the training site is not a trivial point; it is vital to the success of the training. In particular, a CAPI training site should have:

- *The proper type of room.* The room used for training is often determined by what is available. Often, organizations conducting CAPI training have to find an external site that can provide the specified capacity. In general, a conference center with classrooms set up for training is the best option. However, when the numbers of trainees is very large, hotels, which offer more limited training accommodations, may become the only feasible option.
- *Sufficient electrical capacity.* Training sessions are usually too long for trainees to rely on laptop battery power. As a result many trainees will simultaneously need to plug in their laptops, drawing a great deal of

electrical current. It is important to know that the training site can meet this demand.

- *Adequate space per trainee.* CAPI trainees require a good deal of physical space to accommodate computer equipment and training materials. Lack of sufficient space can lead to disruptions and poor morale.
- *Secured storage for computers.* Regardless of the training model, computers need to be delivered to and stored at the training site for some period of time. The amount of space necessary to accommodate a large number of computers should not be taken lightly. It is very easy to underestimate the amount of space required. Furthermore it is extremely important to ensure that access to the storage room be restricted and that proper security measures are in force.
- *Dedicated training space.* Whenever possible, training staff should control access to the training rooms at all times so that material and equipment can be securely stored during nontraining hours. This will minimize the burden of setting up and tearing down the equipment.
- *Access to the proper equipment.* While it is easy to overlook small details, there is usually a great deal of equipment required for CAPI training. It is not uncommon to require overhead projectors, surge protectors, VCR units, slide projectors, or LCD projection panels.
- *Outside telephone lines.* Since the use of data transmission via telephone lines is often one of the least understood steps in CAPI, it is vital that each trainee have the opportunity to practice transmitting data. So, when finding a training site, it is very important to select one that allows multiple telephone lines in the training room. It is also important to ensure that the telephone lines have been activated and that they have direct access to an outside line (rather than requiring the use of a switchboard operator).

17.6.2 Group Size and Structure

As with the early CAPI training, organizations and agencies strive to provide the same level of personal attention that was offered with the early small-scale training. However, since large studies are usually the first ones to adopt CAPI technology, a feeling of personalized attention must be established when training groups that can be as large as several hundred interviewers. Two models have been developed to respond to this need. Each model has advantages and disadvantages.

The first model, used primarily by large government statistical agencies such as the U.S. Bureau of the Census and Statistics Canada, conducts training in regional sites. Instructional development occurs at the main office, but the delivery of the material is the responsibility of each of the regional offices, with each scheduling the required number of training sessions. Group sizes tend to be small, thereby ensuring the personal attention that is so important to CAPI training. At the Census Bureau, regional training is conducted by a single trainer, with group sizes ranging from four to ten people. At Statistics Canada, two trainers are responsible for training groups of about ten. If the number of trainees in a region exceeds these group sizes, another training group is added. Thus, while the overall number of trainees may be very high, the number each office is responsible for is quite manageable.

The main advantage of this model is that training is conducted on a regional level, thereby allowing each office to train the people in its geographic area. Travel costs associated with training may be sufficiently lower than they would be for a centralized training. Furthermore the staff in the regional offices train the interviewers for whom they will be responsible. The greatest disadvantages reported by those who use this model are (1) the difficulty of standardizing the delivery of training, (2) inconsistency in the skill level of the trainers, and (3) the fact that those responsible for the development of the training materials are often one step removed from the trainees and do not receive direct feedback about the materials. Also, while the training is delivered by the regional field supervisors, the skills of these people vary widely. In some cases they may be seasoned veterans with a great deal of training experience, while others may be newly hired with little or no training or interviewing experience.

The second model, adopted in many large organizations, uses a central training facility. All trainees travel to a single location where they are then trained. Often times, with very large surveys, the trainees are divided into two (or more) training sessions, rather than trying to find a location that can accommodate everyone at once. The structure of the training usually begins with large general sessions in which all trainees gather for introductory modules. After the general session the trainees are divided into smaller working groups of 12 to 18 for the remainder of the training. Each of these smaller groups is lead by a training staff of between 2 to 4 trainers. This number of trainers allows one to lead the group and at least one to provide support to the trainees, thereby providing some level of personal attention (see Wojcik, Bard, and Hunt, 1991).

This model too has its advantages and disadvantages. The main advantage is that all training occurs in a single location, making it possible to increase the level of standardization. Trainer meetings are held regularly to discuss issues from the day; this results in uniform decisions that can then be conveyed to all trainees. Another advantage of this model is that experts in specific topic are available to speak to all of the trainees. While this can be accomplished via videotape for regional training, the effect is often not as impressive and does not allow the trainees to ask questions.

The main disadvantages are (1) the size of the training sessions and (2) the costs associated with centralized training. With very large groups, several operational problems arise. First, locating a training site can be problematic. Second, the number of trainers required is quite large; finding enough qualified trainers often presents a challenge. Finally, the cost of bringing everyone to a single location is higher than that required to conduct numerous regional training sessions. In this model most trainees travel good distances to attend the sessions.

17.6.3 Length of Training

The little empirical evidence that exists on interviewer training suggests that there is a correlation between the length of training and interviewer performance. While many factors such as size of the group trained, the complexity of project, and the kind of supervision interviewers receive after training can affect the overall length of training, a reasonable standard is a length of two to five days (Fowler and Mangione, 1990). Training in excess of that amount may actually prove detrimental to interviewer performance.

In current practice a reasonable standard, if it exists at all, is applied only to less complex CAPI applications. The overall length of training varies significantly from organization to organization and even from project to project within an organization. Nevertheless, there is one point on which everyone agrees: The overall amount of time that is spent on the training of CAPI interviewers is increasing. The number of days devoted to CAPI training ranges from three days for the simplest projects, to nine days for very complex projects. Note that this range pertains only to in-person training and does not include any additional self-study work completed before or after the formal training.

17.6.4 Training Staff

We observed two approaches to the selection of trainers, a decision dependent mainly on the organization's structure and whether a training is conducted from a central or a regional location. For example, the U.S. Bureau of the Census and Statistics Canada rely solely on staff in their regional offices to conduct training. These staff, while not usually experts in substantive or technical aspects of the survey, are responsible for the management of field interviewers throughout the data collection period.

There is also a variant used for centralized training when the size of the training group tends to be large. In this appoach groups of 10 to 20 trainees are led by training teams consisting of 2 to 4 trainers, where a lead trainer is assisted by 1 to 3 persons. In each training room the lead and assistant trainers fulfill the following functions (Wojcik, Bard, and Hunt, 1991):

- *Lead trainer.* Must be proficient in many areas, including substantive knowledge about the survey, technical knowledge about the hardware and software, and operational knowledge of proper field procedures. This role is the most difficult to staff.
- *Assistant trainer.* Must be familiar with the same issues as the lead trainer but not necessarily at the same level of detail. The role of these staff is to respond to questions and assist persons experiencing difficulties. They may also take the lead in the presentation of specific training modules.

• *Observers and runners.* The role of these staff is to maintain the pace of the training by assisting persons whose problems cause them to fall behind the rest of the group. These people need to be knowledgeable about the hardware and software but not necessarily the study itself.

There are advantages and disadvantages to each approach. The main advantage to the centralized approach is that more expert trainers are involved in the process. Where the regional training may or may not be conducted by staff with both the technical and substantive knowledge of a project, the centralized approach guarantees that each group is led by a person possessing both these skills. However, this approach may increase the number of trainers needed, increasing the cost of training.

17.7 TRAINING CONTENT

The second broad topic we consider is the content of the training itself. To a large extent the content is determined by the scope and subject of a project. Our discussion thus focuses on broader content areas, including trainers' materials, hardware and operating systems training, the use of paper instruments, and the need to balance training on content with training on technology.

17.7.1 Materials for Trainers

The types of materials that are produced for CAPI training are fairly consistent across all agencies and organizations. Most agencies use interviewer manuals (covering both administrative and technological issues), scripted practice interviews, and Trainer's Guides used for in-person training. There are, however, two different approaches to the development of the Trainer's Guide.

In the first model, trainers are provided with fully scripted training modules. Everything that the trainer has to present is carefully laid out in the text of the Trainer's Guide, which trainers follow verbatim. The main advantages of producing this type of material are (1) the materials encourage a standard delivery and (2) they make the trainer's job easier because the trainer simply needs to be familiar enough with the content to read the text and make sense of it. The main disadvantages of this type of material are (1) trainers often are not well prepared because they do not really know the material (e.g., they are easily stumped by questions) and (2) such materials take considerable time to prepare.

In the second model, the training is not scripted. Rather, trainers are given all of the information to be presented in the form of a detailed outline, with references to the other training materials (e.g., interviewer manuals) where the specific details are contained. Overhead transparencies are often used to help guide trainers through the modules.

The main advantages of this model are (1) the materials encourage preparation on the part of the trainers thereby extending their ability to respond to questions, (2) they can be produced in a shorter period of time than scripted materials, and (3) they allow greater flexibility in accommodating last minute design changes. The main disadvantages are (1) the materials must be sent to trainers in enough time to prepare and (2) if used improperly, they can result in a less standardized delivery of the training material.

Even within each agency or organization the debate about which model produces better results continues to rage. Because it is relatively difficult to gather empirical data about the training of interviewers, it is an issue that is not likely to be settled soon. However, both models have been effectively used for a number of years and have produced satisfactory results.

17.7.2 Training on the Hardware and Operating System(s)

Training on the use of hardware has undergone a great deal of change over the years. Initially, because of the level of apprehension trainees had about the computer, this training was conducted in-person, with a great deal of personalized attention. This ensured that trainees received initial exposure to the laptop in a controlled classroom environment.

As computers have become more widespread, it is no longer necessary to introduce the hardware in a classroom environment. So, the current trend, although not uniformly applied across organizations, is to move the training on the use of the hardware out of the classroom to various self-study media (e.g., hard copy, video, or computer based). While it is not true that all trainees receive a computer prior to their arrival at training, they are doing so in increasing numbers. As an example, Westat sent hardware replacements to their experienced interviewers between rounds of the Medicare Current Beneficiaries Survey, providing at the same time a self-study package on learning the new hardware. No in-person training was required for this task (Edwards et al., 1993). Similarly the Bureau of the Census and the National Opinion Research Center (NORC) have used videotape as part of a larger self-study package to accomplish the same task.

While usually not part of the more formal training program, typing tutorials are sometimes included on the laptop computers. For instance, at NORC, a standard typing tutor is loaded on all CAPI laptops. If a trainer notices that a trainee is experiencing difficulty with typing, he/she can encourage the trainee to use the typing tutorial during the evenings. While no training time is spent reviewing the use of the tutorial, step-by-step instructions are contained in the CAPI Reference Manual which is provided to each trainee.

Very little, if any, time is devoted to training on the use of operating systems (e.g., DOS, Windows) during in-person or self-study training. In fact many organizations prohibit users from having access to the operating system.

17.7.3 Use of a Paper Questionnaire in Training

While the early CAPI training model favored the use of a paper questionnaire, current models avoid this practice for several reasons. First, since the questionnaires for many studies are now being developed for an electronic medium, many studies do not begin with a paper questionnaire that is later converted to CAPI. Although many CAPI authoring systems are able to produce a hard copy questionnaire, these questionnaires rarely resemble a usable instrument.

Second, even if a paper questionnaire is developed, converting it to an electronic one is not an exact process. Designers and programmers often have to make decisions that result in an electronic instrument that is not an exact copy of the paper one. As a result using the paper document to train interviewers can create confusion rather than clarification.

However, there is also a trend to produce materials that can serve some of the same functions of the paper questionnaire. Various organizations have begun using tools such as instrument flowcharts, overview diagrams describing the major characteristics of the instrument, and other graphical depictions of the instrument. These documents are designed to allow the interviewer to become better acquainted with the construction and content of the instrument as well as to provide a road map to gauge the status of the interview.

17.7.4 Balancing Training on Content with Training on Technology

As the instruments and procedures used in CAPI projects continue to become more complex, they require more training time. That leaves training developers with two options: (1) increase the overall time of training or (2) cut the length of training on existing topics to allow for the new topics. Since the former option is expensive, training on technology has often come at the expense of other topics. In most instances training is decreased on issues such as the content of the survey (e.g., training on the content of the questionnaire(s), administrative procedures). This has led, in many cases, to the impression that training on survey content has been abandoned in favor of training on the use of the technology. While there is common agreement that this is a dangerous shift, no one has yet determined a way to adequately balance these two needs, given the constraints of current training.

Even within a given project, there are usually proponents of both types of training. While the instrument designers always want to include more training on the content, the systems designers push for more technical training. In fact an interviewer is unable to properly perform his/her duties without appropriate knowledge of both. Thus it is important that we continue to look for ways to address this issue to the satisfaction of everyone involved.

The future, however, offers many options for resolving this conflict. Many of the technical tasks lend themselves nicely to the use of self-study training via computer-based training (CBT) or videotape. We discuss the application of these newer technologies to the training process later.

17.8 TRAINING STYLE AND METHOD

The last broad topic pertaining to in-person training concerns training style and method. It is in these areas that we are beginning to see the implementation of innovative approaches that use emerging technologies.

17.8.1 Use of Lecture and Demonstration versus Hands-on Experience

There is general agreement among training professionals that the use of varied approaches to training enhances the trainee's experience and level of retention. This is reflected in many of the CAPI training models which include lectures, demonstrations, and hands-on practice sessions. As a rule, the majority of the agencies we contacted preferred to begin training with more structured sessions, like lecture and demonstration modules, and then progress to increased use of less structured hands-on modules. There are some interesting exceptions. For example, Westat's CAPI training incorporates the use of their on-line tutorial very early in the process (see Bittner and Gill, 1996).

17.8.2 Use of Errors and Problem Situations in the Learning Process

There is much discussion about whether common errors and problems should be avoided during training or be incorporated as learning tools (Frese and Altmann, 1989). We see this debate played out in the various approaches to CAPI training. Semmer and Pfäfflin (1978) argue that training should not attempt to restrict the opportunity to make errors but should incorporate "typical" errors, train for them, and use them to improve comprehension of the entire task. This practice is beginning to surface in CAPI training through the development and use of specialized training modules covering such topics as entering comments, navigating through the questionnaire, accessing help screens, diagnosing and correcting system problems, and diagnosing and resolving questionnaire reporting and entry errors. Furthermore organizations that use scripted interviews during training often include typical problems in their designs. Nevertheless, current CAPI training practice allows for training on errors only in a controlled environment. Trainees are not allowed simply to make errors and resolve them on their own. Rather, when errors occur, trainees are carefully led through examples of some of the more common errors and are expected to correct them in a group learning environment. So, while we are beginning to see the advent of training on error, it is not done in any type of a self-exploratory environment.

17.8.3 Use of Self-Study Training

In an effort to decrease the amount of time that trainees spend on in-person training, all organizations we talked to reported the use of self-study materials, some before and some after the in-person training. However, the breadth of the packages and the means of delivery vary widely. Self-study packages has been extensively implemented on major studies such as the National Longitudinal Survey of Labor Market Experience—Youth Cohort (NLS/Y), the Current Population Study, and the Medical Expenditure Panel Survey.

The main goal of these self-study packages is to ensure that everyone arrives at the training with a common base of knowledge, and to allow trainees to become familiar with the computer and associated vocabulary, which is part of the effort to ease the anxiety about the use of the machine (Wojcik and Hunt, 1994). The Office for National Statistics in the U.K. takes this practice a step further and uses data collected in an interviewer self-assessment exercise to group trainees according to their level of computing skills for in-person training.

The delivery of material in these packages can differ greatly even within a single organization. However, the use of exercises, supported by videotape, computer-based practice, and manuals is common.

The most recent development in the approach to self-study training is the increased use of computer-based training (CBT). CBT applications have been used by the Census Bureau, Westat, and NORC, among others. In general, self-study packages can include several components, such as:

- *Training videotape.* Videotape is becoming a popular medium for both in-person and self-study training. The content may vary according to the needs and resources of the organization. Such material included in self-study videotapes has included discussions of organizational and project structure, use of the hardware, introduction to software, and general interviewing techniques.

- *On-line tutorial.* In an effort to improve the efficiency of self-study training, survey researchers are now beginning to implement computer-based training (CBT) programs. CBT programs allow trainees to develop skills by using the computer at their own pace. Like the training video, on-line tutorials can include discussions of organizational and project structure, use of the hardware, introduction to the software, and general interviewing techniques. The advantage of the on-line tutorial is that it is interactive and can be tailored to the needs of individual trainees.

  While the survey research industry is in the early stages of CBT use, there are indications that it can reduce or eliminate the need for in-person training. The challenges associated with the implementation of CBT include the need to provide user support during training, program development time and cost, and the lack of experience with CBT development (see Bittner and Gill, 1996).

- *Written materials and exercises.* These usually comprise the bulk of the self-study package and usually involve the completion of a series of exercises. These exercises, some of which might be completed on the computer (e.g., mock interviews), are usually preceded by either a reading or a viewing of a video segment, or both. Because they comprise the bulk of the self-study training, exercises are often developed for all topics ranging from project background to the use of the computer to procedures for completing interviews.

• *Supervisor monitored interview.* As a final step before beginning work, each interviewer who completed the self-study package is usually required to complete a supervised interview. This can take the form of a scripted mock (practice) interview or an actual live interview. If the trainee does not satisfactorily complete this step, further practice with the self-study program or other retraining is usually required.

The components of any given self-study package will vary according to the requirements of a specific project or the organizational resources available.

17.8.4 Evaluation of Trainees

The primary method cited for the evaluation of CAPI trainees is observation. While there is a tremendous effort toward the development of more objective evaluation procedures, most organizations still rely mainly on subjective observation as their primary means of evaluation.

Furthermore, while some initial observation is conducted to evaluate trainees during their in-person training, much of the evaluation occurs after they return home. This can take several forms. First, some organizations conduct a final telephone mock interview with interviewers after they return home from training (see Couper, Sadosky, and Hansen, 1994). Under this scenario each trainee is called by a field supervisor who functions as a respondent, using a script, and evaluates the interviewer's ability to complete the interview. A second strategy relies on the use of the observation of the interviewer in the field while he/she conducts actual survey interviews. In either case the methods of observation rely upon the observer's ability to make a subjective evaluation of the trainee's ability to perform the job.

17.9 COMMON ISSUES THAT REQUIRE RESOLUTION

We have developed approaches to address each of the issues mentioned in the previous section, but there are still issues that require further attention. While each organization has developed a strategy to respond to these issues, all agree that their current approaches can be improved. Furthermore many of these issues are important to the success of the training effort. Thus these issues offer an opportunity for improvements in future CAPI training.

17.9.1 Training of Trainers

A common problem is being able to produce training materials and deliver them to the training staff well enough in advance to allow adequate preparation. It is usually the case that training materials cannot be finalized until the CAPI management and data collection systems are at least close to final. As a result the development of these training materials is done closer to the actual training than was the case with paper and pencil surveys.

Furthermore, with the growing size and complexity of many current CAPI projects, it may no longer be possible to rely solely on the staff who were responsible for the development of the survey or CAPI applications to conduct all of the training. The number of people who require training may be simply too large to rely solely on the experts as trainers. Thus often it has become necessary to expand the number of available trainers by using staff who are not experts but who have demonstrated some degree of mastery of the tasks.

Since "trainer training" sessions tend to be close to the time of the interviewer training, it is not possible for the trainers to assimilate the material and gain proficiency prior to conducting their training sessions. The obvious solution is to provide more time between the trainer training and the interviewer training to allow for additional assimilation and practice. However, it is more often the case that when this time is available it is used to correct any errors in the systems. The end result is that the trainers become more familiar with systems that are still under revision.

17.9.2   Assuring That Systems Are Ready for Training

Unlike with paper and pencil surveys, there is no longer a date when questionnaires have to be finalized for printing. As a result there is a tendency among questionnaire programmers to continually revise the instrument up to the point when it is loaded on the computers. Even when problems are identified at training, they tend to want to fix them right away. The end result is that training materials are often developed using versions of the questionnaire and management systems that have already been revised.

One practice that has been adopted is to set a date prior to the training to "freeze" the training versions of the systems. While this method ensures that everyone is trained on the same version of the software, that version often differs from the one on which interviewers begin data collection.

17.10   ADOPTING ALTERNATE DELIVERY SYSTEMS FOR SELF-STUDY TRAINING

In discussing the direction we must move in to improve CAPI training, it is useful to look at the trends in training experience outside of our industry. One of the big new developments in the training industry is the rapid improvement of computer-based training. By combining CD-ROM technology and newer operating systems such as Windows 95, leaders in the training field are able to improve training products by using multimedia material. Multimedia-based training is quickly becoming the preferred training method of the big corporations. It saves money, and people tend to remember more accurately, for the training process engages several senses at the same time (Padgett, 1996).

The general model for these improved training packages is not very different from the one developed within our industry. Most of the programs include computerized tutorials, full motion video, and limited animation. The main difference is that the computer is used for all of the training.

Based on a review of the experiences of Fortune 500 companies with CBT there are three important benefits that emerge: (1) continued (and greater) cost savings, (2) shorter training time, and (3) improved comprehension and performance (Padgett, 1996). All of these are goals that we have strived toward since the first implementation of CAPI training.

17.10.1   What Can We Gain by Using Multimedia CBT Products?

Recent efforts in CAPI training (e.g., Bittner and Gill, 1996) are moving the survey industry closer to the CBT programs used in other industries. However, we have to take CAPI training to the next level to realize some of the true advantages of CBT. Naturally, before we make this move toward greater automation of the training process, we must demonstrate the benefits derived from such a move.

Since we have already shown our willingness to move to more self-paced training, CBT seems like the natural next step. Then training programs would take advantage of new and emerging technologies. The CBT modules could automate mixed-media presentations that incorporate videotapes, computer demonstrations, and interactive exercises. Such methods as ACASI are already working on incorporating sound into the survey process. We can likewise improve our training by incorporating similar technology into our training methods.

There are positive reports that people learn better, faster, and retain more of what they learn by using CBT (Padgett, 1996). One of the main obstacles we face in training CAPI interviewers is that automated survey procedures (both administrative record keeping and questionnaire administration) are becoming too complex and difficult to teach. CBT seems to offer a viable means for presenting trainees with the required information, allowing them to practice using it, and then testing their retention.

17.10.2   When Can We Use CBT?

While CBT offers many opportunities for training improvements, it cannot be applied to all situations. The development costs for CBT products can be high. One of the easiest ways to support the cost of CBT development is to re-use the products. Using CBT for modules that apply to general interviewing situations (as opposed to project-specific topics) can help justify their costs. For instance, a CBT module on gaining cooperation skills, or on the basic operation of the data collection software, could be developed and then supplemented with project-specific information during in-person training. Therefore, while we believe that CBT offers exciting ways to improve our current training practices, we recognize that is is not a panacea.

17.11 CONCLUSIONS

The training of CAPI interviewers has undergone significant changes in a relatively short time. It has moved from small and very expensive in-person training, to larger and more efficient in-person training, to the increasing use of self-study training. All these different approaches to training are now converging toward a single acceptable model for training that combines increased self-study with decreasing amounts of in-person training. There are, however, several important points that will allow current CAPI training techniques to be improved and taken to the next level:

- *Maintaining focus on the administration of the survey instrument.* With the quickly emerging changes to the CAPI process, and the inherent complexity that accompanies these changes, it is easy to become bogged down in spending all of the time training on the use of the technology. We must remember that the main purpose in using CAPI is the administration of a survey instrument to obtain the best possible data. Therefore we cannot become so involved in teaching people how to use computers that we no longer offer them the same level of training to be interviewers that was available during paper and pencil (and earlier CAPI) training. We must continue to ensure that people are properly trained to administer the questionnaire(s) and then, as a secondary concern, to use the computer to complete their administrative and record-keeping tasks.

- *Incorporate the same technologies into training now used in surveys.* While we have been very quick to incorporate audio technology into surveys, we have yet to see audio added to a training program. Likewise, as we begin to experiment with the multimedia interview, we should realize the importance of multimedia to the training process and begin to develop computer-based training programs like those currently in use in other industries. We must continue to monitor the newly developing technologies and look for ways to incorporate them not only into our interviewing procedures but also into our training model. No doubt this will force us to incur some initial start-up costs. However, if we take a long-term view of the benefits that are offered, it should be clear that this money is well spent.

- *Do not think of training applications as an afterthought.* Very often, we are so concerned with preparing our electronic questionnaires and management software for production that the preparation of training applications is only considered as an afterthought. We believe that some of the greatest gains in the quality of the survey are the product of adequately trained field interviewers. As a result the development of training applications must be a priority.

It is true that technical resources are often very limited and must be concentrated on the development of the questionnaire. Nonetheless, as we move toward the types of CBT programs that are used by other industries, our options increase. The availability of outside consultants and off-the-shelf products means that we no longer have to rely solely on our own limited resources to develop the tools we require.

- *Continue to develop more objective measures for evaluating the success of our training programs.* Perhaps the biggest problem with our current training model is that we are overdependent on subjective measures of trainee and trainer performance. While we have shown movement toward the use of objective data for evaluation, we still do so on a haphazard basis. If we do nothing else, simply developing means for objectifying the evaluation process will greatly increase the effectiveness of these programs.

We are currently at an important point in the history of training CAPI interviewers. Emerging technology offers options that we could only have imagined a few years ago. However, with these advances comes a responsibility for determining the best and most effective applications.

We must begin to look outside of our own industry to see what is being done by others. We have been reluctant to do so in the past, citing the obvious differences in the availability of resources. However, there is a great deal to be learned from the experiences of private industry. The end result can only be a positive one: better quality training, in less time, resulting in decreased training costs.

ACKNOWLEDGMENTS

The authors would like to thank numerous individuals contacted in the course of gathering material for this chapter, including Debbie Bittner (Westat), Angel Broadnax (U.S. Bureau of the Census), Rolly Jamieson and Jill Bench (Statistics Canada), Beth Ellen Pennell (ISR–University of Michigan), Michael Weeks (Research Triangle Institute), and Jean Martin (Office for National Statistics, U.K.). Finally, special thanks to Robert Wagers, Stephen Smith, and Patrick Bova (NORC), for their assistance in gathering information.

CHAPTER 18

Evaluating Interviewer Use of CAPI Navigation Features

Sandra Sperry, Brad Edwards, and Richard Dulaney
Westat

D. E. B. Potter
Agency for Health Care Policy and Research

18.1 INTRODUCTION

In the early stages of computer assisted personal interviewing (CAPI), designers focused on the feasibility of using computers for in-person interviews. Investigations centered on finding light-weight hardware and fail-proof software that yielded usable, complete data. Usability for interviewers was considered at only the most basic level: Could they carry the computers in the field? Could they read the questions on the screen? (National Center for Health Statistics and Bureau of the Census, 1988). Now that the feasibility of CAPI has been completely established (Baker, 1992; Couper, 1994), designers can focus more fully on the human-computer interface and on better support for CAPI interviewers.

Studying the human-computer interface entails evaluating the way the machine and the user share control of the process. As Norman (1991, p. 51) notes, "Typically human/computer interaction is a dialog in which control is passed back and forth between the user and the computer." In virtually all CAPI systems the computer controls the default (i.e., forward) movement through the questionnaire. Entry of an answer automatically moves the cursor to the next item in a predetermined sequence. CAPI applications also include features that allow the interviewer to assume control of the movement through the questionnaire. Key sequences, pop-up windows, menus, and highlight bars

Computer Assisted Survey Information Collection, Edited by Mick P. Couper, Reginald P. Baker, Jelke Bethlehem, Cynthia Z. F. Clark, Jean Martin, William L. Nicholls II, and James M. O'Reilly. ISBN 0-471-17848-9 © 1998 John Wiley & Sons, Inc.

allow the interviewer to depart from the normal forward flow whenever necessary. To address unanticipated events and occurrences, the design of CAI programs should be flexible enough to allow interviewers to control the flow as unusual interviewing situations arise.

House and Nicholls (1988) note that although one of the greatest strengths of computer assisted interviewing (CAI) is its control of the flow of the interview, the computer's control also "strips the interviewer of the ability to improvise when unusual situations occur." Interviewers need flexibility to recover from errors, react to respondent queries, and verify earlier responses. Navigation features that allow interviewers to control the interview's movement can provide this flexibility.

Controlling the flow of the interview can also help interviewers deal with the "segmentation effect," which is another drawback of CAI that House and Nicholls (1988) describe. Because a CAI questionnaire must be divided into small segments that are presentable on a computer screen, interviewers view the questionnaire one segment at a time. At any point during the interview, an interviewer's attention is focused on the current screen display, which can leave him/her with a disjointed sense of the rest of the questionnaire. Such segmentation can distort an interviewer's orientation in an interview; that is, an interviewer can lose the context of a particular screen, lose the sense of what should logically come next, and in general, lose the broad overview of the interview. Some CAPI features offset the effects of segmentation by providing interviewers with the flexibility to correct for such effects.

Accessing supplemental information, often known as "help," can offset the loss of context for the current screen by presenting probes that define terms used on the screen or by offering an overview of the questionnaire with maps, flowcharts, or lists of items collected earlier in the interview. The feature that allows the interviewer to *break off* an in-progress interview enables interviewers to end the interview at any point deemed necessary or appropriate. Using the *comments* feature, interviewers can enter messages that explain any unanticipated situations. Such information often proves of great use to analysts and data managers.

Some screens consist of more than one element, which allows interviewers to *move around on a screen.* These screens present a set of related questions or a set of entities, like household members, for which the same questions are asked. Presenting related questions in a grid on one screen, with a cursor key that moves freely among answer spaces, allows interviewers to enter responses in an order different from that dictated by the default flow. Because the set of related questions can be seen together on one screen, the interviewer gains a sense of overview, context, and predictability that are usually obscured by question/screen segmentation.

Backing up and returning is used by interviewers to correct an earlier entry. This capability is often needed in unanticipated situations. The interviewer can also regain a sense of both the overview and the context by reviewing previous questions with their associate responses.

Choosing the next questionnaire section to administer provides interviewers with flexibility in multipart interviews. For example, a survey may require that each questionnaire section be asked about each household member. For reasons of respondent availability, the interviewer may need to skip over the sections for a household member who is listed next in the household roster and administer the sections for a different household member. The interviewer may also need to vary the sequence of the questionnaire sections in a record-based establishment survey. To choose the next section, the interviewer must be presented with an overview of the other available sections.

18.2   DATA SOURCE: THE 1996 MEPS NURSING HOME COMPONENT

To understand the way interviewers control the CAPI flow, this chapter examines data collected as part of the 1996 Medical Expenditure Panel Survey, Nursing Home Component (MEPS NHC). The NHC is longitudinal and includes three rounds of interviewing in nursing homes (Potter, in press). Information used in this study comes from the first round of nursing home interviews and includes time stamps, keystroke files, and data from the CAPI database.

For each sampled person (SP) selected in the first round of NHC interviewing, the interviewer first obtained a residence history (RH), and then collected four more sections: health status (HS), prescribed medicines (PM), background (BQ), and insurance (IN). The interview averaged four hours of CAPI administration time per nursing home, collected information on four residents sampled in each facility, and averaged 1300 items per nursing home. The most challenging aspect of the survey, from the standpoint of CAPI design, was the need to interview more than one member of the nursing home staff. In the first round of interviewing, there was an average of 2.5 respondents or informants per facility. Previous nursing home surveys (Anderson et al., 1994) have shown that an interviewer would collect as much information as possible from one respondent and then move to another. (This approach is required in many establishment surveys.) Because it was not possible to predict in advance when respondents would be available or which questionnaire sections a particular respondent could answer, the CAPI questionnaire needed to allow interviewers to navigate to questionnaire sections according to the availability of respondents.

The NHC data provided a wealth of information about a number of different CAPI features that enhance interviewer control of navigation. This chapter focuses on two of these features, backing up and choosing instrument sections.

18.3   BACKING UP AND RETURNING

In the course of conducting an interview, an interviewer may need to return to questions that have been asked and answered previously, or the respondent may decide that an earlier answer needs to be changed. Both interviewer and respondent may want to review the answers to previous questions. With a paper questionnaire, the action is simple. The interviewer flips back through the pages of the questionnaire looking for the questions and answers that need review.

Because computer programs support complex paths, word fills, and on-line data relationships, the use of CAPI may intensify the need for backing up. Special CAPI tools are required to facilitate the backup process, tools that aspire to make backing up as simple as paging through a paper questionnaire.

Most major CAPI systems provide nondestructive backup mechanisms that display the responses previously entered and preserve the integrity of the data. These mechanisms incorporate on-line checks to ensure that data remain consistent with the rules for data collection in normal forward mode. This requires that the keystrokes from the previous forward path be maintained, at least until another path is chosen. By the time the interview session is complete, most backup functions eliminate any inconsistent data that had been entered for questions on invalid paths.

With the NHC application, backup was accomplished in two ways. An interviewer could move backward one screen at a time or "jumpback" over a number of screens. In both backup modes the previously entered values were displayed in reverse video above the entry field. The NHC jump-back function opened a window that displayed a list of descriptions for all questions already asked in the CAPI session. This approach is like that used in most CAPI systems, although the Blaise system also allows developers to choose an option in which the user can move either backward or forward to any point in the CAPI application, including screens that have not yet been viewed. On the NHC, interviewers used the up and down arrow keys to scroll through the descriptors in the jumpback window, moving a highlight bar through the list and selecting from the list by pressing the enter key when the appropriate item was highlighted. If the target item was in an earlier section of the interview, the interviewer could bring up a list of questionnaire sections, choose first to go to the relevant section, and then look for the specific item within that section.

There were several ways that an interviewer could move forward again. Entering a new value or re-entering the old value automatically moved the application forward in the default flow established for that value. To move forward *without* entering a value, interviewers pressed one key sequence to move forward screen by screen or another to "fast forward" to the next screen without an entry. If no entries were changed, the "next screen" was the position where the backup operation had started. If a different value had been entered and there were different questions to be asked, the application went to the first new question. In the NHC application interviewers used the backup features by pressing combinations of two keys, for example, jumpback was invoked by pressing [Ctrl-J], while screen-by-screen backup was invoked by pressing [Ctrl-B].

   To examine how interviewers used the NHC capabilities for backing up and moving forward, we studied records of interviewer keystrokes from all entries made by the 46 NHC interviewers in the course of their first and fifth interviews. The file of keystrokes represents entries made during approximately 400 hours of interviewing and contains more than 800,000 elements, some of which provide context for keystrokes made. Of these 800,000 elements, interviewers pressed the enter key 268,607 times.

   On a complex application like the MEPS NHC, with its many matrices, modules, repeating loops, directory lookups, "code-all-that-apply" items, and long lists, the raw keystroke records are not an efficient means for studying interviewer backup behavior in general. While Couper, Hansen, and Sadosky (1997) examined aggregate keystroke counts, our analysis involved detailed review of a large number of individual backup episodes. We were interested in the context of each episode—at what point in the application it began, how the interviewer initiated it, its duration, and what the interviewer accomplished.

   For this reason we decided to select two completed cases per interviewer. The first interview was chosen because we believed it would reveal what interviewers do shortly after training without having gained much proficiency with the CAPI system. Each interviewer's fifth case was conducted at a point when the interviewer had gained about 16 hours of NHC interviewing. Therefore the fifth case provided a second observation after interviewers had attained a reasonable amount of experience with the backup features.

   To investigate backup behavior, it was first necessary to define episodes of backing up. For this study a backup episode began when an interviewer first pressed one of the backup key sequences. This was the initiation point of an episode. The inverviewer then had to back up at least one screen. He/she might or might not change a data value. The episode ended when the interviewer returned to the point at which the episode was initiated. An extremely simple episode might consist of backing up one screen and coming forward one screen, thereby returning to the initiation point. Although not very complex, these episodes are nonetheless instructive about interviewer behavior. More complex episodes might contain changes to previous responses, which would require different skip patterns, either asking new questions as part of the episode, or skipping over previous questions and deleting the responses. If an episode included changing an answer, and the new answer dictated that the CAPI application skip over the initiation point, the episode ended when the program passed over the field in which the initial key sequence was pressed.

   An episode could also include "nested" backups in which the interviewer backed up and came forward several times before coming all the way forward to the initiation point. When this happened, a new episode was not created. Rather, all interviewer actions from the initiation point until the program returned to or beyond the initiation point were considered as one episode. Times that interviewers jumped back, either at the start of an episode or during an episode, are referred to as jumpback "uses." The number of jumpback uses was greater than the number of jumpback episodes because some jumpbacks were nested in episodes that had already been initiated.

In the first and fifth cases, the NHC interviewers initiated 2167 backup episodes that met our definition, for an average of 24 backup episodes per nursing home interview, or 1 backup episode for every 10 minutes of interviewing. There were nearly 50,000 keystrokes in these episodes, a mean of 23 keystrokes per episode. A majority of the episodes (58%) contained fewer than 10 keystrokes, but some were quite long. For example, 99 episodes (4.5%) encompassed more than 100 keystrokes each.

In 1332 episodes (61.5%), interviewers changed at least one data value on a previous screen. In the remainder of the episodes, the interviewers returned to the initiation point without changing any responses, an activity often called "browsing." Sometimes interviewers used the key sequences for automatic forward movement; on other occasions they re-entered the current values. Four sections of the instrument accounted for 77 percent of the backup episodes with the same four sections yielding 77 percent of the changes made while backing up. This correspondence suggests that whether the interviewer changed data during a backup episode or merely browsed through previous questions without making changes did not relate to the subject matter of the question.

Although the full set of backup episodes in our data set contained nearly 50,000 keystrokes, only about 2000 of these entries were the key sequences specifically designed to move the application forward after backing up. Examination of specific entries made during some of the backup episodes revealed that interviewers frequently re-entered the previous responses displayed above the entry fields rather than using the features that moved the application forward automatically, either one screen at a time or to the next eligible question to be asked. This behavior raises questions about whether the interviewers were comfortable using the "fast forward" feature. Although re-entering the previously recorded data eventually returned the application to the initiation point of an episode, the approach was usually more time-consuming than using fast forward.

There were other notable patterns in the interviewers' backup activities that were related specifically to their use of the jumpback feature. On evaluation questionnaires filled out during training, comments from both trainers and interviewers identified the jumpback feature as the most difficult CAPI feature to learn. The keystroke files support this assessment as only 3.4 percent of the backup episodes were initiated by the key sequence for jumpback.

Other case studies have also found interviewer use of CAPI features like jumpback to be fairly uncommon. For example, Wojcik and Baker (1992) report that interviewers used the *slide* mode, a feature quite similar to jumpback, in only 6.2 percent of interviews for the National Longitudinal Survey/Youth Cohort (NLS/Y). Similarly Edwards et al. (1993) examined the use of a response correction procedure in the Medicare Current Beneficiary Survey (MCBS) and found that this was used in only 9.4 percent of interviews.

Table 18.1. Frequency of Jumpback Use in the NHC

Jumpback Uses per Interviewer	First Interview		Last Interview	
	Number of Interviewers	Number of Jumpback Uses	Number of Interviewers	Number of Jumpback Uses
0	21	0	30	0
1	6	6	5	5
2	10	20	7	14
3	5	15	2	6
4+	4	23	2	11
Total	46	64	46	36

The NHC keystroke data also suggest that the ability to use jumpback varied quite a bit by interviewer, since jumpback use was not distributed uniformly across the 46 NHC interviewers (Table 18.1). The use of the jumpback feature declined from the first to the fifth interview, both in overall use (63 to 36) and in the number of interviewers who used the feature. Fifteen of the 46 interviewers did not use jumpback in either the first or fifth interview.

In the NHC Round 1, use of both jumpback and screen-by-screen backup decreased from the first to the fifth interview. Screen-by-screen backup went from being used an average of 62.7 times during the first interview to 53.2 in the fifth. Jumpback use decreased from a mean of 1.4 times in the first interview to 0.8 times in the fifth interview. This pattern is consistent with interviewer behavior reported by Couper, Hansen, and Sadosky (1997), in which use of backup declined from an average of 8.05 times per interview in each interviewer's first five interviews to an average of 7.30 times for the last five interviews. It is difficult, however, to draw any conclusions about what the declining use of these features means without knowing what motivated the interviewers' actions. The pattern of decreasing use could mean that interviewers became more skilled and needed to back up less, or it could mean that they disliked the features and used them less for this reason.

There was another interviewer behavior that was related to jumpback and that did not create any episodes at all. Of the 126 times in the keystrokes files that interviewers pressed the jumpback key sequence, 27 times (21%) did not include a selection from the jumback list. Interviewers returned to the initiation screen without actually backing up. This action was not the same as browsing, which involved looking at screens used earlier in the interview. In these instances interviewers did not review previous items but looked only at the list of items displayed in the jumpback window. This information suggests that in these cases interviewers started to use jumpback but decided not to use the feature after seeing the jumpback list.

18.4 CHOOSING INSTRUMENT SECTIONS

Recently, there have been reports of CAPI designs that allow the instrument sections to be administered in varying order (Edwards, Sperry, and Schaeffer, 1995; Nusser, Thompson, and DeLosier, 1996; Shepherd et al., 1996; Whelan, Karlsen, and Yost, 1995). Some of these designs provide options that enable interviewers to repeat sections, select new sections, or retrieve missing data. To some extent this work has been grounded in CAPI case management experience, since choosing the next instrument section is similar in many respects to choosing the next case. In general, these designs provide interviewers with more control, echoing a trend observed in other software systems that "allow the user to perform tasks according to the user's plan of action rather than a constrained path of options." (Norman, 1991, p. 25).

Typically the direction in which a CAPI interview moves is controlled by the computer program except when an interviewer interrupts the flow using one of the navigation features designed for special situations. Allowing interviewers to choose the sequence of sections requires a third approach for controlling movement. The program must pass control to the interviewer as each questionnaire section ends. To take control, the interviewer must choose the next section from a menu (or series of menus) that presents a list of the sections that are available.

Designing this type of menu system for the NHC required attention to how the questionnaire would be administered. Previous nursing home surveys had demonstrated the need for flexibility in the order of questionnaire sections (Anderson et al., 1994). This need arose partly because of the wide variety of organizational structures and medical record storage locations among nursing homes and partly because of the demanding schedules of respondents (i.e., the nursing home staff). Sometimes respondents preferred that interviewers abstract data from medical records without the help of a staff member, and then an interviewer might need to return to sections to collect missing data items from a respondent. In other situations one respondent could answer most questions in a section, but a few items had to be collected from a different respondent.

Thus the NHC required a high degree of flexibility in its menu system for selecting sections. These general conditions are not unique to the NHC. Many household surveys collect data on each household member and could benefit from a design that provides the interviewer with the flexibility to administer a questionnaire section for several household members in sequence, or to administer the sections sequentially for each household member that is available.

NHC designers thought that substantial attention to user concerns would be required to create a navigation feature that provides the best possible support to interviewers in choosing the order of questionnaire sections. Many interviewers (especially those with less experience) could be uncomfortable without the precise sequence of sections specified for them. In fact one of the often cited advantages of CAI is that it controls the flow of the interview (House and Nicholls, 1988). The NHC navigation feature needed to strike a balance between program control and the flexibility necessary for interviewing efficiently in a nursing home. Design activities for this feature took place during a CAPI feasibility study that preceded the NHC, as well as in the main study, which included both usability testing and a pretest.

The navigational needs of the NHC suggested that a system with more than one menu should be designed. Although menu systems may seem easy to use and understand, they have grown increasingly complex and can present designers with an assortment of challenging issues. Effective menu structures take advantage of the ways users think about a selection task, that is, how they organize their approaches to making choices. Selecting from a menu uses basic processes in human cognition: receiving a stimulus and encoding it, evaluating options and making a decision, implementing the option by making a response, and evaluating the outcome. According to Norman (1991, p. 1), "The overall process is integrated by an underlying strategy or plan that operates within the context of the user's mental model for how menu selection works."

In the feasibility study the NHC navigation feature used a series of three menus. Interviewers selected the sample person (SP) on the first menu, instrument section on the second, and respondent on the third. This menu structure covered all possible choices and was easy to learn. Its menu lists were short enough that interviewers could scan each screen and select from it very quickly. It seemed that this menu structure would work well for the interviewers because it was so straightforward and easy to use, but in practice, the interviewers disliked it. The design required interviewers to scan and select from three menus to make what they perceived to be one choice. This resulted in interviewers having to repeat the actions of scanning and selecting from menus as many as 100 times in the course of an interview. The experience on the feasibility study clearly showed that this navigation feature needed revision in the main study.

This finding from the NHC Feasibility Study is consistent with several presented in the HCI literature. Shneiderman (1992) reports that frequent users prefer densely packed screens and a minimum of menu levels to reach common choices. Grudin (1989, p. 1166) notes that design priorities must accommodate for and reflect the fact that "ease of learning can conflict with subsequent ease of use." Norman (1991) presents the concept of optimal complexity, which lies somewhere between a menu structure that offers too many choices, requiring users to go through too many levels of decisions to achieve simple movement, and menu systems that offer only a few common paths, failing to support the occasional need for movement along more complicated paths.

In response to the findings from the feasibility study, designers chose to present the information needed to select the next questionnaire section in one menu on one screen. The Navigate Screen (which was developed in a DOS environment) displayed a grid with the names of the four people sampled for a particular facility listed down the left side, defining rows, and abbreviated

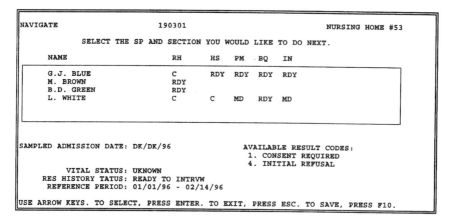

Figure 18.1. NHC navigate screen.

section names across the top as column headings. Figure 18.1 shows an example of a Navigate Screen image.

When using the Navigate Screen, interviewers moved the arrow keys to manipulate the cursor around the 20 cells of the grid, pressing the enter key to bring up the program module for the SP and section combination designated by the cell on which the cursor was located. Respondents for that section were identified by the first question. After each section was administered, the CAPI program automatically returned to the Navigate Screen, and the interviewer arrowed to another cell to choose the next section. The escape key was used to exit the Navigate Screen.

The Navigate Screen also displayed status codes in the cells. The initial status of a questionnaire section was "ready for administration," represented by the code "RDY." When a section was completed for an SP, the CAPI program automatically generated the code "C" for complete. When an interviewer encountered a section that could not be administered for some reason, interim codes for section-level nonresponse (e.g., "consent required" or "initial refusal") could be entered. The code "MD" was used when a section was not entirely complete after it was first administered. This code allowed interviewers to choose the section again to complete it at a different time.

Usability tests conducted with interviewers during the main study suggested additional design touches. These included alphabetizing the SP names rather than listing them in order by the SP number assigned during sampling, setting the cursor on the cell most recently worked instead of putting it on the "next" cell as dictated by a default ordering of sections, and using dynamic displays of the information that varies according to the cell in which the cursor lies, for example, the section level result codes (Figure 18.1).

To analyze interviewers' selection of instrument sections, we chose the case (i.e., the full set of data collected for a nursing home) as the unit of analysis.

Table 18.2. Number of Attempts/Respondents Used to Finalize Instrument Sections

	RH	HS	PM	BQ	IN	All
Sections with more than one attempt						
Number	435	39	266	58	16	814
Percent	19.9%	1.8%	12.2%	2.7%	0.7%	7.5%
Total sections with at least one attempt	2,188	2,176	2.176	2,176	2,176	10,892
Sections with more than one respondent						
Number	91	16	21	18	11	158
Percent	4.2%	0.7%	1.0%	0.8%	0.5%	1.5%
Total sections with at least 1 respondent	2,181	2,174	2,173	2,173	2,174	10,877

We used the actual CAPI database for these analyses rather than using keystroke files. This included 547 completed nursing home interviews with nearly 11,000 completed instrument sections.

Table 18.2 shows the extent to which the Navigate Screen was used to return to particular sections of the questionnaire to collect missing data. Some 814 sections (7.5%) required more than one attempt before they were finalized. The amount of data retrieval varied greatly among the different section types. Relatively straightforward sections, such as those about the SP's health status (HS), background (BQ), and insurance (IN), did not require much data retrieval. Less than 3 percent of each of these section types needed more than one attempt to be finalized. More complex sections, those that included looping and potentially more abstracting from handwritten records, presented more need for data retrieval. The residence history section (RH), which asked iteratively about each time the SP was at a different place during the year, involved more than one attempt almost 20 percent of the time. Prescribed medicines (PM), which asked for detailed information by month about each medicine administered to an SP, was attempted more than once about 12 percent of the time.

Data retrieval was not necessarily conducted with a different respondent. Retrieval of data from the original respondent typically occurred when a respondent broke off the conduct of a section, either to locate records or to continue with the section later in the day. Table 18.2 shows that a second respondent was interviewed for less than 20 percent of the sections that had missing data retrieved. Section nonresponse is counted as an attempt, even though no respondent was selected; therefore there are more sections with attempts than sections with respondents.

The sequence in which interviewers accessed the questionnaire sections for the *first* time was also investigated. Examining the patterns of initial forward flow through the sections can provide a sense of the interviewers' level of comfort with varying the order of sections in the CAPI environment. A review of all first attempts to administer a section in the 547 completed interviews showed that less than one-third (30%) of the interviews were conducted in a highly linear pattern, that is, with all sections or all persons completed in order. This low level of linearity in the ordering of sections implies that the interviewers were, in fact, quite comfortable taking advantage of the options provided by this CAPI feature.

The patterns of initial forward flow were also investigated to determine whether a particular order for administering the sections predominated, suggesting a default path that would have worked in most situations. One route that could have been selected for a programmed default flow presents each section in a predetermined order for each SP, moving horizontally across the rows of the grid and taking the SPs in the order listed on the screen. In first attempts only 9.1 percent of the interviews followed this route. Another possible default route could have presented one section for all SPs before moving on to the next questionnaire section for any SP, that is, moving vertically down the columns of the grid. This route was taken in only 5.5 percent of cases. In addition there was very little or no linearity at all in 57 percent of the routes taken through the sections for first attempts. These results seem to support our expectation (developed from observations of nursing home interviews done with paper and pencil) that a predetermined path through the questionnaire would not have provided the flexibility needed for this data collection.

18.5 DISCUSSION

This chapter has described the design of two CAPI navigation features that pass control of the interview's flow to the interviewer and has looked at the ways in which interviewers used these features during production interviews. The observations of interviewer behavior go beyond training sessions and usability labs to see how interviewers used complex CAPI programs in their actual work environments with no one present to remind them how or when to use the special features. Although the number of interviewers was small (46) and their actions cannot be used to predict how different interviewers would behave on other surveys, the findings provide insights into the success of the designs considered and offer some guidance about how to approach other CAPI development tasks.

There was considerable variation in the extent to which interviewers used the different features. Backing up and returning one screen at a time were used frequently, approximately five times per hour of interviewing. Operating these features was quite straightforward and predictable. Pressing the key sequence for screen-by-screen backup brought up the previous screen, just as making an entry and pressing the enter key brought up the following screen. Only one entry was needed, and that entry could not lead anywhere except to the previous screen.

In contrast, the jumpback and fast forward features were used very little, with a jumpback episode only once every three hours. It may be that these features were not needed so much. The typical need for backing up might have been to a screen that was close enough that using screen-by-screen backup made more sense. Review of the keystrokes made during backup episodes showed that sometimes when interviewers backed up one screen at a time, they backed up far enough that using jumpback would have been considerably more efficient, but there were not a large number of these episodes.

It could be that the interviewers' natural inclination to focus on completing the interview led them to avoid using a feature that resulted in substantial backtracking through the questionnaire. It also seems possible, however, that interviewers would have benefited from greater use of the jumpback feature but found it too difficult to use during actual interviews. This hypothesis is supported by two observations.

In 21 percent of the times that jumpback was accessed, it was not actually used, suggesting that the sight of the jumpback menu led interviewers to decide against using it, and a small number of interviewers used jumpback fairly frequently while a large number (nearly one-half in the first interview and more than one-half in the fifth) did not use it at all, suggesting that only a small number of interviewers knew how to use it smoothly (Table 18.1). If jumpback was avoided because it was too difficult to use in the field, development work with interviewers might improve its design substantially.

Analysis of interviewer use of the Navigate Screen implies that its design addresses the navigational needs of the NHC CAPI questionnaire. Interviewers used it to return to sections and collect missing items from both original respondents and new respondents. When using the screen to move forward through the questionnaire sections, interviewers chose a broad variety of routes. Interviewer opinions about the Navigate Screen were positive: They rated it as easy to learn at training, and in an evaluation form completed after several months of interviewing, they reported a high level of satisfaction with the Navigate Screen in the field.

The Navigate Screen also conforms well to many of the design principles suggested in the HCI literature, in particular, explicitness and comprehensibility (Couper, 1994). Both of these qualities are described by Norman (1988, p. 25) when he writes, "A device is easy to use when there is visibility to the set of possible actions, where the controls and displays exploit natural mappings." The layout of the Navigate Screen is very explicit: a cell for each instrument section and SP, with the current status of the section shown in each cell. The actions to be taken on the screen are easily comprehensible, exploiting natural mappings by having the arrow keys move the cursor in the direction implied by the arrow.

The jumpback menu, on the other hand, does not seem to be either explicit or easily comprehensible. Pressing the jumpback key sequence opens a window in which descriptors of the questions previously asked and answered during that interview are shown with the most recently asked question at the bottom of the window. The user must read from bottom to top, scrolling up through the list toward the first question asked in the session. If the session has been long, the total jumpback menu could include hundreds of items, with only a few of those items visible in the window at one time.

The literature about menus used in computer applications (Norman, 1991, p. 53) notes two ways a menu can be poorly designed. It may be either overcomplex or impoverished. "Menus are overly complex when scores of infrequently used items are constantly displayed." Menus that are impoverished "allow only a limited set of paths... [and] require the user to spend an inordinate amount of time traversing the menu to get to the needed items." The jumpback list seems to be an example of a menu that is *both* complex and impoverished, including many options that are never used and allowing only one path through the list.

A more supportive design for jumpback might be developed using aspects of the successful design of the Navigate Screen. Jumpback could first show a screen displaying the grid of SPs and questionnaire sections that are displayed on the Navigate Screen. With status codes shown in the cells of the grid, such a screen would provide interviewers with an overview of the interviewing session. An interviewer could select a cell for review, and a list of questions that have been asked for that section might appear on a new screen, displayed in the order in which the questions were asked rather than in reverse order. The interviewer could scroll through the list of questions and select the desired question, which would then appear on the screen in the same manner that is currently used to present an item that has been selected from the jumpback menu. The interviewer could then proceed through a series of questions and previously entered responses.

Because this design would require that interviewers select from two separate menus, it might elicit the same type of interviewer criticisms that were expressed about the sequence of three menus used for choosing instrument sections in the NHC feasibility study. Nevertheless, in the absence of usability testing or pretesting, it is impossible to know how interviewers would respond to a new design. A very important aspect of the development of the Navigate Screen was that it was designed and evaluated first in a feasibility study, then redesigned and subjected to more evaluation through usability tests and pretest interviewing in the main study. The design of CAPI features that are used in complex navigational situations would always benefit from this level of attention to their usability.

Involving interviewers in the design of CAPI features can contribute significantly to whether or not the features are used or ignored during interviewing. A goal of CAPI design has been to automate onerous tasks, for example, choosing appropriate wording and following complicated skip instructions. Freed from these responsibilities, interviewers can pay more attention to what they do best, including tasks that a computer will never be able to do, such as establishing and maintaining a positive relationship with the respondent. Well-designed CAPI navigation routines handle the activities that fall somewhere between those that the computer should control, such as choosing the next question depending on the response entered for the current question, and those that only the interviewer can control, like building rapport. The navigation features examined in this chapter require that the programs pass control of the forward flow to the interviewer and that, after completing the needed special navigation task, the interviewer return control of the flow to the CAPI program. The ideal CAPI design will reduce errors by fully automating the interviewers' routine tasks while, at the same time, giving interviewers sufficient control to exercise the judgment required to deal with unpredictable events that occur during a complex interview.

ACKNOWLEDGMENTS

The authors wish to acknowledge the Agency for Health Care Policy and Research (AHCPR) and the National Center for Health Statistics (NCHS), sponsors of the MEPS NHC. The views expressed in this paper are those of the authors. No official endorsement by either of these agencies or by the U.S. Department of Health and Human Services is intended or should be inferred.

CHAPTER 19

Mode, Behavior, and Data Recording Error

James M. Lepkowski
University of Michigan

Sally Ann Sadosky
Intel Corporation

Paul S. Weiss
Trilogy Consulting Corporation

19.1 INTRODUCTION

Interactions between humans and computers have been a subject of research in psychology and engineering long before survey methodology (Carroll, 1991; Falzon, 1990; Shneiderman, 1992). The advent of computer assisted interviewing (CAI) has stimulated research in the nature of the interaction between interviewer and computer as well as respondent and computer (Nicholls and Groves, 1986).

As the mode of data collection has shifted from paper and pencil (P&P) to computer assisted, concerns have been raised about the quality of the data and how it may have changed as mode changed (Groves and Nicholls, 1986). There are a variety of outcomes used to assess the effect of computer assistance on the quality of the data obtained through interviews. Data quality has been assessed through such measures as rates of item missing data, accuracy of reported information, and accuracy of the recording of respondent reports.

The accuracy of recording has received particular attention because of the clear differences between P&P and computer assisted recording modes, recording responses on paper as opposed to recording responses through a keyboard.

Computer Assisted Survey Information Collection, Edited by Mick P. Couper, Reginald P. Baker, Jelke Bethlehem, Cynthia Z. F. Clark, Jean Martin, William L. Nicholls II, and James M. O'Reilly.
ISBN 0-471-17848-9 © 1998 John Wiley & Sons, Inc.

The recording process involves more than simple motor skills to enter an answer. Interviewers must perform complex cognitive processes to record answers correctly. Recording accurately requires the survey interviewers to hear a response provided by the respondent, to decide on an appropriate response category or answer, to probe to clarify responses where necessary, and to record the answer correctly. The accuracy of the recording process depends on the quality of the interaction between interviewer and respondent. It requires the respondent to provide answers that can be coded, and the interviewer to understand the available codes and be able to process the respondent report. It is assumed that the interviewer has the motor skills to record the answer. That is, although accurate data recording depends in the end on interviewer abilities, it also depends on the quality of the respondent's report.

The human computer interaction literature in psychology and engineering usually classifies recording error into two categories: (1) mistakes or errors of judgment, comprehension, and memory, and (2) slips in process involve motor skills or the occurrence of minor cognitive processing errors (Rasmussen, Duncan, and Leplat, 1987; Reason, 1990). As computer assisted data collection became more widespread, survey researchers were most concerned with "slips" (Couper, Hansen, and Sadosky, 1997). Illustrations abound in the survey process: adjacencies such as the interviewer pressing a 2 instead of a 1 to record a "no" instead of a "yes"; transpositions such as entering 15 years for age instead of a 51; and omissions and commissions in entering income, asset, or expenditure amounts, such as dropping a zero, or adding an extra zero.

Programmed skip logic, automatic fills of names and pronouns, range and consistency checks, and other features incorporated into a computer assisted instrument are designed to reduce the frequency of these and other recording errors relative to P&P recording. However, anecdotal reports indicated that recording errors in computer assisted data entry may still be, for some types of data, an important source of error in survey data. For example, survey organizations have observed that interviewers using CAI make decimal location errors in recording dollar amounts for income, asset, and expenditure questions. Interviewers were observed, for instance, entering $10,000 when a $100,000 house value was reported. Such "order of magnitude errors" could bias cross-sectional findings and would seriously increase measurement error in longitudinal analyses when introduced into measures of gross changes over time. The shift from P&P to CAI may have introduced changes in recordings error that may artificially exaggerate real change in the population values under study.

Recording errors due to either mistakes or "slips" ought to be present in both paper and pencil and computer assisted interviewing. Prior to the advent of CAI, little attention was given to recording errors in P&P. Almost nothing was known about the extent to which mistakes or "slips" occurred. An understanding of the importance of recording inaccuracy in CAI should therefore contrast recording errors to those in a paper and pencil mode.

It is easy to imagine that "slips" might be somewhat more frequent in CAI because of the level of motor skills required to record an answer accurately using a keyboard. It is more likely, though, that the interviewer's cognitive processes to determine an answer will more often lead to recording an incorrect answer. What is not immediately obvious is whether computer assistance will lead to more or less frequent recording error compared to paper and pencil.

Survey research studies have examined recording errors in each mode. Rustemayer (1977) examined recording errors in P&P data collection, administering mock interviews to experienced, end-of-training, and new interviewers. She found that experienced interviewers made an entry inconsistent with respondent verbal reports in 4.5 percent of the entries. Kennedy, Lengacher, and Demerath (1990) studied keying errors in CAI by monitoring CATI interviews across four studies. They found only 16 keying errors in 2583 entries, or an error rate of 0.6 percent. Dielman and Couper (1995) used tape recordings of CAPI interviews to assess keying error rates in 16,778 close-ended questions from 116 interviewers. They observed an error rate less than 0.1 percent.

Besides keying or manual recording errors, Nicholls and Groves (1986) suggested CAI would lead to fewer responses to open-ended questions. However, Catlin and Ingram (1988) found no difference in the length of open-ended responses (as measured by the number of characters or words) in a randomized comparison of CATI and P&P modes. Bernard (1988) also found no difference in the length of open-ended responses when comparing CAI with earlier data from P&P.

One area where differences have been found is in following skip logic and failing to complete items. Groves and Mathiowetz (1984) observed five times more skip errors for P&P compared to CAI. Sebestik et al. (1988) reported that more than 90 percent of errors made by P&P interviewers were failures to record an answer; CAPI interviewers made no such errors. Olsen (1991) examined skip errors in the twelfth round of the National Longitudinal Survey/Youth comparison of CAPI and P&P; about one in 100 skips were incorrectly followed in P&P compared to none in CAI. Tortora (1985) and Catlin and Ingram (1988) found fewer edit failures in CATI than P&P, an indication that there were fewer skip logic errors or incomplete items.

These investigations identify recording errors that further investigation ought to examine. None is an experimental comparison that definitively answers the question of whether recording error is consistently different between modes. Further these studies examine the dimensions of recording error and their frequency, but none have attempted to study the underlying behavioral dimensions of recording error. The role of the interviewer and the respondent in the recording process and the extent to which their roles may differ between modes have not been previously investigated.

The purpose of this chapter is to examine the nature of survey interview recording error, comparing recording errors and behavior between P&P and CAI modes of data recording. Section 19.2 presents a model for the survey data recording process to enumerate factors that may affect recording accuracy. Sections 19.3 and 19.4 describe the design and findings of an experimental study comparing recording error by mode. The chapter concludes with a discussion of the findings and their implications for improved design of survey data collection activities.

19.2   A MODEL FOR RECORDING ERRORS IN SURVEY DATA COLLECTION

While the prior research indicates that CAI methods improve data quality and recording accuracy, it has not explored factors that may account for this improvement. It is not clear, for example, whether recording errors are essentially random and unrelated to features of the survey process or whether they are correlated with specific interviewer and respondent behaviors. The nature of the interview process strongly suggests that how interviewers and respondents behave and how questions are written will be related to recording accuracy.

For example, interviewers have been observed skipping the reading of a question because they believe that the respondent has, in a previous response, already given an answer. The interviewer records an answer to the question without reading the question. Skipped questions ought to be subject to higher levels of recording error. Similarly respondent behavior may be related to recording accuracy. Respondents can elaborate an answer to a question with information that is related to but not needed by the interviewer to record an answer. The elaboration may increase the chances that an interviewer will make a mistake in response recording.

While interviewers and respondents may behave in the survey interview at a particular question asking in a manner that increases or decreases the accuracy of the recording, they also tend to demonstrate more often such behavior throughout the course of an interview. For example, some interviewers may give mode-related comments during the interview that indicate some difficulty with recording the answer: "I'm just writing down what you said..." or "The computer won't accept this answer...". Interviewers who make more of these mode-related comments may be, in general, having difficulty with the mode of data recording. One might expect higher levels of recording errors among those interviewers. On the other hand, respondents may consistently behave in ways that could be related to recording accuracy. For example, respondents vary with respect to their ability to provide answers that are within the response frame provided for the question. At a given question asking, an answer outside of the response frame may increase the likelihood of recording error. Alternatively, a respondent who consistently gives answers that are within the response frame for each question makes the recording process easier for the interviewer. We may expect to see lower levels of recording error for respondents who consistently give answers within the response frame for each question.

Of course the question itself may affect recording accuracy as well. Poorly written questions can be expected to generate interviewer and respondent behaviors that may affect recording accuracy. For example, a question that asks about two conceptually different issues at the same time may confuse a respondent. The respondent may then give an answer outside of the response frame. The interviewer should in turn probe the answer to obtain a valid response to record. The original inadequate answer and subsequent responses to interviewer probing create the potential for interviewer confusion about the correct answer to record. Inaccurate recording may tend to increase in these situations. On the other hand, certain types of questions may be more prone to recording errors than others. For example, questions that ask for amounts of income, assets, or expenditures may be more prone to recording slips by omission or addition of zeros than questions that require only a yes/no answer.

We have thus described four classes of factors affecting data recording accuracy (denoted subsequently as A). First, the dynamics of the conversation that is the interview itself, represented by interviewer and respondent behaviors at the question asking, may be related to recording accuracy; conversational dynamics, denoted here as factors C, is composed of both interviewer and respondent behaviors at individual question askings in the interview. Second, the interviewer's behavior throughout the interview may be related to recording accuracy at individual questions as well. Since interviewer behavior, denoted here as factors I, varies across interviews, the recording accuracy may in turn be altered as well. Third, the respondent's behavior throughout the interview can have a corresponding effect on accuracy A. We denote the average respondent behaviors as R. Finally, characteristics of the questions, denoted as Q, may also be related to accuracy A.

These factors and their effects on recording accuracy A may operate differently in CAI and P&P. There is reason to expect, for example, that interviewers will behave differently in the two modes. Interviewers should make fewer question reading errors, especially when fills are required, in CAI. They will probably skip question reading less often as well, since they will not have a sequence of questions and answers displayed on the screen, as in P&P interviews, to use as references to previous questions. If fewer question reading errors and skipped question readings are associated with greater recording accuracy, CAI should have fewer recordings errors. On the other hand, if the computer assisted system is difficult to use, interviewers may make more mode-related comments in the computer assisted mode. A higher frequency of such comments may indicate greater difficulty recording answers, and increase the frequency of recording errors in the computer assisted mode. Differences in interviewer behaviors may be exhibited both at each question asking (conversational dynamics, or C) and on average across all question askings for the interviewer (interviewer behavior, or I).

If the interview is conducted face-to-face, the respondent may likewise be expected to behave differently in the presence of a computer. For example, respondents may tend to elaborate answers less often in CAPI. Their answers may also tend to be briefer, more straightforward, and less qualified or conditional. To the extent that these behaviors are associated with higher recording accuracy, mode will operate through respondent behaviors to reduce the frequency of recording errors in CAI, as compared to P&P. As for interviewers, differences in respondent behaviors may be exhibited both at each question asking (conversational dynamics, or C) and, on average, across all question askings for the respondent (respondent behavior, or R).

Mode may also alter question characteristics, since questions can be displayed differently in the two modes and require different formats for data recording. For example, consider the response frame for a multiple choice question with a time and amount such as "About how much did you make at this job?" In P&P the question response frame may be a box or a line, allowing the interviewer to write the amount and the time period reference (e.g., per hour, per week, per month, or per year) all at one time. A CAI response frame for this item might require two separate entries, one an open response frame to record the amount and a second (perhaps a second screen) to record a numeric code corresponding to the time period reference. These differences may lead to different levels of recording accuracy in the two modes (see Baker, Bradburn, and Johnston, 1995; Bergman et al., 1994).

Figure 19.1 summarizes several potential direct and indirect effects of recording mode on recording accuracy. Mode (M) may be exected to alter all four factors in the recording accuracy model: conversational dynamics (C), interviewer behavior (I), respondent behavior (R), and question characteristics (Q). What is less certain is whether mode will have a separate effect on

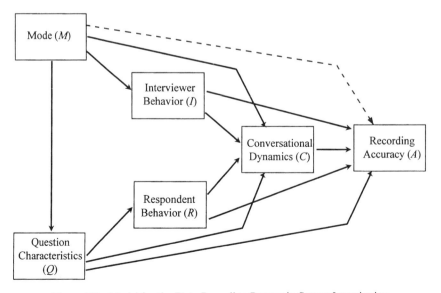

Figure 19.1. Model for the Data Recording Process in Survey Interviewing.

recording accuracy or whether its effects will operate only through these other factors. Thus, in the figure the mode effect on recording accuracy is shown with a dashed line.

Question characteristics (Q) are expected to affect interviewer and respondent behavior at each question asking (the conversational dynamic, C), as well as influencing the interviewer and respondent behavior across question askings (factors I and R). Question characteristics may also be altered through the need to meet constraints of presentation in a given mode (M). Question characteristics are also expected to have a direct effect on data recording accuracy (A).

The remaining factors, conversational dynamics (C) and interviewer and respondent behaviors (I and R, respectively), are expected to have direct effects on recording accuracy. Some dimensions of each of these factors may increase recording accuracy, while others may decrease it. Of less immediate interest here are the effects of interviewer (I) and respondent (R) behaviors on conversational dynamics.

Conversational dynamics (C) can be expected to have quite direct effects on accuracy of data recording (A). Conversational dynamics consist of both interviewer and respondent behaviors at each question asking. The dynamics of the interviewer and respondent might be reflected in terms of separate direct effects of each behavior or in terms of an interaction between them. For the sake of simplicity in subsequent analyses, we assume that the effects of each are separate at the individual question asking.

19.3 METHODS

In order to investigate the appropriateness and adequacy of the proposed model, an experimental comparison of recording modes was conducted in the 1994 Panel Study of Income Dynamics (PSID), the 27th consecutive wave of annual data collection. Sample subjects were respondents for approximately 7200 panel families who have been participating since the first 1968 wave. The PSID interview collects data from the head of the household about basic housing information, current and previous year employment history for the head and the female spouse, food costs for the family, and income for every member of the family (including wages and salaries, self-employment income, and all types of government assistance, obtaining the amount of income from each source). The principal data collection method was computer assisted telephone interviewing (CATI) from a centralized facility. Approximately 150 interviewers (30 with one to two years experience on the project, the remainder newly hired) received three-day training sessions on study specific concepts and procedures using CATI. All newly hired interviewers first received extensive training in basic interviewing techniques.

Interviewers were assigned to nine data collection teams of 8 to 12 interviewers, each of which was assigned a separate sample and led by a specialist. Approximately six weeks into the three-month data collection period, six interviewing teams were randomly selected for the mode experiment. In order to equalize training experience between CATI and P&P groups, all interviewers in these six teams received four hours of P&P training, including a general overview of the experiment, basic P&P data collection instructions, and a practice interview using the paper questionnaire.

Three of the six groups were randomly assigned to paper and pencil data collection, and three to CATI. All 596 PSID interviews conducted by these six teams during a three-week period were tape-recorded. After extensive review of each tape to identify technical problems (inaudible respondents, cutoffs, etc.), 200 tapes were selected randomly from each mode for coding and data analysis.

19.3.1   Coding Accuracy and Behaviors

A simple, replicable coding system was devised to code recording errors. We limit the present discussion to close-ended questions, since nearly all questions in the PSID can be considered close-ended questions of one form or another.

Two types of interviewer-respondent exchanges were identified in the process of developing the coding scheme: simple straightforward and complex. A simple straightforward exchange is one in which the interviewer reads the question and the respondent gives an appropriate response. A highly reliable coding of differences between respondent report and interviewer recorded data is readily made for simple straightforward exchanges. In complex exchanges, interviewers probed a response or the respondent qualified an answer, gave an uncodable answer, or elaborated on an answer.

Simple straightforward answer recordings were classified as "accurate" or "inaccurate." Questions with complex exchanges were classified as "accurate" or, because of the ambiguities in complex exchanges, "may be misrecorded."

Questions were occasionally skipped during question reading, with interviewers still typically recording an answer even though they did not read the question. Skipping question reading occurs most often when the interviewers record an answer they believed the respondent gave at a previous question. From a standardized interviewing perspective such events are recording errors, since every question must be read to the respondent.

For the purposes of investigating the model in Figure 19.1, a second coding operation to obtain data on interviewer and respondent behavior was conducted on the same data. A behavior coding system was used that was originally designed to identify the questions that pose problems for interviewers and respondents. This behavior coding system is based on interaction coding methods used in psychological investigations, and it has a long history of application in survey research to detect questions that are difficult to administer and to monitor the performance of interviewers in a survey setting (e.g., Cannel, Fowler, and Marquis, 1968). A subset of behaviors reliably coded in past studies was used in this study (see Dykema, Lepkowski, and Blixt, 1997). These included the following 18 behaviors:

Interviewers

- made major wording changes that potentially altered the meaning of the question,
- made errors in fill substitutions,
- used incorrect probing techniques,
- used reinforcing feedback,
- made mode-related comments,
- gave affect-related comments about the interview,
- skipped question asking.

Respondents

- gave simple, straightforward response,
- interrupted question reading,
- gave qualified or conditional response,
- expressed uncertainty about an answer,
- gave an answer that was outside the response options,
- gave multiple answers when only one was required,
- refused to answer the question,
- gave a "don't know" response,
- elaborated on an answer,
- gave mode-related comments,
- gave affect-related comments about the interview.

19.3.2 Operationalizing Measures

Data recording accuracy (A) is a simple dichotomy, accurate or inaccurate, regardless of the complexity of the exchange. Conversational dynamics measures (C) are also simple dichotomies that indicate the presence or absence of selected interviewer and respondent behaviors at each question.

Since the influence of the interviewer or the respondent on data recording accuracy is more completely reflected in performance across interviews, the interviewer and respondent behavior measures (I and R) were created by averaging behavior at each question (the conversational dynamic measures presented in the previous subsection) across questions asked by an interviewer or questions asked of a respondent. Thus the interviewer behaviors (I) are the percentage of times the given behavior occurred across all questions asked by the interviewer. Similarly respondent behaviors (R) are the percentage of times the behavior occurred across all questions the respondent was asked in the interview. These interviewer and respondent behavior percentages were assigned to each question asking. That is, for each of the 48,106 records (exchanges) in the study, the percentage of times each interviewer (I) or respondent (R) behavior occurred for the interviewer and respondent was assigned to the exchange record.

Question characteristics (Q) were created as similar percentage measures, but the aggregation was to the level of a question response frame. Each PSID question was assigned to one of seven types which represent different response frames for recording answers:

- *Short answer, simple:* number or other information, other than dollar amount or time (e.g., "How many rooms do you have, not counting bathrooms?")
- *Short answer, amount:* dollar amount (e.g., "How much was from bonuses?")
- *Short answer, time:* amount of time for specified time reference (e.g., "How many years have you been paying on it?")
- *Multiple choice, simple:* multiple responses (e.g., "How is your home heated? With gas, electricity, oil, or what?")
- *Multiple choice, time:* a time amount and period reference (e.g., "About how much did she make at this? ___per/1. Hour/2. Week/3. Two weeks/4. Month/5. Year/)
- *Multiple choice, month:* all months that apply (e.g., "During which months of 1993 did you get this income?/Jan/Feb/Mar/.../Dec/)
- *Yes/No:* yes or no response option (e.g., "Do you have air-conditioning?").

For a subset of 12 conversational dynamic behaviors, the percentage of times the behavior occurred for each of the seven response frames was computed. These percentages were then assigned to the 48,106 exchanges based on the response frame of the question. For example, the percentage of times that an interviewer made major wording changes in a question reading was computed for all of the exchanges involving yes/no questions. This percentage was then assigned to all exchanges for yes/no questions.

A total of 48 different predictors was generated across the four factors. Conversational dynamics (C) consisted of 18 predictors, interviewer behavior (I) consisted of 7, and respondent behavior (R) 11. Question characteristics (Q) contributed another 12 predictors.

19.3.3 Analysis Methods

Preliminary analysis consists of simple two-group comparisons between CATI and paper and pencil results. Subsequent analysis employed logistic regression of a dichotomous outcome (either mode, M, or recording accuracy, A) on various factors (C, I, R, Q, and M). Bivariate models were estimated for each of the 48 behavioral predictors. Multivariate models were then estimated for each of the four factors which included all characteristics represented by that factor. For example, for the 18 conversational dynamics (C) behaviors, there are 18 bivariate models and tests of differences, but two multivariate models, one with the 7 interviewer behaviors and a second with the 11 respondent behaviors. For interviewer behavior (I), there are 11 bivariate models and one multivariate model which includes all 11 behaviors.

Finally, the joint effects of all of these factors on data recording accuracy are explored in two multivariate models to determine whether mode alters in any way the underlying direct effects of the remaining factors on data recording accuracy. The models in this last set of analyses are $A = f(C, R, I, Q)$ and $A = f(C, R, I, Q, M)$.

Statistical tests account for suspected effects of clustering of questions within interview (and thus coder as well as interviewer). All statistical tests were performed using the WesVarPC sampling error software (Westat, 1996).

19.4 MODE AND RECORDING ACCURACY

Since a direct comparison of the frequency of recording error has not been conducted in prior research, there are conflicting expectations about the relative frequencies of recording error between modes. While there is an expectation that slips would be more frequent in CAI because of the somewhat more difficult motor skills required for accurate recording, it was expected that mistakes or errors in judgment or comprehension would be more common in P&P interviewing, since interviewers do not have the aid of features designed to reduce error in the computer assisted mode. The expectation has been that since mistakes would probably be more common than slips, CAI would have fewer overall errors than P&P.

The frequency of recording errors among the close-ended exchanges, simple straightforward and complex, in the experimental mode comparison is presented in Table 19.1. As might be expected, the vast majority of exchanges

Table 19.1. Type of Question and Recording Error by Mode

Type of Question and Recording Error	CAI		P&P	
	n	%	n	%
Simple straightforward	17,490	75.2	18,666	75.1
Accurate	17,471	75.2	18,644	75.0
Inaccurate	19	0.1	22	0.1
Complex	5756	24.8	6190	24.9
Accurate	5495	23.6	5797	23.3
May be misrecorded[a]	261	1.2	393	1.6
Total	23,246		24,856	

[a]Mode difference statistically significant at $p < 0.001$.

(75.2%) are simple and straightforward, with no differences between mode. It is useful to examine the nature of errors in each type of exchange separately, since they are of greatly different frequency and they differ in cause.

The rate of errors in the simple straightforward exchanges was minute: 19 in CATI and 22 in paper and pencil. The difference between the rates in the two modes is not statistically significant. About one-third of these recording errors occurred at yes/no questions, and most of those errors were answers recorded as "don't know" when in fact the respondent had given a Yes or No answer. Another one-third of the 41 recording errors in simple straightforward exchanges occurred when the interviewer incorrectly recorded the time reference, such as recording payment per month instead of payment per week. The remaining errors were distributed among a number of explanations, with no errors due to decimal shifts, transpositions, or other "slips."

No interview had more than two of these recording errors, and none of the 35 interviewers had more than two of these types of recording errors. Thus there does not appear to be any clustering by interview or interviewer. No PSID question had more than one of these recording errors, and there was no difference in the error rates across sections of the questionnaire. As a result, no sequencing of errors was detected. It appears that the recording errors for close-ended simple straightforward exchanges occur haphazardly, probably randomly. And this handful of errors appears to be mistakes rather than slips.

Among the 11,946 complex exchanges, 654 (5.5%) may have been misrecorded. The rate is significantly higher in P&P (6.3%) than in CATI (4.5%). The types of recording errors in the complex exchange is diverse. In order to assist interpretation of the results, the complex recording errors were classified into seven types of error:

- *Implied response.* Interviewer inferred a response from respondent's answer when a valid response was not given (e.g., interviewer assumed a year when the time period was not stated by the respondent).
- *Qualified response.* Interviewer failed to probe for a best estimate when the respondent qualified the answer (e.g., respondent said "Somewhat more than $100," and the interviewer, without further probing, recorded $100).
- *Failure to record answer.* Interviewer left the response blank, marked two or more response options when only one was possible, or wrote illegibly.
- *Failure to repeat question.* Interviewer did not adequately probe or repeat question when required (e.g., interviewer did not probe when a respondent laid off in February also reported being on vacation in that same month, and marked both laid off and on vacation).
- *Simple misrecording.* During a complex exchange, interviewer entered an incorrect response option (e.g., interviewer marked "yes" when respondent said "no".
- *Decimal shift:* Interviewer misplaced the decimal (e.g., respondent reported $1,500 income from a source, but the interviewer recorded $150).

- *Numeric error.* Interviewer wrote incorrect numeric response (e.g., respondent reported $128, but the interviewer recorded $125).

The last three of these error types may be interpreted as slips. The other types are mistakes, including errors in performing the interviewing task.

Distributions across these error types differ significantly between modes (see Table 19.2). While the implied response is the most frequent error in both modes, the relative frequency of implied responses is much higher in CATI (42.6% versus 26.5%) primarily because there are many more failures to record errors in P&P (86 versus 17). Both modes had a very small number of slips (i.e., decimal shifts or numeric errors), but P&P had more than twice as many simple misrecording errors. Thus the higher frequency of recording errors among complex exchanges in P&P is due to a higher frequency of failure to record or simple recording errors. These failures to record or simple misrecording errors did not occur with higher frequency for any types of questions.

The results in Table 19.2 for complex exchanges are unexpected. P&P recording has, according to expectation, a higher frequency of errors. But the nature of those errors indicates that there are more slips in P&P than in CAI and that higher frequency is due almost entirely to many more simple misrecording errors. In addition P&P is subject to higher frequency of failure to record an answer, a type of error more closely related to mistakes than slips.

The results in Tables 19.1 and 19.2 provide quantitative evidence of the effect of computer assisted data recording on the quality of survey data. CAI does indeed have fewer recording errors than P&P, but the source of the improvement is due to a combination of factors, not to one source that can be readily identified as slips or mistakes.

Since there are so few errors among the simple straightforward exchanges, it is sensible from a statistical viewpoint to concentrate further investigations

Table 19.2. Number and Percentage of Complex Interaction Errors by Type of Error and Mode

Type of Error	CAI		P&P	
	n	%	n	%
Implied response	110	42.6	102	26.5
Qualified response	62	24.0	66	17.1
Failed to record answer	17	6.6	86	22.3
Failed to repeat or probe	41	15.9	67	17.4
Simple misrecording	19	7.3	55	14.3
Decimal shift	2	0.8	4	1.0
Numeric error	7	2.7	5	1.3
Total	258	100.0	385	100.0

among complex exchanges. It is important to know whether the errors among complex exchanges are related to various behaviors and question characteristics in the survey interview. If behavior is related to recording accuracy, then the behavior may account for a share of the observed mode effect. However, if behavior does not differ by mode, then there is little reason to expect behavior will account for any of the observed mode difference in recording accuracy. Therefore we examine the association of behavior first with mode and then with recording accuracy, before proceeding to estimate multivariate models that include both mode and behavior.

19.4.1   Behavior and Mode

One could postulate that any number of behaviors in the interview process are related to mode. For example, interviewers are expected to make fewer fill errors, and thus fewer total major question wording changes in CATI due to the automatic fills in the computer assisted interview. We might also expect that interviewers would be less likely to skip question readings in CATI, since each question is displayed one at a time as opposed to multiple questions on a paper and pencil page. For the most part, though, there are not strong reasons to expect differences between mode.

In order to examine potential behavioral differences between modes, bivariate and multivariate models of mode were estimated for the behavioral predictors. Table 19.3 presents these models in a comparison for complex exchanges of conversational dynamics (C) by mode.

Individually three of the predictors are associated with mode: interviewers making fill errors, interviewers skipping question readings, and respondents giving affect-related comments. In the multivariate models, interviewer fill errors appear to be explained by other conversational dynamic predictors. Thus only skipped question reading and respondent affect-related comments continue to remain related to mode. The direction of the mode difference for fill errors, with higher frequency in P&P, is as expected. There is no a priori reason to expect differences in affect-related comments by the respondents.

Similar analyses were conducted for interviewer (I) and respondent (R) behaviors as well as question characteristics (Q) (results not shown here). No evidence of mode differences was found. Thus there are no behavioral or question characteristics related to mode.

19.4.2   Behavior and Recording Accuracy

While there were not strong reasons to expect mode differences in behavior, there are important reasons to expect differences between exchanges that were recorded accurately or inaccurately. The conversational dynamics, interviewer, and respondent behaviors have been used by survey researchers to investigate properties of survey questions. There is reason to believe that several of these

Table 19.3. Frequency of Conversational Dynamics (C) by Mode of Data Recording (M) and Bivariate and Multivariate Tests of Statistical Significance

Conversational Dynamics (C)	CAI % (n = 5756)	P&P % (n = 6190)	p-Value Bi-variate	p-Value Multi-variate
Interviewer made major wording change	4.6	8.5	0.06	0.53
Interviewer made fill error	1.8	5.0	0.00[a]	0.06
Interviewer used incorrect probing technique	15.7	17.7	0.19	0.12
Interviewer used reinforcing feedback	64.3	63.3	0.86	0.97
Interviewer made mode-related comment	4.5	3.7	0.51	0.55
Interviewer gave affect comment	3.8	3.2	0.60	0.91
Interviewer skipped question asking	1.5	2.8	0.01	0.03
Respondent gave simple straightforward response	78.7	78.5	0.91	0.61
Respondent interrupted question reading	3.2	2.9	0.54	0.32
Respondent gave qualified/conditional response	0.4	0.4	0.96	0.93
Respondent expressed uncertainty about answer	19.8	17.5	0.14	0.14
Respondent answered outside response options	23.0	21.8	0.40	0.34
Respondent gave multiple answers	0.2	0.3	0.16	0.13
Respondent refused to answer question	0.5	0.2	0.17	0.19
Respondent gave "Don't know" response	3.7	3.8	0.87	0.69
Respondent elaborated answer	16.7	17.2	0.72	0.46
Respondent gave mode comment	0.2	0.2	0.60	0.71
Respondent gave affect comment	4.2	2.8	0.04	0.02

[a] < 0.01.

behaviors will be related to recording accuracy as well. For example, incorrect probing technique during an exchange may be an indicator that the interviewer and respondent were having difficulty answering the question, and that there may be higher levels of recording errors associated with these question exchanges.

The relationship between the conversational dynamics behaviors and recording accuracy is examined in Table 19.4. These results are presented in a "case-control" format, highlighting differences in behavior between the 11,292 complex exchanges recorded accurately and the 695 that were recorded inaccurately. The comparisons ignore recording mode.

Table 19.4. Frequency of Conversational Dynamics (C) by Data Recording Accuracy (A) and Bivariate and Multivariate Tests of Statistical Significance

Conversational Dynamics (C)	Inaccurate % (n = 654)	Accurate % (n = 11,292)	p-Value Bi-variate	p-Value Multi-variate
Interviewer made major wording change	8.9	6.5	0.04	0.04
Interviewer made fill error	3.8	3.5	0.66	0.15
Interviewer used incorrect probing technique	33.0	19.8	0.00[a]	0.00
Interviewer used reinforcing feedback	61.0	63.9	0.32	0.89
Interviewer made mode-related comment	4.9	4.0	0.23	0.72
Interviewer gave affect comment	3.7	3.5	0.85	0.41
Interviewer skipped question asking	4.1	2.0	0.01	0.24
Respondent gave simple straightforward response	59.2	79.8	0.00	0.00
Respondent interrupted question reading	2.3	3.1	0.33	0.21
Respondent gave qualified/conditional response	0.5	0.4	0.82	0.98
Respondent expressed uncertainty about answer	8.1	19.3	0.00	0.00
Respondent answered outside response options	39.9	21.4	0.00	0.00
Respondent gave multiple answers	0.8	0.3	0.01	0.03
Respondent refused to answer question	0.2	0.3	0.91	0.86
Respondent gave "Don't know" response	5.4	3.7	0.06	0.36
Respondent elaborated answer	17.3	16.9	0.86	0.05
Respondent gave mode comment	0.2	0.2	0.97	0.93
Respondent gave affect comment	6.0	3.3	0.00	0.00

[a] < 0.01.

Methodologically interesting and statistically significant differences are present for a number of behaviors in the bivariate analyses. Among interviewer behaviors at the question asking, major wording changes, use of incorrect probing techniques, and skipped questions are much more frequent among inaccurately than accurately recorded exchanges. For example, in nearly a third of the inaccurately recorded exchanges the interviewer used an incorrect probing technique, compared with less than one-fifth of the accurately recorded questions. In the multivariate analyses the difference for skipped questions is no longer statistically significant, but the differences for major wording changes and incorrect probing remain.

There is a larger number of respondent behaviors for which differences exist between accurate and inaccurate recordings, and all of these are statistically significant in both the bivariate and multivariate tests. Simple straightforward responses are much less frequent among inaccurate recordings. The reasons for this difference are reflected in the other significant respondent behaviors at the question. For example, respondents are much more likely to give answers outside of the response frame, give multiple answers when only one is required, and give affect-related comments when the answer is inaccurately recorded. The relative frequency of these respondent behaviors is at least twice as large among inaccurately recorded questions for almost all of these behaviors. On the other hand, respondents who tend to express uncertainty about answers during an exchange are associated with higher levels of recording accuracy.

Thus, while several of the differences for interviewer behavior at the question are significant, there are more behaviors and larger differences for the respondent behaviors. That is, it appears that recording accuracy is more strongly associated with what the respondent does than with what the interviewer does. Further, suppose that the largest differences for interviewer behavior at the question — use of incorrect probing technique — is actually an interviewer behavior made in response to a respondent behavior. Then clearly respondent behaviors at the question are driving interviewer behavior, which in turn effects recording accuracy. That is, question askings recorded inaccurately are particularly difficult exchanges for the respondent and the interviewer, reflecting a complex, confusing conversational dynamic around the question. These difficult (as a subset of complex) exchanges lead to significantly higher levels of the kinds of behaviors traditionally associated with poor interviewing technique and with inaccurate recording of data.

The multivariate model results in Table 19.4 indicate that this process has several separate features that are not merely interrelated dimensions of the same problem. The persistence of the associations between recording accuracy and respondent uncertainty about answers, answers outside the response frame, multiple answers, and affect-related comments in the multivariate model indicates that these behaviors may be acting separately in causing recording errors.

The relationships among recording accuracy (A) and interviewer behavior (I), respondent behavior (R), and question type characteristics (Q) showed only one statistically significant association (results not shown here). In the multivariate analysis involving respondent behaviors, only the association for simple straightforward responses remained statistically significant at the $\alpha = 0.05$ level. In particular, the respondents who consistently gave simple straightforward responses were less likely to have inaccurately recorded responses.

These tabular presentations show that there are only a few direct effects across all of the different conversational dynamics (C), interviewer behavior (I), respondent behavior (R), and question type characteristics (Q) examined in the bivariate and multivariate models. The most dramatic differences between inaccurately and accurately recorded complex exchanges are for respondent behaviors in conversational dynamics, and those appear to operate separately.

Two principal questions remain to be addressed: the extent to which these behaviors and question characteristics explain recording accuracy simultaneously in a multivariate model across factors, and the role of mode in recording accuracy when these behaviors are taken into account.

19.4.3   Mode, Behavior, and Recording Accuracy

The empirical results so far provide support for some aspects of the model presented in Figure 19.1. There is a strong association of mode with recording accuracy, but there are no strong associations of behaviors or question characteristics with mode. Yet there appears to be a relationship between recording accuracy and some behaviors in the conversational dynamics set (C) and one behavior in the respondent behavior set (R). None of the behaviors represented in the interviewer behavior (I) and question type characteristics (Q) sets appear to be associated with recording accuracy.

The model in Figure 19.1 suggested that the mode effect may disappear after behavior and question characteristics have been accounted for. A multivariate model of recording accuracy on behavioral factors and questions characteristics needs to be developed first. When mode is added to this multivariate model, we can determine whether mode differences in recording accuracy are truly separate from these behaviors and question characteristics or merely act through them.

As noted previously, there are 48 different predictors across conversational dynamics (C), respondent behavior (R), interviewer behavior (I), and question characteristics (Q). A single multivariate model is not feasible, as the dependent variable, recording accuracy, has only 695 instances of recording inaccuracy to be explained. A forward variable selection procedure was thus used within each collection of predictors (i.e., conversational dynamics, C, respondent behavior, R, interviewer behavior, I, and question characteristics, Q) with respect to recording accuracy to obtain a subset of predictors to be used in the final regression model for recording accuracy among the complex exchanges. The four forward selection procedures yielded 16 total predictors, which were then used in a multivariate logistic regression model to examine their joint association with recording accuracy and the added contribution of mode to the model. The results are summarized in Table 19.5.

Model 1 is the regression of recording accuracy (A) on the 16 predictors selected from the 48. Ten of the 16 predictors are not significantly related to recording accuracy when the other predictors are present in the model. Among conversational dynamics behaviors, respondent elaboration of an answer is not significant, and none of the respondent behaviors (R) is. Further, one of the two interviewer behavior (I) predictors, fill errors, also fails to achieve statistical significance in the multivariate model. The single question type characteristic fails to achieve statistical significance in the multivariate model as well.

Table 19.5. Multivariate Model Odds Ratios and *P*-values for Regression of Data Recording Accuracy (*A*) on Conversational Dynamics (*C*), Respondent (*R*), Interviewer (*I*), and Question (*Q*) Characteristics (*n* = 48,102)

	Model 1		Model 2	
Predictors	Odds Ratio	*p*-Value	Odds Ratio	p-Value
Conversational dynamics (C)				
Interviewer used incorrect probing technique	1.50	0.00[a]	1.50	0.00
Respondent gave simple straightforward response	0.44	0.00	0.44	0.00
Respondent expressed uncertainty about answer	0.40	0.00	0.40	0.00
Respondent answered outside response options	1.44	0.00	1.44	0.00
Respondent elaborated answer	0.71	0.08	0.71	0.07
Respondent gave affect comment	1.92	0.00	1.94	0.00
Respondent (R)				
Respondent tended to give simple straightforward responses	0.38	0.35	0.38	0.36
Respondent tended to interrupt question reading	1.02	0.24	1.02	0.21
Respondent tended to express uncertainty about answers	0.99	0.56	0.99	0.62
Respondent tended to answer outside response options	1.01	0.22	1.01	0.23
Respondent tended to give multiple answers	0.75	0.25	0.74	0.23
Respondent tended to elaborate answers	0.99	0.37	0.99	0.35
Respondent tended to give mode comments	0.80	0.10	0.79	0.11
Interviewer (I)				
Interviewer tended to make fill errors	1.06	0.10	1.05	0.35
Interviewer tended to use incorrect probing techniques	1.02	0.00	1.02	0.00
Question type (Q)				
Interviewers tended to give reinforcing feedback for some response frames	1.00	0.38	1.01	0.38
Mode (M)	–	–	0.89	0.54

[a] <0.01.

The odds ratios corresponding to each of the Model 1 predictors are also given in Table 19.5 to indicate the direction of the association. The respondent (R), interviewer (I), and question characteristic (Q) predictors have all been scaled to a percentage, while the conversational dynamics (C) variables are all dichotomies (behavior present or absent). Thus care must be taken when comparing odds ratios across the predictors.

For statistically significant conversational dynamics predictors, the strongest effects on recording accuracy are respondents' giving simple straightforward responses, expressing uncertainty about answers, and giving affect related comments. Simple straightforward responses lead to a 56 percent decrease in the odds of recording inaccuracy relative to questions without simple straightforward responses. Expressing uncertainty also leads to decreased odds of recording inaccuracy, a decrease of 60 percent. This may reflect a part of a negotiation between the respondent and the interviewer to obtain the most accurate information possible, and a more carefully recorded response because of the negotiation process. Respondents' giving affect-related comments leads to nearly a doubling of the odds of recording errors over questions where no such comments occurred. The single interviewer behavior that remained statistically significant, using incorrect probing techniques, leads to a 2 percent increase in odds of recording error for each percentage point increase in the use of incorrect probing techniques.

This first multivariate model in Table 19.5 reinforces the findings of the previous analyses. The most important behaviors in predicting recording accuracy continue to be respondent behaviors at the question. The only interviewer behavior at the question that remains significant in the model is the use of incorrect probing techniques. To the extent that the use of incorrect probes is related to the presence of the indicated respondent behaviors, it is a behavior that has a separate and statistically significant contribution beyond that attributable to the respondent behaviors at the question (i.e., conversational dynamics).

Model 2 is the same set of 16 predictors as Model 1, but now recording mode (M) has been added to the predictor set. While mode was strongly associated with recording accuracy when examined alone in Table 19.1, it is, in the presence of the other predictors, no longer related in Table 19.5. Further the relationships of the other 16 predictors to recording accuracy in Model 1 do not change. That is, mode does not account for any of the explanatory power of these other 16 predictors. It was speculated that mode would operate independently of any observable behaviors or other characteristics of the interviewing process. The results of Model 2 indicate quite clearly that mode does not have any separate observable effect; the other behaviors are now accounting for whatever initial differences in recording accuracy had been observed for mode. That is, the recording accuracy differences between CAI and P&P are attributable not to mode but to differences in the behaviors of interviewers and respondents between modes. In particular, mode (M) differences in recording accuracy (A) can be explained by conversational dynamics (C), especially respondent behaviors, as well as an interviewer behavior averaged over question askings (I).

19.5 CONCLUDING REMARKS

The shift from P&P interviewing to CATI has largely occurred without any substantial research on mode differences in recording accuracy. Given that CAI was touted as a means to improve data quality, it is perhaps not surprising that so little research into mode differences occurred before changes were made. Still, for ongoing survey operations, the impact of a change in mode needs to be carefully assessed to understand whether mode changes have any substantial effect on the quality of data.

Much of the research on mode of interview has been limited to simple differences between modes. There are compelling reasons to suspect that observed mode differences can be attributed to other factors in the survey process. The purpose of this chapter has been to shift the focus of mode comparisons from simple differences to examinations of the features of the survey interview that explain those differences.

One set of reasonable explanatory factors for the mode differences in recording accuracy is behavioral: what the interviewer and the respondent do during the survey interview. The investigation reported here has used standard behavior coding measures of the interviewer and respondent behaviors to provide a set of behavioral characteristics of the survey interview as explanatory factors. The findings indicate that these behaviors do explain the difference in recording accuracy between P&P and CAI. That is, mode differences are a reflection of changes in how interviewers and respondents behave when the recording mode changes.

It is important to keep in mind, though, that the first and most important finding has been that mode is clearly associated with recording accuracy. Thus recording accuracy is higher in CAI than in P&P. There is some association of mode with interviewer and respondent behaviors at the question asking (C), and several question type characteristics as well. The associations are in an expected direction, with behaviors and characteristics that indicate some difficulty in the interview being more frequent in P&P than CATI.

There are some strong associations of behaviors with recording accuracy among complex question asking exchanges. Respondent behaviors at the question, such as answers outside the response frame or elaborating an answer, generate higher levels of recording error, while a respondent expressing uncertainty about an answer actually appears to suppress recording errors. Only one interviewer behavior at the question generates higher rates of recording error, use of incorrect probing technique. The majority of the incorrect probing is a failure to probe when a probe is needed. One interpretation of these findings is that when respondents have difficulty providing an answer at a question (reflected in answers outside the response frame and elaboration) interviewers fail to probe adequately and tend to record an answer that is incorrect. There appears to be a subset of complex exchanges where respondent difficulty in providing an answer leads to less accurate recording.

There is an import of this particular finding for the survey researcher. Interviewers clearly have greater difficulty handling complex exchanges and recording the answers to them accurately. Greater emphasis should perhaps be placed in general interviewer training on how interviewers should handle these complex exchanges. Most survey organizations spend considerable training time on the use of good probing techniques. What should accompany that training is interviewer preparation to carefully examine the response that is recorded from such exchanges. Further the interviewers can be trained to detect and correct the kinds of recording errors identified in Section 19.4.1.

Finally, when mode is introduced as a control in the behavior and recording accuracy setting, it has no effect. This finding ought to encourage survey researchers who detect a simple mode difference to explore more fully the nature of that difference. The mode difference finding is important, but, as indicated here, it is not necessarily explanatory of the underlying process that generates the difference. Behavioral and other aspects of the survey process ought to be examined. There appears to be a new role for traditional behavior coding: providing behavioral data on the process that can be used to explain mode or other experimental findings.

While the expected and observed mode differences have been explained by behavior, this is still not a complete and satisfactory explanation of the mode difference. That is, while behavior does explain mode differences, it is not clear how the interviewer and respondent behaviors operate to create differences between modes in recording accuracy. CAI clearly has a beneficial effect on recording accuracy. The exact mechanisms by which it operates through interviewer and respondent behaviors remain unresolved.

ACKNOWLEDGMENTS

Support for this investigation was received from NSF grant and NIA supplemental award number SES 9022891 and from the Survey Research Center, University of Michigan. The advice and counsel of Greg Duncan of Northwestern University, the Panel Study of Income Dynamics Board of Overseers, and the staff of the Panel Study is gratefully acknowledged. Steve Blixt of MBNA was responsible for the development of coding schemes and training of coders. Marshall Cummings of the Survey Research Center developed computer applications. We also thank Beth-Ellen Pennell, Lesli Jo Scott, Stephanie Chardoul, and Lisa Carn, together with the staff of the Survey Support Laboratory, at the Survey Research Center for their numerous contributions to the research investigation.

CHAPTER 20

Computerized Self-Administered Questionnaires

Magdalena Ramos, Barbara M. Sedivi, and
Elizabeth M. Sweet
U.S. Bureau of the Census

20.1 INTRODUCTION

While the replacement of paper and pencil (P&P) interviewer-administered surveys with computer assisted methods (CATI and CAPI) has been underway for some time, the replacement of mailed paper questionnaires with computerized alternatives has been slower. Since the mid-1980s, however, electronic questionnaires have been appearing in most areas of data collection. In essence, computerized self-administered questionnaires (CSAQs) bring classic CAI functionality to mail surveys, and hence solve some of the traditional data quality problems associated with P&P surveys. Still we do not expect that CSAQs will fully replace their P&P equivalents. Instead, we expect they will play a key role in specialized survey applications, including those on the Internet and the World Wide Web (WWW), and become one option in surveys that employ multimode or multitechnology data collection.

20.1.1 CSAQ Definition

Computerized self-administered questionnaire is the general term we use to identify all computerized questionnaires that request information electronically from respondents without an interviewer present and where respondents use their own (or their organization's) personal computer (PC) to respond.

There are two main forms of CSAQ: (1) disk-by-mail (DBM) and (2) electronic mail surveys (EMS). In DBM, a diskette mailout package is sent to

Computer Assisted Survey Information Collection, Edited by Mick P. Couper, Reginald P. Baker, Jelke Bethlehem, Cynthia Z. F. Clark, Jean Martin, William L. Nicholls II, and James M. O'Reilly. ISBN 0-471-17848-9 © 1998 John Wiley & Sons, Inc.

and received from the respondent via postal service mail. Because computer diskettes are mailed in place of paper questionnaires, DBM CSAQ applications are the closest computerized alternatives to traditional mail questionnaire surveys. Electronic mail surveys differ in that respondents receive the request and return the survey data by electronic means.

The CSAQ instrument controls the flow of survey questions, provides instructions, including help screens, and performs consistency and completeness edit checks as the respondent enters data. In some CSAQs, respondents may import existing data from other computer applications, such as a predefined ASCII file, a spreadsheet or database.

With these capabilities CSAQ technology offers several potential advantages over P&P questionnaires. Timeliness may be increased by eliminating a separate data entry step at the survey organization. Data accuracy may be increased by built-in edits, reducing respondent transcription and reporting errors. Respondent burden may be reduced by automatic branching, data import capabilities, and electronic transmissions. With CSAQ the survey organization avoids the printing and handling of paper questionnaires, and with EMS even the preparation and the handling of computer diskettes may be eliminated. However, as of 1998 the usefulness of CSAQ data collection has been limited by PC availability and compatibility. For those forms of CSAQ employing electronic transmissions, CSAQ acceptance also has been limited by issues of confidentiality and security (Sedivi and Rowe, 1993).

The method we call CSAQ has been given other names by other investigators. For example, Keller, Bethlehem, and Metz (1990) call it CASAQ for *computer assisted self-administered questionnaires,* while Baird and Walker (1995) refer to it as EQ for *electronic questionnaires* (see also Synodinos and Brennan, 1988). We do not equate CSAQ with computer assisted self-interviewing or CASI. We consider CASI a broader category of computerized, self-administered data collection methods. CSAQ is a subcategory of CASI. CASI also includes some applications where interviewers are present. For example, it may be employed as part of an in-home CAPI interview, as described in Tourangeau and Smith (Chapter 22) and Turner et al. (Chapter 23). CASI also includes applications where interviewers are not present, but the respondent does not use their own PC. These forms of CASI include kiosk applications (e.g., Synodinos, Papacostas, and Okimoto, 1994) and the telepanel (see Saris, Chapter 21).

20.1.2 CSAQ Origins and Evolution

The CSAQ technology had its origins in several disciplines, including academic research, government survey research and applications, and commercial or market research. Each of these areas recognized the potential advantages of using the computer for data collection of self-administered questionnaires.

Academic studies of CSAQ concentrated from the beginning on the psychological effects of using the computer and its potential effects in bias reduction.

Kiesler and Sproull (1986) were the first to compare the effects of computerized questionnaires with P&P self-administered methods, including effects on the reporting of sensitive or stigmatized behaviors. Higgins, Dimnick, and Greenwood (1987) similarly compared CSAQ and P&P in the reporting of alcohol consumption and drinking behavior. Olson and Schneiderman (1995) explored the use of DBM CSAQ to collect information from physicians, while other academic researchers examined the use of EMS CSAQ and the potential biasing effects of self-selected respondents in electronic surveys (Fisher, Margolis, and Resnick, 1995; Fisher et al., 1995; Walsh et al., 1992; and Werner, Maisel, and Robinson, 1995).

The goal behind government CSAQ research was to develop faster survey completion methods that also produced higher-quality data. A pioneer in government CSAQ applications was the U.S. Energy Information Administration (EIA) which in 1985 began development of its electronic reporting system called PEDRO, the Personal Computer Electronic Data Reporting Option. Other early U.S. government developers were the Internal Revenue Service, with its application for electronic transmission of tax returns, the Department of the Navy, the National Center for Education Statistics (NCES) and the National Science Foundation (NSF) (see Federal Committee on Statistical Methodology, 1990b; Kindel, 1992; Somer and Murphy, 1989; Trumble et al., 1995.)

Since 1989 additional electronic reporting systems have been developed both in Europe and North America. Applications by the U.S. Bureau of the Census and the U.S. Bureau of Labor Statistics (BLS) are discussed in Sweet and Ramos (1995), Ramos and Sweet (1995), Harrell et al. (1996), and Clayton and Werking (Chapter 27). These and most other recent government applications have primarily focused on the collection of economic data in surveys of businesses or other establishments. They usually collect payroll, employment, and sales data at the company or establishment level and other economic data particular to the survey (Blom, 1993; Jamieson, 1996; Keller, 1994; Keller, Bethlehem, and Metz, 1990).

In market research, Goldstein (1987) and Morrison (1988) first reported tests of DBM CSAQ methods in surveys of marketing executives and corporate buyers. A variety of CSAQ market research applications have subsequently been developed. CSAQs are currently used in the commercial sector for product registration, new product design and follow-up, product pricing and positioning, brand image research, and customer satisfaction surveys. Commercial applications also have developed complex interviewing technologies, such as adaptive trade-off analysis, said to have been extremely difficult to implement on paper (Gershenfeld et al., 1989, Gum, 1989; de Leeuw and Nicholls, 1996; Porst, Schneid, and van Brouwershaven, 1994; Witt and Bernstein, 1992).

This chapter summarizes the interdisciplinary development and research of both DBM and EMS CSAQs. Section 20.2 examines a broad range of CSAQ issues including choice of software for questionnaire design, distribution and retrieval methods, confidentiality and security, universe definition and sampling, response rates and data quality, costs, and organizational considerations. A brief review of the future direction of CSAQ is also presented, including options for Web/Internet CSAQ data collection (discussed further by Clayton and Werking, Chapter 27).

20.2   ISSUES IN CSAQ

20.2.1   Software and Questionnaire Design

Four types of development software have been used for CSAQ over the years. The earliest applications typically employed general programming languages, such as BASIC, CLIPPER or C. Government agencies that already used CATI for telephone interviewing sometimes employed the same CAI authoring language (e.g., CASES, Blaise) for their CSAQ needs. Market research interest in CSAQ was partly stimulated by commercial off-the-shelf electronic questionnaire software. With the growth of the Internet in the 1990s, additional CSAQ authoring choices using Internet software options also became available. These include direct use of e-mail, Hypertext Markup Language (HTML), Java and JavaScript (Curry, 1990b; Federal Committee on Statistical Methodology, 1991; Keller, 1994; Pitkow and Recker, 1994; Schuldt and Totten, 1994; Sweet and Ramos, 1995; Sweet and Russell, 1996; Witt and Bernstein, 1992).

The optimal CSAQ authoring system would have the flexibility to allow users to choose and combine the different screen layout designs, with forms-based (matrix style) design preferred for business survey questionnaires. It would allow most types of question and preferably would provide flexible navigation style capabilities, a graphical user interface (GUI), and all of the functional capabilities common to CAI systems (Nicholls, 1988; Carpenter, 1988; Bethlehem and Schuerhoff, 1994; Federal Committee on Statistical Methodology, 1990b).

Each type of CSAQ authoring software has specific strengths and limitations in questionnaire structure and functionality, although with sufficient custom programming, all survey authoring software can be adjusted to satisfy most CSAQ requirements. The question is how much time and money should be invested in customization for every new application.

General programming languages require a great deal of customization and a highly skilled computer programmer, while most CAI authoring languages have much of the functionality already built in (data import and printing capabilities may be exceptions). CAI authoring languages generally do not require a programmer, but they do need an extensively trained and capable author. Advanced CAI systems also provide flexible screen layout and navigation options, but they are currently weak in user interfaces. The move toward GUI interfaces is expected to be slower for CAI systems because of their size, complexity, and DOS based architecture. However, various GUI applications software are already used for custom programming of CSAQs (Carpenter, 1988; Federal Committee on Statistical Methodology, 1990b; O'Muircheartaigh and Murphy, 1991).

Commercial off-the-shelf (COTS) electronic questionnaire software often lacks capabilities like complex skip patterns, error checking, randomization of response categories and question blocks, complex fills, and arithmetic functions which are available in CAI authoring languages. This makes COTS packages easier to learn, although they typically require a dedicated staff that follows set standards for CSAQ development. If the desired functionality is lacking, it may be necessary for a programmer to custom code the extra requirements. Another common drawback is that most commercial packages do not accommodate form-based questionnaires, but evidence of change in this area is seen in recent versions of COTS systems (Downes-LeGuin and Hoo, 1994; Schneid, 1995; Smith and Behringer, 1992).

Alternatively, there is another genre of COTS software called "forms design software" that does allow form-based, as well as item and screen-based, designs. One example is JetForm which originally was developed for electronic business form completion but has evolved to include most functional capabilities required for CSAQ applications. While currently expensive, such software may become a more feasible option in the future.

Internet authoring options for surveys are in their infancy. Both e-mail and HTML have limited instrument design capabilities. A screen of questions generally requires scrolling to view the complete questionnaire; navigation is point and click with the mouse, and question types are restricted. E-mail has the added disadvantage of being based on unstandardized, proprietary software. Since HTML is a standard language for the WWW, HTML surveys have fewer problems with survey layout across different interfaces. Due to the simplicity of the authoring, a subject matter specialist could probably produce a simple questionnaire independent of a programmer or author. But to address such CSAQ requirements as branching, filling, data validation, and data retrieval from a user's PC, a programmer would need to add Common Gateway Interface (CGI) scripts, Java applets, or JavaScript to the HTML questionnaire. As with other CSAQ authoring options, the further development of e-mail and Web software should reduce the amount of customization required to meet CSAQ requirements in the future.

CSAQ software also varies in cost. If a survey organization already uses a general programming language, e-mail, or even a CAI programming language for other purposes, then it might seem logical to use it for CSAQ authoring also, resulting in no additional costs. Nevertheless, the cost of customization to bring the software up to CSAQ requirements must also be taken into account. Most CAI languages, for example, require software enhancements for CSAQ use. Enhancing software can result in substantial costs that must be balanced against the costs of procuring software already adapted for CSAQ use. If software must be purchased for CSAQ applications, commercial packages are generally less expensive than CAI authoring systems, and HTML is available gratis as a standard language for the WWW. The best value is a function of both costs and capabilities.

20.2.2 Distribution and Retrieval

Until the early 1990s, almost all CSAQs were distributed and retrieved via mailed diskettes. The advantages of CSAQ transmission by modem, e-mail, the Internet, or the Web have since began to attract the attention of the CSAQ developers. If CSAQ distribution and retrieval could be conducted using pre-established electronic networks, all mailout activities could be eliminated.

Diskette technology and PC operating systems have evolved, bringing changes that affected CSAQ software. These included changes in both diskette size (5.25 to 3.5 inch) and density (low to high), and the changes from DOS/Macintosh to Windows operating systems, which reduced the number of diskettes required, simplified the installation instructions, and reduced the size of the CSAQ mailing package.

Modem communication currently is a much less common form of CSAQ distribution and retrieval than DBM. Distribution by modem is especially uncommon because it entails transmitting the CSAQ software, the CSAQ instrument, and historical data (if any) to the respondent's PC. Modem retrieval is simpler, since only the respondent's data need be transmitted back to the survey organization. Respondent access to modems and communication software also is currently limited. If respondents initiate the connection to the survey organization with their own communication software, protocol incompatibilities may arise. To avoid these difficulties, most survey organizations provide communications software to the respondents in a mailed diskette before any modem communications are attempted. Even then, when respondents who have modems are given the choice of a diskette or a modem return, most have chosen to reply by diskette (Sweet and Ramos, 1995; U.S. Department of Energy, 1992).

E-mail has also been used for CSAQ transmissions and seems most valuable when all respondents to a survey use the same e-mail system within the same organization. Recently CSAQs also began to use wide area networks (e.g., Blom, 1993; Schuldt and Totten, 1994; Walsh et al., 1992), and more recently the Internet. Even with the problems associated with this emerging technology, the Internet apparently is becoming an increasingly popular medium for CSAQ distribution and retrieval.

When the organization conducting a survey and the survey respondents are under different e-mail systems (or system managers), there is a risk that those systems may differ in message length limitations, use of attachments, compression and encryption options, or in other ways. A lengthy text file containing a CSAQ instrument may be left intact by one e-mail system, converted to an attachment by another, and rejected by a third. E-mail CSAQ transmissions also can face additional implementation problems from outdated e-mail addresses and concerns about the security or confidentiality of e-mail communications. Still e-mail remains an option when multitechnology surveys are planned, even when e-mail is appropriate only for a limited segment of the target sample.

Large government agencies in the United States are beginning to consider the Internet, even for CSAQ surveys of the general public. While access by the general population to the Internet and Web is limited, the network interface is much more easily understood than that of modem communication (Fisher, Margolis, and Resnick, 1995; Fisher et al., 1995; Oppermann, 1995; Pitkow and Recker, 1994; Quarterman, 1994; Sweet and Russell, 1996).

20.2.3 Confidentiality and Security

Confidentiality and security remain vital concerns for CSAQ data collection. Respondents need assurance that their survey responses will remain confidential and will not be revealed or distorted by electronic communications. They also need to be assured that installing and using the CSAQ questionnaire and communications package will not negatively affect their computer hardware and software.

Appropriate confidentiality and security steps for CSAQ depend on the mode used for CSAQ distribution and retrieval. For DBM CSAQ, security measures are similar to those for a P&P mail questionnaire. Quality control checks prior to mailout should ensure that confidential information is not disclosed and that the software contains no computer viruses. The secure delivery and return of the disk is then entrusted to the postal service whose postal security measures are believed to be effective. Although not commonly used for DBM, CSAQ technology also offers capabilities for data encryption before mailout that are not available for P&P methods. Data encryption uses algorithms and separately sent keys to make the transmitted information useless to anyone but the intended recipient.

For EMS CSAQ the communications travel through telephone and other publicly accessible data lines, usually considered less secure than the postal service. To ensure confidentiality, data encryption prior to transmission is recommended. When the data are transferred between the respondent and the survey organization, authentication, access control, nonrepudiation, and data integrity checks should be used to restrict access and prevent damage to the information or computer system. When CSAQ applications employ the Internet, greater complexities and security threats arise because this environment is outside the survey organization's control. Because Internet communications are broken into packets that pass through a multitude of routers before reaching the target computer, encryption is generally essential. A brief definition of these security measures is provided in Table 20.1.

With the escalating popularity of the Web, groups such as the World Wide Web Consortium (W^3C) are eager to resolve the issues of WWW infrastructure and security standards. Numerous forms of encryption, authentication, and digital signatures are currently under development (see Cain, 1995). On the Internet it is now possible to avoid the three types of risks described in Table

Table 20.1. Security Measures

Authentication	Verification of the claimed identity of a computer or computer network user.
Access control	Verification and enforcement of the authorized uses of a computer network by a user, subsequent to authentication.
Data integrity	Verification that the contents of a data item (e.g., message, file, program) have not been accidentally or intentionally changed in an unauthorized manner.
Data confidentiality	Protection of the information content of data from unauthorized disclosure, generally using some form of encryption.
Nonrepudiation	Protection against denial of sending (or receiving) a data item by the sender (or receiver).

Sources: McNulty (1994), Cain (1995).

20.2. Depending on the degree of sensitivity of both the data being sent and those residing at the survey organization's computer, a security system could be designed by combining the methods described in Table 20.2. Of these, source verification is the most difficult to achieve, given the current infancy of standards for signatures, certificates, and such.

20.2.4 Target Universe and Sampling Concerns

Coverage error is defined as the loss of survey data "through the failure to include some number of persons in the list or frame used to identify the target population of a survey" (Nicholls, Baker, and Martin, 1997). Coverage error is a major concern for CSAQs, since this technology requires that the respondent, rather than the data collection organization, has access to and operates the data collection device, that is, the computer. Current CSAQ software is not hardware independent. In other words, it will not always work with different

Table 20.2. Methods to Safeguard against Internet Security Risks

Organization security	Use of firewalls, authentication, and access control.
Confidentiality of data-in-transit	Cryptography protects the confidentiality of the data in transit to and from the respondent.
Source verification	Data integrity checksums, digital signatures, digital certificates, and smart cards can be used to ensure the validity and source of the data (nonrepudiation).

Sources: McCarthy (1995), Cain (1995).

combinations of system configurations, monitors, and operating systems. For a survey to be conducted exclusively with CSAQ, the target population must meet two requirements. First, respondents need access to a PC equipped with the particular hardware and software that allow interaction with the survey CSAQ. Second, they have to be willing to use that PC to complete the survey. Given these requirements, PC penetration is of major concern (Pilon and Craig, 1988; Carpenter, 1988).

As of 1995, estimates of PC penetration of U.S. homes range from 26 to 31 percent. For business or establishment surveys, PC availability is not as limiting. Although 98 percent of U.S. large businesses were estimated to have PCs as of 1993, that figure dropped to 63 percent for businesses with less than 100 employees and to 20 percent for those with less than 10 employees (Oppermann, 1995; Ogden Government Services, 1993).

Additionally not all PCs can run all CSAQ instruments. Access may be limited by the PC's operating system, such as DOS, MS-Windows, OS/2, Macintosh, or Unix. Before a 1994 CSAQ test, the Census Bureau conducted a screening survey of potential business respondents to learn which operating systems they used. At that time about 70 percent used MS-Windows with DOS, and an additional 15 percent used DOS only. Less than 2 percent reported a Macintosh system. A DOS-based CSAQ instrument was therefore designed, although some Windows users said they would have preferred a questionnaire in Windows. Two years later a similar screening survey was conducted for a Web-based survey. Of those business respondents saying they were willing to participate in a Web-based CSAQ test, 93 percent ran their PCs under Windows 3.1 or Windows 95, about 4 percent used a Macintosh system, and less than 3 percent ran under OS/2, Unix, or some other operating systems. The CSAQ was written for Windows 95 and a Netscape Navigator 3.0 browser. These results apply only to Census Bureau surveys of business and industry establishments. For new CSAQ surveys now being planned for surveys of schools and of households, larger proportions of Macintosh systems and other operating systems may be found (Sweet and Ramos, 1995).

Limited modem and Internet access by the general population also affects EMS surveys. In 1994 the Times Mirror Center estimated that only about 16 percent of U.S. households had a modem equipped PC and that only 6 percent had e-mail or Internet access. For businesses the Census Bureau estimates that as of 1995 approximately 72 percent of large companies have modem connections, but only 10 to 25 percent would be willing to report via the Internet (Oppermann, 1995; Quarterman, 1994; Sweet and Russell, 1996).

Given these estimates, substantial coverage and nonresponse bias would be introduced if the groups who do not meet the PC requirements and those that meet the requirements but are not willing to participate in an electronic survey are different from those willing to participate. Studies have shown that the more computer experience people have, the more likely they are to complete at least some parts of an electronic survey. Those with less experience are less likely to participate (Allen, 1987; Couper and Rowe, 1996; Walsh et al., 1992;

Witt and Bernstein, 1992). There is a danger that restricting the survey to those willing and able to respond via a CSAQ might bias the estimates. In fact many CSAQ surveys have been restricted to respondent populations with PC access and knowledge.

The coverage limitations of CSAQ applications are expected to improve as computers continue to penetrate the workplace and home, but currently these problems appear insurmountable for household survey applications. In the meantime CSAQ researchers have several options, including limiting the scope of the universe to cases expected to have PCs, and thus generalizing the conclusions only to populations of similar composition. Another option is indiscriminate distribution, but this can result in lower than acceptable response rates, since many may not have the required equipment. A better option is to prescreen the potential universe or to send respondents alternative data collection modes, such as both a CSAQ and P&P questionnaire (Sawtooth Software, 1989; Walsh et al., 1992).

While prescreening cannot guarantee a representative sample, it serves to identify those willing to respond via CSAQ and those that meet the hardware/software requirments for the particular application. For those that either do not respond to the screener or do not have the needed capabilities, a paper questionnaire can be used. Prescreening adds time and cost to the survey process. This may still be worthwhile for ongoing surveys that can establish a permanent CSAQ panel. Although prescreening usually reduces the burden on respondents who do not fit all requirements of the survey instrument, it is not always completely effective. Even in the prescreened sample, some respondents will cite incompatible hardware as a reason for CSAQ nonresponse (Gum, 1989; Sweet and Ramos, 1995).

Mailing both the CSAQ and a paper questionnaire to prospective respondents eliminates the need for prescreening. Still the cost of developing and sending both forms of the questionnaire and the survey organization's desire to receive the reported data electronically often reduce the attractiveness of this option. This practice still may be useful for continuing surveys to create a separate, permanent CSAQ subpanel for future data collection.

20.2.5 Response Rates, Timeliness, and Respondent Reactions

When respondents have the required hardware and software to complete a CSAQ, other factors come into play in their decision to complete the CSAQ or not. Factors specific to CSAQ applications include their knowledge about computers, their perceptions about security, the ease of installation and use of the CSAQ, company policies, and past experiences with electronic reporting. More general factors, which influence both P&P and CSAQ response rates, include the relevance of the survey topic; whether prenotices, prescreening, reminders, or incentives are used; and perceptions of the project sponsor (Saltzman, 1992; Witt and Bernstein, 1992).

Market researchers suggest that the more influential factors are the respondents' perception of the project sponsor (when disclosed) and their interest in the survey topic. These may not be the more important factors in academic and government applications where follow-up methods often have proved most effective in P&P self-administered questionnaires. One clear difference between a CSAQ and a P&P questionnaire that may affect the response rate is that respondents cannot guess the survey length with CSAQ until the software is loaded into the computer. Research has been undertaken on the consequences of disclosing the expected completion time when respondents are asked to complete a CSAQ survey, but results are not conclusive (Saltzman, 1992; Witt and Bernstein, 1992; Zandan and Frost, 1989).

O'Brien and Dugdale (1978) and Messigner (1989) have suggested that CSAQ use could result in poor cooperation levels due to confidentiality concerns and to computer or technology phobia. A common speculation is that the novelty of the task may motivate respondents to load and complete a CSAQ. These views suggest that ease of use is of paramount importance in gaining and maintaining respondent participation for a CSAQ. The CSAQ package should be easy to load, easy to use, free of errors, provide help screens and help desk assistance, and ensure that security measures are highlighted and followed (McGarr, 1995; Pilon and Craig, 1988; Sawtooth Software, 1989).

The market research literature frequently claims that DBM response rates exceed those of P&P questionnaires; but such claims remain largely unsupported by scientific studies or experiments. The few comparative response rate studies report higher response rates for CSAQ than for P&P methods, but theses studies usually have very small sample sizes. The CSAQ response rates they cite as improved also appear lower than would be acceptable for many academic or government survey organizations (Quirk's Marketing Research Review, 1987; Curry, 1990a; Goldstein, 1987; Gum, 1989; Machrone, 1992; Pilon and Craig, 1988; Saltzman, 1992; Wilson, 1989).

The burden of rigorous experimental evaluation has rested on universities and government agencies, which have often reported mixed results. Smith and Behringer (1992) compared three panels of sales agents using a telephone interview, a P&P mail survey, and a DBM CSAQ (all sales agents had portable computers). The telephone panel had the highest response rate of 93 percent, the mailed P&P survey achieved a 77 percent response rate, and the response rate of the DBM panel reached only 63 percent. In their split panel design Sweet and Ramos (1995) also found that the response rate to P&P, at 73 percent, was significantly higher than the 53 percent response rate obtained for CSAQ.

Other researchers have reported higher response rates for DBM panels, although not necessarily from fully controlled experimental comparisons. Higgins, Dimnick, and Greenwood (1987) observed a 63 percent response rate for paper, significantly lower than the 78 percent for DBM. Kindel (1992) found that states using the P&P questionnaire in an NCES library survey had a lower response rate than states receiving the CSAQ survey, but the recipients of the NCES diskette surveys were not selected in an experimental design. Similarly the report by Higgins et al. (1987) of a higher response rate for their DBM panel could have been influenced by their sampling only IBM PC users.

Response rate research for EMS surveys is even more limited. Prior to 1996 there had been no official government surveys undertaken with EMS methodology and no recorded comparisons of EMS and P&P response rates. The Kiesler and Sproull (1986) study, which involved direct terminal linkages among a population of "recently active computer mail users," observed higher response rates for paper (75%) than for the electronic survey (67%). Similar research suggests that e-mail surveys have relatively poor response rates. Schuldt and Totten (1994) reported a 56.5 percent response rate for mail compared to a 19.3 percent response rate for e-mail in a university setting, using a nonprobability selection of professors for whom an e-mail address could be obtained.

A number of researchers have observed that the novelty effect of CSAQ, especially EMS CSAQ, apparently draws an earlier pattern of returns. This can be important to surveys with short data collection periods, but the effect may not persist if the novelty factor wears off with time or as the technology is more widely used. Timeliness of results for DBM CSAQ also has been mixed. Some studies have reported responses to DBM CSAQs two days to one week quicker than P&P methods (Higgins et al., 1987; Slatzman, 1992; Witt and Bernstein, 1992). Sweet and Ramos (1995), who followed response rates over a three-month period in a Census Bureau study, found that after the first two weeks the P&P panel had a consistently higher response rate than the CSAQ panel. Although this study offered modem return options, almost no respondents chose modem return.

Timeliness of results is also mixed for EMS. Kiesler and Sproull (1986) observed that the paper questionnaires took longer to return on average (10.8 days) than EMS CSAQ (9.6 days), while Schuldt and Totten (1994) concluded that e-mail responses were not more timely than P&P mailback responses. In any event major improvements in timeliness were not demonstrated by these studies. Still EMS has the potential to provide more timely data than either DBM or P&P (see Saris, Chapter 21).

Both overall CSAQ response rates and the promptness of response may be partly a function of respondent preferences, which may in turn depend on PC experience. Respondents who report via CSAQ instead of paper usually are favorable to CSAQ surveys and indicate a willingness to respond via CSAQ in the future. When asked, most CSAQ respondents understimate the time spent working on a CSAQ. For example, Higgins et al. (1987) observed an average of 30 minutes recorded by the instrument, while respondents estimated an average of 23 minutes. MacBride and Johnson (1980) suggest that this implies greater respondent involvement during the interview, which could result in better quality responses (see also de Leeuw and Nicholls, 1996; Morrison, 1988; Smith and Behringer, 1992; Sweet and Ramos, 1995; Witt and Bernstein, 1992; Zandan and Frost, 1989).

20.2.6 Data Quality

One of the most important gains of CSAQs over P&P mailed questionnaires is improved data quality. CSAQ methods enhance data quality through controlled routing to avoid item omissions and eliminate skip pattern errors, and from range and consistency checks to prompt respondent corrections where needed. In addition CSAQ methods eliminate the survey organization's need to enter data from P&P forms with its attendant transcription and keying errors.

The significant improvements CSAQs bring to survey data quality in comparison with P&P methods are well documented. In the United States the EIA attributes the higher-quality data produced by their PEDRO system to its embedded edits. The IRS found that only 6 percent of the electronic tax returns had errors compared with 20 percent of paper returns in the 1988 filing season. In the 1994 Census Bureau split panel test, the CSAQ panel had a 23 percent postinterview edit failure rate, which was significantly lower than the 38 percent for the P&P panel. The investigators attributed the difference to the built-in edits in CSAQ. In the early Kiesler and Sproull (1986) study, only 10 percent of the CSAQ cases "failed to complete or spoiled one or more items" versus 22 percent of the P&P cases (see also Sweet and Ramos, 1995; Weeks, 1992).

Because CSAQ instruments can be programmed to require a response to each question before moving to the next, they can improve item response rates. The Kiesler and Sproull (1986), Olson and Schneiderman (1995), and Sweet and Ramos (1995) studies all found supporting evidence that they do. For the Sweet and Ramos (1995) study, the CSAQ panel averaged 38 completed items per respondent, while the comparable P&P panel averaged 18.

CSAQ methods may improve data quality in additional ways. The EIA PEDRO system allows the respondent to extract data from other computer files and import them directly into the CSAQ application. Data import capabilities thus contribute to data accuracy by avoiding potential respondent keying errors. In addition, since the CSAQ respondent cannot always read ahead in the questionnaire, its length should not bias respondents into typing shorter answers. Higgins et al. (1987) found evidence that open-ended responses by DBM respondents included more distinct ideas and contained more words than those of P&P respondents. In this study open-ended CSAQ responses averaged 39.2 words compared with 31.0 words for the P&P responses.

Use of CSAQ methods also may encourage more candid responses to survey questions compared with P&P questionnaires. When Kiesler and Sproull (1986) compared an e-mail (EMS) questionnaire to a mail P&P questionnaire on health issues, they found that the e-mail respondents were more honest on sensitive topics, made greater use of the scale extremes, and were more thoughtful in their responses. Finegan and Allen (1994) and Smith and Behringer (1992) found no meaningful differences between response distributions of P&P and CSAQ methods. However, when Smith and Behringer (1992) made a three-way comparison among DBM, P&P, and CATI in their study of salesperson work satisfaction, they found the CATI satisfaction ratings substantially higher than those for DBM or the P&P questionnaire. They attributed the higher (less critical) CATI ratings to the respondents' reduced privacy in speaking to an interviewer. Tourangeau and Smith (Chapter 22) and Turner et al. (Chapter 23) further examine the effects of self-administration and survey automation in respondent reporting of sensitive topics.

20.2.7 Costs

Systematic cost comparisons of CSAQ with P&P methods are rare, and summary judgments based on broader evidence are not consistent. Saltzman (1992) and Wilson (1989) conclude that the costs of DBM methods are generally higher than those of P&P mailed surveys. Higgins et al. (1987) conclude that once initial program development costs are absorbed, CSAQs are not more expensive than other data collection methods.

If a survey uses only CSAQ methods, cost savings are realized by eliminating questionnaire typesetting, handling and storage, and data keying. When on-line edits are embedded in the CSAQ instrument, further savings should result from reductions in clerical edit review and telephone follow-up of edit failures. When electronic communications are used for both distribution and retrieval of CSAQs, savings in postage may be realized. Unlike CATI and CAPI no major hardware purchases are necessary for interviewers, since the respondents answer on their own home or office computers. Additional office hardware may be required to prepare, send, and receive CSAQs, and some significant new survey costs may be added; these are listed in Table 20.3.

Most of the activities listed in Table 20.3 are necessary for highest-quality DBM or EMS CSAQ surveys and can result in substantial costs. If CSAQ is

Table 20.3. Additional Costs of CSAQ Data Collection

- Purchasing a questionnaire authoring system or designing a CSAQ system from scratch
- Programming and design testing of the questionnaire
- Careful debugging
- Addressing universe PC availability and other sampling issues
- Gaining respondent cooperation
- Protection of confidentiality
- Purchase, preparation, and checks for incoming and outgoing diskettes
- Establishment of telephone help lines
- DBM CSAQ package mailing costs
- Staffing and training of help desk personnel for technical assistance
- Setup and maintenance of communications hardware
- Development of attractive software and enhanced capabilities
- Implementation of incentives when appropriate

employed as one part of a multimode or multitechnology design to ensure proper coverage of a broad population, the increased costs of maintaining and coordinating multiple collection methods must also be considered.

In one detailed examination of CSAQ costs, Sweet and Ramos (1995) found that elimination of data keying and other postcollection processing of P&P forms could not make up for the increased mailing and material costs incurred in that study with CSAQ. Their test application replaced a one-page paper questionnaire per company with a CSAQ package that included three diskettes and an instruction manual. CSAQ has the potential to prove more cost effective in surveys that can replace boxes of paper questionnaires (like those used for large companies) with a one-diskette package. Further cost savings are likely as CSAQs evolve from DBM to EMS methods, since savings would accrue from discontinuation of diskette acquisition, handling, and mailing.

The Internet currently seems not only the largest and most promising but the least expensive means of CSAQ distribution and retrieval. To date, however, no detailed cost estimates of Internet EMS CSAQs based on production experience have appeared in the literature. Only the expectation of greatly reduced costs is currently documented, such as the projections presented by Clayton and Werking (Chapter 27).

In general, paper surveys still seem more cost effective than CSAQ for one-time short surveys. The extra expense of CSAQ may be more justifiable for long or complex questionnaires used in panel, periodic, or repeated surveys.

20.2.8 Organizational Considerations

The integration of new data collection technologies into the everyday activities of a survey organization deserves careful consideration. New methods require new resources and new skills, for they must work in concert with established data collection methods. Their introduction can affect both organizational structure and performance.

Costs and staffing requirements are usually the first factors considered. Organizations experienced in conducting large mailed surveys with P&P methods generally have well-established procedures for mailout, mail receipt, clerical editing, data entry, and associated quality control activities. These are often labor-intensive, time-consuming, and costly operations with long traditions and committed staffs. The introduction of CSAQ methods may eliminate, reduce, or greatly change most of these activities. The greatest effects would occur through the elimination of data keying for CSAQ returns, reduced mailout and mail receipt in EMS methods, and reduced office editing and failed edit follow-up from the use of CSAQ instruments with embedded edits.

With these changes substantial savings in costs and personnel could result for reallocation to new CSAQ equipment, new technical specialists, and staff retraining. The feasibility and nature of this reallocation will partly depend on the extent to which CSAQ replaces P&P methods. If CSAQ technology is employed only for a subset of an organization's self-administered surveys while traditional P&P methods are used for the remainder, procedures and staff required for traditional P&P activities must be retained. Savings in cost and staffing for traditional methods may then be more difficult to achieve (Federal Committee on Statistical Methodology, 1990b). This is equally true if CSAQ is only employed for a subset of cases (respondents with their own PCs).

At the same time the new resources required for CSAQ (listed in Table 20.3) need to be acquired, planned, and developed. All may have an effect on the structure of the organization. A small, dedicated research and development staff is recommended to coordinate the changes necessary for CSAQ implementation. The same staff could participate in a continuing program of research on new data collection methods and in methodological comparisons with traditional methods. The Federal Committee on Statistical Methodology (1990b) recommends that this staff have a strong interdisciplinary background in statistics, methods test design, computer system design, questionnaire development, and human-computer interaction. Such investments can only be made by the larger survey organizations.

The development of CSAQ production applications also typically requires a technical staff to author (program) the CSAQ questionnaires and to design an efficient distribution and retrieval system. Some organizations may elect to contract out these tasks rather than develop the necessary technical expertise internally. The choices between heavy investments in internal development, contracting out, or mixed strategies are strategic ones for a survey organization and should be made after careful evaluation of the organization's CSAQ requirements, projected workload, and overall resources.

The current coverage limitations of both DBM and EMS CSAQs make it likely that they will be employed in multitechnology surveys using both CSAQ and P&P data collection methods. Only in that way can these surveys reach all parts of the target population. Organizational efficiency would therefore be enhanced by software that coordinated the design of multiple collection technologies. For example, the ideal software might consist of an executable file that would be mailed out on diskette, sent out over the Internet, printed on P&P forms for mailing, FAXed to respondents, and perhaps even used by CATI for telephone follow-up. Such common software would facilitate survey documentation and postcollection processing as well as survey collection. It would also assist in standardizing the collection process across technologies which in turn would reduce the need for custom programming.

The Census Bureau is examining one such trans-technology software, a "forms design" system that offers the capability of producing a generic form in one design session that can be distributed via paper, diskette, or the Web. Other software options are emerging as possible candidates for multitechnology distribution. Further investigation is required to determine how successful these and other software alternatives are in producing generic instruments that are portable across multiple collection technologies.

Since multitechnology survey methods are relatively new, difficult design issues abound and further methodological studies are clearly needed. Should the design of mixed technology applications be satisfied with the lowest common denominator approach suitable for all technologies—CSAQ, P&P, and FAX? Or should it take full advantage of a particular mode's special capabilities, such as CSAQ's use of embedded edits? Even without varying capabilities, the differing form layouts of these technologies could affect responses, yielding different results from different parts of the sample. Perron, Berthelot, and Blakeney (1991) have warned that technological enhancements may bring questionnaire and graphical design factors into play, which could introduce differential biases. But if the CSAQ instrument is required to emulate a paper form to maintain consistency across the sample, it may forgo the very features, such as embedded editing and data importing, that make it valuable and attractive to sophisticated respondents.

There are no simple answers to these questions, nor even the beginnings of a research literature to address them. Further development in the fields of visual graphical components of questionnaire design (Jenkins and Dillman, 1997) and of human computer interaction may provide some guidance (Federal Committee on Statistical Methodology, 1990b; Sweet and Ramos, 1995). But in the interim, survey designers must focus on feedback from potential respondents and employ careful pretesting to optimize the balance between standardization and respondent acceptance.

20.2.9   Likely Future Directions of CSAQ

At least for some applications, CSAQ data collection methods have well documented advantages over P&P mail procedures. When enough of the public have access to computers and electronic communications, and when salient confidentiality and security issues are satisfied, CSAQ methods may be more widely used. The optimal system should work well in a mixed-technology data collection environment (as described in Section 20.2.8). The CSAQ instrument should be easily accessible for respondents, without the need for mailed diskettes, so that the survey organization can prompt the respondent to answer with a simple postcard or e-mail reminder. The respondent would then retrieve the CSAQ questionnaire, import needed data with a few keystrokes, complete the requested information, correct items that failed the edit checks, and return the completed data electronically (Sweet and Ramos, 1995).

Since the Internet is currently the fastest growing communications and marketing medium, CSAQ may have its most promising future on the Internet and Web, assuming the coverage and security issues can be solved. One of the most attractive features of the Internet (and especially the Web) is its ability to overcome many of the system and communication incompatibilities that impede CSAQ activities in the DBM environment. The Java programming language is able to deliver interactive information over the Internet with fewer platform dependence issues (users' hardware, operating system, or browser). Although Java provides a means of avoiding some hardware, software, and communications incompatibilities, at this point it does not support the full functionality of CSAQ. JavaScript can be used to add client-side CSAQ interactivity to plain HTML instruments, but this solution creates CSAQs that can only be accessed by specific browsers or operating systems. These restrictions may be overcome with time. As another option, Web forms can be made interactive with forms software accessed through a helper application or browser plug-in. These COTS forms software products can be used in this way to develop Web-compatible CSAQ instruments. Since they also can produce comparable questionnaires for distribution by diskette or paper forms, this may provide a solution to multitechnology applications as well. Such forms of software with even more advanced capabilities are likely to become available in the future.

The coverage problems of the Internet and Web may be reduced by new products, just coming on the market, that offer more affordable Internet access. Although cheaper than PCs, these costs still may be significant for many households, and the appeal of these options to the general public remains to be assessed.

The future of the Internet, with or without low-cost access, is not easily predicted. The exponential growth of the Internet may or may not be maintained, and the security concerns that discourage its use may persist or be quickly resolved. Within a decade, computers may become as common as telephones and television sets, and some futurists predict that they will eventually merge into the same appliance. It is also possible that the willingness of the public to use the Internet to respond to surveys may prove disappointing. Even under the most discouraging scenarios, we anticipate that both DBM and EMS CSAQ software will still greatly improve over that currently available, perhaps even including multimedia capabilities on CD-ROMs (Clayton and Werking, Chapter 27; Johnson, 1992; Saris, Chapter 21).

Although many survey organizations are already employing the Internet and the Web for survey data collection, we urge caution in assessing their survey results. Appropriately designed Internet surveys may provide reliable inferences about the population of Internet users, and the Internet may be legitimately used in multitechnology surveys where all members of the target sample have the option of responding by one medium or another. Problems arise when survey organizations attempt to draw conclusions about the general public from samples reached only through the Internet. While PC ownership and Internet/Web access are growing rapidly, they are still confined to a self-selected minority who are not representative of the general public.

20.3 CONCLUSION

The general purpose of CSAQ methods is to replace mailed P&P questionnaires. To do so, they must be superior to P&P methods in important ways without having major disadvantages of their own. Much of the CSAQ literature is devoted to summarizing their potential advantages over P&P questionnaire methods in response rates, return time, survey data quality, and respondent preference or burden. There are also suggestions of cost savings with EMS. Our review of the literature suggests that the empirical evidence for most of these claims is limited and sometimes contradictory. The one exception is data quality, where the evidence of CSAQ superiority is compelling. These data quality gains by CSAQ methods occur primarily from minimizing respondent errors through controlled routing and the inclusion of range and consistency checks as part of the data collection process.

The primary disadvantage of CSAQ methods is population coverage. CSAQs have been used effectively for specialized applications, but currently the penetration of PCs is not high enough to guarantee a sample that is representative of the population as a whole or even of most subpopulations of interest. Legitimate applications of CSAQs in single method surveys are currently limited to specialized populations of known PC users, such as LAN managers, subscribers to on-line services, or employees connected to a common e-mail system. As the penetration of PCs continues to expand, the number of legitimate single-method surveys will continue to increase, especially for surveys of businesses, governments, schools, employees, student bodies, purchasing agents, and so on. CSAQ methods also have been employed as a supplemental option in multitechnology surveys, such as those offering respondents a CSAQ or P&P questionnaire.

The coverage problems of CSAQ methods are not limited solely to PC ownership. The variety of operating systems, equipment configurations, and communications alternatives among PC owners presents continuing problems of compatibility for the designers of CSAQ surveys. Although software designed for the Internet and Web may assist in overcoming potential compatibility problems, for Internet CSAQ to be a major alternative, its security and confidentiality issues must be resolved to the satisfaction of potential respondents.

CSAQ applications cannot be expected to completely replace P&P mail methods until PC ownership approaches the levels achieved by telephone ownership in the industrial nations, hardware and software compatibility problems are solved, and respondents feel comfortable with Internet security. Rather, CSAQs are likely to replace P&P questionnaire methods in selected surveys where CSAQ coverage problems are minimal or where their use is combined with that of other survey methods in multitechnology applications.

There is one additional major consideration in the future of CSAQ methods: their costs. Major expenditures are typically required to acquire attractive and fully functional CSAQ software, to obtain (or train) staff to design and manage CSAQ activities, to establish telephone help lines, to ensure careful debugging of questionnaires, to check incoming and outgoing diskettes, and for mailing procedures. Cost deterrents are likely to impede DBM development, while EMS options, including those through the Internet, still hold the promise of cost savings. These savings remain to be documented.

As with other technologies we suspect the initial costs will be unlikely to dampen CSAQ's future. In fact we believe that CSAQ software development and enhancements will continue at a rapid pace, driven by advances in computer technology and the rising penetration of the Internet, perhaps spurred by new low-cost options affordable by most households and the possible integration of the PC, television, and telephone into a common household appliance. A future in which all households are linked by flexible, interactive communications is clearly feasible. In such a future CSAQ data collection would surely have an important role.

CHAPTER 21

Ten Years of Interviewing Without Interviewers: The Telepanel

Willem E. Saris
University of Amsterdam

In 1984 the European Society for Marketing and Opinion Research (ESOMAR) organized a conference under the title: "Are Interviewers Obsolete?" At that conference Clemens (1984) introduced the first completely automated system for survey research without interviewers. Respondents were asked to answer survey questions interactively displayed on terminals in their homes or offices which were linked via telephone modems to remote central-computers in a research agency. Clemens used the British Videotex system for this purpose. French research agencies later employed the Minitel terminals that FRANCE TELECOM distributed to telephone subscribers for similar survey uses (Gautier, 1995). In the Netherlands, Saris and De Pijper (1986) of the Sociometric Research Foundation (SRF) began development in 1984 of the first Telepanel for use in computer assisted panel research of a representative sample of Dutch households.

The British Videotex System did not prove economically viable, and survey uses of the French Minitel system have been limited by its screen, entry ergonomics, and by the uneven distribution of Minitel computers across the French population. The first Dutch Telepanel has been functioning successfully since 1986 and has the longest, continuous history of nationwide interviewing without interviewers. Four additional Telepanels have been established since that time. This chapter describes (1) the historical development and special features of the Telepanel data collection method, (2) the social requirements of its operation, (3) the nature and quality of its sample, (4) the quality of its survey data, and (5) possible future developments.

Computer Assisted Survey Information Collection, Edited by Mick P. Couper, Reginald P. Baker, Jelke Bethlehem, Cynthia Z. F. Clark, Jean Martin, William L. Nicholls II, and James M. O'Reilly.
ISBN 0-471-17848-9 © 1998 John Wiley & Sons, Inc.

21.1 BACKGROUND

21.1.1 Origins

Computer assisted panel research (CAPAR) using computers in respondents' homes became feasible for nationwide surveys at about the same time as computer assisted personal interviewing (CAPI). The first portable microcomputers with keyboards and screens suitable for CAPI came on the market in 1985 (see Couper and Nicholls, Chapter 1). About two years earlier, inexpensive modems appeared on the market, making it feasible for home computers to receive and return information from remote central computers. Computer hobbyists used modems to reach central bulletin boards to exchange information and computer programs without charge, and commercial ventures, such as Videotex, began to sell subscription services to provide requested information for a fee.

In 1984 Saris and De Pijper recognized that these new capabilities of home computers could be used for survey research. If potential respondents had computers, telephones, and modems, a questionnaire could be sent from a remote central computer through telephone lines. Respondents could then respond to the questions, and their answers could be transmitted back to the central computer by telephone. We called this system the "Telepanel" because it used the telephone for the communications and the television as the computer "monitor."

The SRF field tested CAPI and the Telepanel at about the same time. We found that the Telepanel had two major advantages: First, since we generally preferred respondents to answer in self-administered mode, CAPI interviewers would have little to do other than bring the computer to resondents' homes and instruct them in its use. Second, since the Telepanel did not need to hire and pay interviewers, it should be much less expensive than CAPI.

The primary problem in using home computers for data collection is coverage. Most (or at least many) potential respondents in populations of interest will not possess a computer. This was especially true in 1985 when the first Telepanel was tested, and it remains true in 1998 as this chapter is written. This coverage problem can be solved in three basic ways.

The first is to restrict use of this data collection method to those specialized populations who own computers and modems or who have access to a shared computer system, such as the college population studied by Kiesler and Sproull (1986). Since very few populations meet these criteria, research opportunities are limited.

A second possibility is to take advantage of hardware placed in homes for commercial purposes and use it for survey research. For example, French market and opinion research companies have made use of the Minitel system that FRANCE TELECOM provided to many French telephone subscribers as a substitute for telephone directories.

The third possibility is for the survey organization to provide the necessary computing equipment to a sample of households who agree to participate in the survey. Clemens (1984) was the first to try this approach by providing Videotex equipment to a sample of doctors to obtain their views on medical issues. The same strategy was independently developed by SRF for its first experimental Telepanel of 100 Dutch households chosen in 1985 (Saris and De Pijper, 1986).

In 1986 an expanded Telepanel of 1000 households representative of the Dutch population was selected by the Dutch Gallup Organization (NIPO) for its opinion and commercial research. This original panel continues to serve that function. In 1991 two additional Telepanels were established in the Netherlands by the Telepanel Foundation (STP). The first is a 2000 household sample representative of the Dutch population. The second is a 1000 household sample drawn from the highest Dutch income decile. Telepanels of at least 1000 households also have been established in Finland and Italy. Thus a total of five Telepanels are currently known to be active.

21.1.2   Standard Telepanel Approach

All Telepanels other than the Dutch high-income panel were designed to be representative of their national household populations. They are probability samples of households who have agreed to be members of a continuing panel, participating each week in a survey that takes about 20–30 minutes per person. All members of panel households are asked to participate because the weekly surveys may target different demographic subgroups of the population. A maximum of about 2300 persons may be reached in the Netherlands with a 1000 household panel. Telepanel households are not paid for their participation although they are reimbursed for related telephone charges and other costs. The primary benefit of their participation, is having the Telepanel computer placed in their homes which may be used for other purposes. Panel households are given games and other software, and they can also arrange additional options such as access to the Internet. Supplementary rewards, such as participation in lotteries, are offered for unusually long or difficult questionnaires.

Each weekend, when a household member turns on the computer and selects the interview option, a new questionnaire is automatically downloaded via the modem. Data collection begins on Saturday. The first household member can then answer the questions or postpone survey completion until later that weekend. The system prompts for other household members as needed. When all have answered, the system automatically dials the central computer, uploads the answer files, and clears the diskette or hard disk for the next week. The household telephone is free for other uses except during downloading and uploading. Households who have not responded by Monday are telephoned by a member of the Telepanel staff. Data collection ends on Wednesday; and the analysis files and response reports are automatically generated. The next round of research begins the following day.

21.1.3   Comparisons with Other Methods

The Telepanel method is a form of computer assisted self-interviewing (CASI) for households in which the questionnaires are delivered by an electronic mail system (EMS). It differs from the CASI methods described by Tourangeau and Smith (Chapter 22) and Turner et al. (Chapter 23) because respondents answer the Telepanel questions on their own at their own pace. An interviewer does not have to be present when the questions are answered. The Telepanel also differs from the computerized self-administered questionnaire (CSAQ) and Internet/Web methods described by Ramos, Sedivi, and Sweet (Chapter 20) and Clayton and Werking (Chapter 27) in two important and related ways. First, because the Telepanel provides households with computers and modems, it avoids the coverage biases that CSAQ and Internet/Web surveys face through loss of potential respondents who lack the necessary equipment. Second, the Telepanel conducts surveys of households and household members while CSAQ and Internet/Web surveys have most often been used for surveys of organizations or businesses.

Because Telepanel respondents are committed to weekly participation for an extended period of time, the system obviously lends itself to panel surveys in which the same respondents are asked questions on the same topics in repeated waves. The STP Telepanel has been used for panel studies on consumption, income, savings, victimization, political party preference, and the effects of advertising. It differs from the large household panels described by Brown, Hale, and Michaud (Chapter 10) because they typically collect data only once or twice a year, while the Telepanel members participate weekly. Since many Telepanel weeks are not scheduled for panel surveys, ad hoc or cross-sectional studies are frequently conducted as well. Cross-sectional studies have been conducted on such topics as life satisfaction, current opinions, life histories, reactions to government decisions, and use of time. The total amount of information collected is very large, and it is all obtained from the same respondents. Individual studies can omit demographic details on the respondents who are already known from prior weeks.

The Telepanel resembles TV panels and consumer panels in its frequency of data collection but covers a much broader range of topics and interests. TV and consumer panels tend to concentrate on only one aspect of life. The Telepanel does share with TV and consumer panels the problem of avoiding a heavy response burden.

21.2   SOCIAL REQUIREMENTS

The social requirements of the Telepanel are as important as its technical requirements. Its success depends both on its ability to recruit, retain, and encourage panel members and on the quality of the software which should allow people of all backgrounds to participate. These two sets of requirements converge in the designs of the software and fieldwork.

21.2.1   Automated Operation

The process of downloading and uploading information between the central computer and the household's PC has been fully automated. If the PC is switched on and the interview option selected, the computer automatically makes contact with the central computer and downloads the interviews. Similarly the program detects when all household members have completed their interviews and automatically makes the connection to upload the responses to the central computer.

The survey questions must be designed for easy self-administration. The various types of questions and response options are shown to the respondents in an initial session when the computer is installed. In addition all instructions are available to the respondents at all times on each interview screen or in easily accessible help screens.

21.2.2   Help Desk

When technical problems arise, respondents forget the procedures, or they feel the need for personal contact, they may telephone a help desk which is staffed during each data collection period. Under normal circumstances about ten queries are received per four-hour shift in a sample of 2000 households. Most involve hardware failures and missing interviews or diskettes. These queries are easily handled by a single staff member who can perform other tasks between calls.

The help desk also serves a crucial function in alerting research staff to questionnaire errors. Such errors prompt many telephone calls and must be corrected as soon as possible with an updated interview file installed on the central computer. To avoid such problems and the irregular staff hours and respondent irritation they produce, questionnaires are extensively tested in advance.

21.2.3   Respondent Comments

In the STP Telepanels, respondents have the opportunity to enter comments or remarks at the end of each question and more general remarks at the end of each session. Respondents may report that they did not understand a question or could not correct an answer. The comments also are used for more general information, such as to explain that the household will be on vacation the following week, that they will be moving to another residence, or that they no longer want to continue in the Telepanel.

It is important that participants be given the opportunity to make comments and that these comments are read and acted on. Comments play a key role in maintaining two-way communication between respondents and the survey organization.

21.2.4 New Study Announcements

The Telepanels differ in the way they inform panel members that a new questionnaire is ready to be answered. The most common approach is by asking panel households to participate on a weekly basis, starting on the same day each week. In that way most households learn that it is their responsibility to initiate contact with the central computer each week. All Telepanels except the Dutch high-income panel use this as their primary method.

Although, in principle, the same procedure could be used for monthly rather than weekly surveys, a monthly schedule has several disadvantages: A day of the month is more difficult to remember (and control from the office) than a day of the week, monthly surveys could result in lengthier fieldwork periods, and new surveys would have to be announced through the system. This would be effective only if respondents left their computers on all the time. Most households will not do that because of the electricity costs. Therefore a mailed announcement is used for monthly surveys.

21.2.5 Panel Management and Personal Contact

All panels need a system to monitor questionnaire completion and to encourage the cooperation of panel members. The Telepanel, like other computerized systems, can quickly record and summarize questionnaire returns to identify participants who have not yet replied. Although electronic reminders can be sent to late respondents, they will only be effective for households who use their computers frequently enough to receive prompting messages or leave them on all the time.

To maintain high response rates and continuing panel participation, STP has found that personal relationships are beneficial, if not essential, between panel households and survey staff. Regional managers are employed for this purpose. Each is responsible for about 200 sample households and receives a database containing general information about these households, their participation over six months or more, and their entered comments about planned vacations, illnesses, or other matters that could affect participation.

The regional managers play a key role in nonresponse follow-up. They begin a weekly routine on Monday by telephoning all households who did not answer the latest questionnaire and have not previously communicated an explanation. If there is at technical or logistic reason for nonresponse, the manager will try to solve it immediately. Otherwise, the household is asked to respond by Wednesday morning. If the household cannot respond by Wednesday when fieldwork terminates, the manager asks the household the reason, which is included in the response reports.

In the evenings the regional managers contact those households who have entered messages in the past weekend's questionnaire about problems with questions or other matters. Respondents become irritated and may terminate their participation if such messages are not answered. In total, regional managers spend about six hours a week overseeing 200 households. With time, regional managers get to know all their respondents and the respondents get to know their regional managers. This makes participation less anonymous for the respondents and encourages their continued participation.

21.2.6 Work Schedule and Staffing

Weekly research requires an efficient team and a tight time schedule as the following typical schedule suggests:

Day	Task
Wednesday	Select people for the next study.
Thursday	Test the final questionnaire and install it on the central computer.
Friday	Test the cleaning tasks and questionnaires.
Saturday	Data collection with help desk.
Sunday	Data collection with help desk.
Monday	Identify nonrespondents; call to remind delinquent panel members; conduct extra data collection.
Tuesday	Conduct extra data collection.
Wednesday	Create data files and response reports; identify those who will no longer participate; and select substitutes.

If these tasks are not completed on time, the workload builds up leading to serious problems. These tasks are primarily the responsibility of three people: one responsible for the questionnaires, one for the regional manager system, and one as systems manager. All other tasks, such as preparing the questionnaires for special studies, testing the questionnaires, and completing detailed analyses are carried out by additional staff. Some tasks, such as solving technical problems, choosing replacement households, and the placing of computers in their homes, can be spread over a period longer than a week. However, to maintain a stable sample size, the replacement tasks must not lag far behind.

21.2.7 Panel Administration

All panel studies face problems of maintaining current addresses of their panel members and of learning the new addresses of those who have moved (see Brown, Hale, and Michaud, Chapter 10). The Telepanel can often circumvent this problem as it maintains weekly contact with survey households. Households often provide information about a coming change of address in their weekly comments, and even if they fail to do so, their absence is immediately detected and results in a call by the regional manager. Locating respondents who have moved is simple under these conditions and very few households are lost in this process.

At the same time the use of PC equipment in the Telepanel generates problems that other panel studies do not have. This is not simply a matter of re-installing the equipment for households who move. Any household may experience technical difficulties with the computer, modem, or monitor. The technical staff first attempts to correct such problems by telephone. If that fails, a member of the technical staff goes to the household to exchange the computer or to make other repairs. Furthermore, to maintain management control of the whole process, a current inventory of all equipment and its status is needed. This requires more work than is necessary in paper and pencil panels but is similar to procedures for interviewer-administered CAPI. Fortunately the number of broken or stolen Telepanel computers is very small.

21.3 SAMPLE QUALITY

The quality of survey data collection depends on the quality of its sample and the quality of its responses. This section examines sample quality while Section 21.4 examines data quality.

21.3.1 Sampling Objectives

The objectives of the Telepanel sample design are based on its use for both panel studies and for ad hoc cross-sectional surveys. The Telepanel has many advantages for cross-sectional research, some of which are that a new sample does not have to be chosen each time; the research can be rapidly implemented; and interviews can be shorter because much information is already available about panel members. To be suitable for both cross-sectional and panel studies, the sample must be and remain representative of the population from which it is drawn. The sample requires adjustment over time to compensate both for panel attrition and for changes in the population (Duncan and Kalton, 1987).

The quality of the final sample for any specific application depends on four components: the sample design, the initial response rate in recruiting sampled households to the panel, panel attrition over time and substitution for those losses, and the response of the panel members in the individual study.

21.3.2 Sample Design

The sample must be continuously updated to match the relevant population characteristics. This could be accomplished with a systematically rotating panel. But such a design would complicate the study of individual change. Another alternative is the combination of a fixed and rotating panel to facilitate both panel and cross-sectional studies. Various cohorts could rotate in and out of the sample as needed. Alternatively, substitutes could be found for individual households who cease to participate. This can be done in such a way that the sample remains representative of the population for the most relevant characteristics. This last option was chosen for the Telepanel. The STP National Telepanel used a two-stage stratified cluster sample for the initial selection (Geldrop, 1993).

21.3.3 Initial Nonresponse

The cooperation of sampled households in the Telepanel is obtained in stages. The households are first asked to participate in a very short telephone interview that obtains basic demographic and other background data. The telephone interview concludes by asking appropriate households whether they are willing to participate in a longer interview. If they agree, the next interview takes place in the respondent's home. This second interview is a computer assisted self-interview (CASI) with an interviewer present (CASI-IP). The interviewer brings the computer to the respondent's home. One purpose of this interview is to demonstrate how simple and easy self-interviewing using a computer can be. Only after this proves a positive experience are households asked to join the panel.

Not all households who complete the second interview agree to join the panel. Very few give technical reasons for refusing, such as saying that the computer is difficult to work with. The most common reasons for refusal are "no time" or "no interest" as in other panels or surveys. The initial response rate in the STP Dutch National Telepanel (slightly over 30 percent) is comparable to that of the Statistics Netherlands' Family Expenditure Survey. We should note that other surveys in the Netherlands also have low response rates. We hypothesize that response rates for Telepanels in other countries may be comparable to those of other high burden surveys such as expenditure surveys. Households with children are more likely to participate in Telepanels than those with no children (Geldrop, 1993). Cooperation in the Telepanel is also lower in large cities and among persons under 35 or above 65 years of age.

A two-stage sampling procedure can be used to compensate for these nonresponse deviations. In the first stage, a representative sample of approximately 10,000 potential households is asked to participate in the panel. In the second stage, households are randomly selected from the pool of households who have agreed to participate with probability inversely proportional to their likelihood of cooperation (Geldrop, 1993). In this way the unequal chances of participation can be corrected while retaining a probabilistic selection procedure.

21.3.4 Attrition

All panels face the problem of attrition as respondent participation diminishes over time. The size of the attrition problem is strongly related to the frequency of data collection and the response burden associated with each questionnaire. With the fixed automation costs of the Telepanel, cost effectiveness requires frequent use, and all Telepanels face continuing problems of attrition.

Sikkel (1994) showed that within 100 weeks after the start of the STP National Telepanel, only about 50 percent of the original households continued to participate. The tasks the respondents were asked to perform were so burdensome that most households were willing to participate only for a limited time. Sikkel also found that the hazard rate, or probability of dropping out, was highest after a year, but attrition naturally depends on the type of research, the number of questions per study, and the frequency of the data collection. After five years of experience, the STP National Telepanel has lowered the response burden, and the dropout rate fell by about half.

Several demographic characteristics have a significant effect on the hazard rate (Oppenhuisen, 1994), such as:

- *Age*: Young people drop out sooner. Although older people are more difficult to recruit initially, when they cooperate, they remain loyal longer than younger people.
- *Family size*: Small families drop out sooner.
- *Occupational status*: The retired and unemployed drop out more slowly. This suggests that people busy with jobs drop out sooner.
- *Income*: The highest income group drops out first.

The equipment the research organization provides can also affect participation. Households given a monitor remain longer in the panel. A separate monitor means that the computer does not have to be connected to the household TV so that answering survey questions does not interfere with TV watching. Respondents also are more satisfied and remain in the panel longer if they use their own home computer rather than the relatively simple computer provided by the research organization. Since the standard Telepanel computer had no hard disk, the interview process was slower and more tedious.

Differential attrition will bias the panel over time if lost households are not replaced by substitutions, thus taking account of any deviations between the panel and the target population. This is a problem all continuing panels face, especially those with frequent data collection. The STP Telepanel has solved this problem by selecting substitute households from the pool of potential respondents with a probability weight that is a function of the stratum the household was drawn from, the size of the stratum in the sample, and the stratum size required to represent the target population.

By applying this procedure on a continuous basis, the distributions of the stratifying variables in the sample will be comparable to those in the population (Geldrop, 1993). This does not mean that the sample is unbiased for other variables. That depends on the strength of the relationship between the stratification variables and the variables of interest.

21.3.5 Nonresponse by Wave

The last component of sample quality is the cooperation rate for individual weekly surveys. This has proved surprisingly high across the various Telepanels. In the initial period of the first Telepanel, NIPO reported a response rate of 95 percent or more of those who were able to participate (Saris, 1988). The latter qualification is necessary because there is a fluctuating number of people who cannot participate due to technical problems, illness, or vacations. With time these exceptionally high response rates decline, but even after five years the STP National Telepanel still reports weekly response rates of around 85 percent for households able to participate. Those not participating may have forgotten to notify the Telepanel of an absence or may be committed to other activities on that particular weekend. Since these events are approximately random, higher response rates are found for multiple weeks. In a typical two-week period of the STP Telepanel, the combined response rate was 91.2 percent, and for a three-week period, it was 93.4 percent of panel households.

Because the weekly response rates are so high, the representativeness of the weekly samples depends mostly on the representativeness of the continuing panel from which the households are drawn. Thus a great amount of effort must be spent on construction and maintenance of the panel and on procedures to avoid attrition and nonresponse.

Some of the most important methods of minimizing attrition and nonresponse are basic elements of the fieldwork design previously mentioned. They include:

- Well-designed questionnaires that are clear, informative, and cover a variety of topics over time to hold the interest of diverse respondents.
- A system of regional managers to maintain personal relationships with respondents, expedite the solution of technical problems, answer respondent questions, and follow up on respondent comments.
- The provision of useful computer hardware and interesting software to panel households to motivate their continued participation.

In addition lotteries can be used as an added incentive for households who are not interested in the hardware or software. As requested by the panel members, the lotteries emphasize one or two large prizes (e.g., video camera, or tickets to a major sports event) rather than many small prizes.

A newsletter for panel members also seems to be appreciated (see Sudman and Ferber, 1979). Newsletters present the results of completed studies, include photos of lottery winners, and describe other Telepanel activities, emphasizing that the panel members are participating in an important but pleasant activity.

21.4   DATA QUALITY

This section summarizes evidence on the quality of Telepanel data drawn from methodological studies conducted over the last ten years, focusing on three types of studies in particular. The first compares the quality of Telepanel data with that of other collection modes under a multitrait-multimethod design. The second compares the quality of Telepanel data with that of other collection modes based on independent samples of respondents for each mode. The third examines the use of Telepanel data from complex survey instruments based on internal checks made during data collection. The complexity of these instruments would make cross-mode comparisons prohibitively difficult and expensive. This section concludes with a discussion of the effect of the Telepanel's high response burden on underreporting in survey panels.

21.4.1   Multitrait-Multimethod Studies

Multitrait-multimethod (MTMM) designs are perhaps best known for assessing alternative formats of attitude questions, for example, 100-point scales, 10-point scales, or ordinal categories. The same sample of subjects (respondents) is asked to answer the same set of attitude questions (traits) in each of the three formats. Using MTMM models and assumptions, the test-retest reliability and "true value validity" of each format may be estimated from the study data. The reliability and validity components also may be combined into a summary measure of data quality by format. (See Andrews, 1984, and Saris and Andrews, 1991, for a full description of these methods.)

The same MTMM design can be applied to comparisons of alternative data collection modes, such as CAPI, CASI, and their paper and pencil equivalents, rather than to question formats. The key requirements are that the same questions are asked of the same subjects (respondents) in all the modes being compared. The quality (reliability * validity) of each mode may then be estimated based solely on the observed data and MTMM model assumptions. The quality coefficients normally vary between 0 and 1 and can be interpreted as standardized regression coefficients. The coefficient squared is the variance in the observed scores explained by the latent trait(s) of interest.

Several MTMM studies have compared the estimated quality of Telepanel data with that of paper and pencil (P&P) or other computer assisted data collection modes (see Saris and van Meurs, 1990). One study, which focused on satisfaction questions asked in 15 different regions in Europe (Scherpenzeel, 1995), reported the following quality coefficient (in parentheses) for four collection methods: Telepanel (0.90), P&P personal interviewing (0.88), mail questionnaire (0.83), and P&P telephone interviewing (0.81). Using graphical line scales, which were only available in the Telepanel, the Telepanel quality coefficient was even higher (0.95) (Scherpenzeel, 1995). In a similar study of satisfaction questions in the Netherlands, Scherpenzeel and Saris (1995) report quality coefficients of 0.85 for the Telepanel, 0.68 for CATI, and 0.61 for CASI-IP. Another large MTMM study in the Netherlands found Telepanel data quality coefficients for several opinion items generally exceeding those from CASI-IP, which in turn were higher than those from CATI (see Scherpenzeel and Saris, 1997).

The Telepanel does not always obtain the highest-quality coefficients in these MTMM studies. For respondent evaluations of the seriousness of crimes, the quality coefficients were CASI-IP (0.67), Telepanel (0.57), and CATI (0.56). The evaluation of crimes was a difficult task for the respondents, as shown by the generally lower-quality indexes. The presence of the interviewer and the use of show cards in the personal interviews led to better performance (Scherpenzeel, 1995).

21.4.2 Mode Comparisons with Independent Samples

In the MTMM studies the same survey questions were asked of the same respondents in each mode and estimates of quality were obtained from their combined results. In traditional mode comparison studies, an independent sample is used for each collection mode, and comparable estimates of quality are not directly available. One can only examine such mode differences at an aggregate level. With independent samples, additional sources of variation, such as differential response rates by mode, may confound the comparison as may specific procedures necessary to adapt the questions or study objectives to the specific mode. The results are less pure mode comparisons than system comparisons. Several such studies have nevertheless provided useful information about the data quality of Telepanels.

Kalfs (1993) compared estimates of media use reported in Telepanel and CATI surveys. She found that reported daily television watching by highly educated people was about 30 minutes higher in the Telepanel than in CATI. This difference seems attributable to the social desirability bias when the questions were asked by a CATI interviewer. Similarly Bon and van Doorn (1988) found that drug use was between 15 and 20 percent higher when reported in a Telepanel than in face to face interviews.

The special capabilities of computer assisted data collection might also make important contributions to mode effects. In comparative time budget studies, Kalfs (1993) found that 50 percent of the trips reported in Telepanel data collection were left unreported in paper and pencil diaries. In the Telepanel time budget study, a consistency edit looked for a trip report whenever a change of location was reported. Such interactive edits were not possible in P&P diaries.

21.4.3 Data Quality in Complex Questionnaires

The design of complex questionnaires for the Telepanel and other forms of CASI requires special care because respondents must answer on their own. Typically no interviewer is present to guide the process, correct errors, and provide help when needed. All assistance must be built into the interview program. The screens should be simple, the instructions clear, and help screens readily accessible.

Assistance with navigation, or movement from screen to screen, is especially important when no interviewer is present. One important means of avoiding navigation problems in CASI is the use of summary and correction (S&C) screens (Saris, 1991). Respondents inevitably make occasional mistakes in answering survey questions, and if those mistakes occur in questions later used for routing (or branching), the consequences can be serious. If the erroneous routing paths lead to requests for information about income sources mistakenly chosen or persons no longer in the household, the respondent may be sufficiently annoyed to terminate the self-interview. S&C screens located at strategic points throughout the instrument summarize data already entered and ask the respondent to verify the information and make any necessary corrections.

In practice, Telepanel respondents frequently make corrections in the S&C screens and thereby avoid many potential errors. The S&C approach is more efficient than asking respondents to back up in the questionnaire to the original entry and make the correction there, something that even experienced CAI interviewers often find difficult (see van Bastelaer, Kerssemakers, and Sikkel, 1988).

The effectiveness of S&C screens has been demonstrated in constructing calendars in life history research. STP Telepanel respondents were asked to list in chronological order major events in various domains of their lives and to provide the dates for each of those events. The Telepanel program then constructed a S&C calendar summarizing these events and dates. The respondents were then asked to make any necessary corrections. About a fifth (21%) of the respondents corrected the order or events and almost four-fifths (79%) corrected the dates after seeing the information summarized in a single calendar (Vis and Wouters, 1996). Without a S&C screen these errors would have remained undetected. As shown by Holt, McDonald, and Skinner (1991), errors in dates can have especially devastating effects on parameter estimates in event history models.

The data summarized on S&C screens may include information from prior panel waves as well as from the current one. In the Telepanel, past information can be efficiently stored on the household computer's hard disk for use in later waves. Data from prior waves are especially important for income surveys where they can be used for sophisticated edits in dependent interviewing.

Verwey et al. (1989) has described a study in which the reported income of the household head fluctuated substantially from month to month. The median income of household heads in the Netherlands at that time was about 2200 Guilders (NLGs) per month (NLG 1.60 ≈ 1 U.S. dollar at the time). Monthly fluctuations of more than NLG 150 were found for 12.7 percent of the respondents, while 5.6 percent had monthly changes of more than NLG 500. In an initial study no special methods were used to check the quality of the data because such large fluctuations were not anticipated. In a subsequent study edits were added using the income reported in the prior wave.

If a respondent reported a monthly income NLG 50 greater than in the previous month, he or she was asked whether the reported income was correct and shown the previously reported amount on the same screen. This probe or check question was asked only under the specific condition that

$$I(t) > I(t - 1) + 50 \quad \text{or} \quad I(t) < I(t - 1) - 50$$

where $I(t)$ is the income just reported and $t - 1$ is the income reported on the previous wave. The screen appears as follows:

> Perhaps you made a mistake because last time you reported that your income was "$I(t - 1)$." If the amount was incorrect, return to the previous question by pressing the F1 key, otherwise press ENTER.

Brown, Hale, and Michaud (Chapter 10) have called this the "reactive" approach to dependent interviewing. In this case the check was not very effective because the month-to-month fluctuations in reported income were not reduced (Verwey, 1992).

Since we were still not convinced that such large monthly income changes were possible, a third study added another procedure called "scheduling" (Saris, 1991). Before the current month's income was asked, last month's reported income from 23 possible income sources was displayed on the screen, and the respondent was asked to correct any of those figures that were wrong. In this way an effort was made to prevent as many typing errors as possible while allowing for large changes to be reported (Hartman et al., 1991). This approach is called "proactive" by Brown, Hale, and Michaud (Chapter 10). The question appeared as follows:

> Our records indicated that your income last month from {Source S} was "$I(t - 1)$."
> If this is correct, press ENTER. If this amount is incorrect, type the proper amount here _____ .

Then after the current month's income was entered, the reactive check described above also was used. Scheduling and checking were used consecutively. The results of checking alone and of scheduling with checking on reported income of the household head are shown in Table 21.1.

The result of these scheduling and checking procedures was that minor typing errors were reduced and reported monthly income fluctuations decreased. The revised estimate of monthly income fluctuations may have been partly an artifact of the procedures used. We accepted this possible bias to see whether the large monthly income changes would disappear. Although the monthly changes became smaller, rather large fluctuations still remained, as can be seen in Table 21.1. Once all possible checks were built in and major monthly changes remained, we concluded that large fluctuations in monthly income do occur (Saris, 1995).

Table 21.1. Changes per Month in Income for the Head of the Household (in %)

	Checks in Data Collection		
	No Checks	Reactive	Proactive and Reactive
Change in NLG			
Over 500	5.6	9.9	3.3
150–500	7.1	12.6	4.8
50–150	10.1	6.9	6.6
5–50	13.6	6.9	5.7
0–5	61.6	63.7	79.7
(Sample size)	(950)	(950)	(1500)

Similar procedures have been developed to check the quality of prices reported in family expenditure studies (Hartman et al., 1991; Saris, Prastacos, and Marti-Recober, 1995). In these studies respondents are asked how much they had purchased of various commodities and the total amount they paid. For most goods the national average price per unit is known and can be stored in the instrument as a check for the price reported by the respondent. This is somewhat inefficient because prices vary rapidly over time and any given household can pay a variety of prices for a single good that they purchase on a regular basis.

To address these concerns, a procedure was developed that takes into account both variations in the families' buying patterns and the price changes those families experience over time. This is done by using the following condition to evoke the edit check question:

$$P(t) > P(t-1) + a(t-1) \quad \text{or} \quad P(t) < P(t-1) - a(t-1)$$

where $a(t-1) = [a(t-2) + \{P(t-1) - P(t-2)\}]/2$, $P(t)$ is the price per unit calculated from the price specified in the interview, $t-1$ is the time of the previous interview, and $t-2$ is the time of the interview before the previous one.

The use of $P(t-1)$ in the check makes this test family specific by using the last price this family paid. The use of the interval $a(t-1)$ makes the edit dynamic, since the interval is a function of fluctuations in the price *this family paid* over the two prior time periods. When past price fluctuations have been large, so is the interval. If the price does not fluctuate from wave to wave, the interval is halved for each interview until a minimum is reached. Too narrow an interval could annoy respondents. When the condition for the edit check is satisfied, participants see the following screen:

You may have made a mistake. We have recorded that a (kilo/liter) of {goods} costs NLG "$P(t - 1)$" while you paid NLG "$P(t)$." Could you please indicate which of the following was the case?

1. This is now the normal price.
2. The price is correct but I got a special offer.
3. The price was correct but this time I paid more.
4. The price entered was incorrect.
5. The amount entered was incorrect.
6. Both were incorrect.

If the respondent indicates that the wrong price or amount was entered, he/she is asked to correct it. If the entry is correct, the respondent is asked to indicate whether the entered price is now the normal price. If it is, the new price is stored for further use, and the interval $a(t - 1)$ is adjusted on the next wave. If the respondent indicates that the family paid more or less this time only, no change is made to the price database.

The stored information can be used to impute missing information as well as for consistency checks. For example, if the family reports the total price of purchased goods but not the amount, the computer can suggest the amount, as in the following screen:

Given the price per kilo/liter and the total price you paid, we think that you bought {*estimate*} kilos/liters. Do you think that this is correct?

1. Yes
2. No
3. Don't know.

If the respondent answers "yes," the estimate provided is recorded as the answer. If the respondent answers "no" or "don't know," a missing value is recorded. Total costs can be similarly imputed and suggested from amounts entered. In this way many missing values can be avoided. And since the suggested amount or total cost is based on the family's most recently reported purchases of those goods, it is more useful and accurate than a missing value.

With such procedures to ensure data consistency and impute missing values during self-interviewing, Maartens (1995) concluded that it was no longer necessary to clean the data after collection. Postinterview corrections led only to nonsignificant and substantially irrelevant changes in consumption estimates at the aggregate level. This clearly shows the benefits of dependent interviewing and dynamic range checks as employed in the Telepanel. In family budget research, postinterview data editing typically takes a lot of time and effort and still may not provide better data than obtained directly from the Telepanel.

This does not mean that the Telepanel data are without error. Respondents will at times report very deviant total costs or other quantities and fail to correct them in summary and correction screens. The Telepanel also is subject to other biases that are common to all panel surveys that collect data frequently.

21.4.4   Limitations of the Telepanel

Panels that ask respondents for a significant portion of their time each week create a high response burden which may have unfortunate consequences. In addition some studies, such as consumer research, may be seen as more burdensome than others. Consumer research requires a major effort from respondents; the topic is less interesting than social or political subjects, and these studies often are highly repetitive. If the purchase of consumer goods is the topic too frequently, panel attrition will increase.

Underreporting can be an even more serious consequence of high response burden in panel studies. The effect of panel participation on underreporting has

Table 21.2. Recorded Amounts of Selected Goods over a Period of Six Months Expressed Relative to the Base Month (Base = 100)

Goods	Month					
	Jan	Feb	Mar	Apr	May	Jun
Red meat						
Expense	100.0	99.3	91.0	93.8	89.7	88.9
Volume	100.0	95.4	87.4	90.8	82.8	85.1
Number of products	100.0	99.4	92.9	92.9	89.6	90.3
Meat products						
Expense	100.0	101.0	92.8	91.0	90.6	82.3
Volume	100.0	103.0	90.9	84.8	84.8	78.8
Number of products	100.0	102.1	95.9	92.8	87.2	84.1
Poultry						
Expense	100.0	87.5	82.1	80.2	79.0	75.5
Volume	100.0	80.8	76.9	73.1	69.2	76.9
Number of products	100.0	90.9	87.9	81.8	81.8	75.8
Eggs						
Expense	100.0	91.8	86.6	90.7	73.2	70.1
Volume	100.0	91.3	85.4	92.1	75.7	75.7
Number of products	100.0	91.3	85.4	92.1	75.7	75.7
Total						
Expense	100.0	98.0	90.2	91.1	87.8	84.5
Volume (excluding eggs)	100.0	93.8	87.0	87.0	80.8	82.2
Number of products	100.0	99.8	93.3	91.3	86.7	84.8

Source: STP panel, *n* = 1750.

been treated in many studies (Kemsley, 1961; Bailar, 1975; Olivier, 1987; Silberstein and Scott, 1991; van Meurs, van Wissen, and Visser, 1989). The Telepanel provides an additional example.

Table 21.2 summarizes reported purchases of a variety of products over a six-month period in 1994 in the STP National Telepanel (Kaper and Saris, 1996). Total purchases of all four commodities decreased substantially over time, except in April which included Easter when the Dutch traditionally purchase and consume more eggs. It is implausible that the downward trend represents a true decline in consumption; these figures were collected before the onset of mad cow disease caused an actual decline in meat consumption. It is also not likely that these effects are the result of telescoping. The respondents were asked to report their purchases weekly. The only obvious explanation for this decrease is respondent satisficing as a consequence of this study's heavy response burden. It is also clear that in this form the resulting estimates cannot be used for any practical purpose.

Underreporting due to response burden is likely to occur in any panel with frequent measurement and requiring active respondent participation. There are several possible solutions to the underreporting problem.

One option is to screen the respondents more carefully, only retaining those who seem to respond in a serious and stable way. The difficulty with this strategy is that people retained in the panel may no longer be representative of the total study population. In one study of this approach, Oliver (1987) found that those retained in the panel disproportionately reported the purchase of inexpensive products.

A second solution is to estimate the effect of nonresponse, refusals, and underreporting on the aggregated data and to adjust for these factors (Van den Oord and Saris, 1994; Kaper and Saris, 1996). This procedure requires extensive knowledge of the interrelationships among the variables of interest, which can only be developed over an extended period of time.

A third solution is to attempt to control the data quality by personally contacting each household when it reports deviant results. This strategy is relatively expensive in staff time and requires a staff with exceptional ability, dedication, and continuity. If the staff efforts in contacting deviant households vary, this may cause fluctuations in the data that have nothing to do with real change.

A fourth approach to controlling data quality is through the use of the proactive and reactive edit checks described above. For example, if a given household normally buys meat but then stops reporting meat purchases, an additional question can ask why meat was not bought. If the probes are sufficiently elaborate, omitting a purchase will no longer reduce the interview time, eliminating a motivation for underreporting. This strategy has been investigated by van den Oord and Saris (1994).

In view of the costs or limitations of all four alternatives, it may be better to attack the problem directly by reducing the response burden. Hogendoorn and Sikkel (1994) completed a statistical analysis that indicated that measuring consumer purchases every two weeks, instead of every week, would reduce underreporting. A reduction in the total response burden of the STP household Telepanel also seemed to alleviate attrition rates. When the total response burden was reduced by approximately half, the attrition also dropped by half. Thus the best solution to the problems associated with a high response burden seems to be a reduction of that burden.

21.5   EVALUATION AND DISCUSSION

This chapter summarizes 10 years of Telepanel development and research. It identifies several favorable features of the Telepanel, including the high weekly response rates and the quality of its data in comparison with other data collection modes. These data quality benefits may not be realized unless study directors invest the resources to develop the necessary edit checks. The effort required to design and test data quality improvements is most cost effective in repeatedly used questionnaires (Saris, 1991). The high quality of Telepanel data partially stems from its use for panel and repetitive cross-sectional surveys where data quality improvements are more easily supported.

This chapter also identifies a major limitation of the Telepanel. Its high response burden can result in attrition and the underreporting of behaviors measured frequently, especially in consumer research. These problems are common to all panels that collect frequent observations on the same topic, unless their data collection procedures are fully automated and make little or no demands on the respondents' time.

In view of these common problems, an appropriate future direction for survey methods should include the further automation of data collection for frequently measured factual behaviors. This has already happened in TV audience research where the "People Meter" almost completely automates the measurement of household TV viewing (Buck, 1987; Saris, 1989). The use of bar code scanners to record purchases, both in stores and in respondents' homes, has brought high levels of automation to consumption and expenditure surveys. Similar procedures, possibly involving the use of smart cards, may in time be extended to the measurement of doctor and dental visits, hospital stays, and other recurrent behaviors.

The need to ask respondents questions about the reasons for their behaviors and about their plans, intentions, life satisfaction, attitudes, and opinions will remain. For these purposes the Telepanel and other forms of CAI will have a major role, although they may take new forms. The experience of the Telepanel suggests some limits on these future systems.

Many writers on survey methods have suggested that the Internet will be the primary medium of future data collection. Whether this happens depends on whether access to the Internet will be free or will provide other general benefits the public is willing to pay for. Thus far all data collection systems that households have had to pay for have failed to achieve a sufficient penetration of the population to yield a representative sample of households. This applies to the Videotex system and to all tests to date of paid Interactive TV. But, if at some point Internet access is made possible at minimal cost through a TV or other common household appliance, the Internet may expand very rapidly as a vehicle for survey research. Should that happen, research organizations will be able to undertake interviewing without interviewers without making large investments. If the Internet does not achieve broad household penetration, the Telepanel system will remain an attractive option for continuous research on representative samples of the population.

In either event, the general kind of system this chapter describes may become the standard model of opinion research in the future, whether that system functions through the Internet, through computers placed in respondents' homes, or in other ways. The requirements for frequent survey research based on a continuing panel will still apply, including largely automated operation, simple, clear questions, help desks, regional managers to foster two-way communication, systems to announce new studies, continuous panel management, and so on. The procedures for data quality improvement discussed in Section 21.4 also could be used under a variety of possible systems, but to be effective, they will require as large an investment in time as they did in the Telepanel.

ACKNOWLEDGMENTS

I am grateful for the useful comments on an earlier draft of this chapter from the following colleagues: Robert Voogt and Corrie Vis of the STP Telepanel; William Nicholls, Magdalena Ramos, Barbara Sweet, and Elisabeth Sweet of the U.S. Census Bureau; Richard Clayton of the U.S. Bureau of Labor Statistics; and Roger Tourangeau and Tom Smith of the National Opinion Research Center.

CHAPTER 22

Collecting Sensitive Information with Different Modes of Data Collection

Roger Tourangeau and Tom W. Smith
National Opinion Research Center

22.1 INTRODUCTION

The search for better methods to gather information on sensitive topics—questions on private, embarrassing, or criminal matters—has been a long-standing methodological quest. Surveys have always included questions that made some respondents uncomfortable, and since the onset of the AIDS epidemic, the need for data on some very sensitive topics, including sexual behavior and injection drug use, has dramatically increased. Typically such data have been collected in personal surveys, those in which interviewers contact the respondents at their homes or by telephone and ask the questions themselves or provide the respondent with a questionnaire to complete. The survey questions—whether read by the interviewer or by the respondent—are increasingly in electronic form. But, despite the widespread use of these methods to collect sensitive data, the accuracy of the data obtained is still very much in doubt.

With most sensitive topics, these doubts concern underreporting. For example, since its outset, the National Survey of Family Growth (NSFG) has asked respondents in face-to-face interviews about the outcomes of their pregnancies. Responses to these questions are the basis for projections about the number of abortions performed in a given year. Comparisons between these estimates and those derived from data from abortion providers suggests that

Computer Assisted Survey Information Collection, Edited by Mick P. Couper, Reginald P. Baker, Jelke Bethlehem, Cynthia Z. F. Clark, Jean Martin, William L. Nicholls II, and James M. O'Reilly. ISBN 0-471-17848-9 © 1998 John Wiley & Sons, Inc.

the survey respondents report fewer than half of their abortions (Jones and Forrest, 1992). There also is evidence of underreporting in studies of drug use. To cite just one example, Fendrich and Vaughn (1994) show that some respondents who admitted illicit drug use in one round of a longitudinal survey deny or minimize their drug use in subsequent rounds. Discrepancies in the opposite direction are rare.

Underreporting in these cases is likely to reflect both motivational and cognitive factors. Many respondents are reluctant to report these embarrassing or illegal behaviors, but even if they were willing to report them, they would still be susceptible to such additional sources of reporting error as failure of comprehension, lapses of memory, and flawed judgments that could lead to underreporting. Although it might seem unlikely that respondents would simply forget incidents of illicit drug use, say, several considerations suggest that such events may be quite forgettable. These events may happen quickly, evoke little emotional reaction, and trigger few discussions with others afterward. Each of these characteristics of the events is likely to produce poor recall.

Underreporting of sensitive behaviors may be the most common problem, but it is not the only one. Reports about sexual behaviors illustrate a more complex pattern of reporting errors. Within a closed population, men and women should report the same aggregate number of opposite-sex sex partners, since members of both sexes are reporting the same sexual pairings. As Smith (1992) has demonstrated, men consistently report more opposite-sex sex partners than do women, a difference that persists even when differences in population sizes are taken into account. The simplest explanation for this difference is that men and women differ in the direction of their reporting errors, with men overstating their partners and women overlooking theirs. These findings suggest that the validity of self-reported data on sexual behavior is far from perfect.

22.1.1 Effects of the Mode of Data Collection

Over the last 20 years the use of computers in the collection of survey data has revolutionized the survey industry and greatly expanded the roster of methods for collecting data in personal interviews. At least six methods can be used to collect self-reported data on sexual behavior, illicit drug use, or other sensitive behaviors:

- Paper and pencil personal interviews (PAPI).
- Paper and pencil self-administered questionnaires (SAQ).
- Walkman-administered questionnaires (ASAQ).
- Computer assisted personal interviews (CAPI).
- Computer assisted self-administered interviews (CASI).
- Audio computer assisted self-administered interviews (ACASI)

We use each of these terms to refer to methods in which an interviewer contacts the respondents and elicits their cooperation in the study. The interviewers either administer the questions themselves or provide the respondents with a paper or electronic questionnaire and transmit the data back to central office. Our research has examined in-person methods of data collection, and this chapter focuses on these methods. As the present monograph makes clear, there are several additional data collection modes available, including those that employ the telephone (Turner, Chapter 23) or home computer (Saris, Chapter 21). We believe that many of our conclusions will generalize to other methods.

Self-Administration
These six modes of data collection differ in several ways. Perhaps the major difference is whether the questions are self-administered or interviewer-administered. A hypothesis guiding much of the survey literature on reports about sensitive topics is that a major source of error is more or less deliberate misreporting. Respondents doubtless want to cooperate by providing accurate information, but they are not immune to other considerations — such as the desire to avoid embarrassment. Sudman and Bradburn (1974) conclude that personal interviews are more affected by self-presentation concerns than are self-administered questionnaires and that this difference is likely to have an especially large effect on reports about sensitive behaviors. Self-administered questionnaires generally obtain higher levels of reporting of sensitive behaviors than do face-to-face interviews (Bradburn, 1983). The advantages of SAQs have been demonstrated in studies of sexual behavior (Boekeloo et al., 1994), illicit drug use (Aquilino, 1994; Aquilino and LoSciuto, 1990; Schober et al., 1992; and Turner, Lessler, and DeVore, 1992), alcohol consumption (Aquilino and LoSciuto, 1990; Hochstim, 1967), and abortion reporting (London and Williams, 1990; Mosher and Duffer, 1994; Mott, 1985).

Table 22.1 summarizes the results from a series of similar studies examining different methods of collecting data on illicit drug use. The table displays the ratio between the percentage of respondents admitting to a given behavior (e.g., lifetime use of marijuana) under interviewer administration of the questions to the corresponding percentage obtained in self-administered questionnaires. The table indicates, for example, that in the Aquilino (1994) study, respondents who completed an in-person paper and pencil interview (i.e., via PAPI) were only 0.67 times as likely to report they had used cocaine in the past month than were respondents who completed a self-administered paper questionnaire (SAQ). This suggests that at least one-third of the recent cocaine users denied their cocaine use in the in-person interview. Doubtless, some cocaine users also denied their recent use in the SAQ.

Several of these same studies also examined reporting of alcohol consumption (e.g., Aquilino, 1994; Aquilino and LoSciuto, 1990; Schober et al., 1992). Generally, the trend for reports on alcohol consumption is similar to that for

Table 22.1. Mode Effects and Reports of Illicit Drug Use: Proportions of Respondents Reporting Drug Use Relative to SAQ

Study	Modes of Data Collection	Drug	Time Frame		
			Past Month	Past Year	Lifetime
Aquilino	In-person vs.	Cocaine	0.67	0.75	0.88
(1994)	SAQ	Marijuana	0.83	0.77	0.98
	Telephone vs.	Cocaine	1.00	1.00	0.76
	SAQ	Marijuana	0.67	0.62	0.96
Aquilino and	Telephone vs.	Cocaine	0.83	0.85	1.10
LoSciuto	SAQ — Whites	Marijuana	1.00	0.96	1.00
(1990)	Telephone vs.	Cocaine	0.60	0.82	0.83
	SAQ — Blacks	Marijuana	0.41	0.72	0.80
Gfroerer and	Telephone vs.	Cocaine	—	0.49	0.70
Hughes	SAQ	Marijuana	—	0.65	0.75
(1992)					
Schober et al.	In-person vs.	Cocaine	0.60	0.75	0.89
(1992)	SAQ	Marijuana	0.75	0.83	0.99
Turner, Lessler,	In-person vs.	Cocaine	0.42	0.68	0.94
and Devore	SAQ	Marijuana	0.62	0.77	0.95
(1992)					

Note: SAQ refers to self-administered paper questionnaires; in-person refers to face-to-face (paper and pencil) interviews; telephone refers to paper and pencil telephone interviews.

drug use — less drinking is reported when the questions are administered by an interviewer than when they are self-administered, although the pattern is less striking.

Another characteristic of self-administration is that it reduces the salience of the interviewer in the data collection process. Interviewers may divert the respondent's attention from the task of answering the question, or they may help maintain respondent motivation. Regardless of the overall direction, one might expect the interviewer's effect to be reduced when the questions are self-administered relative to when the interviewer administers them (Tourangeau et al., 1997). However, the effect of the interviewers on the answers may be modest (Sudman et al., 1977).

Computerization
A second dimension that distinguishes the six methods of in-person data collection is the computerization of the data collection process. Computerization can have several effects on the data collected. Most obviously, the programs typically prevent errors of administration, thus reducing the propor-

tion of answers that are missing, outside the permissible range, or logically inconsistent with other answers. Computer assistance may also have more subtle effects on data quality. Laptop computers are still a novelty for many respondents, and their use may impress respondents, making the survey seem more important, more objective, or more scientific than when paper question-naires are used. At the same time computers may make some potential respondents nervous, increasing their reluctance to participate. Resistance to the computer may be especially high when respondent must interact with the computer directly, as with CASI and ACASI (Couper and Rowe, 1996).

Auditory or Visual Presentation of the Questions
A final difference among the methods is whether the questions are presented aurally, visually, or through both modalities. Methods of data collection, such as CAPI or ACASI, in which the questions are read aloud reduce the overall cognitive burden imposed on the respondent and eliminate the requirement that respondents be able to read. This may be especially important within subgroups where literacy problems are common. On the other hand, ACASI may impose cognitive requirements of its own, such as numeracy and minimal typing skills, especially when open-ended items are used. In addition, when the questions are only read aloud, the respondent has less control over the pace of the interview and may be prone to "recency" effects, favoring options presented at the end of the list of permissible answer categories over those presented at the beginning (Krosnick and Alwin, 1987). Auditory presentation (without concurrent visual display) may also overtax the respondent's listening ability, and the comprehension of even moderately long or complicated questions may suffer as a result. By itself, auditory presentation does not directly affect privacy, since the questions may be presented via earphones or out of the earshot of others.

Particular combinations of these three underlying variables may produce additional differences among the six modes. For example, self-administered paper questionnaires allow respondents to look ahead and go back to earlier items, reducing the effects of question order (Bishop et al., 1988). In principle, CASI and ACASI applications may also allow respondents to "leaf" through the questionnaire, but as a practical matter respondents are less likely to take advantage of this capability on a computer than with a paper questionnaire. Because CASI and ACASI present the items on the screen one at a time, they may increase the time respondents spend on each question relative to an SAQ. For example, respondents may be prevented from rushing through a block of similarly formatted items displayed individually on the screen rather than presented as a block in a paper questionnaire.

Overall Model of the Effect of Mode
Figure 22.1 depicts an overall model of the effects of the different methods of in-person data collection on responses to sensitive questions. The figure shows the main hypothesized causal links among the features of the different modes,

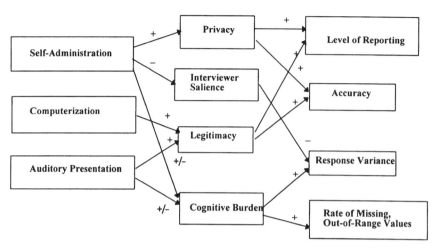

Figure 22.1. Model of mode effects.

the psychological variables that are affected by these features, and the consequences of these psychological variables for the quality of the data that are ultimately collected. According to the model, four variables mediate the effects of data collection mode on data quality—privacy, interviewer salience, cognitive burden, and sense of legitimacy fostered by the mode. These psychological variables are affected by the method of data collection and in turn affect the properties of the data obtained. The model has not been systematically tested, but it does provide a useful summary of much of what is known about how the method of data collection affects the data that are collected and suggests a number of specific hypotheses for future research. For example, the model incorporates the hypothesis that self-administration increases the perceived privacy of the data collection process and that this effect mediates mode differences in reporting sensitive behaviors. Each link in the diagram corresponds to a similar specific hypothesis about the effects of the mode of data collection on the data that are ultimately collected.

The model is not an attempt to depict all the variables that can affect reporting about sensitive topics, only those that are affected by the method of data collection. Poorly worded questions, difficulties in recall or estimation, and numerous other problems affect reporting without being much affected by the method of data collection. Such sources of mode-invariant error are omitted from the model. The model also omits other aspects of data quality, such as coverage or nonresponse errors. Nicholls, Baker, and Martin (1997) provide a useful summary of the findings on the effect of computerization on data quality.

Figure 22.1 is based on our work with in-person methods of data collection and does not include methods of collecting data by telephone. Some prior research suggests that telephone data collection is seen by respondents as intermediate in privacy between self-administration and face-to-face interviews (Aquilino, 1994; Hochstim, 1967; Sudman and Bradburn, 1974). However, Smith (1984) reviewed a dozen studies and found no clear indication of any reduction in social desirability effects from telephone data collection (see also Johnson, Hougland, and Clayton, 1989). In addition, telephone interviews clearly rely on auditory presentation of the items. Although the model could be expanded to examine telephone modes, we have chosen to focus on modes of data collection that can be used in the context of in-person data collection.

22.1.2 Types of Items

Our discussion of mode effects has implicitly focused on responses to sensitive questions. Questions about voting, sexual behavior, abortion, sexually transmitted diseases, exercise, smoking, drinking, and illicit drug use have all been considered sensitive. Such questions clearly have socially "correct" answers. In the studies summarized later, we included questions on most of these topics.

The direction of the reporting errors is likely to vary by the topic of the question. Questions about embarrassing or illicit behaviors (e.g., excessive drinking or abortion) are likely to produce underreporting; those about obligatory or widely lauded behaviors (e.g., voting or exercise) are likely to produce overreporting. The findings on sexual behavior are particularly interesting because the direction of the errors seems to vary by the sex of the respondent. The hypothesis that men and women distort their answers in opposite directions (Galbraith et al., 1974; Herold and Way, 1988; Kinsey et al., 1953; Klassen, Williams, and Levitt, 1989; May, Anderson, and Blower, 1989; Ornstein, 1989) is consistent with sex differences in sexual values. Women's attitudes toward sexual behavior are less permissive than men's, and respondents of both sexes show greater disapproval of sexual activity by women than by men (e.g., Klassen et al., 1989).

With nonsensitive behaviors the privacy of the interview will presumably have little or no effect on reporting, so differences among modes related to privacy will be greatly attenuated. Many of the prior reviews of the mode effects literature examine a broader range of items and therefore come to a picture quite different from the one conveyed by Figure 22.1 (e.g., see deLeeuw and van der Zouwen, 1988). In fact one index of a question's degree of sensitivity may be its susceptibility to mode effects, especially to differences between self-administered and interviewer-administered. The data in Table 22.1 show reporting ratios closer to 1.0 (indicating the absence of a self-administration effect) for lifetime drug use than for drug use in the past month. We suspect that recent drug use is more sensitive than drug use in the distant past. Most of the reporting ratios involving lifetime marijuana use are near 1.0. Even political candidates have admitted to trying marijuana.

22.2   STUDIES ON MODE EFFECTS AND REPORTS ON SENSITIVE TOPICS

We have recently conducted a series of studies on sensitive topics, particularly reports about sexual behavior. The first, the Women's Health Study, was carried out as part of a larger effort to test new methods for the National Survey of Family Growth (NSFG). The second examined the effect of several variables, including the mode of data collection, on the disparity between the number of sex partners reported by men and women. The third attempted to determine whether the main source of this disparity was motivated misreporting. A bogus pipeline procedure was used in an attempt to increase motivation to answer truthfully.

22.2.1   Women's Health Study

Since it began in 1971, the NSFG has obtained information on a number of sensitive issues, such as contraceptive practices, pregnancy histories (including fetal and infant deaths), sexually transmitted diseases, and infertility. Our study examined three strategies for improving the accuracy of self-report data in the NSFG.

The first strategy involved increasing the privacy of the data collection. We compared self-administration of the questions with administration by an interviewer. In addition, we investigated the effects of moving the interview to a site outside the respondent's home, away from other family members. Relatively few studies have recorded whether face-to-face interviews on sensitive topics were conducted in private or with other household members able to overhear the respondent's answers. Aquilino (1993) provides an exception. In the studies that have examined this variable, the effects of the privacy of the interview setting have been unclear.

A second strategy we investigated was the use of computer assisted data collection. Early evaluations suggest that computer administration of the items may produce gains similar to those from conventional self-administration (e.g., Waterton and Duffy, 1994) and that computerization itself might have an effect on respondents' willingness to report truthfully (Baker, Bradburn, and Johnson, 1995).

The final strategy we examined was that of placing the interview as a whole, and especially the questions on abortion, in a medical context. We believed that respondents might be more willing to provide truthful answers when the setting underscored the health-related purposes of the interview. (For more detailed descriptions of this study, see Jobe et al., 1997, and Tourangeau et al., 1997.)

Sample
The Women's Health Study comprised a large field experiment, with more than 1000 women being interviewed. (A small comparison sample of 100 men was

included but their data are not presented here.) The sample was selected from an area probability sample of the city of Chicago and from the rosters of two cooperating Chicago-area health clinic. The area probability portion of the sample was a stratified, multistage sample of dwellings in the city of Chicago, selected using standard methods. Screening was used to identify persons in the 15 to 35 age. The clinic component of the sample was drawn from lists of names of women between the ages of 15 and 40 who had had abortions during the preceding year. We selected 2296 persons — 1564 from the area sample and 732 from the clinic sample — for the experiment and randomly assigned them to an experimental condition. The study incorporated an after-the-fact permission form procedure in which women were asked to sign a release form giving us access to their medical records at their sources of gynecological care. Forty-eight women from the clinic sample who refused to sign the permission form were dropped from the analysis, and their data were eliminated from the data files.

Study Design

Questionnaires based on the one used in the NSFG were administered to the sample. They included items on abortion, sexual behavior, and illicit drug use and took about an hour to complete. The experiment examined five variables: (1) whether the questionnaire began with a series of questions on medical procedures and conditions or with a series of questions about pregnancy, (2) whether the interview was conducted by a nurse or by a regular NORC interviewer, (3) whether the interview was done at the respondent's home or at a site outside the home, (4) whether the interviewer asked the questions or the questions were self-administered, and (5) whether the data were collected via computer or on paper. The latter two variables were crossed yielding four methods of data collection — CAPI, PAPI, CASI, and SAQ. In an attempt to keep the response rates comparable across conditions, we offered respondents $40 to complete the interviews outside their homes. This incentive proved over-generous; the outside-the-home group ended up with a higher response rate than the in-home group. The analysis examined a number of outcome variables, including unit and item nonresponse rates under the various experimental conditions, level and accuracy of abortion reporting, and level of reporting on other sensitive topics (e.g., number of sex partners). The main findings involved the effects of self-administration and computerization on reports about sexual behavior, on item nonresponse, and on interviewer variability.

Number of Reported Sex Partners

The questionnaire asked respondents to report the number of their opposite-sex sex partners during the past year, the past five years, and over their lifetimes. The sex partner data are counts, and the distribution of the responses for all three time periods is highly skewed. To compensate for this departure from normality, we added 0.5 to the values and then carried out a logarithmic transformation prior to performing statistical analyses. For ease of interpretation, the means reported are generally based on untransformed data. The analyses examined the number of sex partners reported by the women respondents as a function of the five experimental factors, controlling for differences across the experimental groups introduced by nonresponse. The standard errors and significance tests took the clustering of the sample into account.

For all three time periods, women who completed self-administered questionnaires reported more sex partners than women who responded to questions administered by an interviewer (see Table 22.2). The effects of self-administration were significant for all three time periods. For the past year, the women who answered self-administered questions reported a mean of 1.72 sex partners versus 1.44 for those who answered interviewer-administered questions. For the five-year period, the corresponding means were 3.87 sex partners versus 2.82. For the lifetime item, the means were 6.51 versus 5.43. No other main effects were significant.

Table 22.2 also displays the proportion of respondents who reported ever having used illicit drugs. The absence of mode differences on drug reporting may reflect the lifetime reference period for the illicit drug use item. As we noted earlier, recent drug use seems more sensitive than drug use in the distant past.

Computerization interacted with the site of the interview to affect the number of sex partners reported. During home interviews, more sex partners were reported by women interviewed using computer assisted questionnaires than by those responding to paper and pencil questionnaires. For women

Table 22.2. Mode Effects in the Women's Health Study

Experimental Group	Mean Reported Sex Partners			Percent Admitting Illicit Use
	Past Year	Past 5 Years	Lifetime	
Self-administered questions	1.72	3.88	6.54	40.9
Conventional (SAQ)	1.56	3.37	6.88	42.5
Computer assisted (CASI)	1.89	4.40	6.25	39.3
Interviewer-administered questions	1.44	2.82	5.43	40.7
Conventional (PAPI)	1.56	2.86	4.58	39.3
Computer assisted (CAPI)	1.36	2.79	6.27	42.2

Source: Tourangeau et al. (1997).

Note: Each mean or proportion is based on approximately 240 interviews; rows for self- and interviewer-administration are based on nearly 500 interviews.

interviewed outside the home, more sex partners were reported on the pencil and paper questionnaires. Bringing computers into the respondents' homes may have fostered a sense of the importance or objectivity of the survey, promoting more complete reporting of sex partners. Outside the home, especially in public places, the computer may make respondents feel conspicuous, inhibiting reporting. Any conclusion must remain tentative at best, since the site variable is confounded with the incentive offered to the respondents.

Item Nonresponse

We analyzed responses to 43 of the questionnaire items administered to the women respondents. These items covered a range of topics — demographics, sexually transmitted diseases, pregnancy history, medical procedures for inducing abortion, illicit drug use, and sexual behavior. Because of the skip patterns in the questionnaire, few respondents were actually administered all 43 of the items: on the average, the respondents were supposed to have answered about 30 of them. Across the experimental treatments, the women answered almost all of the items they were supposed to — an average of 29.4 items, or about 97.4 percent of the total.

Both computerization and self-administration appeared to affect the completeness of the data, as shown in Table 22.3. Computerization increased the proportion of the questions that were answered (98.5% vs. 96.2% for paper questionnaires), and self-administration decreased the proportion (96.2% vs. 98.5% for interviewer administration). In an analysis that fit a simple additive model including the five experimental factors, both of these main effects were significant. A fully saturated five-way model that incorporated all the interactions among the experimental variables revealed an additional interaction effect involving the type of interviewer — the difference between self- and interviewer administration was more striking for the regular field interviewers than for the nurses.

Table 22.3. Proportion of Questions Answered by Method of Data Collection

Group	Mean Proportion	(n)
PAPI	97.8	(261)
SAQ	94.6	(256)
CAPI	99.2	(244)
CASI	97.9	(244)
Self-administered	98.5	(505)
Interviewer-administered	96.2	(500)
Computer assisted	98.6	(488)
Paper and pencil	96.2	(517)

Source: Tourangeau et al. (1997).

Interviewer Effects

One of the potential benefits of self-administration is a reduction of the effect of the interviewers on the answers. When the interviewers merely hand the respondent a questionnaire to be completed (or set up a laptop computer to administer the items), they seem likely to have less effect than when they read the questions and record the answers.

Computerization may also reduce the interviewers' effect on the answers by enforcing a certain level of standardization on the interviewers. We explored these hypotheses by conducting one-way analyses of variance across interviewers for 19 sensitive items, including questions on sex partners, sexually transmitted diseases, specific sexual practices (e.g., oral sex), abortion, and illicit drug use. Separate analyses were carried out for respondents interviewed under each of the four modes of data collection. Because the interviewer assignments were geographically clustered, the ANOVA estimates represent both interviewer effects and the effects of the overall area in which an interviewer's cases were concentrated. The findings are thus suggestive at best, and we make no attempt to test their statistical significance.

Despite these limitations a clear pattern emerged in the data. Using the estimated variance components from the ANOVAs, we estimated values of ρ, the intraclass correlation, for each variable and mode of data collection. (When the estimate of the between-interviewer variance component was negative, we set $\hat{\rho}$ to zero.) Table 22.4 displays the median value of the ρ estimates across the 19 variables, as well as the results for six specific items. Both computerization and self-administration appear to reduce variation across interviewers. For most of the variables, there is no evidence for interviewer effects when the questions were self-administered; the estimate for ρ was typically zero. Similar results were reported in a comparison of CATI and paper and pencil telephone interviews by Mathiowetz and Groves (1984), which found lower estimates of ρ for CATI than for paper and pencil interviews on 18 of 24 variables. Although consistent, these differences were not significant (see also Groves and Magilavy, 1980).

Table 22.4. Estimates of ρ by Method of Data Collection and Variable

	PAPI	SAQ	CAPI	CASI
Median across 19 variables	0.072	0.000	0.016	0.000
One-year sex partners	0.158	0.036	0.000	0.000
Five-year sex partners	0.072	0.000	0.000	0.000
Lifetime sex partners	0.178	0.000	0.021	0.000
Used illicit drugs	0.108	0.003	0.002	0.000
Used marijuana	0.100	0.000	0.013	0.000
Used cocaine	0.100	0.044	0.035	0.000

Source: Tourangeau et al. (1997).

22.2.2 ACASI Experiment

We completed the second experiment on the effects of mode of data collection in the fall of 1994 (Tourangeau and Smith, 1996). This study compared three methods of computer assisted data collection for face-to-face interviewing — CAPI, CASI, and ACASI. The three modes were compared in a cross-sectional sample of respondents from Cook County, IL. The key outcome variables were nonresponse rates and levels of reporting, particularly reporting of illicit drug use and sexual behavior. Our hypotheses were that relative to CAPI data collection, both CASI and ACASI would increase the proportion of respondents admitting illicit drug use and decrease the disparity between the average number of sex partners reported by men and women. More generally, we predicted that by offering greater privacy to the respondents, the two self-administered modes would reduce the effects of social desirability on the answers. Because a large number of partners is seen as undesirable for women and a small number of partners is seen as undesirable for men, we expected self-administration to have opposite effects on the number of sex partners reported by men and women. We did not anticipate a difference in response rates by mode.

The design included two other variables thought to affect the number of sex partners reported. These were the format and context of the sex partner questions. Work described in Bradburn, Sudman, and Associates (1979) had shown that closed-ended items tend to elicit lower levels of reporting of alcohol consumption than open items, at least among those who report drinking at all. To test the effect of the answer options used in the closed questions, we developed two versions of each closed item — one with categories mainly at the low end of the range and a second version with categories concentrated at the high end. As Schwarz and Hippler (1987) have shown, the response categories offered can effect the overall level of the answers (see also Schwarz et al., 1985). Thus we prepared two closed versions and also included an open version of each of the sex partner questions.

Sample

The sample was a multistage area probability sample of dwellings in Cook County, IL. Seventeen of the 32 sample segments (individual blocks or adjoining blocks) were located in the city of Chicago, and the remainder were drawn from the balance of the county. Within those segments we selected more than 1100 housing units (HUs) for a short screening interview to determine whether anyone living there was in the 18–45 age range. Screeners were completed at 975 of the 1069 occupied HUs, for a completion rate of 91.2 percent.

Of the 643 potential respondents selected in the screener, 365 completed the main interview for a final completion rate of 56.8 percent. Unfortunately, the study used a new version of the software, and because of a design flaw in the program, some 79 of these initial interviews were lost. Subsequently the error was corrected, and 53 of the original cases were reinterviewed under the same experimental conditions to which they had been assigned originally. The results presented here are, except where noted, based on 339 completed cases (including the 53 reinterviewed cases). The results are not appreciably altered if the reinterviews are excluded from the analysis. Although the response rates may seem low, they are comparable to those achieved within large metropolitan areas by major national surveys (e.g., the NSFG).

Experimental Design

Prior to screening, each dwelling was randomly assigned to an experimental condition; the data collection mode variable was crossed with the context and format of the sex partner items, producing a total of 12 conditions. The random assignment was done within segments to ensure that the experimental conditions were not confounded with geographic areas. The main experimental variable was the mode of data collection: Respondents were assigned to data collection by computer assisted personal interview (CAPI), computer assisted self-administered interview (CASI), or audio computer assisted self-administered interview (ACASI). In all cases an interviewer was present throughout the interview. In the CASI and ACASI conditions, interviewers were instructed not to look at the screen but to busy themselves with administrative chores while the respondents completed the questionnaire.

We also examined the effects of prior items on the reporting of sex partners. Just before the sex partner questions, respondents answered one of two sets of questions about their sexual attitudes. One set of items was designed to encourage respondents to report large numbers of sex partners; these questions consisted of statements expressing "permissive" views about sexual activity. The other set was designed to discourage reporting large numbers of partners; these statements in this set expressed more "restrictive" views about sex. We thought that because reports about sexual behavior may be affected by attitudes toward sex, the permissive items would encourage female respondents to admit how many partners they had and that the restrictive items would encourage male respondents to admit how few they had. The open version of the sex partner question simply asked for the number of partners. For example, the one-year question administered to women read: "During the last 12 months, that is, since August/September, 1993, how many men (if any) have you had intercourse with? Please count every partner, even those you only had sex with once." A number was entered as the answer. The closed versions of this question presented the same question followed by a set of response options. The response options presented to one group of respondents were 0, 1, 2, 3, 4, and 5 or more; the options presented to the other group were 0, 1 through 4, 5 through 9, 10 through 49, 50 through 99, and 100 or more. A follow-up item asked respondents in the closed format groups the exact number of partners (when that could not be inferred from the response category they had selected). Respondents received the same version of the sex partner item for all three recall periods and the same set of response options.

Aside from the items on sex partners, the questionnaire included items on a range of sensitive topics, including sexually transmitted diseases, pregnancy, condom use, and illicit drug use. The questions were mostly drawn from existing sources, including the NSFG, the sex supplements to the General Social Surveys, and the National Household Survey of Drug Abuse.

Response Rates

The mode of data collection did not appear to affect the response rates; 55.5 percent of those assigned to ACASI, 57.7 of those assigned to CAPI, and 57.1 percent assigned to CASI data collection completed the interview. We examined whether different sorts of respondents took part under the three modes, comparing the composition of the three groups on sex, age, marital status, and educational attainment. The three groups differed only on educational attainment, with the ACASI respondents reporting a somewhat higher level of education than those in the other mode groups. The differences in reporting by mode described below remain significant even when these differences in educational attainment are taken into account.

Reporting of Sex Partners

Each respondent was asked to report the number of sex partners over three time periods: past year, past five years, and lifetime. We examined the final exact answers to these questions here. Few discrepancies were found between the final and initial answers to the closed questions. Again we transformed the raw counts (by adding 0.5 and taking the log) to reduce the skewness in the data. We also dropped five outliers—two respondents who reported 50 partners in the prior year, one who reported more than 100 partners during the previous five years, and two who reported more than 100 partners in total. Their omission had little effect on the results. We analyzed the transformed reports for each time period by sex, data collection mode, question format, and item context.

For all three time periods, there were significant main effects for the sex of the respondent. As has been found in prior studies, the male respondents reported significantly more partners than their female counterparts (2.9 vs. 1.6 for the past year; 5.2 vs. 2.7 for the past five years; and 8.4 vs. 4.8 over the lifetime). In addition there were significant main effects for all three of the experimental variables (see Table 22.5). For all three time periods, ACASI elicited the highest mean number of reported sex partners. The effect of data collection mode was significant for the lifetime reports and marginally significant for the five-year reports. An additional contrast revealed that for both lifetime and five-year reports, ACASI yielded significantly higher numbers of reported partners, on average, than CASI did. For the one-year and five-year periods, the closed item with the high-response categories elicited the highest level of reporting and the closed item with the low-response categories the lowest. In both cases, the open item produced intermediate levels of reporting. The effect of item format was also significant for both the one-year and

Table 22.5. Mean Reported Sex Partners and _F_ Values by Experimental Variables

	One-Year Partners			Five-Year Partners			Lifetime Partners		
	Raw	Log	(n)	Raw	Log	(n)	Raw	Log	(n)
Females	1.60	0.51	(187)	2.73	0.83	(188)	4.79	1.20	(188)
Males	2.92	0.80	(106)	5.17	1.12	(104)	8.35	1.62	(104)
_F_ value		9.29 (<0.01)			7.69 (<0.01)			9.18 (<0.01)	
ACASI	2.26	0.67	(86)	4.52	1.01	(87)	7.05	1.45	(88)
CAPI	2.14	0.59	(103)	3.44	0.95	(103)	5.51	1.33	(102)
CASI	1.87	0.60	(104)	2.99	0.85	(102)	5.75	1.28	(102)
_F_ value		0.93 (ns)			2.19 (ns)			3.03 (<0.05)	
Open	1.65	0.50	(102)	3.12	0.85	(102)	7.20	1.45	103
Closed-low	1.43	0.43	(105)	2.62	0.78	(103)	4.73	1.23	103
Closed-high	3.38	0.98	(86)	5.33	1.21	(87)	6.28	1.37	86
_F_ value		14.05 (<0.001)			7.03 (<0.01)			2.35 (<0.10)	
Permissive context	1.88	0.57	(148)	2.95	0.85	(145)	5.77	1.34	(146)
Restrictive context	2.28	0.66	(145)	4.24	1.01	(147)	6.34	1.36	(146)
_F_ value		2.92 (<0.10)			4.03 (<0.05)			0.52 (ns)	

Source: Tourangeau and Smith (1996).

Note: _F_ values based on log transformed data.

five-year data. Finally the restrictive context elicited higher numbers of sex partners than the permissive context for all three time periods; however, this "backfire" effect was significant only for the five-year data, and the effect of context was qualified by higher-order interactions.

The data in Table 22.5 show the usual discrepancy between men and women in the reported number of sex partners. We expected this discrepancy to be reduced when the questions were self-administered. Moreover we expected the effect of self-administration to differ by sex — with males reporting fewer partners and women more partners under the two self-administered modes than under the interviewer-administered mode. Consistent with this hypothesis, the disparity between the number of sex partners reported by men and women was largest under CAPI for all three time periods; in addition CAPI yielded the highest level of reporting for the men and the lowest level for the women.

Figure 22.2 displays the means of the transformed sex partner reports for the past year by sex and mode of data collection. The pattern is similar for five-year and lifetime reports as well. However, none of the mode by sex interactions were significant; the pattern apparent in Figure 22.2 was qualified by higher-order effects involving the context variable. The convergence in the reports of men and women under the self-administered modes of data collec-

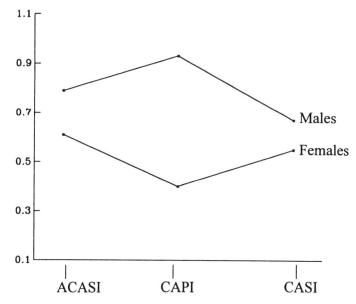

Figure 22.2. Mean reported sexual partners (log transformed) in past year, by mode and sex.

tion — CASI and ACASI — was far more pronounced among respondents who received the restrictive context items than among those who received the permissive items. Within the restrictive context groups, the mode of data collection had a large effect on the reported number of sex partners and the gap between men and women was widest under CAPI. Within the permissive context groups, the mode of data collection had relatively little effect on the number of partners reported and the disparity between the reports by men and women persisted under all three modes of data collection.

Other Sexual Behavior
In addition to the sex partners items, respondents were asked about some specific sexual behaviors, including how frequently they had had oral and anal sex over the past five years. We analyzed the proportion of respondents who reported that they had oral and anal sex at least some of the time as a function of data collection mode. A significantly larger percentage of respondents indicated they had oral sex under ACASI (75.3%) than under CAPI (51.5%) or CASI (64.6%) data collection. Similarly, a significantly higher percentage of respondents indicated they sometimes had anal sex under ACASI (20.2%) than under the other modes (4.8% and 9.8% under CAPI and CASI).

Reporting of Illicit Drug Use
The mode of data collection also had some effect on the proportion of respondents who admitted using illicit drugs. We examined reported marijuana

or cocaine lifetime use, within the past year, and during the past month. For both drugs and all three reporting periods, ACASI yielded the highest percentage of reported drug use. In general, CAPI yielded the lowest level of reporting with CASI between the other two modes. The mode difference was significant for lifetime marijuana use and marginally significant for one-year marijuana use and lifetime cocaine use. The difference between levels of reporting under ACASI and CASI was not significant for any of the drug use variables.

22.2.3 Bogus Pipeline Study

We sought to test the hypothesis that reports about sexual behavior and illicit drug use increased so strikingly with self-administration because of more or less deliberate misreporting of information to interviewers. In our final study we attempted to increase respondents' motivation to provide accurate information through a very powerful manipulation—the "bogus pipeline" procedure (Jones and Sigall, 1971). In the bogus pipeline procedure, respondents are led to believe that inaccurate answers can be detected by a physiological recording device (e.g., a polygraph) or a test performed on a biological specimen (e.g., a saliva sample). The crucial feature of the method is not that the device or test can actually detect invalid answers but that the respondents believe that it can. Reviews of the literature indicate that the technique can produce more accurate self-reported data than those obtained under standard conditions. For example, in a review of 11 studies on the effect of the bogus pipeline procedure on adolescents' self-reports about smoking, Murray et al. (1987) find an overall effect for the method.

The respondents in our third study were adult volunteers from Chicago, paid $20 for taking part. A total of 120 were interviewed, 62 under the bogus pipeline procedure and the other 58 under control conditions. The respondents completed a preliminary "screening questionnaire," covering such basic items as sex, age, and race. After respondents had completed this questionnaire, they were led to a second room, where the main interviewer introduced herself. The interviewer then administered the main questionnaire. The main interview covered a range of health-related topics—the respondent's smoking, drinking, and exercise habits, as well as illicit drug use and sexual behavior. Three of the items asked how many opposite-gender sex partners the respondent had in the usual three periods—the past year, past five years, and over his or her lifetime.

In the control treatment, the interviewer and the respondent sat at a desk across from each other, and the interviewer, after a brief introduction, conducted a conventional face-to-face interview. In the experimental group, the procedures were more complicated. The interviewer seated the respondent in a chair, facing a battery of machinery that included a desktop computer, a polygraph, and a microphone. The interviewer then attached four electrodes to the respondent and adjusted the microphone so that a light displayed in response to changes in the volume of sound. The interviewer then demonstrated the equipment's apparent ability to detect lies, telling the respondent that it was necessary to "calibrate the machine" and the respondent should lie in answering some test questions. These were drawn from the screener. Without the respondents' knowledge, the experimenter in the reception area had forwarded to the interviewer their answers to the preliminary questions electronically; these were displayed on the interviewer's computer monitor. The interviewer was thus able to detect any discrepancies. After this demonstration, the main interview was conducted.

The results of the study were simple and quite dramatic—overall, 7 of the 19 sensitive items we examined showed significant differences between the bogus pipeline and control respondents, and three others showed marginally significant differences. Altogether on 15 of the 19 items the differences between groups are in the expected direction. Clearly, when they are under enough pressure, respondents are willing to admit to a number of behaviors they would otherwise deny. (For a more detailed account, see Tourangeau, Smith, and Rasinski, 1997.)

22.3 CONCLUSIONS

The most consistent finding from the studies we have summarized is the overwhelming effect of self-administration on levels of reporting sensitive behaviors. Table 22.6 displays ratios comparing the levels of reporting in interviewer-administered and self-administered conditions from the Women's Health and ACASI experiments. These ratios are analogous to the ones presented in Table 22.1. Almost without exception, respondents report fewer sex partners, less sexual activity, and less drug use when they must report their answers to an interviewer. The major exception—reduced reporting of sex partners by men—is consistent with the hypothesis that self-administration reduces self-presentation concerns. Our results are also consistent with the results of a series of studies by Turner et al. (1996a; Chapter 23). Respondents appear to withhold embarrassing information on certain topics when interviewers administer the questions. Self-administration may not eliminate misreporting due to social desirability, but it may be the single most effective means for minimizing this source of error.

The magnitude of the effect of self-administration on levels of reporting may depend on several variables, including the sensitivity of the behaviors, the level of trust of the respondents, the privacy of the interview setting, the context and format of the questions, and the respondent's race. We have already noted the effect of the degree of sensitivity. When the behavior in question is very sensitive, self-administration can have dramatic effects, doubling or tripling the proportion of respondents willing to report the behavior; when the behavior is not especially sensitive, the effect of self-administration seems negligible. Differences in sensitivity seem to account for the variations in reporting ratios

Table 22.6. Ratios Comparing Levels of Reporting by Mode of Data Collection

	Contrast and Ratios	
Women's Health Study	*PAPI vs. SAQ*	*PAPI vs. CASI*
Average number of sex partners (females)		
Past year	0.97	0.72
Past five years	0.85	0.64
Lifetime	0.67	1.03
% Reporting STD	0.77	0.77
Average % of time using condom	0.84	0.70
ACASI experiment	*CAPI vs. ACASI*	*CAPI vs. CASI*
% Reporting anal sex	0.24	0.49
% Reporting oral sex	0.68	0.80
% Reporting marijuana use		
Past month	0.60	0.84
Past year	0.62	1.01
Lifetime	0.68	0.78
% Reporting cocaine use		
Past month	0.57	1.05
Past year	0.35	0.73
Lifetime	0.55	0.99
Average number of sex partners (males)		
Past year	1.33	1.85
Past five years	1.00	1.43
Lifetime	0.99	1.11
Average number of sex partners (females)		
Past year	0.61	0.68
Past five years	0.53	0.86
Lifetime	0.60	0.81

Sources: Jobe et al. (1997); Tourangeau et al. (1997); Tourangeau and Smith (1996).

Note: Sex partner data are untransformed counts.

across behaviors that are apparent in Tables 22.1 and 22.6. For a given respondent the level of sensitivity of a behavior may reflect its perceived prevalence, the degree of social approval (or disapproval), and whether the behavior is illegal or not. Sensitivity can also vary across subgroups of the population; for example, smoking is probably more sensitive for adolescents than for adults.

If the effects of self-administration reflect fears about disclosure or about the reactions of the interviewer, then they may be heightened among respondents with chronically low levels of trust in people. This hypothesis receives some support from a study by Aquilino (1994), in which responses to an item on trust interacted with mode of data collection to affect reports of illicit drug use. Similarly one might expect self-administration to affect answers less in private settings than when other persons are present during the interview. To date, however, few studies have been able to demonstrate the effects of the privacy of the data collection setting. The Women's Health Study attempted to vary the setting of the interview experimentally and found few effects. At least two prior studies had varied the mode of data collection and recorded whether other people were present during the interview, but neither study found either main effects for the privacy measure or interactions with self-administration (Shober et al., 1992; Turner et al., 1992; see also Mosher and Duffer, 1994). In theory, the effect of the presence of other people would depend on whether the respondent has anything embarrassing to admit, whether the others present are already aware of what the respondent might report, and how the others are likely to react to the respondent's admission of something they were previously unaware of. It is one thing to admit having smoked marijuana in the presence of one's spouse, who is probably already aware of it, quite another to make the same admission in front of one's parents or children.

Our study of ACASI indicated that the effect of mode of self-administration on the reported number of sex partners depended both on the format and context of the sex partner questions. Self-administration had relatively muted effects with the open-ended versions of the sex partner questions compared with its impact with the two closed-ended versions of the items (Tourangeau and Smith, 1996). The context variable had an even clearer effect. When the restrictive items preceded the sex partner questions, the convergence between the number of sex partners reported by men and women respondents under self-administration was more marked than when the permissive items preceded them. It seems likely that the restrictive context items heightened self-presentation concerns and therefore increased the effect of self-administration on the reported number of sex partners.

The effect of self-administration may vary by demographic subgroup. For example, Aquilino and LoSciuto (1990) found that reported levels of alcohol and marijuana use were generally lower when the data were collected by telephone than in an SAQ, but as previously shown in Table 22.1, this mode difference was more pronounced among black than among white respondents. By contrast, Schober et al. (1992) found the opposite, concluding that non-minority respondents were more affected than minority respondents by self-administration in their reporting of cocaine and marijuana use. These contradictory findings about the interaction of race and mode of data collection may reflect the complex underlying relationships involved. Subgroup differences in the impact of self-administration are likely to result from differences across subgroups in the sensitivity of the behaviors in question, the perceived privacy of the different modes of data collection, or in the trust inspired by the interviewers. These differences in turn may vary over time or by region of the country.

Self-administration has its drawbacks. An interviewer is trained to recognize when respondents need help and to give such help. Self-administration can increase item nonresponse. It may also reduce the length or quality of open-ended material respondents provide. It eliminates the interviewer comments that can clarify the meaning of an answer. The great virtue of self-administration is that it permits some respondents to make admissions that would otherwise be too embarrassing.

22.3.1   Effect of Auditory Presentation

Our second experiment allows us to compare two modes of data collection — CASI and ACASI — that differ mainly in whether the items are presented only visually or both visually and aurally. The results in the lower panel of Table 22.6 suggest that ACASI may enhance the advantages of other methods of self-administration. For several items, ACASI had greater effect on reporting (relative to CAPI) than CASI did. The use of earphones may induce a greater sense of isolation and privacy than CASI does.

ACASI also helps overcome the literacy barriers to self-administration. There are several distinct sources of literacy problems. Some people never learn to read because of visual impairment or severe mental disability. Others learn to read other languages but cannot read English. By itself, ACASI does little to address such problems. Instead, ACASI — and auditory presentation more generally — is probably of most benefit with respondents who fail to learn to read because of a learning disability or lack of education.

22.3.2   Effect of Computerization

The effects of computerization on the reporting of sensitive information seem far less dramatic than those of self-administration or auditory presentation of the items. There are a few scattered effects reported in the study by Baker, Bradburn, and Johnston (1995). In the Women's Health Study, computerization reduced the rates of missing data and the variability attributable to interviewers. Our studies may have missed the full effect of computerization because our samples consisted of relatively young people, under 45 years of age. As Couper and Rowe (1996) observed, older respondents, those with little education, and those with limited prior experience with computers may be reluctant to use the new methods of data collection, which require direct interaction with the computer. Computerization may lead to increased nonresponse rates for such sample members.

Whatever its other effects, computerization has one clear virtue: It is a necessary feature for ACASI and other new technologies for collecting data. Even in the absence of clear evidence of the benefits of computerization, computers will, without doubt, continue to play a growing role in the collection of sensitive data.

ACKNOWLEDGMENTS

Some of the research reported here was supported by Contract 200-91-7099 from NCHS to NORC and by an NSF grant (SES 9122488) to Tom Smith and Roger Tourangeau.

CHAPTER 23

Automated Self-Interviewing and the Survey Measurement of Sensitive Behaviors

Charles F. Turner
Research Triangle Institute

Barbara H. Forsyth
Westat

James M. O'Reilly, Phillip C. Cooley, Timothy K. Smith, Susan M. Rogers, and Heather G. Miller
Research Triangle Institute

23.1 NOMENCLATURE: THREE FLAVORS OF CASI

The early 1990s witnessed the first ripples of what is likely to become a tidal wave of surveys using automated self-interviewing systems. Those systems currently fall into three main groups: video computer assisted self-interviewing (CASI), audio-CASI, and telephone audio-CASI (T-ACASI).

23.1.1 Video Computer Assisted Self-Interviewing

In video-CASI, respondents view the questions on a computer screen and enter their answers using the computer keyboard. This technique has precursors from the age of large mainframes and minicomputers (e.g., Evan and Miller, 1969), and it provided the earliest demonstration that automated self-inter-viewing might reduce underreporting of sensitive or stigmatized behaviors (Waterton and Duffy, 1984). Video-CASI, using laptops, has been explicitly tested in several methodological studies (e.g., O'Reilly et al., 1994; Tourangeau

Computer Assisted Survey Information Collection, Edited by Mick P. Couper, Reginald P. Baker, Jelke Bethlehem, Cynthia Z. F. Clark, Jean Martin, William L. Nicholls II, and James M. O'Reilly.
ISBN 0-471-17848-9 © 1998 John Wiley & Sons, Inc.

and Smith, 1996); video-CASI also is available as an alternative interviewing mode in most audio-CASI systems.

23.1.2   Audio Computer Assisted Self-Interviewing

This technique adds audio features to those provided by video-CASI. Using laptop computers, respondents listen to questions through headphones and enter answers by pressing labeled keys. The recorded audio component has voice quality sound; it does not rely on synthesized voices and presents no significant delays in playing back audio-delivered questions. Early researchers and developers saw this technology as a way of providing complete privacy to *all* respondents (including those with poor reading skills) when reporting on sensitive topics. Although audio-CASI bears a superficial resemblance to early attempts to use a Sony Walkman to read survey questions (Camburn, Cynamon, and Harel, 1991), it is in fact fundamentally different. Because audio-CASI is computer controlled, it is capable of executing skip patterns, checking for out of range responses and inconsistencies across similar questions, and generating electronic data.

Today the use of this technology is experiencing explosive growth. A number of large-scale probability surveys using audio-CASI have been tested in the field (e.g., Mosher and Duffer, 1995; Udry, 1995; Turner et al., 1996a). In addition the federal government recently announced that its National Household Survey on Drug Abuse (NHSDA) would be converted to audio-CASI.

23.1.3   Telephone Audio Computer Assisted Self-Interviewing

In 1995 audio-CASI was adapted for use in telephone surveys, thus providing a promising new mode of survey administration. The technique was developed by adding telephone capabilities to a standard audio-CASI system. A successful T-ACASI pilot study has been completed, replicating the National AIDS Behavioral Survey (Catania et al., 1994) and comparing standard telephone survey administration with T-ACASI administration (Turner et al., 1996c). In addition implementation of T-ACASI in a large-scale methodological experiment (Turner, Miller, and Catania, 1995) began in November 1996 with the survey of a large sample of gay men (target $N > 6000$) residing in major cities in the United States. A parallel experiment is under way in the winter of 1997–1998 using a probability sample ($N = 3000$) drawn from the telephone accessible population ages 18 to 49 in the United States.

23.2   THEORETICAL ADVANTAGES OF CASI TECHNOLOGIES

Automated self-interviewing differs fundamentally from other technologies that have automated aspects of survey procedures, such as CAPI and CATI. In those applications, the respondent-interviewer interaction was not fundamentally altered by the introduction of the new technology. In CAPI and CATI surveys the measurement process still relies on an interviewer asking questions and a respondent providing answers. Thus, not surprisingly, there is only weak and inconsistent evidence that these technologies affect the willingness of respondents to provide accurate responses to survey questions. Of course, the new technologies may have altered the error structure of *some* measurements.

All CASI methods remove or minimize the role of the human interviewer in the measurement process, a change that can have numerous effects. The key role of the interviewer in survey research has been well-recognized in the social sciences for more than half a century (e.g., Hyman et al., 1954). Indeed it has been argued that all survey data arise from a complex interpersonal exchange, that thus they embody the subjectivities of both interviewer and interviewee, and their interpretation requires a multitude of assumptions concerning, among other things, how respondents experience the reality of the interview situation, decode the "meaning" of survey questions, and respond to the social presence of the interviewer and the demand characteristics of the interview (Turner, 1981). By removing the requirement that respondents divulge sensitive, stigmatized, or counternormative behaviors to another human, CASI procedures may substantially reduce the extent to which response accuracy for such measurements is compromised by the social presence of the human interviewer.

Audio-CASI and T-ACASI technologies also offer other important methodological advantages over standard survey methods, such as interviewer-administered questionnaires and paper self-administered questionnaires (SAQs). Most notably, audio-CASI and T-ACASI:

- can be used with any respondent who can hear and speak — without the requirement of literacy in any language;
- may permit efficient multilingual administration of surveys without requiring multilingual survey interviewers;
- offer the traditional advantages of computer assisted survey technologies (i.e., computer controlled branching through complex questionnaires; automated consistency and range checking; automatic production of data files; etc.); and
- provide a completely standardized measurement system — every respondent (in a given language) hears the same question asked in exactly the same way.

23.3   IN-PERSON AUDIO-CASI

The earliest video-CASI systems were designed for mainframes and minicomputers (Kiesler and Sproul, 1986; Griest, Klein, and Erdman, 1976; Lucas et al., 1977; Waterton and Duffy, 1984; see also Saris, Chapter 21). With the advent of personal computers (PCS) in the early 1980s came proposals for the development of PC-based video-CASI (e.g., Dubnoff, Kiesler, and Turner, 1986). At about the same time the growing epidemic of the human immunodeficiency virus (HIV) generated a widely recognized argument for better measurements of sexual and other sensitive behaviors (see Turner, Miller, and Moses, 1989; Miller, Turner, and Moses, 1990).

CASI measurements have emerged as a methodological response to the need for improved data on such behaviors. The addition of audio to the CASI interviewing mode was accomplished in 1991 by groups at the University of Michigan and Research Triangle Institute (RTI) (see Johnston and Walton, 1995; O'Reilly et al., 1994). Development of RTI's audio-CASI interviewing system was stimulated by the suggestion of David Celentano that a "voice-administered CAPI" interview might reduce measurement bias in the collection of data on AIDS-related behaviors from samples of intravenous drug users, many of whom possess limited literacy skills (Project Light, 1991; Turner, 1991).

Early tests established the feasibility and advantages of using audio-CASI to administer complex questionnaires in personal interview surveys. The RTI audio-CASI pilot test conducted in 1991–1992 indicated that the technology was stable and could be used with a minimum of disruption to typical survey and research routines (O'Reilly et al., 1994). Soon thereafter Turner, Lessler, and Gfroerer (1992a) recommended testing the use of audio-CASI in the NHSDA. The first small-scale methodological tests embedded within the context of that survey were not completed until the winter of 1996, and the results are not yet publicly available. Nevertheless, a variety of other evidence accumulated since 1992 demonstrates the soundness of the above recommendation. In the next section we present some examples derived from our research program. Further data are presented by Tourangeau and Smith (Chapter 22).

In January 1995, RTI's in-person audio-CASI technology was field tested in two major national surveys: the National Survey of Family Growth, or NSFG (Cycle V: $N = 10,000$ females, ages 15 to 44) and the National Survey of Adolescent Males, or NSAM (new cohort: $N = 1741$ males, ages 15 to 19). Preliminary data from the surveys indicate that field interviewers and survey respondents had few problems accepting the audio-CASI technology and that audio-CASI produced substantial increases in the reporting of sensitive behaviors.

23.3.1 National Survey of Family Growth

The 1995 NSFG collected repeated measurements using interviewer-administered questioning (IAQ) and audio-CASI. Women in the 1995 NSFG first reported on their history of abortions, number of sexual partners, and a wide variety of other topics (e.g., history of pregnancies and contraceptive practices) to a female interviewer, who used CAPI technology to record their answers. In a second phase of the interview, a small number of the same questions were repeated, but this time the respondent used the audio-CASI technology. Thus all of the women who completed the NSFG's face-to-face IAQ (10,847 respondents) had a second opportunity to provide responses in the audio-CASI mode, which we hypothesized would encourage more complete reporting of sensitive events such as abortion.

Preliminary analyses of the data from the 1995 wave (Miller et al., 1997) confirm that among sexually active women, more women reported a history of abortion in the more private audio-CASI mode than in the interviewer-administered mode (see Table 23.1). Of the women who did not report an abortion in the original IAQ mode, 4.5 percent reported one or more abortions in the audio-CASI mode. (All percentages are weighted to account for variation in the probability of selection and nonresponse.) The preponderance of discrepant reports of abortion across modes occurred in the direction of increased reporting with Audio-CASI. Thus, among women who reported one abortion in the IAQ mode, only 1.8 percent decreased the number of abortions they reported while 5.8 percent increased the number in audio-CASI. It should also be noted that the effect of audio-CASI was greater among black women than among whites: 7.3 percent of black females who reported no abortions in

Table 23.1. Estimates of Percent of U.S. Women Who Have Had One or More Abortions by Mode of Questioning and Race (Estimates for Sexually Active Women Ages 15 to 44 Interviewed in the 1995 NSFG)

Original Report in IAQ	Subsequent Report in Audio-CASI			Unweighted N
	No Abortions	1 Abortion	2+ Abortions	
All women				
No abortions	95.4	3.5	1.0	7,827
1 abortion	1.8	92.5	5.8	1,265
2 abortions	0.4	2.2	97.4	582
White women				
No abortions	95.8	3.3	0.9	5,675
1 abortion	1.1	93.9	5.0	850
2 abortions	0.6	2.5	96.9	343
Black women				
No abortions	92.8	5.3	2.0	1,742
1 abortion	4.4	85.3	10.3	355
2 abortions	0.0	1.8	98.2	217

Source: Miller et al. (in press).

Note: Of 9674 sexually active respondents who were interviewed in the IAQ mode, 2121 reported no history of pregnancy, and consequently interviewers did not administer the questions concerning abortion. In this analysis those women are considered as reporting no abortions. All women, regardless of pregnancy history, received questions on abortion in the audio-CASI mode. From this analysis 230 cases were excluded due to missing data.

the IAQ mode reported one or more abortions when using audio-CASI; for white participants the comparable percentage was 4.2 percent. In addition 10.3 percent of black females reporting one abortion in the IAQ mode reported two or more abortions in the audio-CASI mode, compared with 5.0 percent of white women. Overall, the odds of reporting an abortion were approximately 1.3 times greater when information was collected using the audio-CASI technology rather than an IAQ.

We suspect that these NSFG results understate audio-CASI's actual effect. Because of the design of the NSFG experiment, all women completed the IAQ before the audio-CASI reinterview; thus the order of presentation was not randomized. If respondents were motivated to provide consistent responses — or to avoid admitting that they had deceived the human interviewer — we would expect the audio-CASI mode to have less effect on the results than might have been obtained in a simple randomized experiment. In that kind of design, respondents would experience no pressure to provide consistent responses.

The second major audio-CASI survey conducted in 1995, the National Survey of Adolescent Males, used such a simple randomized design — but with a quite different population.

23.3.2   National Survey of Adolescent Males

Since 1988, NSAM has tracked the sexual, contraceptive, and HIV risk-related behaviors of a national probability sample of U.S. men ages 15 to 19 in 1988. A follow-up of the original cohort was conducted in 1991. In 1995 a third round of interviews with the original cohort, as well as interviews with a new cohort of young men ages 15 to 19 in 1995, was carried out.

In the 1988 and 1991 rounds of NSAM data collection, behavioral and demographic data were collected in face-to-face interviews. Questions about sensitive behaviors, such as use of illicit drugs, male-male sexual contacts, and violence were presented in a self-administered questionnaire that followed the interview and that the respondent sealed in an envelope before returning it to the interviewer (Sonenstein, Pleck, and Ku, 1991). The data contain some perplexing puzzles, the most troubling of which involve measurements of male-male sexual contacts, the most common mode of HIV transmission during the 1980s. The estimates are considerably lower than one might predict, based on the retrospective reports of national samples of adult men (see Turner, Danella, and Rogers, 1995).

Concerns about potential underreporting bias in the 1988 and 1991 rounds of NSAM and the desire to increase the actual and perceived privacy of the interview prompted the addition of a methodological experiment to the 1995 round (Turner et al., 1996a). Respondents were randomly assigned to receive the most sensitive NSAM questions in either the format employed in 1988 (personal interview followed by self-administered questionnaire) or the audio-CASI mode. Preliminary results are available from the first 928 respondents, or approximately one-half of the total sample. These are reproduced in Table 23.2.

Table 23.2. Percentages Reporting Different Types of Male–Male Sexual Contact, and Odds Ratios, in a National Sample of Males Ages 15 to 19, by Mode of Data Collection (Self-Administered Questionnaire or Audio-CASI)

Measurement	Estimated Percentage		Odds Ratio	*p*
	Paper SAQ	Audio-CASI		
Ever masturbated another male	1.1	2.3	2.07	0.29
Ever been masturbated by another male	0.6	3.0	5.44	0.03
Ever had insertive oral sex with another male	0.6	2.5	4.42	0.07
Ever had receptive oral sex with another male	0.6	2.1	3.67	0.13
Ever had receptive anal sex with another male	0.0	1.2	—	0.05
Ever had insertive anal sex with another male	0.6	1.6	2.93	0.23
Any male–male sex	1.1	4.7	4.26	0.01
(*N*)	(176)	(728)		

Source: Turner et al. (1996a).

Notes: Preliminary (unweighted) results from the 1995 NSAM. The *p*-values are for likelihood ratio chi-square for fit of independence model to the two-way table of mode by reporting of behavior.

Table 23.2 shows the percentage of respondents who reported engaging in each of six types of male-male sexual contacts. The final line of the table shows the results for a composite measure that indicates whether the respondent reported at least one type of contact. As the table shows, there were substantial and statistically reliable differences in the reports given in the audio-CASI interview and the paper and pencil SAQ. Indeed respondents were more than four times more likely to report some male-male contact in the audio-CASI interview. Those preliminary results suggest that audio-CASI reduced the underreporting of male-male sex that occurred in the prior NSAM surveys.

23.4 TELEPHONE AUDIO-CASI

23.4.1 Motivation for Implementing Audio-CASI in Telephone Surveys

Because of the substantial cost of sending field interviewers to tens of thousands of households across the country, some of the largest surveys of HIV-related and other sensitive behaviors have used telephone survey techniques (e.g., Catania et al., 1992; ACSF Investigators, 1992). Large samples are particularly important in such surveys because they permit analyses of sub-populations that are crucial to understanding the dynamics of the HIV epidemic (e.g., the population with multiple new sexual partners in the past year). These telephone surveys have interviewed substantially larger samples than typical in-person surveys of HIV-related behaviors (e.g., Tanfer, 1993; Laumann et al., 1994).

Nevertheless, the cost advantage that argues for use of telephone survey methodology to collect sensitive data ignores the effect and importance of other factors. Researchers must also take into account that traditional telephone surveys of sensitive behaviors are vulnerable to serious reporting biases (Aquilino, 1994; Gfroerer and Hughes, 1992; Tourangeau and Smith, Chapter 22; Turner, Miller, and Rogers, in press). In an attempt to provide a technology that would reduce such biases, we extended our audio-CASI system for use in telephone interviewing.

23.4.2   Early Development Efforts

Our goal in modifying the audio-CASI technology was to enable the conduct of complex call-in or call-out telephone ACASI surveys. In a call-in survey the respondent initiates the interview by calling a number that is answered by the T-ACASI system. In a call-out survey a human telephone interviewer calls the respondent and subsequently transfers the call to the T-ACASI system. We developed a software platform that fully integrated audio-CASI and T-ACASI capabilities and that could be implemented on a wide array of hardware. In our T-ACASI system, PCs equipped with a hardware interface handle incoming and outgoing calls; as a result the administration of a survey can be painlessly transferred from one environment to another by cloning the relevant software and digitized voice files.

We use the term *T-ACASI* for these applications both for consistency with the nomenclature for other CASI modalities and to distinguish them from the more circumscribed data collections that have been carried out with earlier touchtone data entry (TDE) systems. Early experimentation with TDE was begun at the Bureau of Labor Statistics (BLS) during the late 1980s (Werking, Tupek, and Clayton, 1988; Phipps and Tupek, 1990). These early TDE applications have been limited to quite simple data collection tasks, typically involving only 5 to 10 questions asked without complex skip patterns or other tailoring of the survey instrument (see Weeks, 1992).

23.4.3   Major Effects

Our T-ACASI system became fully operational in February 1995. Testing began with a pilot study using the questionnaire from the National AIDS Behavioral Survey (Catania et al., 1992). The study used a cross-over experimental design to test respondents reactions to the new technology (see Turner et al., 1996 for a full description). Half of the sensitive NABS questions, which dealt with heterosexual and same-sex encounters, HIV serostatus, and drug use, were asked by using standard telephone interview methods; the other half were administered with the T-ACASI technology. The order of presentation was balanced across the experiment. At the end of the interview, a human telephone interviewer queried respondents about their experience with each mode of interviewing.

The pilot test not only evaluated the feasibility of the new T-ACASI technology, but it also tested two hypotheses: (1) whether respondents would feel more comfortable reporting sensitive sexual behaviors to a computer compared with a human interviewer and (2) whether they were more likely in the T-ACASI mode to report engaging in stigmatized or sensitive behaviors and less likely to report normative behaviors than in the standard telephone interview mode. The quota sample for the experiment comprised people 18 to 49 years of age recruited either from households with listed telephones in Cook County, IL, or from the Wake County Sexually Transmitted Disease (STD) Clinic in Raleigh, NC. Quotas were set to produce approximately equal numbers of males and females in each stratum. Preliminary data from the first 142 cases in the study are encouraging and suggest that T-ACASI surveys are not only feasible but that they may improve the quality of data on sensitive topics (Turner et al., 1996c).

Touchtone phone ownership was quite common among respondents, a finding that supports the feasibility of T-ACASI. Data on touchtone ownership from the entire sample indicated that among the 306 screened households that reported having an adult ages 18 to 49 and that also reported on the types of telephones in their households, 302 (99%) said they had a touchtone phone. Touchtone telephone ownership was considerably less common among elderly households (i.e., those without an 18- to 49-year-old). Only 153 of 203 households (75%) in that group reported having a touchtone phone.

Secondary analysis of data from a national telephone survey conducted in 1993 (Frankovic, Ramnath, and Arnedt, 1994) indicate a similar pattern for the telephone accessible portion of the U.S. population. Approximately 90 percent of respondents ages 18 to 49 reported that they were speaking from a touchtone phone. For 85 percent of respondents in this age group, the interviewer verified that the phone was *currently switched* to generate the tones required for T-ACASI. In 5 percent of cases, the phone appeared to be switched to generate pulses; in those instances the telephone mode selector would have to be reset for a T-ACASI interview. Touchtone telephones, however, are less widely available among people age 50 and older, and in both age groups they are less widely available among people with lower levels of income and education.

Also supporting the finding of T-ACASI's feasibility was the stability of the system and the relative ease with which interviewers used it. In general, respondents were enthusiastic about the system as well. Preliminary results from questions to respondents about their experiences with T-ACASI and standard telephone interviewing indicated that respondents thought T-ACASI was better at protecting privacy, provided a more comfortable environment for answering sensitive questions, and was more likely to elicit honest reporting of sexual and drug use behaviors.

Even with the small sample sizes available for analysis (maximum n's for each condition were 79 and 67), the preliminary data from this study indicated significant differences, or differences bordering on significant, in the responses given to many of the most sensitive questions asked in the survey (see Table 23.3). For example, among respondents who reported engaging in anal sex during their lifetime, preliminary data indicated a 17 percentage point difference between the two modes of administration (25.4% for interviewer administered questioning vs. 42.0% for T-ACASI). That result appears to be attenuated in the full study; however, Table 23.3 shows other results indicating that, compared with standard telephone interviewing, T-ACASI substantially increased the likelihood that respondents would report sensitive behaviors or experiences.

Although such preliminary results from a small sample must be approached with caution, they suggest that subjects prefer T-ACASI for answering sensitive questions. They also suggest that T-ACASI increases the likelihood that subjects will report sensitive behaviors and decreases the likelihood that they will overreport normative ones. Of course much work remains to be done to secure a broader range of evidence on the effects of T-ACASI on respondents' reporting. Part of that effort is likely to take account of the cognitive demands of such new technologies present.

23.5 COGNITIVE ASPECTS OF AUTOMATED SELF-INTERVIEWING

With the rapid development of enhanced CASI technologies, including audio-CASI and T-ACASI, early methodological work has generally focused on preliminary questions of feasibility and general data quality (O'Reilly et al., 1994; Turner et al., 1996b, 1996c; Hendershot et al., 1996; Tourangeau and Smith, 1996, Chapter 22). In these early stages less attention was given to the new cognitive demands that accompany the technologies. In addition CASI technologies are frequently used for surveys about highly sensitive or stigmatized behaviors in research that focuses on the cognitive demands of the question answering tasks rather than the cognitive demands that are unique to the data collection technologies.

Because of the limited research and knowledge base and the tentativeness of the conclusions that can be drawn, we present below—in a very abbreviated fashion—some of the emerging findings from the limited cognitive research that has been done on CASI interviewing.

Offering both audio and video questioning modes improves respondent acceptance. O'Reilly et al. (1994) found that respondents rated audio-plus-video CASI as easier to use and more interesting than the video-only mode. Rogers

Table 23.3. Estimates of the Prevalence of Sensitive Behaviors Obtained from Telephone Interviews Using Human Interviewers and Telephone Audio-CASI (Preliminary Results)

Measurement	Estimated Percentage		Odds Ratio	p
	Human Interviewer	T-ACASI		
Anal intercourse				
Ever had anal intercourse	25.4	42.0	2.13	0.03*
Had anal intercourse in past 6 months	3.0	12.0	4.43	0.03*
Oral sex				
Given oral sex (since age 18)	79.7	79.5	0.99	ns
Received oral sex (since age 18)	89.8	89.0	0.92	ns
Limited sexual experience				
Had no sex partners since age 18	1.6	7.6	4.93	0.09
Had no sex in last 5 years	4.8	11.4	2.53	0.15
Did not have sex in past 6 months	1.5	8.0	5.74	0.01*
Had sex fewer than 10 times in past 6 months	22.7	41.3	2.51	0.01*
Condom use				
Never used a condom in lifetime	8.1	18.4	2.57	0.07
Used condom every time had sex in past 6 months	14.8	6.8	0.42	0.14*
Used condom almost every time had sex in past 6 months	27.8	15.9	0.49	0.14*
Stability and quality of relationships				
Most recent sexual relationship lasted less than 6 months	5.8	21.3	4.42	0.01
Never discussed sex life with most recent partner	1.9	14.8	8.83	0.03*
Discussed sex life less than once a month	28.8	49.2	2.39	0.03*
Ever had a one-night stand since age 18	59.0	64.4	1.26	ns
(Approximate N)	(67)	(79)		

Source: Turner et al. (1996c).

Note: p-value marked with asterisk do not apply to test of individual odds ratios but are derived from testing of multi-category response distribution for a mode effect of trend (see source for details).

et al. (1996) found similar results for audio-plus video CASI compared with audio only. These results suggest that offering multiple CASI modes enhances respondents' favorable reactions to CASI methods. In cognitive testing with audio-plus-video CASI administration for NSAM and the NHSDA, respondents had the option of turning off either the audio or the video presentation. In both studies all cognitive interview participants completed interviews with both the audio and the video presentations turned on. During debriefing interviews several cognitive interview participants indicated that they thought the audio presentation was unnecessary for them. However, none of them exercised the audio-off option, which suggests that the redundant mode was not sufficiently annoying to cause respondents to turn it off.

Even in sex surveys, the gender of the voice is unimportant. Most of our methodological studies have involved administering potentially sensitive items about sexual behavior. Thus we hypothesized that the gender of the audio-CASI voice might affect how respondents reacted to the audio-CASI interview and how they answered audio-CASI items. In cognitive testing for NSAM, we used three voices to present the audio-CASI interview: one male voice and two female voices. Follow-up debriefing items asked respondents about their reaction to the audio-CASI voice they had heard, and most respondents were unable to report whether their interview was administered by a male or a female. Rogers et al. (1996) included voice gender as an experimental factor in their methodological study and found no effects of voice on either ratings of the audio-CASI methods or on survey response distributions, even when controlling for effects of the respondent's gender and race.

Respondents will back up to change answers during CASI interviews, but backing up is likely to occur less frequently than in interviewer questioning. Rogers et al. (1996) calculated how often respondents repeated questions or backed up to re-hear earlier questions. Overall, a small proportion (3.2%) of survey questions were repeated; that is, very few respondents listened to the question or a portion of the question again before entering a response. Not unexpectedly, respondents repeated questions more often in the audio-only than in the audio-plus-video mode. In either mode, respondents backed up over less than 1 percent of the questions one or more times. In T-ACASI, respondents rated interviewer questioning as overwhelmingly superior to T-ACASI for changing answers (Turner et al., 1996c).

Elderly people, individuals who are unfamiliar with computers, the less educated, and some other segments of the population may hesitate to participate in a CASI interview. However, most people in those segments appear able to complete interviews and to provide reasonably reliable data. Couper and Rowe (1996) have reported that interviewers read questions or entered answers for a substantial proportion of respondents (21%) in the video-CASI segment of a 1992 Detroit area survey. Even among respondents with "prior computer experience," interviewers provided assistance in 16 percent of cases. Interviewers in the study had discretion in providing such assistance, although it was intended that "these options were to be used as a last resort in the case of refusal or incapacity" (p. 92).

In contrast to that experience, the NSFG and NSAM did not permit such direct interviewer involvement, and there were only a handful of cases in which an audio-CASI interview could not (or would not) be completed by a respondent. Similarly Hendershot et al. (1996) reported relatively little difficulty in conducting multilingual audio-CASI interviewing with a sample of elderly, monolingual Koreans (mean age of 71.5 years)—even though the field interviewers who conducted the survey spoke no Korean themselves. The addition of the audio mode might explain the observed divergence from Couper and Rowe's results; however, Saris (personal communication, 1996) reports similar findings for his video-CASI surveys, which include interviewing of the elderly. These results lead us to suspect that allowing interviewers to use their discretion in providing assistance may transform signs of respondent "hesitancy" into interviewer perceptions that "help" is necessary.

The foregoing conclusions should probably be treated as interim speculations. Relatively little is known with certainty about the cognitive aspects of automated self-interviewing, and much work remains to be done.

We are currently conducting a broadly focused program of research on the applications and effect of audio-CASI and T-ACASI on survey and other measurements of sensitive behaviors. Among the activities under way are a number of experimental comparisons of these techniques in both general samples and clinic studies. Over the next several years we hope that this research program and those of other investigators will provide a firmer empirical foundation than now exists for drawing inference about the cognitive aspects of CASI and its effect on the quality of survey and other measurements of sensitive behaviors.

23.6 HARDWARE AND SOFTWARE CONSIDERATIONS

The audio-CASI systems currently used in survey research have been developed by a few large survey organizations to meet the needs of their own research programs and those of major clients. These organizations have drawn on internal programming staff with specialized skills to develop either an audio-CASI subsystem for their existing computer assisted interviewing (CAI) package or a custom audio-CASI interviewing system. In the following sections we describe how audio-CASI systems have been designed and implemented. We also discuss some of the technical factors that make DOS implementations of audio-CASI inferior to Windows-based implementations, and we consider how future Windows-based CAI packages will offer audio and other multimedia capabilities.

23.6.1 In-Person Audio-CASI

In 1996 virtually all audio-CASI interviewing in survey research was based on DOS platforms despite the difficulty of implementing audio and other multimedia features in DOS. In the DOS environment the interfaces between the audio component and the central processing unit are achieved by using audio-specific device drivers. However, the design of the drivers that operate on DOS platforms is based on Intel 16-bit architecture assumptions, whereas the CAI systems currently available are based on Intel 32-bit architecture. As a result compatibility between the audio and CAI parts of the audio-CASI system can be directly achieved only by developing a CAI routine that does not utilize the full capabilities of the 32-bit architecture used by new personal computers.

In order to use audio routines from CAI systems, researchers have developed two distinct methods: terminate and stay resident processes (TSRs), which are invoked by a "hot key" or a software equivalent, and external DOS-executable routines, which are invoked by system calls from the CAI software. In both methods, either an external custom audio device connected to the PC's parallel port or an internal chip added to the PC's motherboard provides the audio capability.

This rather cumbersome and somewhat fragile architecture in DOS audio-CASI systems is unlikely to change. Audio services are most widely used on PCs in conjunction with multimedia applications. In the DOS environment audio and other multimedia are implemented only in a limited, specialized manner, that is, for computer games. However, the success of Windows in the early 1990s created what has become a growing consumer market for multimedia. Today most vendors of multimedia devices and software tools have shifted their business focus almost entirely to Windows, which means that audio device drivers are now based on Intel 32-bit architecture assumptions.

Although the mass market has moved strongly to Windows, survey research organizations continue to use DOS for the great bulk of their interviewing systems, in large part because the CAI systems they employ were, and continue to be, DOS-based. A second reason, we suspect, is that prior to the development of audio-CASI, there was no serious claim that a Windows CAI application would add significant value to the automated interviewing process.

23.6.2   Telephone Audio-CASI

In T-ACASI the sound card used for audio-CASI is replaced by a telephony card. That change in hardware and associated software makes it possible to ask questions using audio sent over the telephone line and to collect a respondent's answers to those prerecorded questions by decoding the tones generated when the respondent presses the keys on a touchtone telephone. The telephony card provides audio features for T-ACASI applications in a manner that is completely analogous to the functioning of the sound card in audio-CASI applications. Similarly, in T-ACASI, the telephone keypad replaces the PC's keyboard. There are, however, obvious differences between T-ACASI and audio-CASI applications that arise from the distinct technological features of the two media. For example, T-ACASI keypad responses are limited to 10 digits (0 to 9) and two special characters (# and *). In addition, T-ACASI systems have no special screen handling needs.

Automated touchtone response systems, known as interactive voice response (IVR) systems, have proliferated in the past two years and are used in a large number of commercial applications. Many vendors market such systems. A major difference between IVR and T-ACASI systems is that the latter are designed to support complex survey operations, and they have at their core a CAI routine that manages the survey component of the application. In contrast, IVR systems use relatively simple data collection instruments to capture limited amounts of data from a large number of respondents. Consequently the design of T-ACASI systems focuses on maintaining the survey processing features of the CAI component of the package, whereas the design of IVR systems focuses on handling many respondent calls simultaneously (i.e., on supporting channel multiprocessing features).

For T-ACASI systems implemented under DOS, which is a single-user operating system, the multitasking features have to be built directly into the IVR software. Consequently event-based programming procedures must be used to develop channel multitasking capabilities. That constraint is a fundamental problem for T-ACASI designs. T-ACASI systems have a more complex CAI core than IVR systems; the core frequently supports a separate language (with its own grammar and syntax) to describe edit and skip logic. Such systems are more difficult to convert from a procedural style of programming, which is typical of DOS-based applications, to an event-based style typical of Windows-based applications. Hence, on the one hand, the migration of T-ACASI applications to Windows is not a trivial task. On the other hand, moving the T-ACASI application to Windows enables T-ACASI to make use of a channel multitasking environment. That means that different T-ACASI processes can use different channels on the same telephony card simultaneously — in other words, the same machine can handle multiple interviews. (Such simultaneity cannot be achieved by running multiple DOS-based T-ACASI applications on a Windows 95 machine.)

Recently we moved both IVR and T-ACASI applications to a Windows NT platform, and the performance of the two applications improved significantly compared with the DOS versions. In addition the multitasking capabilities of the NT operating system, though not totally free of bugs, have been successfully demonstrated in both IVR and T-ACASI applications. Improvements in the NT drivers are expected to resolve the few remaining problems.

23.6.3 Emerging Availability of Audio-CASI in Standard Systems

Currently three major CAI system vendors (CASES, Blaise, and Surveycraft) are developing Windows versions of their DOS-based CAI systems. Many market research software vendors have already made the move to Windows. Because multimedia features are available in the languages and systems that reside on Windows platforms, one can expect that audio-CASI capabilities will soon become standard. In the meantime some vendors are developing audio-CASI capabilities for existing DOS-based systems.

Our experience in developing audio-CASI systems in DOS and Windows suggests that the difficulties in implementing a general-purpose CAI system with audio features in the Windows environment should not be under-estimated. A key design consideration is that the system's target user must be assumed to have little or no computer experience and only a few minutes of training before being let loose to complete a complex interview. That demand-ing usability requirement means the system must provide a high level of speed, simplicity, and robustness along with the desired graphical and audio features (Cooley et al., 1996; Cooley and Turner, 1996). The system must also be able to work well with people who may never have used a computer or even a keyboard. Thus the system must permit respondents to engage the instrument with full audio accompaniment without confusion, delays, or intrusions that disrupt their concentration. Even small extra demands on the respondent may undercut support, participation, and data quality.

23.6.4 External Sound Device versus Built-in Audio Capability

Cooley et al. (1996) present a thorough review of technical issues and document their experience in implementing audio-CASI for PC laptops. They conclude that for (non-Windows) DOS-based audio-CASI applications using laptop PCs, the only viable options are the use of one of a limited but increasing number of laptop PCs with built-in Sound Blaster-compatible audio, or the use of an external audio device connected to the laptop's parallel printer port. PCMCIA cards are not feasible for DOS platforms because the audio interface is typically designed for Windows.

Both built-in audio and external devices provide equivalent functionality including acceptable voice quality audio and performance that is adequate for most low-end PCs. The main differences between these options are the comparative level of their reliability and their effects on the field interviewing process. External devices add complexity, increase weight, and take time to set up. Thus an interviewer in a respondent's home must not only set up the laptop but also attach the audio device to it, although extensive and demanding field experience of the 1995 NSFG and NSAM indicates that external audio devices are a minor inconvenience that field staff are usually able to handle relatively easily. Indeed, in the NSFG pretest and main study, RTI field staff reported infrequent malfunctioning of the audio devices and few difficulties in managing the equipment (Kinsey et al., 1995).

In NSAM the same equipment worked reliably after a problematic start-up phase with older model laptops. Interviewers did, however, complain that the external audio equipment was somewhat cumbersome to use. After the early spate of equipment problems, those complaints dwindled. Nevertheless, survey managers reported that they preferred laptops with built-in audio capability.

The National Opinion Research Center reported some significant problems using another external audio device. In the Adolescent Health Study, Grilley, Kean, and Nichols (1996) cited the following:

- Because of malfunctions, an estimated 10 percent of the field interviewers probably did not use the audio equipment.
- Twelve percent of documented support calls to the central office "involved problems with the programming and hardware associated with the Audio-CASI portion of the interview."
- The audio device had to be carefully screwed into the parallel port before raising the laptop's cover, or the device would not function properly in the audio-CASI section of the interview, which occurred after a length CAPI section.
- Interviewers had to be careful not to attach the audio device too tightly or they would be unable to remove it at the end of the interview.

Those experiences suggest that a number of factors determine whether an external audio device is robust and reliable, including the design of the device itself, the laptop, the software, organizational experience, and so forth. As organizations gain experience, serious difficulties are reduced to a low level. Overall, despite problems and shortcomings, external audio devices have established a strong, successful track record in demanding field situations. Therefore, although internal sound chips promise the highest level of reliability and ease of use, at least for the interim, external sound devices can be a capable alternative.

23.6.5 Prospects for General Audio-CASI Availability

The key factor in determining general audio-CASI availability will be the arrival of general-purpose CAI software for the Microsoft Windows and Windows 95 graphical environment. Our experience in transporting a CAI system from DOS to Windows 95 demonstrated the difficulties of moving a sophisticated software package from a linear procedural design to an event-based design (Cooley and Turner, in press). Nevertheless, the resulting benefits proved well worth the effort. They included true 32-bit performance without any of the side effects associated with accessing external audio processes, a single program that works with all Windows-compatible sound systems, and flexible screen handling characteristics such as font and bitmap manipulation. All of those capabilities are now available without any apparent loss of performance.

We can look to such Windows-based CAI systems for other features as well. They should be able to interface easily with virtually all types of audio devices, including built-in audio chips, PCMCIA cards, and external parallel port devices. Almost all PC laptops sold since 1995 are capable of running such software. The key remaining obstacle is the release of Windows versions of CAI systems, such as CASES, Blaise, and Surveycraft, that offer the other features required for large-scale complex surveys. Given the actual or imminent availability of suitable software and hardware, plus the growing body of evidence that audio-CASI is a powerful, if not essential, survey research method for addressing sensitive topics, much wider adoption of audio-CASI can be anticipated in coming years.

23.7   FUTURE DEVELOPMENTS

The growing use of CASI — and of audio-CASI in particular — has sometimes led us to speculate that these technologies could well revolutionize the way all types of social measurements will be made in the future. Key elements of that speculation will, we believe, stand the test of time. Nevertheless, the more general claim is likely to be found wanting. The ultimate result for those concerned with social measurement may still be dramatic, but the drama will play itself out in scenes and with nuances that are only vaguely anticipated or appreciated today.

The great rush of enthusiasm that fueled the adoption of in-person audio-CASI interviewing in research (our own and others) has arisen in part from the fact that the technology elegantly solved a problem of great import for *some* areas of the social and statistical sciences. In particular, the idea of audio-CASI interviewing emerged to meet a pressing need in the field of behavioral research on the spread of HIV — that is, a measurement method that would reduce bias in the reporting of highly stigmatized behaviors (such as male-male sexual contacts) that risked HIV transmission. The need was particularly apparent among subpopulations, such as injection drug users, for whom literacy levels were inadequate to permit confident use of paper and pencil questionnaires. Indeed that specific problem was the initial stimulus for our current program of research and development of audio-CASI interviewing systems.

In considering the application of audio-CASI technology to social measurement, we believe nonetheless that the range of areas in which one might expect large improvements in response validity is not sufficient to sustain a true "revolution" in measurement methods, nor is the target population extensive. Literacy may be a problem for a notable fragment of the U.S. population, but that fraction is considerably smaller than a majority of potential respondents. Since audio-CASI is required only for that segment of the population that is incapable of responding to written questions (or of following written instructions), the theoretical advantages of in-person audio-CASI are delimited both in terms of the target population and in terms of the content areas in which major improvements in measurement validity might reasonably occur. Furthermore, for typical in-person surveys, the economic benefits of audio-CASI, if any, will not be large because interviewers typically must wait in the respondent's home while the respondent completes the audio-CASI interview.

So, one might ask, where is the "revolution" coming from? We offer two replies. First, CASI and audio-CASI are fundamentally altering our understanding of the prevalence and patterns of certain hidden and highly sensitive characteristics of the U.S. population. Even preliminary data indicate that those alterations are not always occurring in ways that would have been expected. Thus the first revolutionary aspect of the adoption of these technologies is the substantial rethinking of the realities we and others have described in our past attempts to assess the prevalence and patterns of sexual, drug using, and other sensitive behaviors (e.g., see Catania et al., 1994; Laumann et al., 1994; Turner, 1989; Turner, Miller, and Moses, 1989; Miller, Turner, and Moses, 1990; Fay et al., 1989; Rogers and Turner, 1991; Turner, Danella, and Rogers, 1995). In that regard we would offer the additional observation that what is called the "mode effect" in survey measurements (i.e., the difference in responses obtained from in-person versus CASI questioning) is likely to become a useful instrument for judging the social sensitivity of survey questions and topics and the variation in that sensitivity over time.

The second revolutionary aspect of the use of CASI technology arises from a technological afterthought that will, we believe, induce fundamental change in the way social measurements of sensitive aspects of human behavior are carried out in the future. That afterthought was the extension of audio-CASI technology to the telephone, an almost serendipitous move that has fundamentally redefined what is possible when conducting telephone surveys. Before this merging of technologies, there was no way to offer respondents in a telephone survey a completely private mode of data collection. Although in-person surveys have long used paper and pencil questionnaires — and can now use CASI — telephone surveys have required respondents to report on even the most private aspects of their sexual and other behaviors to a human interviewer. Now the advent of T-ACASI has transformed the potential of telephone surveys by providing a fully private mode of data collection within the more economical telephone survey process. This development should allow researchers to feel more comfortable in using telephone methodologies for future surveys of sensitive behavior.

ACKNOWLEDGMENTS

Preparation of this chapter was supported by the U.S. National Institutes of Health through grants to the first author from the National Institute of Child Health and Human Development and the National Institute on Aging (grant R01-HD/AG31067-04) and from the National Institute of Mental Health (grant R01-MH56318-01). The opinions expressed in this chapter are the sole responsibility of the authors.

CHAPTER 24

Response Rates, Data Quality, and Cost Feasibility for Optically Scannable Mail Surveys by Small Research Centers

Don A. Dillman and Kent J. Miller
Washington State University

24.1 INTRODUCTION

Optical scanning has long been used to capture responses to self-administered questionnaires, but that use has been limited primarily to large surveys conducted by large survey organizations. Unless a survey received thousands of replies to the same questionnaire, manual keying of responses was clearly more economical than reading pencil marks with a photosensitive device. The technical requirements of printing and processing optically scannable surveys placed formidable limits on the situations where their use could be justified. Dramatic improvements in scanning technologies may be changing this situation. Powerful stand-alone scanners and microcomputers are now available at relatively low costs that may provide smaller survey organizations access to technologies once reserved for large organizations. At the same time optical character recognition (OCR) capabilities are improving the range of survey answers for which scanning may be used beyond the traditional optical mark recognition (OMR).

In this chapter we explore whether the changes now occurring in scanning technologies are in fact making it possible for small survey organizations to efficiently conduct small-scale surveys using scannable questionnaires. We are interested in whether an organization that conducts no more than 15–20

Computer Assisted Survey Information Collection, Edited by Mick P. Couper, Reginald P. Baker, Jelke Bethlehem, Cynthia Z. F. Clark, Jean Martin, William L. Nicholls II, and James M. O'Reilly.
ISBN 0-471-17848-9 © 1998 John Wiley & Sons, Inc.

self-administered surveys per year, each with a unique questionnaire and averaging less than a thousand respondents (ranging from perhaps 200 to 2000), can effectively convert all of these surveys from manual data entry to optical scanning, thus eliminating traditional keypunching. The figures are only illustrative. The defining element is that the organization does not have any large survey with tens of thousands of replies to depreciate the costs of scanning technology and leave it available for small surveys a significant amount of time. This is a situation faced by many private and university survey centers throughout the world, including our employer, the Social and Economic Sciences Research Center (SESRC) at Washington State University.

24.2   ORIGINS AND RECENT DEVELOPMENTS

24.2.1   Historical Development and Limitations

The earliest uses of optically scannable questionnaires were generally limited to situations where very large numbers of identical questionnaires were collected from specialized audiences, such as students or employees, with a high motivation to complete the survey request. Another early use was the U.S. Decennial Census, to which residents are legally required to reply.

The early technologies typically required that questionnaires be printed only on one side of a page. When two-sided printing was allowed, answer spaces were not permitted back-to-back on the page. Some formats limited answer choices to small areas of each page, for example, columns on the right side. All questions had to fit specified page layouts. A soft-leaded pencil was usually required for answers, and people were asked to completely darken small circles or ovals, rather than mark answers with crosses or check marks. Prominent equipment guide markers, sometimes called "skunk" marks, were necessary for scanning alignment. These marks often dominated the questionnaire's appearance, visually distracting respondents (Jenkins and Dillman, 1995, 1997). Flat mailings were preferred because folding the questionnaire interfered with machine reading of answers.

These early optical questionnaires were mostly used in controlled settings, for example, in testing students and collecting information from employees. In these situations soft-leaded pencils, instructions, and supervision could be provided. Moreover, since the respondents had much at stake in the correct interpretation of their answers, they could be assumed to mark their answers carefully.

Among the leaders in the development of optical mark recognition systems was the U.S. Bureau of the Census which developed the Film Optical Sensing Device for Input to Computer (FOSDIC) system. FOSDIC was used to count people in decennial censuses from 1960 to 1990, pioneering the use of optical questionnaires with the U.S. general public. National Computer Systems (NCS), a private organization founded in 1962, was a leader in scanning of optical mark recognition forms for student tests, personality assessments, and later for surveys.

Smaller survey organizations faced formidable barriers to using these technologies. Early scanning systems required large, mainframe computers, which were too substantial an investment for users with a small volume of questionnaires to process. The optical mark recognition (OMR) technology then available also was limited to reading the presence or absence of marks, such as filled ovals or circles. Survey items requiring words for answers had to be sent through an additional keypunching operation for these responses to be captured.

24.2.2 Recent Developments

Barriers to the use of scannable surveys may be diminishing for all organizations (large and small) for three reasons. First, scanning technologies now have a much greater capacity to record survey answers. Most surveys require that respondents write numbers, letters, or words as well as mark answer choices. Optical character recognition (OCR) capabilities have been developed making it possible to capture numbers and letters as well as marks, diminishing the need for dual processing of questionnaires. Although many OCR systems can recognize individual digits and letters with reasonable accuracy, text answers to open-ended questions cannot be read well without the letters being separated. The reading of cursive writing remains a formidable challenge to the best OCR technologies.

Second, the development of imaging technologies has fundamentally changed the structure of data capture operations. The entire image of the questionnaire is scanned into the computer for the recognition process. The recognition engines of the most sophisticated imaging software can assign a confidence level (e.g., 98%) that the image of the respondent's answer (a letter or number) corresponds to a unique character which the computer has been programmed to recognize. Entries the recognition engine cannot identify at that confidence level, called rejects or spurious marks, are separately stored and displayed on a computer screen where a human operator interprets their meaning. In this way common problems such as double marks, erasures, or attempts to obliterate an answer with a heavy mark and replace it with another, can be easily and quickly corrected. If the operator cannot determine the respondent's intent from the visual display of a specific answer, the full questionnaire page, or even the complete questionnaire, can be displayed on the screen to see the ambiguous response in context. For example, an operator attempting to decide whether a respondent meant to enter an "n" or an "r" can see how those letters were written in other answers.

Imaging technologies are dramatically changing the organization of processing procedures for self-administered forms as illustrated in Figure 24.1. In traditional keypunching operations, questionnaires were clerically preprocessed to

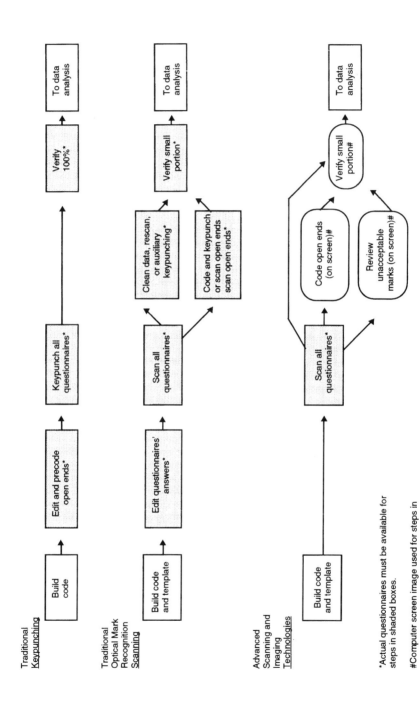

Figure 24.1. Differences in processing for three types of data capture.

Traditional
<u>Keypunching</u>

| Build code | → | Edit and precode open ends* | → | Keypunch all questionnaires* | → | Verify 100%* | → | To data analysis |

Traditional
Optical Mark
Recognition
<u>Scanning</u>

Build code and template → Edit questionnaires' answers* → Scan all questionnaires*

Clean data, rescan, or auxiliary keypunching*

Code and keypunch or scan open ends scan open ends*

Verify small portion* → To data analysis

Advanced
Scanning and
Imaging
<u>Technologies</u>

Build code and template → Scan all questionnaires*

Code open ends (on screen)#

Review unacceptable marks (on screen)#

Verify small portion# → To data analysis

*Actual questionnaires must be available for steps in shaded boxes.

#Computer screen image used for steps in rounded boxes.

identify unclear answers, to determine which of two double-marked answers to enter, to code open-ended answers, and generally to facilitate the keypunching process. The key-operator often was asked to make additional decisions while keypunching, such as entering the highest educational level if educational attainment was double-marked. To ensure keying accuracy, questionnaires typically were keypunched twice, a process known as verification. If an error was discovered after the keypunching (e.g., a person age 5 with a driver's license) the original questionnaire was retrieved to resolve the inconsistency. Thus every step of processing, from preliminary editing through verification, required access to the questionnaire document.

The processing of OMR questionnaires where respondents had darkened ovals or circles to record their answers also began with an editing process. An operator visually examined the entire questionnaire to further darken marks too light to scan and to erase or "white-out" duplicate or inappropriate marks. After a template was designed for the mark recognition engine and the preliminary editing was complete, the questionnaires were scanned. Open-ended questions and questions requiring coding (e.g., occupation) were sent through a separate process which led to another round of scanning of coded marks or to auxiliary data keying. Initial data runs (or batch editing) of the scanned data often identified double entries or other problems missed by the preliminary editing. This necessitated retrieval of the original questionnaire for corrections. As with keypunching, this process required continuing access to the physical questionnaire.

With advanced imaging capabilities, the entire questionnaire is scanned into memory at the start. The questionnaires need not be edited prior to scanning. Corrections traditionally made in preprocessing are made afterward, using the screen image of the questionnaire. The first step is to build a recognition template, after which questionnaires can be scanned at a rate of several hundred per hour. Components of the questionnaire may then be sent to different workstations where (1) all marks which have not been accepted at the specified level of confidence are reviewed by retrieving the questionnaire image to the screen, (2) open-ended answers are entered or coded from a screen image of the questionnaire, and (3) a randomly selected portion of the scored marks are reviewed for scanning accuracy. Each of these processes can occur simultaneously because none of the operators need access to the physical questionnaire. Under the most sophisticated scanning systems, it is unlikely that physical retrieval of the respondent's completed questionnaire will ever be needed once it has been scanned.

With advanced imaging, further changes in the organization of work are possible, and further improvements seem inevitable. Instead of making corrections one questionnaire at a time, corrections can be made on the same item across all questionnaires that exhibit a particular problem. This adds the potential of greatly speeding up the correction process and helping to ensure that corrections are made consistently across all respondents. Much of the potential for efficiency stems from making scanning the first step of processing and freeing the processing personnel from dependence on the physical questionnaire. Capacity can then be expanded or speed increased simply by adding to the number of computer workstations. These are major advantages for organizations processing large surveys under severe time pressures. The advantages of advanced imaging for small survey organizations remain unclear.

A third major scanning development that reduces barriers to current survey use is improved forms design permitting questionnaires that are more respondent friendly than in the past. It is now possible to design questionnaires which permit either ×'s or checks instead of darkened circles. Ballpoint pen marks can be read as well as pencil marks provided that the pen color is not similar to that of the questionnaire page. Colored drop-out inks are often used for the printing of scannable questionnaires. These inks "disappear" when a light of the same color, typically red or green, is shone on the questionnaire page. Answer choices can be placed virtually anywhere on the page, and two-sided printing is allowed. Whereas in the past, technology constraints dictated many aspects of questionnaire layout and design, this is no longer the case.

The most promising systems combine advanced imaging technologies with forms design permitting respondent friendly questionnaires. Our observations of one of the most advanced systems now available lead us to conclude that it could be efficiently used for small surveys (less than 1000 questionnaires), as well as large ones because of its ease of template building and correction efficiency. However, this system is too expensive to be cost effective for small survey organizations. The scanner alone costs over $30,000. Further developments with these technologies, especially in cost effectiveness for small-scale operations, need to be watched closely.

24.2.3   Demands for Scanning

The new developments in scanning technology are timely inasmuch as new expectations are being placed on all survey organizations, large and small. Customers typically want surveys completed more quickly than in the past; imaging the entire questionnaire at the start of processing has great potential for shortening mail survey production time. The demand for surveys with very large samples also is growing. Businesses and other organizations are increasingly managed on the basis of customer feedback not just to the total organization but also to small units, such as dealerships, within it. Many large businesses and public agencies also conduct surveys of all their employees. Survey organizations are being asked to design surveys that obtain large numbers of responses so that feedback can be provided to the smallest work units of the organization. One result of this interest in total organization feedback is that the demand for mail surveys appears to be increasing. The ability of small survey organizations to meet this expanding demand for mail surveys may depend in part on their ability to master and effectively utilize advanced scanning and imaging technologies.

A related question is whether scanning and imaging technologies may wholly replace keypunching. The advanced technologies now available to large organizations seem to offer much promise for switching their entire mail operations to scanning technologies. It remains to be seen whether these technologies offer equal promise to small organizations.

24.2.4   Do Scannable Questionnaires Elicit Lower Response Rates?

Scannable questionnaires have long had a reputation for eliciting low response rates to surveys. For example, recent research at the U.S. Census Bureau found that changing a form designed almost entirely to meet OMR scanning requirements of the Census Bureau's FOSDIC system to one based on visual principles of respondent friendly design improved response rates at least modestly (Dillman, Sinclair, and Clark, 1993). However, it is unclear whether scanning's reputation for low response rates is deserved or even applies to scanning's newer forms.

Research on improving mail survey response rates suggests that the most powerful inducers of better response rates are repeated contacts (prenotices, reminders, replacement questionnaires, etc.), prepaid financial incentives, personalization, stamped reply envelopes, and special outgoing postage (certified, special delivery, two-day priority mail, etc.) (Heberlein and Baumgartner, 1978; Dillman, 1978, 1991). These procedures are routinely used by academic survey centers for small surveys of general populations, and they have been responsible for achieving respectable response rates of 60 percent or higher for general public populations.

Initial conversations with the managers of large scannable mail surveys conducted by large private organizations suggested that they frequently did not use such response rate enhancing procedures. Surveys that could be scanned generally had very large samples conducted at low per unit cost. Their objectives, such as total organizational feedback, emphasized other considerations than response rates and nonresponse error. The use of response rate enhancing procedures would dramatically increase their per unit costs.

Only two experimental studies examining how scannable formats might influence response rates could be located. Neither obtained response rates different from traditional questionnaires. Klose and Ball (1995) sent a folded 11 by 17 inch four-page questionnaire to electric utility customers in a northern great plains state. The traditionally constructed survey obtained a 20 percent response rate compared to 22 percent for the OMR survey which required that respondents fill in bubbles with a pen or pencil. However, they also found that 26 percent of the OMR questionnaires had incomplete responses compared to 9 percent of the traditional questionnaires. Taking these data into account, they concluded that the effective response rate for the OMR questionnaire was 17 percent compared to 18 percent for the traditional questionnaire. A study by Jacobs (1986) of school teachers compared a two-page questionnaire designed for OMR with a traditional questionnaire and also reported no difference in response rate, with 81 percent responding to the OMR question-naire versus 79 percent for the traditionally constructed questionnaire.

Nonetheless, an image of low response rates persists for scannable question-naires. This image comes in large part from the "mass survey" culture associated with their use. Sending out tens of thousands of questionnaires, which encourages the use of scannable surveys, also encourages other efforts to reduce costs—bulk rate mail, lack of personalization, printing letters on the questionnaire itself, crowded formats to keep the page count down per questionnaire, no prenotices or follow-ups, no token financial incentives, and so forth. Ink colors, chosen for their drop-out value in scanning, a standard $8\frac{1}{2}$ by 11 inch page size (more efficient for scanning), and prominent scanning guide marks contribute to instant identification as some type of test or questionnaire, limited visual appeal, and a lack of respondent friendliness.

To better understand the response rates associated with scannable surveys, we systematically contacted organizations who conducted such surveys to learn about their methods and response rates. We asked about the nature of the population, the sample size, the response rate, and whether more than one mailing was used as a means of increasing response. Group administered surveys were left out of our analysis because virtually everyone who received a questionnaire filled it out or response information was not maintained. Information was provided for 44 individually administered surveys mailed from 10 sponsoring organizations. The questionnaires ranged in length from 2 pages to 28 pages.

Summary results are presented in Table 24.1. Sample sizes ranged from 100 to 4,000,000 with an overall mean of over 300,000. However, the only surveys under 1600 respondents were special situation surveys; they probably would not have been conducted had the client not already contracted for other, much larger surveys.

We divided the surveys into those of individuals (mostly customer satisfac-tion surveys), of businesses, and of employees, since different factors influence their response rates (Heberlein and Baumgartner, 1978; Paxson, 1992). For the surveys of individuals, response rates averaged 20 percent when only one contact was used and 53 percent when two or more contacts were made. Similarly employee response rates averaged 43 percent with one contact and 79 percent for two or more contacts. For surveys of businesses, extra contacts did not produce higher response rates, but only two of the eight business surveys used more than one contact. Overall, response rates for the 44 surveys averaged only 33 percent.

The data in Table 24.1 tend to confirm the association of mass delivery methods with sample size. The average sample for surveys using only one contact (about 460,000) was almost four times larger than the average sample for surveys using two or more contacts (about 125,000).

Although the response rates of scannable surveys reported in Table 24.1 tend to be low compared to those mail surveys can achieve, these data are not evidence that scannable surveys necessarily produce lower response rates.

Table 24.1. Sample Sizes and Response Rates for 44 Studies Using Scannable Questionnaires

Population	Number of Studies	Sample Size		Response rate (%)	
		Range	Mean	Range	Mean
Individuals					
1 contact	15	1017–4 million	331,559	10–60	32
2+ contacts	12	168–736,221	63,843	29–68	53
Totals	27		207,998		42
Businesses					
1 contact	6	14,715–2.3 million	986,786	14–42	23
2+ contacts	2	600,000–900,000	750,000	18–21	20
Totals	8		927,589		22
Employees					
1 contact	3	5500–300,000	160,167	34–48	43
2+ contacts	6	100–60,000	37,168	55–84	66
Totals	9		78,168		58
Table totals	44		314,702		42

Based on the details available to us, we concluded that the vast majority of these surveys did not employ many of the techniques that past research has shown effective for improving response rates. The data do suggest, that when additional contacts are made, response rates improve. This conclusion is consistent with the Klose and Ball (1995) experiment which reported that response to their OMR questionnaire doubled (from 22% to 44%) when a prenotice, reminder, and $500 sweepstakes notification were included in the treatment.

We had hoped to obtain systematic information on item nonresponse and other data quality measures for these scannable surveys, but these efforts were abandoned because such information was frequently unavailable or considered proprietary. However, conversations with the managers of the scannable surveys suggested that the "mass survey" culture often extended to processing as well as to mailing. Although a few managers reported meticulous efforts to review and correct identified errors (previewing each form to remark light entries and whiting out duplicate marks), they appeared to be in the minority. In other surveys the choice between double marks was programmed into the computer processing instructions, but minimal efforts were made in a large number of surveys to identify and correct errors. Questionnaires that would not scan due to poor marks or other reasons were simply ignored in the analysis. Their survey managers explained that time pressures to report survey results were often great and did not allow for either pre- or postprocessing steps to improve data quality. Our impression is that the "mass survey" culture associated with scanning effects both response rates and data quality.

24.2.5 Influence of Visual Design on Survey Response

If response rates or data quality are lowered by the choice of scannable questionnaires over traditional ones, graphical design may be part of the explanation. In the past, scannable surveys have been instantly recognizable, in part, by the use of colored inks. Bands of color have been used to guide respondents across the page from questions to answer choices, while successive questions may employ alternating background colors. Words printed in black on light red bands is a popular choice. Choosing colors that lend themselves well to the drop-out functions of scanning is a further help. Red lights are easily available with scanning, and few people use red pens when answering questionnaires. The use of these bright colors, small type, standardized layouts, and the common requirement that respondents completely darken circles contribute to an image of scannable questionnaires as less than respondent friendly.

Research at the U.S. Bureau of the Census has attempted to assess the effect of respondent friendly design and other variables on responses rates to decennial census questionnaires. A series of experiments revealed that improved graphical design can increase response rates but will do so only modestly. Tests of the effects of 13 techniques hypothesized to influence response rates revealed that only a prenotice, reminder, replacement questionnaire, statement that census response was mandatory, and respondent friendly design seemed to have independent effects on response. Together these variables improved response an estimated 36 percentage points, from 42 to 78 percent, but only about 3.5 of the 36 points were due to a significant improvement in respondent friendliness (Dillman, Clark, and Treat, 1994; Dillman, Sinclair, and Clark, 1993).

Additional research at the Census Bureau suggested that principles drawn from the vision sciences might help explain that difference (Jenkins and Dillman, 1995, 1997). It was theorized that a visually prominent navigational path should be established for respondents to follow when completing a questionnaire. Limited confirmation of these ideas was provided through extensive cognitive interviews (Dillman et al., 1996).

Among the principles posited as helping to define the navigational path was one concerning the figure-ground format, that is, how the foreground (usually words) is displayed on background (Jenkins and Dillman, 1997). Use of a consistent figure-ground format (e.g., black ink printed on the same background color) should make it easier for respondents to answer the questionnaire than alternating bands of color. Colored bands also may cause respondents to group answers in an inappropriate manner. These same principles suggested that the use of different colors for questions than for answers, varying colors to highlight different questions, printing special instructions in distinct colors, and making the questionnaire superficially more attractive may make it more difficult for respondents to answer. Another undesirable effect is that the design may then obscure the navigational path. To better understand the influence of visual design on response quality and the factors affecting the response rates of scannable questionnaires, we undertook an organizational experiment.

24.3 ORGANIZATIONAL EXPERIMENT

The experiment had two equally important objectives. The first was to determine whether scannable questionnaires and their associated procedures would affect survey response rates or response quality. The second was to identify the challenges and consequences of introducing optical scanning technologies into the regular operating procedures of a small university survey research center. Each year our center conducts no more than 20 mail surveys, each with a unique questionnaire and with sample sizes between 200 and 2000 questionnaires.

24.3.1 Options for the Small Survey Organization

Small survey organizations appear to have three scanning options. The first is to acquire questionnaire setup and scanning services from one of the many organizations that have now provided them for many years. These organizations can also mail the questionnaires so that in essence the contractor takes over most of the normal functions of the survey research center, except for analysis and reporting of the results.

We requested a cost estimate from one company for a general public survey with an initial sample of 1000 households, a common size for our center. The immediate advice we received was that unless the sample was much larger, it would not be cost effective either to contract with them to set up and process the questionnaire *or* to purchase the scanning equipment and attempt the scanning ourselves. Their rule of thumb was that at least 16,000 returned questionnaires were necessary to make contracting the proposed survey cost effective.

Despite this admonition we then asked the approximate cost for conducting a hypothetical survey of 10,000 people. We assumed two questionnaire mailings and a fairly high response rate of 70%. We also assumed that the questionnaire would use standard $8\frac{1}{2}$ by 11 inch pages, the cheapest to process. The amount proposed for setting up, printing, mailing, and processing the questionnaires, using only OMR, was approximately $17,000. The cost of a scanner and software from the same company to set up and process this questionnaire ourselves was estimated at approximately $9600. This included a scanner with an automatic feeder capable of processing 2000 sheets per hour. Our total costs with about $14,000 for printing and mailing would be about

$23,600 to do it ourselves. We concluded that contracting for complete setup and processing services could hardly be justified unless a very large survey was being conducted or unless the cost of scanner and software could be depreciated over many small surveys.

A second option might be to purchase one of the services, either questionnaire setup or scanning, but not the other. In some ways these costs may not be as separable as they at first seem. If one has decided to invest in a scanner, the marginal costs of obtaining software may be fairly insignificant.

A third option is to purchase everything and do the processing ourselves, and this was the alternative we chose to test. We did not feel that contracting out would meet our continuing needs in conducting small mail surveys.

24.3.2 Essential Performance Criteria for Use in Small Survey Organizations

Before purchasing the needed equipment, a concise list of performance criteria was developed through discussion with our center staff. The reasonableness of these criteria was further validated through discussions with personnel of other small survey organizations. They included:

- *Both OMR and OCR capabilities.* Virtually all mail questionnaires employed by our center in the previous five years required writing of both numbers and letters and several open-ended questions. A scanning system would not meet our needs unless it handled both: (1) answers marked by filling a bubble, marking an ×, or circling a numbered answer, and (2) entered numbers and characters associated with open-ended questions. Maintaining both scanning and manual data entry capabilities was deemed undesirable in the long run. Our goal was to fully replace (and thereby eliminate) keypunching operations.

- *No adverse affects on response rates.* Our usual questionnaire formats and mailing procedures typically produced general population response rates from 55 to 75 percent when all response enhancing methods were employed. We were unwilling to accept even a slight reduction in response rates below our accustomed standards.

- *Low item nonresponse rates.* We set our item nonresponse standard at less than 5 percent, our usual expectation. Item nonresponse can be affected by the graphical layout of questionnaires, although it also can result from a question being intrusive (e.g., income) or difficult (e.g., rank these 20 items from top to bottom).

- *An efficient data repair procedure.* Many of our clients require (and our center's standard is to provide) the best possible data set from the information provided by respondents. We would not accept a scanning method that made it necessary to ignore mismarked or unrecognized answers or to identify and correct them at great cost. If a scanning and imaging system did not provide efficient handling of marking or recognition errors, it would be unacceptable for general use.

• *Cost effective between 500 and 1,000 questionnaires.* Since most of our surveys receive between 500 and 1000 questionnaires, it was essential that the scanning operation be cost effective at that volume. We recognized, however, that it was difficult for the smallest surveys to achieve full economic efficiency either with scanning or keypunching.

These criteria made it clear that the conversion of all questionnaire processing to optical scanning would be a formidable task for small survey organizations. Proven off-the-shelf OCR technology is not yet available for reading cursive answers, or even most hand-printed answers without boxes to separate individual characters. Keying operations could not be eliminated, although imaging the full questionnaire and keying text answers from the screen might produce some gains in efficiency over keying from paper questionnaires.

24.3.3 Design of the Experiment

We purchased an inexpensive flatbed scanner and the Teleform software package because that package had both mark (OMR) and character (OCR) recognition capabilities as we required. The Teleform software also provided considerable flexibility in the design of individual questionnaire pages, which allowed us to maintain our traditional questionnaire booklet format as part of a survey design and implementation system successfully used by our center for many years (Dillman, 1978). This scanner and software together cost about $2000, considerably less than other options we considered. Thus, if the technology worked, it would be affordable by a small research center that conducted only occasional mail surveys.

Although the Teleform software provided on-line correction of recognition failures and the manual entry of open-ended text answers from a screen image, the version we used did not offer the power and efficiency of the advanced scanning and imaging technologies described earlier and costing more than ten times as much. For example, the Teleform correction process proceeded only one questionnaire at a time. It did not permit correction by item across all questionnaires as in the most advanced and expensive system we observed.

Questionnaires designed by our center are usually printed on legal size sheets of paper, which are folded into $8\frac{1}{2}$ by 7 inch booklets, and have graphically designed, interest-getting covers (Dillman, 1978). The finished booklets can be folded once again (vertically) to fit into normal sized business stationery envelopes. This minimizes costs for first class return mail.

The specific questionnaire chosen for our scanning experiment was one previously designed to adhere to the principles of visually prominent navigation developed by Jenkins and Dillman (1997). This questionnaire had been tested and revised in two full-scale pilot studies and had been employed in several prior response rate experiments (Miller, 1996). The intended respondents were new applicants for drivers licenses from outside Washington State. This population was chosen for the experiment because of the challenge such a relatively young group presented for attaining a high response rate. A previous study had produced response rates of about 60 percent using four personalized contacts, a stamped return envelope, and a $2 prepaid financial incentive (Dillman, 1997). Available records provided information on the sex, age, and residence (rural vs. urban) of both respondents and nonrespondents. Age was of special interest, since younger people were thought more likely to be familiar with scannable forms through their educational experiences.

A sample of 1500 names of new Washington State residents was divided systematically into three subsamples of 500 each. One sample received the booklet questionnaire normally used in our center's research. Thus Treatment 1 received a prenotice letter, a 12-page booklet questionnaire using an attractive color cover, a reminder postcard, and a replacement questionnaire. Correspondence was sent by first class mail containing a return envelope with a 32 cent stamp affixed. Each of the cover letters in this treatment was personalized, that is, individually produced on letterhead stationery with respondent names inserted and real signatures. Personalization is a standard part of our mailing procedures but rarely if ever used in the large-scale scannable surveys previously described. We decided to include personalization as a separate factor in the experiment as a possibly significant consideration in the decision of small survey centers to adopt scannable surveys.

For Treatment 2, identical procedures (including personalization) were used except that the scannable form was substituted. On the outside the questionnaires looked the same. Inside however, circles were used for marking answers, colored (gray) background bars were used to align horizontal questions and answer choices, and the required scanning guide marks were used. The Teleform software did not dictate the first two design decisions. Check boxes could have been used instead of fill-in circles, and dotted line connections could have been used instead of color bars, perhaps enhancing the respondent friendliness of the questionnaire. However, we thought it important to test the darkened circle and colored lines procedures used in virtually all of the 44 optically scannable questionnaires included in our previous response rate analysis.

The questionnaire included 124 fixed-response items answered by filling circles. Twelve open-ended questions requiring 22 separate fields of numbers and names (e.g., years and addresses) were set up for optical scanning by printing segmented answer spaces with one box per digit or character as illustrated in Questions 1, 3, and 4 in Figure 24.2. An additional 15 questions requiring narrative open-ended answers (e.g., Question 7 in Figure 24.2) were set up as large unsegmented boxes. These were questions where it was nearly impossible to anticipate how many character boxes would be needed. A previous study on another questionnaire had indicated that without boxes to write in, people were much more likely to use cursive handwriting which the software would not be able to recognize (Dillman, 1995). We believed it would be more efficient to image those words onto the screen and have an operator enter those answers directly into the data file.

Treatment 3 was identical to Treatment 1, except that none of the correspondence was personalized. Sampled individuals received a letter on copied (black-and-white) letterhead stationery containing a preapplied signature and salutation of "Dear New Washington Resident" along with the standard booklet questionnaire. These procedures were repeated through each of the multiple mailings.

24.3.4 Findings

Unit Nonresponse

The response rate is the proportion of the mailed sample who returned a usable, completed questionnaire. People who wrote back saying they were ineligible for the survey (e.g., not a Washington resident) were deleted from the response rate calculations. Undelivered envelopes marked "return to sender" were counted as nonresponse.

The Teleform optically scannable questionnaire in this experiment achieved a response rate of 59.9 percent compared to 60.0 percent for the conventional format. Thus the format changes undertaken to make our questionnaire scannable had no detectable influence on the response rate. This result is very encouraging for the use of traditionally formatted (fill-in circles and alternating colored bars) scannable questionnaires.

The third treatment group, which used the conventional format, but did not personalize its mailings, attained a somewhat lower response rate of 55.0 percent. This difference, although suggestive, was not statistically significant with this sample size (chi-square $= 2.26$, $df = 1$, $p = 0.13$).

Effectiveness and Efficiency of the Scanning Technologies

Our ability to evaluate the efficiency and effectiveness of the scanning and imaging process in relation to keypunching was hampered by two considerations. The first was that the Teleform software we acquired was designed to run under Windows 3.1, and our office had upgraded to Windows 95. We were led to believe that the 3.1 version would work on Windows 95, information that proved later to be incorrect. The second impediment was our decision to maintain our usual $8\frac{1}{2}$ by 7 inch booklet format. This meant we had to process pages of a unconventional size and therefore could not use a scanner with an automated document feed. The combined result of these factors was that scanning was quite slow and errors occurred that could not be corrected.

After considerable developmental time in learning how to operate the scanning equipment, the scanning of questionnaires required approximately 4.5 minutes each. An additional 8 minutes were necessary to retrieve the questionnaire image on the computer screen and manually enter in answers which, according to the software, had not been correctly read. In addition the entire questionnaire was reviewed at this stage to check for inconsistencies as done

Q1. In what <u>month</u> and <u>year</u> did you move to (or back to) Washington State?
(Please fill in circle for month and write in year)

○ Jan	○ May	○ Sep
○ Feb	○ Jun	○ Oct
○ Mar	○ Jul	○ Nov
○ Apr	○ Aug	○ Dec

`1 9 [ ] [ ]` YEAR

(If you moved before 1990, you do not need to complete the rest of the questionnaire. However, please mail it back so we can remove your name from the mailing list.)

Q2. Are you a <u>first time</u> Washington resident? *(Please fill in circle of your answer)*

○ Yes

○ No ————→ **Q3. (If No) What year did you leave here to live someplace else?** `1 9 [ ] [ ]` YEAR

Q4. All together, how many years have you lived in Washington? `[ ] [ ]` NUMBER OF YEARS

Q5. Did you move here alone or with others? *(Please fill in circle of answer)*

○ Alone

○ With other(s) → **Q6. Who did you move here with?** *(Fill in circle of all who moved here with you.)*

○ Spouse or partner
○ Child or children ————→ `[ ] [ ]` NUMBER
○ Other (please specify)

Q7. What was the most important reason you moved here?

Q8. Did any of the following job-related considerations influence your decision to move to Washington? *(Please fill in YES or NO for each item.)*

	YOURSELF Y N	YOUR SPOUSE OR PARTNER (If you had one) Y N
To accept a job with a new employer	○ ○	○ ○
To look for new work/job	○ ○	○ ○
To start/take over a business	○ ○	○ ○
A military transfer	○ ○	○ ○
A transfer by current employer (except military)	○ ○	○ ○

Figure 24.2. Comparison of scannable versus conventional TDM questionnaire format used in experiment.

Q1. **In what month and year did you move to (or back to) Washington State?**

_____ MONTH and _____ YEAR

(If you moved before 1990, you do not need to complete the rest of the questionnaire. However, please mail it back so we can remove your name from the mailing list.)

Q2. **Are you a <u>first time</u> Washington resident?** *(Please circle number of your answer.)*

1   Yes
2   No ⟶ **Q3.** **(If No) What year did you leave here to live someplace else?**

_____ YEAR

Q4. **All together, how many years have you lived in Washington?**

_____ NUMBER OF YEARS

Q5. **Did you move here alone or with others?** *(Please circle number of answer.)*

1   Alone
2   With other(s) ⟶ **Q6.** **Who did you move here with?** *(Circle number of all who moved here with you.)*

1   Spouse or partner
2   Child or children   _____ NUMBER
3   Other *(Please specify)*

_____

Q7. **What was the most important reason why you moved here?**

_____
_____

Q8. **Did any of the following job-related considerations influence your decision to <u>move to Washington</u>?** *(Please circle YES or NO for each item.)*

	YOURSELF		YOUR SPOUSE OR PARTNER (if you have one)	
To accept a job with a new employer	YES	NO	YES	NO
To look for new work/job	YES	NO	YES	NO
To start/take over a business	YES	NO	YES	NO
A military transfer	YES	NO	YES	NO
A transfer by current employer (except military)	YES	NO	YES	NO

Figure 24.2. *(Continued)*

at the preprocessing stage for keypunching operations. Afterward, three additional minutes were required to enter answers for the remaining open-ended questions into a data file. Thus about 15.5 minutes were required for the scanning and data entry process for each questionnaire. This number is a carefully considered estimate. During the initial scanning efforts, the correction process took somewhat longer. We based our estimate on the levels achieved during later stages of scanning. With our current equipment we could not achieve any additional speed with the scanning activities, but might achieve some slight additional gains in the correction phase. In contrast, keypunching required 3.1 minutes of preprocessing for coding in preparation for keypunching, 5.5 minutes to keypunch (including open-ended questions), and 3.2 minutes to verify for a total of 11.8 minutes, or about 3.7 minutes less than scanning. Thus scanning did not improve the speed of data entry.

The questionnaire contained 146 scannable variable fields, 124 requiring marks and 22 requiring the writing of numbers or characters. Only 1.5 percent of the marked (OMR) question fields required manual corrections, an average of about 1.8 per case. By comparison, about a third (31.8%) of written character (OCR) fields required manual correction, an average of 7.0 per case.

We measured item nonresponse for the questionnaires in two ways, first by manual keypunching of the scannable questionnaires to evaluate respondent intent, and second by scanning to evaluate the recognition technology. The first indicates whether switching to the scannable format (circles, character boxes, and colored bands) encourages or discourages respondents from answering each question. Even if a mark misses a box slightly and cannot be read by the computer, the keypunched results show what the respondent intended.

The item nonresponse rates based on keypunching both sets of forms were quite similar: 6.3 percent for the conventional format versus 7.1 percent for the scannable questionnaire, both slightly above our 5.0 percent standard. The conventional questionnaire had a lower item nonresponse rate for 70 items compared to 62 for the scannable questionnaires. The larger differences in item nonresponse, however, favored the conventional questionnaire. There were 25 instances in which the conventional questionnaire had more than a three

Table 24.2. Comparison of Item Nonresponse for Scannable and Regular Questionnaires in Experiment

	Scannable Questionnaire	Regular TDM Booklet Questionnaire
Mean item nonresponse (%)	7.1	6.3
Items with lower item nonresponse	62	70
Items with item nonresponse 1–3% less	34	28
Items with item nonresponse +3% less	9	25

percentage point item nonresponse advantage, but these occurred disproportionately for one type of question.

Item nonresponse rates were calculated in the manner normally used by our center. Each field is counted independently except for "mark all that apply" questions. A "no answer" for a "mark all that apply" question counts as an omission for each of its options, giving such items extra weight. Of the 25 fields where item nonresponse of the scannable form exceeded that of the conventional form by 3 percent or more, 18 were from Q48, shown in Figure 24.3. This pair of "mark all that apply" items asked respondents to check each kind of computer equipment they used and that their spouse or partner used. The percent of respondents indicating no use of any of these kinds of equipment

TELEFORM FORMAT

Q48. **Does this person use any of the following types of equipment or services in his/her work, regardless of work location?** *(Fill in all that apply.)*

	YOURSELF	YOUR SPOUSE OR PARTNER (If you have one)
A computer with a keyboard	O	O
A computer modem	O	O
Fax machine or fax modem	O	O
Overnight or courier delivery of materials, products, or information you produce	O	O
Overnight or courier delivery to you of things you need for work	O	O
Electronic mail, e.g., Internet	O	O
Answering machine or voice mail	O	O
Cellular telephone	O	O
Conference telephone capability	O	O
Other equipment (please explain)	O	O

CONVENTIONAL FORMAT

Q48. **Please circle the number of each type of equipment or service that is used by this person in his/her work, regardless of work location.** *(Circle all that apply.)*

	YOURSELF ▼	YOUR SPOUSE OR PARTNER (if you have one) ▼
A computer with a keyboard	1	1
A computer modem	2	2
Fax machine or fax modem	3	3
Overnight or courier delivery of materials, products, or information you produce	4	4
Overnight or courier delivery to you of things you need for work	5	5
Electronic mail, e.g., Internet	6	6
Answering machine or voice mail	7	7
Cellular telephone	8	8
Conference telephone capability	9	9
Other equipment (please explain)	10	10

Figure 24.3. Higher item nonresponse on scannable versus conventional questionnaire, format (12.9% versus 5.8% for yourself; 11.4% versus 5.6% for spouse or partner).

was 5.8 percent on the conventional form and 12.9 percent on the scannable questionnaire. A similar difference of 5.6 versus 11.4 percent was observed for the spouse/partner items.

Item Q48 was the only one that asked respondents to mark all items that applied both for themselves and for their partner/spouse. We believe that the reason for the scannable form's higher nonresponse on this item derives from its use of horizontal bars to align questions with answer categories. The visual effect of the bars is to "group" responses across people (respondent and spouse/partner) as if each equipment option must be answered first by the respondent about her/himself and then about the spouse. It would seem much easier for the respondent to answer the question for all equipment she/he uses, as encouraged by the visual design of the conventional questionnaire, and then switch to the spouse or partner. This interpretation is consistent with the grouping effects of visual layout hypothesized by Jenkins and Dillman (1997). The scannable form also received a somewhat higher proportion of multiple marks to a question on the educational attainment of the respondent (26%) than on the traditional form (20%). This seems to be the result of the horizontal bars visually separating categories from one another and encouraging people to mark each level of school completed.

Except for these questions, no noteworthy differences in item nonresponse or multiple marks occurred between the two questionnaires forms. With these exceptions, the filling of circles and the use of lightly colored bars to align items with answer choices do not seem to influence either unit or item nonresponse, based on the keypunched results of the scanned forms.

The attempt to develop comparable measures of item nonresponse based on actual scanning of the scannable forms was thwarted by a series of events. First, for a number of items we learned that we had mistakenly set up the scanning template to export the data for several items as one variable (e.g., 111) rather than as separate variables (e.g., 1, 1, 1). This occurred, for example for the question, "Did you move to Washington for any of the following reasons?" This was followed by three yes/no items, which should have been separate variables. We were unable to find a means of correcting that mistake either from the software manuals or by talking with a company representative. We also discovered frequent errors in the process of assigning scanned page output to case IDs. We originally thought these were human errors from the need to manually insert and turn booklet pages one at a time for the scanner, but when the scanning process was repeated, exactly the same result occurred. While attempting to find an answer to this problem, we were informed that technical support for the Windows 3.1 version of the Teleform software was no longer provided because it was known to have problems and the version for Windows 95 had been released. Although we subsequently purchased the Windows 95 version, it would not read questionnaires set up under the previous version.

To circumvent these problems, we prepared a reduced data set consisting only of questionnaires whose pages were known to be linked to the appropriate case IDs and limited to those variables where we were confident that we could determine the correct scanning rates. This reduced data set consisted of 30 variables from 53 percent of the scannable questionnaires. For this data set, item nonresponse based on scanned values was three time higher than for the same items when keypunched, 14.1 as opposed to 4.6 percent. The main reasons for the difference, based on visual examination of answers successfully keyed but missed in scanning, appeared to be light marks and those slightly outside the answer circles. We also found a few reasonably dark marks apparently filling answer circles completely that for unknown reasons were missed by the scanning software.

The incompatibility of our Windows 3.1 optical scanning software and Windows 95 operating system, which in the end could not be resolved, makes the results of the experiment less valuable than might otherwise have been. Nevertheless, we can conclude that scanning in this manner was both slower and less accurate than manual data entry.

24.4 CONCLUSIONS

The development of advanced imaging and scanning with OCR as well as OMR capabilities is dramatically increasing the options for mail survey design and administration. The most advanced imaging and recognition systems described in Section 24.2.2 offer the promise of achieving the data capture quality possible with manual keypunching with verification and to do so with great efficiency. This is accomplished with new, albeit expensive, hardware and software that, in essence, turn the data entry process upside down. Questionnaire scanning proceeds with minimal preprocessing; and once the scanning is complete the actual questionnaires are no longer necessary. Corrections, coding, and verification can proceed using images of the questionnaire.

The experiment we conducted to test a much less expensive scanning and imaging system as a replacement for manual data entry was not successful. Our decision to purchase Teleform software resulted in technical problems that could not be resolved. Nonetheless, several important things were learned from the experiment.

First, the experiment demonstrated that common questionnaire scanning formats thought to reduce respondent friendliness did not have the hypothesized negative effects on response rates. Specifically we found that the use of shaded (or colored) bands to connect question stems to response choices, the filling of small circles to select answers, and the entry of numbers and characters in response boxes generally did not result in higher unit nonresponse or in higher item nonresponse, as measured by manual data entry of respondent intent.

Second, the experiment and the analysis of the 44 large scannable surveys summarized earlier suggest that the same factors that influence response rates of traditional mail surveys (e.g., multiple contacts and personalization) also influence the response rates of optically scannable surveys.

Third, the experiment also suggests that the visual layout of optically scannable questionnaires may in some situations increase both item nonresponse and inappropriate extra responses to single response items. Although the use of colored bands to guide respondents across the page generally has no effect on item response, they probably should not be used if they encourage respondents to answer more than one question at a time or make it more difficult for respondents to perceive an entire set of response categories as a group.

Our experiences in conducting the experiment also suggest some important next steps. One should be a revised approach to general questionnaire formats. In the current study we attempted to apply the scanning technologies to our existing questionnaire procedures. We retained our traditional questionnaire booklet rather than adopting the $8\frac{1}{2}$ by 11 inch page format standard in scanning. Thus the scanning process became a slow manual operation which was subject, we believe, to considerably more error and a slower pace of work than would have occurred with a standard paper size permitting use of an automated document feeder.

Choice of a standard paper size also would permit printing questionnaires in-house on a standard laser printer rather than outsourcing the printing and stapling of the booklets. In-house printing would allow us to pretest all aspects of the formatting, scanning, and correction process before the printing production run, a step that is prohibitively expensive with outsource printing for small surveys.

Our reluctance to make these changes initially was partly based on concerns about response rates and about the costs for paper and larger envelopes. The absence of response rate differences between the conventional and scannable forms in the current experiment has now encouraged us to test more radical changes in questionnaire design and printing methods. Perhaps the most important lesson this study has for small survey organizations is to recognize that a *switch from manual data entry to scanning requires a change in the entire system of survey activities, not just in the questionnaires.*

Despite the difficulties encountered in this experiment, we remain optimistic about the future of optical scanning for small sample surveys conducted by small survey organizations. In the short run we doubt there will be a great rush among such organizations to eliminate manual data entry in favor of scanning. A major effort is required to redesign an entire system of survey procedures, and cost and quality arguments do not warrant an immediate change. In the long run we see no prohibitive barriers to scanning for most or all mail surveys. The technology exists to conduct small sample surveys with efficient scanning. At present this technology remains very expensive, limiting its use to large organizations that can amortize its purchase over very large surveys or continuous rather than intermittent use. Since new computing technologies historically decrease in cost with time, we anticipate that this will be true of advanced scanning and imaging as well, facilitating increased use by small organizations in the future.

One of the frequently mentioned final barriers to the full use of scanning for mail surveys is the inability of all but the most advanced and custom-tailored recognition systems to accurately read cursive handwriting (Wilson et al., 1996). We no longer see this as a major problem. The ability of advanced imaging systems to display answers by item rather than by case promises greater efficiencies for processing open-ended responses than could be achieved with manual data entry. Thus we expect that the processing of text entries will steadily diminish as a barrier to the replacement of manual data entry with optical scanning. Our conclusion about the future of optically scannable mail surveys is that it has already arrived for some large survey organizations who conduct large surveys. Its availability for small organizations whose emphasis is small surveys may not be far behind. Costs, rather than technology, constitute the true final barrier.

ACKNOWLEDGMENTS

Research on which this chapter is based was supported by the Washington State University Social and Economic Sciences Research Center, the Department of Rural Sociology, Western Region Project W-183, Rural and Agricultural Surveys, and a grant from the USDA National Research Initiative to Don Dillman and Priscilla Salant. We acknowledge with much appreciation the information on response rates for scannable questionnaires provided by many individuals and organizations in both the public and private sector. We also wish to thank others for demonstrating and explaining operating procedures associated with particular scanning and imaging technologies. Appreciation is also expressed to several individuals who read and commented on previous drafts.

CHAPTER 25

Scanning and Optical Character Recognition in Survey Organizations

Evert Blom and Lars Lyberg
Statistics Sweden

25.1 INTRODUCTION

Data collection and processing are frequent bottlenecks in survey production. When done manually these operations are time-consuming, costly, and even boring. CASIC methods seek to automate the process and make it more efficient, as has been demonstrated in applications across survey agencies. Computer assisted interviewing (CAI) eliminates the need for a separate data entry step and integrates data editing with data collection. In computer assisted self-interviewing (CASI) respondents do the keying themselves. In the case of electronic data interchange (EDI), the respondent's database is linked directly to the data collection application. These and other CASIC methods share the common goal of reducing or even eliminating the paper flow of traditional data collection, making the process faster and less expensive.

Despite CASIC's success in moving toward paperless data collection, paper is still the dominant medium for recording and transferring information. Estimates run as high as 95 percent for the amount of information worldwide that is still on paper (Goodyear, 1995). The new technologies of document image processing (DIP) now make it possible to automate data capture from paper rather than eliminate it altogether. DIP technology and especially devices for scanning and optical character recognition (OCR) change the entering of data from the keyboard to automatic recognition, making it

Computer Assisted Survey Information Collection, Edited by Mick P. Couper, Reginald P. Baker, Jelke Bethlehem, Cynthia Z. F. Clark, Jean Martin, William L. Nicholls II, and James M. O'Reilly. ISBN 0-471-17848-9 © 1998 John Wiley & Sons, Inc.

possible to achieve significant gains in efficiency while relying on traditional paper-based data collection methods.

This chapter focuses on applications of new scanning technologies to survey processing from paper forms. Scanning (or image technology) is sometimes used as a general expression to describe a sequence of activities including both imaging and the interpretation (the OCR process) of survey forms and information. The wider concept of document management, which includes workflow, is also used but is largely outside the scope of this chapter, since it refers to the processing not only of paper documents but also of a spectrum of multimedia documents from a variety of different sources. We normally use OCR as a term for the interpretation process, which in turn is generally synonymous with intelligent character recognition (ICR), a term used by some vendors to describe the most advanced techniques for the interpretation of large sets of multifont characters and unconstrained handwritten characters.

A DIP system for surveys has two main components: (1) a scanning process that produces a digital image of the document which can be stored, retrieved, displayed, and printed, and (2) an OCR process that interprets document information automatically and converts it to computer readable format (e.g., ASCII text). The two elements can be conceived as the data capture components of a CASIC system, producing a complete image of the questionnaire itself and interpreting handwritten or typed numerical information (digits 0–9), handwritten or typed alphabetical characters (e.g., occupation), bar codes, and various other types of markings (crosses, lines, etc.). By integrating edit rules and document case management, the scope of functions involved in data capture increases and the process can be enhanced and streamlined, including archiving and easy retrieval of documents, and the registration of incoming documents.

The chapter begins with a brief review of the history and development of scanning technology followed by a discussion of the potential advantages and disadvantages to survey processing. We describe the basic capabilities of scanning technology and present a process flow through the system along with essential concepts and terminology. A short overview of the hardware configuration is also presented. From there we move to an overview of the use of scanning technology and review findings reported by survey organizations, mainly within governments. We conclude with guidelines for technology selection and some comments on the future.

25.2 HISTORY AND DEVELOPMENT OF SCANNING TECHNOLOGY

25.2.1 History of Scanning

The history of image technology and especially scanning and OCR begins with simple pattern comparisons and evolves to unconstrained "intelligent" systems called intelligent character recognition (ICR) that use neural network technology. The vision of machine reading has become real, and today many applications for performing OCR exist in a wide range of organizations. The origins can be traced back to 1870 when the retina scanner was invented, using a mosaic of photocells (Eikvil, 1993). Modern versions did not appear until the 1940s with the development of digital computers. Business applications were the driving force, and the first reading machine was installed at Reader's Digest in 1954. These first generation OCR machines were characterized by constrained (pre-established) letter shapes that could be interpreted but were limited to a library of fonts available for comparison to the character image.

Second generation OCR machines appeared in the 1960s. They were capable of recognizing machine printed as well as handwritten characters. Typically the market provided specialized high-speed OCR readers that were very expensive. They could be run only by specialized personnel, knowing the special hardware and software languages needed for building the applications. Also during this period the first standards emerged including OCR-A (American) and the OCR-B fonts (European). An example of an early application using OCR is the coding of the Swedish Population Census in 1970 (Eiderbrandt and Lyberg, 1974). About the same time applications for the recognition of different types of marks (crosses, lines, etc.) appeared. Such recognition was called optical mark recognition (OMR). Special reading machines were used for interpretation; the survey form was designed so that the position of the mark signaled the content of the response. An early use of optical mark recognition (OMR) was the innovative and successful application of film optical sensing device for input to computer (FOSDIC) for the 1960 U.S. Censuses of Population and Housing. In these censuses a high-speed electronic device was used for direct transfer of data from microfilm of the enumerated schedules to magnetic tapes (U.S. Department of Commerce, 1965). Users at that time were very optimistic about the new technology, and they were looking forward to processing normal human handwriting in the near future, even in the format of continuing script. Unfortunately, this never happened. In the 1980s the advent of high-performance personal computers brought new opportunities for development. At the same time specialized hardware and software platforms for imaging continued to evolve and play an important role for specialized high-speed and large-volume applications.

Today scanning equipment is based on much faster and less expensive hardware than ever before. The costs of OCR have decreased dramatically, and it now is supported by many standard software packages that are easy to adapt and run (see Dillman and Miller, Chapter 24). Systems have become more interactive and are able to handle a large variety of printed and handwritten characters. They are also able to integrate data capture and editing with document imaging, work flow, and modern case management tools to make it easier to speed up and control the process. Unrecognized characters can be entered interactively, instead of in separate batch processes used before. Systems can be built from various standard components and run on different platforms across networks. These developments have produced a viable technology for a wide range of applications, not only in commercial establishments but also in survey organizations.

Despite the new technology and various improvements in the function and capacity of modern OCR systems, there are still limitations. At present there are no systems in practical use that can interpret free form handwritten letters, namely continuous joined-up script. However, controlled handwriting in block letters can be interpreted. Free figure panels can also be interpreted by the more highly developed systems so that no special boxes or letter fields are required for the various characters. This makes it possible to use a more user friendly layout of forms and achieve a better use of paper space.

25.2.2 Advantages and Disadvantages to Survey Processing

One of many goals of a DIP system is to free professionals from the burden of document searching and handling. This is important since much time is currently devoted to the preparation of documents, entering data, and copying, faxing, handling, and filing documents. Many processing activities have in fact been made more efficient. However, we have also shifted an enormous amount of clerical work to professionals and managers. The latter not only perform data entry and word processing, they are now often responsible for handling the documents as well.

Data collection and data capture are particularly resource and staff intensive parts of statistical production. Image technology can simplify the capture and processing of paper bound information. While traditional data entry is time-consuming, costly, and boring, scanning and OCR can result in higher throughput, and there is less need for moving mounds of paper. Once documents have been scanned into the system, their images can be instantly accessed by many users, throughout the production process, and data can be captured on the spot, ready for immediate processing. Manual search of archives is eliminated.

According to Gartner (1992), the most compelling reason to invest in DIP technology is simply to put information in the hands of users faster. The second most important reason is productivity. Reduced workforce is the third, and better management control over the system is the fourth.

The specific advantages of scanning and OCR can be summarized as follows:

- Improved survey efficiency and timeliness via shorter production time.
- Time-consuming and monotonous manual data entry (heads down) can be eliminated.
- Images are immediately available and can be accessed by many users simultaneously, which is crucial since most of the actions associated with survey forms normally take place in parallel immediately after the forms have been received.

- Time spent on sorting and searching in archives disappears. Electronic archives make it easier to retrieve a specific questionnaire for checking and, if necessary, recontacting respondents in order to complete inadequate responses.
- Costs are reduced not only for data entry but also for case management and follow-ups.
- Work flow generates information about the process and makes it easier to establish the status of the various tasks as they progress. Performance rates can be calculated.
- Integrated editing and correction is done more efficiently. There is less need for listing the errors and conducting any batch oriented re-editing.

Of course there are also problems and limitations associated with image technology, and it is necessary to consider both pros and cons when planning to use it. Some of the crucial issues are:

- costs for investment in the development of systems and training in their use,
- time needed to build applications,
- hardware and storage costs compared to savings in workload,
- the total throughput of a survey,
- the costs to correct errors,
- quality problems due to accuracy rate,
- implications for questionnaire design,
- uncertainty about storage media.

Given these advantages and disadvantages, what are the typical survey applications for potential use of imaging? Such applications should most certainly include large volume censuses and other surveys comprising large numbers of forms, including preprinted identifications, marks, and handwritten numeral responses. But under what conditions could other surveys profit from this technology? To answer this question, more information and practical experience must be gained.

25.3 DESCRIPTION OF THE TECHNOLOGY

25.3.1 Steps in the Scanning Process

Document image processing is a completely computer-based approach to document scanning, storing, retrieval, and interpretation. The sequence of key steps in this process begins with scanning, which is then followed by character recognition, correction of rejects, verification of interpreted characters, possibly batch quality control, regular statistics production, and—parallel to this flow—image storing on an image server and document image processing. A work flow system also can be integrated into the process. The process is complex and a standard terminology is not yet agreed upon. In this section we present the essential concepts and terminology, delineate the scope of image technology and distinguish the systems needed for statistics production from document management technologies in general. System architecture is reviewed, as well as the organization of the related data collection and input processing (data flows). Systems components are described in terms of hardware and software needed.

There are five main steps in the image process: (1) preparation, (2) scanning and imaging, (3) data capture and validation, (4) reject repair, search for substitutes, and editing, and (5) transfer to the target application. We describe each in more detail below.

25.3.2   Work Flow through the System

Figure 25.1 depicts the scanning and OCR workflow in detail, where the flow goes from the left to the right, from application setup and planning through all phases of the process.

Application setup is when the survey manager starts preparing the use of image technology in a survey and defines the steps to be followed through the production phase. It contains the form definition: that is, how to scan filled-out forms, how to interpret the information, how to define error procedures by the validation routines and how to transfer information to the target system for further processing. The form definition typically is based on a scanned image

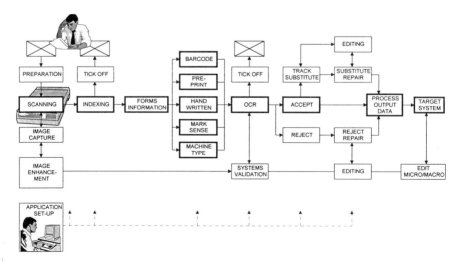

Figure 25.1. Scanning and OCR process.

of an empty form and includes determining the field positions, types of information to be captured in the different fields, validation checks, and so forth.

Preparation of source documents involves making pages ready for scanning. It includes postal handling and the separation of forms from letters and notes, for example, which must be taken care of in a separate flow of activities. At a minimum, forms must be removed from folders, binders, or other containers. Folded pages must be unfolded. Torn pages should be mended or copied if necessary. Scanners equipped with automatic stack feeders require removal of staples, paper clips, and so forth.

Scanning and image capture is the capture of images of documents and forms in a scanner. The device used can be a flat bedded or stack-fed high-performance model with the ability to feed a variety of documents. The images produced by the scanner, still not interpreted, are copies of forms with their individual data, including common text and layout. They are stored in a standard format, usually tag image file format (TIFF), and compressed to save storage and speed up transmission time. Some systems make use of a subsequent "form out" procedure to further reduce the volume of data that must be stored and to enhance the OCR efficiency. This means the extraction of the fixed part of the form, leaving only the reading areas and variable data to be stored. In a subsequent reconstruction process the original form can be created by combining the saved template with the various parts.

Image enhancement is necessary to support the later process of interpretation. It is quite common to pass images through a postscanning process for further enhancement, such as (1) removing the stationary content of a form using a feature known as color drop-out, (2) removing shaded areas and erroneous marks irrelevant to data capture, and (3) enhancing image output through the use of sophisticated thresholding techniques.

Indexing involves assigning document and form identifiers necessary for storage and subsequent retrieval. This information should be acquired at the time of scanning with as little effect on throughput as possible. The index values can be typed manually at the input workstation. Other more automated features include OCR-driven indexing by preprinted identity numbers on the form, automated full text indexing, or other auto-indexing techniques. The registration or tick-off process can be integrated with the indexing process, allowing for automated mailing of nonrespondent reminders at a predefined time of the data collection process.

Tick off (or check in) is the procedure of registering incoming forms, counting the number of responding objects, matching against a list of units included in the actual survey, and identifying units for reminders. This can be done using index information at the time of scanning or following the OCR procedure having recognized the form's identity.

Forms information is the filled-in data, the areas of which have been declared in the application setup phase. Data can take the form of a *bar code*, information that is *preprinted* such as identifying numbers, *handwritten* numericals or alphabetical strings, *mark sense* such as crosses, or *machine-type* alphanumerical information.

OCR is the automated recognition of the different marks and characters. In this context we include the following: (1) bar code recognition (BCR), (2) optical mark recognition (OMR; recognition of specifically delimited marks, small marked areas, and ticks), and (3) intelligent character recognition (ICR; recognition of machine printed and handwritten characters using neural network technology). The characters may be alpha, numericals, or alphanumerics. The character recognition can be preceded by a "formout" procedure, which is the separation of an image of the original document from data filled in by respondents. The recognition process converts characters on images to ASCII format data.

The first step of the OCR process is the identification of the form and location of the fields containing the information that is to be interpreted. In this step the application uses the information provided by the users at the time of setup, especially the form definition. The next step is segmentation, that is, the identification of the set of pixels (picture elements) needed to build a single character. The final step is feature extraction, the task of assigning a list of numerical values (a feature vector) to the shape of a character using some kind of classifier. The classifier uses well-known mathematical methods, for instance, neural networks. Three key factors influence the accuracy of the OCR output, namely (1) image quality, (2) the strength of the recognition vehicle, and (3) application-specific features.

Systems validation is an integrated part of the OCR process itself. Using a variety of complex algorithms, the system computes a measure of confidence for the correctness of its interpretation of each character. Characters with low confidence are presented on the operator's screen for a final decision, accepted or corrected, when compared to the image. The system can propose a possible interpretation or just an asterisk to declare that the character is rejected. In some systems several OCR engines are involved in the decision, presenting the character with the best score as the output.

Rejects are characters that are not interpreted by the system. Rejects can be flagged on the operator's screen for manual keying. We call this operation *reject repair*. The rejection rate is defined as the number of times the system failed to classify a character as a percentage of the total number of characters read. This rate can also be calculated on the field (variable) level. The balance between rejects and substitutes (misinterpreted characters) can be influenced and adjusted through a confidence level preset in the interpretation process. A high confidence level produces few substitutions or flagged characters and many rejects, increasing the overall quality of the interpreted data. Conversely, a low confidence level produces many substitutions or flagged characters and few rejects. With a high confidence level the manual intervention increases, and with a low confidence level this intervention decreases.

Tracking substitutes is the search for misinterpreted characters. Substitutes therefore constitute a serious error source in image processing. Examples of substitutes are 3 interpreted as 8 or 1 as 7. The substitution error rate is the number of characters incorrectly interpreted by the system as a percentage of the total number of characters read. This rate too can be calculated on a field (variable) level. The OCR system may use additional information, such as lookup tables, checksums, and spell checkers to increase accuracy and thus to some degree avoid substitutes.

The survey manager has to carefully consider strategies for detecting substitutes. A simple but cumbersome procedure is to inspect all critical fields on the operator's screen and correct any errors found. Another approach is to apply ordinary edit rules, which can be integrated in the actual application (where the edit rules are decided already in the setup phase) or managed as a separate step downstream in the target system (see Granquist, 1995).

A cumbersome proofreading of a sample of forms is the obvious method to estimate the level of errors, but how should these errors be found and displayed on the operator's screen to allow corrections? A new approach to tracking substitutes and performing *substitute repair* is the following. For a batch of forms the images of all single characters are displayed on the screen, grouped in their interpreted classes of characters, 0, 1, 2, etc. This allows for quick mass inspection of each character that will help the operator detect and correct substitution candidates. Such visual inspection is not error free. For instance, it is a well-known fact that it is very difficult to accurately count the number of specific characters among a large total number of characters (like counting the number of times the letter "e" appears on a book page) (see Tortora and Faulkenberry, 1979).

Editing is the process of reviewing and adjusting the scanned data to control quality. Its key steps consist of survey management, data capture, data review, and data adjustment (United Nations, 1994). The edit process deals primarily with errors and inconsistencies in the data generated during the collection phase, whatever the collection mode might be. However, when OCR and reject repair processes are part of a survey system, it is important to have the opportunity to track and adjust for substitutes caused by erroneous interpretation (as is the case for discovering keying errors when using traditional data entry). Editing can use auxiliary information, and can include sophisticated kinds of checks, such as ratios and combinations of variables, which may not be included in the recognition software verification and validation routines. Thus editing should be distinguished from the earlier reject management and systems validation stages.

Process output data is the final stage in the process. At this point the data can be transferred to the target system for subsequent production, possibly including additional editing and analysis to fit the purpose of the survey. The image of the form captured in the scanning phase can be retrieved and re-used at this stage and if necessary even further in continuing surveys. Work flow systems can distribute documents into specific branches in the organizational structure according to rules prescribed for document management. The work flow system can also monitor the present status of documents in the work flow and remind people to act if deadlines are approaching.

25.3.3   Hardware Configuration

The configuration of image systems can vary considerably. A basic system for scanning and OCR consists of a scanner, a workstation for interpretation and verification, and a database server for storage of images and data. A configuration run on a network is illustrated in Figure 25.2.

Large volume productions normally require a stack-fed high-performance scanner with high throughput. A simple flat bed scanner can be used as a backup device to manually feed forms that do not fit into the automated stack feeder of the production scanner, although these significantly reduce throughput. Actual throughput depends on the scanner's ability to handle the physical condition and the size of the document to be captured, as well as its ability to deliver the quality of image output required to meet the needs of the application (forms processing, OCR, archiving, etc.). Throughput rates vary according to the document type to be scanned. For automated batch feed of international A4-size documents (210 × 297 mm) at 200 dpi, scanning speeds usually start at 40 documents per minute; the fastest performers deliver a throughput of around 100 to 150 documents per minute, capturing both sides in a single pass (Gawen, 1996). There are high-speed scanners suitable for large-volume applications that can achieve speeds of 30,000 documents per hour or more. Many scanners offer optional plug-in cards that improve image

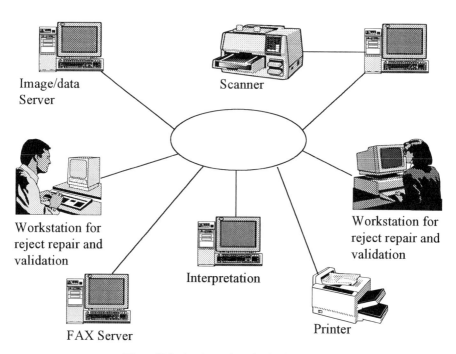

Figure 25.2. Configuration of a basic system.

quality. Other valuable features include built-in bar code readers, endorsers, and colored lights for form template drop-out.

The interpretation device performing OCR can be a PC with OCR software (or other standard platform) or a dedicated hardware system. Today's best image processing board may be eclipsed by software running on a high-performance Pentium PC.

The server for image and data storage must meet the user's need for capacity, depending on the actual application, at least until the recognition processes are complete. A typical questionnaire page held as a graphic image requires about 30k bytes of storage. Elder and McAleese (1996) give the example of a recent survey carried out by Social and Community Planning Research (SCPR) for the U.K. Department for Education and Employment. The survey produced approximately 17,000 questionnaire booklets each consisting of 20 A4 pages. The total disk requirement for having all page images on-line would have been in the order of 10.8 gigabytes. The images were actually written to optical disk in order that they could be called up and reviewed subsequently if necessary.

Workstations for reject repair and validation should be high performance PCs with large screens (at least 17-inch diagonal size).

25.4 OVERVIEW OF ORGANIZATIONAL EXPERIENCES

Scanning and OCR systems have developed extensively over the last decade. Today such systems are frequently used by market research companies, various governmental institutions, service bureaus, the pharmaceutical industry, health care institutions, mail order companies, bookmakers, financial institutions, manufacturing industries, tourism organizations, and the like. The users appear to benefit from the new technology because they have to manage large quantities of information, improve customer satisfaction through faster response time, and accomplish real cost savings in their data capture processes. This section provides a brief overview of how image technology is used in National Statistical Institutes (NSIs) and similar survey organizations, and the experience gained so far. We do not cover the total propagation of applications, but we do know from our investigation that many NSIs, including our own, are exploiting the new technology. Specific systems or applications are not evaluated, rather we try to describe the current status of scanning and OCR in these organizations. Applicable surveys are discussed first and then experiences are presented along a number of dimensions that are important to keep in mind when implementing the technology. The dimensions we have chosen are limited to what we consider important drawing on different presentations and our own experience. They include types of interpretation, recognition rate, substitutions, quality control, forms design, paper quality, organizational issues, costs, response rate, coding, and FAX input.

We have brought together and reviewed information from 13 NSIs about their experiences with scanning and OCR techniques. The organizations are Statistics Canada, Statistics New Zealand, Israeli Central Bureau of Statistics (ICBS), Central Bureau of Statistics in Croatia, Swiss Bureau of Statistics, Statistics Bureau and the Statistics Center of Japan, U.K. Office for National Statistics (ONS), Australian Bureau of Statistics (ABS), Statistics Norway, Statistics Sweden, U.S. Bureau of the Census, Federal Statistical Office in Germany, and the Statistical Institute of Italy (ISTAT). We also included reports from the U.K. Employment Department. Most of these agencies have benefited from the new technology, but negative experiences are reported from Italy. Also the early use of scanning in the Swiss case was more problematic than expected.

25.4.1   Examples of Applications

DIP technology has not yet reached a real breakthrough among NSIs in general. One reason could be the problem of cost efficiency for small surveys (see Dillman and Miller, Chapter 24). Compared to many document production systems in the business sector, surveys often process a relatively small number of documents, often multipage documents (including booklets), featuring various formats. Only censuses and some large surveys have such volumes that processing time exceeds more than a few weeks. Furthermore many surveys are one-time efforts, or they are at least relatively infrequent, say, quarterly or yearly collections. An inventory at Statistics Sweden during the fiscal year 1992–93, before the decision to explore image technology was made, is illustrative of the situation at one organization:

- A total of 274 surveys together representing 5.7 million pages, with an average of 3.6 pages per questionnaire.
- Sixty percent of the surveys have an annual volume less than 3000 pages.
- About half of the surveys have less than 500 respondents, about 20 percent have more than 3,000 respondents, and only four surveys have more than 50,000 respondents.
- The type of information most frequently recorded on questionnaires is handwritten numerical characters, which occur on approximately 80 percent of the annual page volume.
- Different types of marks, such as crosses and lines, occur on approximately 70 percent of the annual page volume; handwritten text occurs on approximately 70 percent of the annual page volume.

Table 25.1 shows the typical use of scanning in the countries covered by our review and the type of interpretation used. It should be noted that the surveys mentioned by no means constitute a complete list of applications or types of interpretation employed at each institute.

Table 25.1. Use of Scanning in Different Countries, Distributed by Type of Survey and Interpretation

Country of National Institute	Type of Survey	Type of Interpretation
Canada	1996 Census of Agriculture	Bar code
New Zealand	1996 Census of Population and Dwellings	OMR and handwritten numerals
Israel	1995 Census of Population	Bar code, preprinted numbers, marks, and handwritten numerals
Croatia	1991 Census of Population	Handwritten numerals and alpha
Switzerland	1991 Population Census	Handwritten numerals and alpha
Japan	1990 Population Census	OMR
United Kingdom	1993 Census of Employment; New Earnings Survey	Bar code, preprinted numerals, handwritten numerals and alpha
Australia	Population Census trials; 1991–92 Retail Census	Numerals and alpha characters
Norway	Annual Activity Surveys (from 1992)	Bar code, OMR, handwritten numerals
Sweden	Survey on Child Care; some 20 other surveys 1996–97	OMR, preprinted and handwritten numerals
United States	Survey of Manufacturer's Shipments and Inventories (M3)	Handwritten numerals (FAX)
Germany	Extra-community Trade	Typed numerals and alpha
Italy	Agricultural, demographic, and legal statistics, a number of forms	Information not available

As seen from the table, large-volume scanning tends to be favored among the NSIs, which is why censuses are reported with such high frequency (from seven agencies). Smaller surveys are also present, but they depend heavily on off-the-shelf software and the existence of medium-sized low-cost scanners. Also the application setup must be so easy that nonspecialized EDP personnel can work with it.

25.4.2 Type of Interpretation

As Table 25.1 shows, the reported surveys cover the whole range of interpretation types, both numerical and alpha data, machine printed, and handwritten. Marks (crosses, check marks, etc) are most commonly used, but often in combination with other interpretation types within the same application, which means that the newest technology integrates OCR and OMR. A distinct advantage of the OCR and ICR technologies compared to the older OMR technology is the software's ability to recognize real numbers instead of marks for numerical intervals. According to Tozer and Jaensch (1994) the OMR option requires collection of numeric data in ranges and therefore is not recommended because of the inaccuracies introduced. Most organizations are currently in transition from OMR to OCR/ICR techniques.

25.4.3   Recognition Rate

Recogniton rate is often considered an important factor when evaluating the efficiency of different OCR systems. Recognition rate is frequently reported, but the level varies considerably, even considering the fact that calculations are made on different bases such as a digit basis, a field basis, or form basis. Among the major explanations of this variation are the different kinds of characters that are interpreted and the different capacities and performances of the systems used. Printed numbers are more accurately interpreted than handwritten numbers; the latter are in turn more accurately interpreted than printed or handwritten alpha. None of the individual results reported deviate from this general trend. Other factors affecting the recognition rates are form design, navigation instructions for questionnaire completion, and adjustments and calibration of hardware and resolution (dots per inch) for scanning. Last but not least there is a balance problem between rejects due to recognition rate, and the level of substitutes that must be taken into account when measuring the efficiency of an OCR engine.

Vendors' information about system performance is often very optimistic. They typically predict very high level of interpretation, on the order of 95 to 100 percent depending on type of information. In practice, however, performance can be much weaker.

The experience of the NSIs and other organizations investigated deviated substantially from the numbers claimed by vendors. Tozer and Jaensch (1994) report that the recognition rates (presumably on a digit basis) achieved across their trials at ABS were 90 percent for numeric characters in Retail and Agriculture Censuses, 93.7 percent in Population Census trials, and 86.5 percent for alpha characters in Population Census trials. In the image trials at Statistics New Zealand, accurate recognition rates were 99.4 percent for marks and 74 percent for numerics. Six percent of the numerics were recognized incorrectly (substitutions), and 20 percent were not recognized at all (rejects) (Scott, 1995; Archer and Scott, 1995). A survey on child care conducted by Statistics Sweden estimated a 98.9 percent overall recognition rate including mark fields. For numerical handwritten characters only, the recognition rate was 94.3 percent (Blom and Friberg, 1995).

The U.K. Employment Department reported that 95 percent of all forms were sent to scanning and OCR, while only 4 percent had to be referred to keying. The final 1 percent was not scanned due to damaged forms. The recognition rates are reported to be 80–87 percent for numerical and alphanumerical fields (Springett, 1994; Thomas, 1994). Statistics Norway has reported a recognition rate of 70 percent on a form basis, namely 30 percent of the forms contained at least one illegible digit. A batch control revealed that approximately 35 percent of the forms were registered and edited without any human intervention, while the remaining 65 percent had to be reviewed and edited by subject matter staff using imaging as a support tool (Pellerud, 1994).

The 1992 ICR Conference at the U.S. National Institute for Standards and Technology discussed research and development on machine recognition of handwritten information. The main issue was to evaluate test results submitted, and it was concluded that the software tested was at least 95 percent accurate at recognizing handwritten numbers, at least 90 percent accurate at recognizing uppercase handwritten letters, and 84 percent at recognizing lowercase handwritten letters. It was emphasized that the software used to recognize handwritten alphanumerical text is indeed usable but needs further development (Rowe and Appel, 1994; Geist et al., 1994).

Finally ICBS reports experience from a total of 5.1 million census forms. The total error rate, after editing, was 0.5 percent. The character interpretation rate ranged from 100 percent for preprinted numbers and 98 percent for bar codes to 57.3 percent for marks and 60.9 percent for handwritten numbers. The overall OCR average error rate (substitution rate) for numerical data was 1.2 percent on a digit basis (Nathan and Givol, 1996; Blum, 1995; Blum and Ben-Moshe, 1996).

25.4.4 Substitution

OCR accuracy can be measured by the substitution rate given the actual recognition rate. Tozer and Jaensch (1994) report that the substitution rate in the Retail Census conducted by ABS was estimated at about 1 percent on a digit basis. According to Springett (1994) and Tomas (1994) the level of substitutions is almost the same at the U.K. Employment Department. Vézina (1994) reported a substitution rate of 1 percent for numeric characters in a test using a subset of questionnaires from the 1991 Census of Agriculture in Canada. It was 2 percent and 5 percent for alphabetic and alphanumeric characters, respectively. Finally Dumicic and Dumicic (1994) report from the Central Bureau of Statistics, Croatia, that the relative number of errors (which we assume to be substitution) made during the recognition process are reported to be 1.6 to 3.1 percent for alphabetic characters, 0.8 to 1.5 percent for alphanumerical characters, and 0.2 to 0.3 percent for numeric characters (the intervals are minimum and maximum error rates over the data processing period). The error rates increased slightly during the period.

25.4.5 Quality Control

In most administrative applications such as financial transactions, insurance, trade, and social security systems, scanning and OCR activities are supported by a battery of quality checks. Typically such checks include check numbers, checksums, and controls associated with comparing input/output figures. It is easy to see why such a battery is necessary; errors can generate substantial bad will and financial loss.

In statistical applications of scanning and OCR, quality control systems are not as common. They are only superficially, if at all, described in references on scanning and OCR. Often concepts such as "batch control" and "mainframe batch editing" are mentioned without much detail. Obviously there is room for improvement here. Statistics production can be viewed as a chain of activities where scanning and OCR constitute one link. For some links, dependable quality control systems have already been developed. Examples are coding, keying, interviewing, and frame construction. For others, many involving the use of new technology, the development has not yet come as far. This state of affairs is understandable. The first step in introducing new technology such as scanning and OCR in surveys is to check whether it is at all feasible. Once feasibility has been established, the new link can be strengthened by developing quality control measures (see Lyberg et al., 1997). To the best of our knowledge, scanning and OCR are currently in that first phase. Dependable quality control systems involving development of methods for error detection and error correction will appear eventually.

25.4.6 Forms Design

Several studies focus on the effect of image technology on forms design. In order to ensure sufficiently high outgoing quality of the estimates, the forms design has to be considered at an early stage. Scott (1995) mentions this crucial issue but adds that "the impact on questionnaire design is not as significant as initially thought. Even results from entirely unmodified 1991 Census forms produced over 99 percent tick box recognition and 74 percent unconstrained numeric recognition." In the U.K. Employment Study mentioned above, the size of the boxes and the larger spacing between fields has resulted in the expansion of survey forms from two pages to three pages. Dillman and Miller (Chapter 24) discuss the importance of the visual layout of opscan questionnaires in order to obtain accurate answers, and they conclude that factors that influence response rates of traditionally constructed surveys also influence the response rates of opscan surveys in approximately the same way.

25.4.7 Paper Quality

The scanning process is highly dependent on paper quality. The paper must have the right reflectivity, opacity, texture, and fiber content. The U.S. Bureau of the Census found that thickness of paper used in scanning may need to be increased from the usual 70 grams to 80 or 90 grams. Statistics Canada used imaging to support the 1996 Census of Agriculture by means of having the respondent-completed questionnaires available at different stages of processing. The questionnaire was made of 70-gram bond paper, which was folded and stapled along the spine to form booklets of sixteen pages using double-sided printing. A cutter was used to remove the staples and $\frac{1}{4}$ inch of the spine from the booklets. In total, the imaging of 274,000 questionnaires resulted in slightly more than 1 percent of questionnaires being affected by double feeds (Duggan, 1996).

25.4.8   Organization

The willingness of individual NSIs to become involved in high-risk projects, which technology projects undoubtedly are, varies among countries. Two major trends can be distinguished. Some countries choose to outsource the development of image management systems to private companies. Statistics New Zealand, Statistics Canada, ISTAT, and the U.K. Employment Department belong to the set of producers of statistics who prefer this strategy. Others, such as Statistics Norway and Statistics Sweden, prefer to manage a major part of the development projects in-house. There are also examples of assigning just the scanning process to a service bureau, which was the case in one of the applications at Statistics Norway. The same strategy is mentioned by Baird and Walker (1995).

25.4.9   Costs

Cost savings are claimed in most reports, even though no detailed calculations are given. Pellerud (1994) reports dramatic achievements, mentioning a 40 percent staff reduction, although there are numerous reservations associated with this figure. He provides little detail because of the difficulty of estimating exactly the actual savings and because of the different approaches and methods of calculation. Crude estimates on savings are reported from ABS. Direct savings are estimated at 78 percent of "keyboard hours." Other examples of savings stem from the reduced number of computer workstations required for data entry (Tozer and Jaensch, 1994).

Statistics Canada reports data capture savings estimated at 60–75 percent, depending on the system. The agency claims that the ICR solution will reduce significantly the number of staff required to capture Census of Agriculture data. There are also savings due to a faster process, the ability to feed the data directly into the agency's Generalized Data Capture System, and the ability to use the images of the questionnaires throughout the remainder of the process. Furthermore the new environment is expected to reduce significantly the time lost in paper handling, saving the time of subject matter specialists, time that can be devoted to analysis and improved data quality. In Statistics Sweden's experience, a midrange scanning system needs a yearly volume of at least 200,000 pages to break even, given the investments in hardware and software.

25.4.10   Response Rate

The possible effect of the image technology on response rates is mentioned briefly in one reference. Tozer and Jaensch (1994) report that forms become less user friendly if reporting areas are defined with individual boxes. The actual effect of this phenomenon has not yet been evaluated. However, Population Census studies relating to average item nonresponse for the bureau's OCR trial form, compared to the OMR form, revealed minor differences only. Dillman and Miller (Chapter 24) find no strong evidence that opscan surveys necessarily produce lower response rates than surveys with regular forms.

25.4.11   Coding

Overall very little information is available about coding operations conducted in conjunction with OCR. However, the connection of back-end automated coding packages to interpreted alphabetic information appears to be beneficial, as mentioned by the Central Bureau of Croatia and the U.K. Employment Department. Population Census trials conducted by ABS show that significant additional savings were obtained through more extensive use of automated coding in conjunction with OCR.

25.4.12   FAX Input

The U.S. Bureau of the Census reports on the development and use of paperless FAX image reporting. This is similar to the imaging technology discussed elsewhere in this chapter. The main feature of the FAX image technology is that it is a kind of "distant scanning." The scanning of documents is carried out on the respondent's FAX machine. An electronic image is submitted to the agency, and the information is automatically recognized and may be optionally verified on screen, much like data scanned and interpreted from hard copy documents at the agency. To test the technology, the Census Bureau assembled a proof of systems and began testing on the survey of Manufacturer's Shipments and Inventories. Of 523 forms received in the first six months of testing, 3 percent were not recognized. The average unconditional success rate, defined as characters interpreted correctly by the software, was 84 percent during this period (Appel, Petunias, and Russell, 1994).

25.4.13   Summary

The studies of NSI applications suggest that scanning is a valuable new tool for data capture in surveys. The information we have obtained, however, is hard to interpret and compare due to the lack of a common vocabulary in which information about processes and measurement methods could be expressed. This is not uncommon for areas in which emergent technologies are tested and applied. Since new technologies seem to be frequently used and widely accepted, there is now a need to focus more on quality problems.

25.5 GUIDELINES FOR TECHNOLOGY SELECTION

Given that scanning and OCR seem favorable and worthwhile using, this section provides some guidelines for planning and implementing an application. The process includes the procurement of software and hardware components and building the actual application. A general recommendation is that work should be done according to some kind of structured concept, such as business process re-engineering (BPR) or total quality management (TQM). One must precisely identify the problem, select evaluation criteria, collect data, and set priorities among technology features. Baird and Walker (1995) provide a model for the relative speed of implementation of new techniques, based on a benefit and risk analysis, where OCR appears midrange on a high-to-low scale. The importance of early tests and evaluation programs is also stressed by Blom and Friberg (1995), and more or less defined requirements are provided in the references cited in this chapter. For instance, Schaadt (1992) recommends that potential users develop a series of checklists of their own operations adjusted to their own environments, namely to look at how the system is going to correct errors and how these errors can be displayed.

We start with some general recommendations for implementing a scanning system and then continue with some requirements as they can be formulated in a typical biding process. We end up focusing on the necessity of a testing and evaluation phase when a system is considered.

25.5.1 General Recommendations

The following points may be helpful for beginners or organizations without substantial OCR experience.

- The adaptation of image technology cannot be separated from other possible technology choices. Consider the balance between different technologies and the need for mixed mode data collection. Standard packages are always preferable to tailor-made if they satisfy the requirements. Because of the rapid development of new products on the market, one should not engage in time-consuming and extensive development of tailor-made solutions. Normally standard packages can be easily adapted to any IT environment.
- Make sure that the new system is able to produce information of a quality at least similar to that of the current system at a lower cost, or of a higher quality at approximately equal costs.

- The system must be able to record forms, and interpret, register, edit, correct, and store information. It must also be able to transfer information and images to other systems, and control the flow of material, including case and folder management, retrieval, and printout capabilities.

- Priority should be given to the equipment's manageability and its functional, environmental, and ergonomic properties; the availability of programs, operative and development aids; and possibilities for extension, including an image database/document image processing system.

- The system must fit into the existing EDP environment, have replaceable components, and be able to expand so that future increasing volumes can be managed. It must be a multiuser system having an interface with the IT platform and be capable of handling multipage forms with double-sided print.

25.5.2 Systems Requirements

- As for *recording/scanning*, the system must be capable of recording forms in any format from international standard A5 up to A3, read double-sided print, handle variation in the number of pages in a set of forms, simultaneously record different types of forms, index automatically, and read at least 30 A4 sheets per minute in continuous production.

- As for *definition of forms*, the system must (1) be easy to handle when it comes to defining forms for recording, including a simple check function, without special EDP skills being needed, (2) have functions that allow an easy modification of the field formats, positioning of fields, and so forth, and (3) have a function that filters irrelevant information such as separation and other special characters.

- The system must be capable of *interpreting* TIFF files. Final requirements depend very much on the specific application and should be *verified by testing*. Of course it is possible to give priority to some specific type of characters. Table 25.2 can serve as a reasonable set of required recognition rates.

- The system should include *editing, correcting, and coding capabilities* comparable to those of the normal data-entry/correction system used in surveys. This includes duplicate controls, validation of values and intervals of values, checks against external tables, controls between all fields in the form, execution of equations, acceptance of values that initially were considered suspicious, among other things. The system must also handle reject correcting and other editing/correcting with multiple simultaneous users, and have functions for manual supplementary keying of fields that are not interpreted. In the process of editing/correcting, the system must use no more than 1.5 seconds for changing an image when operated in stand-alone mode.

Table 25.2. Recommended Recognition Rates per Character

Type of Characters	Recognition Rate per Character (in %)
Marks in boxes	>99
Numerical preprinted (like OCR-B)	>95
Numerical machine printed	>90
Numerical handwritten	>90
Machine typed text	>60
Handwritten text constrained with boxes	>80

- On the *output side* the system must be able to transfer images to an electronic image database adapted to the EDP environment, deliver data in ASCII format, and create images in the standard TIFF format.
- For *work flow information and production statistics*, the system must include a process control and have statistical functions showing the recognition rates on a character, field, and form level.
- For *safety and protection against unauthorized access*, the system must be able to store information and images while offering adequate protection against unauthorized access, authorize the access to information and images at the user level, have backup procedures that are simple to handle, allow for backup of information and images without having to close down, allow simple re-creation of all information in the system after interruptions, and maintain a system log of all occurrences.

25.5.3 Test and Evaluation

Selection of technology should be guided by results from various testing activities. First, it is important to check whether general user requirements are met. This test should include, at a minimum, volume and performance checks for scanning, OCR, editing, and correction. Second, this first test could be conducted on parts of the organization, parts of the material, parts of the staff, and so forth. Errors and other problems are identified and proper changes made. The improvements are introduced and the test expanded to include new parts of the organization, material, or staff. Third, all test forms should come from the users' own applications. Fourth, the vendor should be informed about the testing and given time to prepare for it. Finally a few users' practical work should be observed and evaluated.

25.6 TRENDS AND FUTURE NEEDS

Taking into account all the options for data collection and capture that today's technology offers, imaging is just one of many methods that should be included in the survey manager's toolbox. We hope this chapter has shed some light on the issue of choosing scanning and OCR for the data capture process. However, the question of the extent of the future use of this technology is still unclear. We expect electronic forms to be distributed by modern data communication media replacing traditional paper questionnaires, a process that has in fact already started. Electronic data interchange (EDI) will probably play an increasingly important role as the direct mode for data collection, and system-to-system communication will decrease the need for paper conversion. But those who are resonsible for today's data collection will need more efficient tools for carrying out their tasks while they wait for the paperless society. We expect that for at least another five to ten years, paper media will dominate the processing of censuses and sample surveys while image technology is being more fully developed.

Admittedly, the technology still needs improvement. Scanners will become cheaper with higher speeds and better image quality. Personal computers and networks will become even more powerful and allow for more efficient and flexible OCR products. Resources for interpretation will probably be available in global networks like the Internet, which will also give small organizations the opportunity to use image technology. Today's limitations due to constrained handwriting will be taken care of by a more powerful and flexible OCR technology that allows for a more user friendly layout of forms. This means that there will be less need for special and space-consuming boxes for the different characters and drop-out colors for field frames and fixed background forms data. We do not know how long it will take to develop an OCR technology that can interpret free format continuous handwriting, but OCR accuracy will definitely increase. For instance, new techniques for tracking substitutions will be developed.

Perhaps in the not too distant future imaging will be commonplace in survey data collection. Distant scanning via FAX and OCR will appear as FAX mail-systems. We also have to consider the problems and possibilities with different combinations of collection methods for surveys and connect technology research with mixed mode data collection research.

CHAPTER 26

Pen CASIC: Design and Usability

John Bosley, Frederick G. Conrad, and David Uglow
U.S. Bureau of Labor Statistics

26.1 INTRODUCTION

Pen computers provide an attractive option for the CASIC developer. They are uniquely suited for standing field data collection; they facilitate natural conversation between interviewer and respondent, and can be easier to operate than conventional, keyboard-driven laptop computers. Pen computers may be especially appropriate for surveys requiring observational data collection or complex interviews in which interviewers may need to navigate freely around data collection forms.

This chapter reviews the advantages and drawbacks of pen computing for CASIC. It presents a brief history of "pen CAPI" and some initial guidelines for designing and implementing pen-based CASIC. It is based on our experience developing and testing the usability of a pen-based application for the Consumer Price Index (CPI) conducted by the U.S. Bureau of Labor Statistics (BLS). In addition the chapter has two subsidiary goals: (1) to discuss the pros and cons of using graphical user interfaces (GUIs) in CASIC and (2) to introduce the notion of CASIC usability and its evaluation.

26.2 ADVANTAGES AND DISADVANTAGES OF PEN CASIC

The primary advantages of pen computers for survey data collection are ease of use, support of a "natural" conversation between interviewer and respondent, and the compatibility of the pen with graphical user interfaces.

Using pen computers involves a talent even inexperienced computer users have—writing and pointing with a pencil or pen. As a consequence training

Computer Assisted Survey Information Collection, Edited by Mick P. Couper, Reginald P. Baker, Jelke Bethlehem, Cynthia Z. F. Clark, Jean Martin, William L. Nicholls II, and James M. O'Reilly. ISBN 0-471-17848-9 © 1998 John Wiley & Sons, Inc.

data collectors to use pen-based applications may be significantly easier than training for conventional CAPI where typing experience is almost a must. In addition pen input is naturally one-handed, an advantage for standing data collection where two-handed touch typing or use of a mouse may not be possible. (For further discussions of the benefits of direct manipulation devices—pens, mouses, trackballs and touchscreens—for nonexpert users, see Shneiderman, 1997.)

Pen computers may also foster more natural interaction, helping the interviewer to retain eye contact and focus attention on the respondent. MacNeill (1995) speaks to this point in his enumeration of personnel benefits from using pen-based systems in the field: "Pen devices allow natural communication.... Sit down with a keyboard computer on your lap and you have set up a wall between you and your contact. The best ... meetings are friendly and conversational, face-to-face. Field workers find that pen tablets and slates disappear like a notepad or clipboard, and that the devices help them maintain the eye contact they desire."

One reason that pen computing may promote eye contact with a respondent is that it simply requires fewer eye movements than its keyboard equivalent: When the interviewer is not looking at the respondent, he or she needs only look at one place—the screen—to both read the computer's output and enter his or her input; even the best typists need to move their gaze to the keyboard on occasion. Thus the interviewer can more easily restore eye contact after breaking it to view the screen. However, there may be interactional costs associated with this design: When pen input requires close attention and fine hand-eye coordination, when text is input by tapping a virtual keyboard, or when very small objects on the screen must be activated by a precise pen movement, most or all of this naturalness is lost.

Finally pen computers are especially appropriate to graphical user interfaces (GUIs) and the advantages this kind of interface offers. While pointing devices such as a pen can be used with text-based systems, GUIs optimize their use through screen objects known as *controls*. Compared to screen objects created with combinations of text characters, a graphical control (e.g., a radio button) stands out better on the screen. For example, it can have the visual attributes of a three-dimensional object and display an appearance as "pressed" or "not-pressed," providing instant feedback to the user as to the success of an action using a pointer. In addition controls can be manipulated in relatively complex ways (i.e., more than just on and off). An example would be a listbox, where pointing to the box causes a list of entry choices to drop down from the box, allowing the user to select from that list. (This and several other controls are shown in Figure 26.1.) This ability to expand when needed permits more functions on a single screen without the need to scroll or paginate to a new display. The features of GUIs that particularly affect CASIC design are presented in Section 26.4.1.

Pen CASIC also has its drawbacks, many of which are inherent in portable computers. All full-function portable computers are still heavier than users

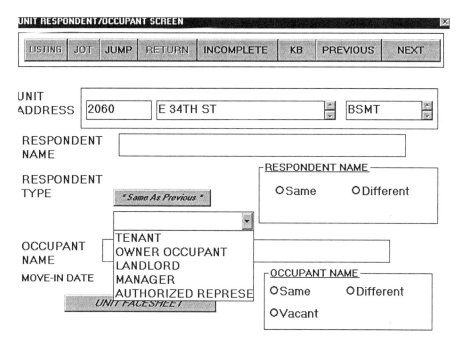

Figure 26.1. Sample of GUI controls from the CPI housing instrument.

would like for them to be. Battery life is still too short to support a full day's work. In addition the typical portable computer screen, usually an LCD, is not as easy to read as a desktop monitor. To the extent that weight and battery life problems have been partially overcome in devices such as Personal Digital Assistants (PDAs), this has come at the cost of reducing desirable features such as screen size and memory capacity.

A particularly important problem for pen computers has been the slow development of handwriting recognition. The original proponents of pen computing believed that handwriting would be reliably recognized and stored as text files. Today most handwriting "recognizers" (the software that translates pen strokes into ASCII text) are neither accurate nor versatile enough to fully support development of pen-based applications. For example, LaLomia (1994) argues that users will not accept recognition accuracy below 97 percent. Gibbs (1993) reviews 13 recognizers from seven different vendors and cites vendor estimates of "untrained" accuracy in the 85–95 percent range.

Frankish, Hull, and Morgan (1995) tested recognizer performance and user satisfaction across different types of applications. They found that a minority of characters accounted for a very high proportion of recognition errors. Error rates were as high as 38 percent for the most-frequently misinterpreted character. This prompted researchers to propose that recognizer accuracy be improved by focusing individualized recognizer "training" on only the most troublesome characters, ignoring the remainder with low error rates (see Blom and Lyberg, Chapter 25; Dillman and Miller, Chapter 24, for the role of handwriting recognition in other CASIC technologies). This same research found no evidence that recognition improved as users acquired more experience with the system. Furthermore the most troublesome characters varied widely across users. These findings led to a recommendation that a user should identify her or his own subset of troublesome characters and train to improve their recognition.

Given the current capabilities of handwriting recognition software, pen CASIC may only be suited to surveys where most of the data entered are numeric (since the recognition of numeric characters is significantly better than for alpha), and where text responses can be reduced to multiple choice objects or items in pick lists.

The ergonomics of current pen computers continue to be problematic. Full-size pen computers today still weigh several pounds. The machines used by BLS weigh about 3.5 pounds, and they are relatively light compared to others currently on the market. This means that all except the strongest users will quickly tire if they try to hold the machine like a paper tablet or clipboard. Steadying a device of this weight is tiring even when the user braces it against the body. Furthermore, when one end of the computer is supported by the torso, a fairly wide strip along the bottom of the screen is hidden by the user's clothing or body, and this screen region becomes unusable.

Unfortunately, lighter machines usually have smaller screens. Using the BLS machines again as an example, the screen is slightly less than 8 inches diagonally—already smaller than would be ideal. This small screen area is quickly consumed by standard-sized controls and icons, partly nullifying the cognitive and ergonomic advantages gained through forms-based design and often forcing the designer to compromise on screen layout. Some of the ways of circumventing problems learned in the process of dealing with this disadvantage will be discussed in the design Section 26.4 below.

Finally, with the exception of some "convertible" models (pen and keyboard), most pen computers are designed so that the screen is not itself physically covered by the case when closed. This makes these devices more susceptible to accidental breakage than standard laptops (whose screens are covered when not in use).

26.3 DEVELOPING FOR PEN CASIC

The origins of GUI and pen CASIC applications can be traced to the overall expansion of CAPI surveys in the late 1980s, and the need for *forms-based* CASIC authoring systems. As opposed to many commercial data collection applications that used the paper form as a design metaphor, CASIC instruments were typically designed as *question-based* systems—presenting one or two literal questions to the interviewer at one time.

The development of forms-based CASIC occurred under pressure from users who needed to conduct field surveys. Where it is difficult to establish the exact order of items in advance, multiple item display can be an aid to the data collector (Uglow, 1991). Such studies are often targeted at business establishments rather than households. Forms-based authoring systems were developed beginning in the late 1980s but were built on text-based software.

The need for forms-based CASIC systems led in turn to an interest in pen, particularly where the interview or collection was conducted standing and using conventional laptops was difficult. An early CASIC study by Couper and Groves (1992, p. 209) found that neither a pen computer using handwriting recognition nor a touchscreen machine "performed as well as 'traditional' keyboard computers in terms of speed and errors." However, they also noted that "such differences could well disappear given sufficient familiarity with and training on these machines." Data collectors were evenly split between "happy" and "unhappy" with the pen machine, which in this respect scored lower than one conventional laptop but above another keyboard machine and the touchscreen (Couper and Groves, 1992, p. 208). They also concluded that pen machines would likely find their best application for surveys heavy in numeric and light on text input.

Several U.S. federal agencies have experimented with pen CASIC. Researchers at the National Agricultural Statistical Service (NASS) tested pen computers for collection of field agricultural data (Ecklund, 1991). The National Center for Health Statistics (NCHS) funded a pilot project using pen computers for collecting survey data in 1993–94. Twenty-five field representatives collected data for the National Hospital Discharge Survey. Pen-based collection was ultimately not implemented because of hardware costs (Noce, personal communication; Burt, 1994).

The Bureau of the Census and BLS began research and development work on pen computing in 1990. The Bureau of the Census began its testing of pen computers with a field test for the Current Population Survey (Harvell, 1990). In 1993 the Census Bureau tested a pen-based application for sample listing data and followed up with some production collection. This application was similar to the BLS CPI listing instrument that was used to list the CPI housing sample in 1997. The Census Bureau also is involved in preliminary research combining pen computing with audio-CASI.

Because the CPI is heavy on both numeric data collection and standing collection, it is a good candidate for pen CASIC by the Couper and Groves (1990) criterion. Each month the CPI collects data from nearly 100,000 residential and establishment reporters to create a measure of price change aggregated across a predefined market basket of purchases. As noted above, BLS began its pen CASIC research and development work in 1990. In late 1991 and early 1992 nearly all of the CPI's field data collectors tested several mock instruments, and the feedback from those sessions encouraged CPI planners to continue pen development work. The tests also helped generate some early design guidlines for pen instruments (Uglow, 1992). Two pen

CASIC applications have been used to produce the CPI Housing Survey since 1997, one to list the housing sample and the other to collect price data.

While pen CASIC is an embryonic enterprise, pen computers have achieved a firmer foothold in commercial data collection. The most intensive use of pen computers is for observational collection of organizational records, particularly for applications that stress the collection of numeric data (Giangrasso, 1995; MacNeill, 1995; MacNeill and Giangrasso, 1996). Examples include on-site disaster insurance assessments by the Federal Emergency Management Agency, computerized medical records, and patient charts in hospital settings; warehouse inventory management; and tracking data for the transportation and shipping industries. These examples and those CASIC applications for which pen computing seems particularly well suited normally involve capturing a limited range of data elements, such as case identification data and other case management information, and respondents' selections from among a few structured responses to well-defined questions about a narrow domain of inquiry. These applications, then, are *vertical*, as opposed to *horizontal* applications such as desktop office software. Meyer (1995, p. 74) defines vertical applications as "designed for clearly defined problems. They are very strict in their structure and are usually developed for a specific client or market. Additionally he says, "[E]ach field has a very small input domain, i.e., it is well known what data can be entered. Therefore [vertical applications] need not rely on difficult free-form handwriting... recognizers, but use radio buttons, pop-up lists, etc. instead. Forms are easy to set up and easy to use" (1995, p. 73).

26.3.1   Hardware

In the initial years of pen computing, pen computers were designed and engineered as a separate product line. Today most pen computers are developed from standard laptops, with the addition of a pen-sensitive (digitizing) screen and the pen itself. Most do not have built-in keyboards. As a consequence pen machines have tended to lag conventional laptops by a generation in terms of adopting processors, screen features, and so on. For example, Pentium chips were in laptops about two years before they first appeared in pen computers. Other features, such as screen display types (including color) and battery and storage options, mimic those available in conventional laptops.

There are currently several PDAs available which use pen input. Some are being used for CASIC work where the interface can be implemented on the extremely small screen (e.g., Nusser et al., 1996). A related development is Windows CE, an operating system that makes it easier to port software to PDAs. Windows CE is currently being implemented on new hardware using small keyboards. These platforms, while potentially promising, are outside the scope of this chapter.

26.3.2 Development Platforms for Pen CASIC

There is currently no CASIC development system specifically designed for the pen. Pen CASIC computing will become generally available, however, as each of the major systems becomes available in Windows versions. Generally speaking, Windows applications are able to become "pen aware" through the addition of special drivers. This software allows the pen to function similarly to a mouse and the interviewer to enter handwritten text in text box controls. As an alternative to instrument building in traditional CASIC authoring systems, instruments can be built in a visual development environment, such as Visual Basic, Visual C++, Delphi, CA Objects, and PowerBuilder. The drawbacks are the same as those confronting any CASIC designer looking to develop a CASIC instrument without benefit of a true authoring language. Mechanisms peculiar to CASIC must be written to implement and control movement, data integrity under irregular movement, and special data-driven conditions such as survey rosters of persons and conditions.

For relatively simple data collection tasks, there exists a number of versatile, easy to use *forms creation* software tools, that use the paper forms metaphor while providing quick development. Many of these packages are pen-enabled, particularly those that have been developed for PDAs. One use for such packages is rapid prototyping of instruments that may be completed in a richer programming environment.

26.4 DESIGNING FOR PEN CASIC

The combination of GUI platforms and pen computing to produce CASIC applications requires rethinking our models of data entry methods, instrument sequencing, and user interface design. Some issues have to do with capitalizing on GUI features in the application, while others are grounded more in pen system characteristics.

26.4.1 CASIC Design in the GUI Environment

Before approaching particular pen-specific design issues, we need to address the more general challenges and opportunities of the move from text-based to GUI-based CASIC instruments. These issues will become increasingly important over the next several years as CASIC moves to GUI platforms and older, text-based instruments are redesigned and rewritten.

In traditional text-based CASIC interfaces, individual questions generally are presented one at a time in a linear succession. This mode of presentation works well if the survey has a relatively simple structure. Many surveys, however, are organized around concepts, not specific questions, and so question order is not the same for all respondents. It is hard to represent such conceptually based designs in text based applications but GUIs are well suited

528 PEN CASIC: DESIGN AND USABILITY

to this. In some situations, though, the interviewer wants to know "Where am I?" relative to that general framework. The wider *visual* scope of forms-based GUIs gives the interviewer a way to maintain some awareness of his or her location in the overall conceptual framework. For example, the system might display multiple items on one screen, or change layout from one context to another.

With GUIs and pointing devices it is possible to design interactions so that they more naturally reflect users' thinking. For example, GUIs enable designers to capitalize on the distinctive properties of the pen as an input device. Buxton et al. (1986, p. 475) offer the concept of "phrasing" to describe an input task that is "woven together by a thread of continuity similar to that [which] binds together a musical phrase." The type of phrasing is contingent on the particular input device—so that keyboard "phrasing" would not necessarily facilitate mouse, trackball or pen inputs. A simple example of an efficient "phrase" that can be employed in a pen environment is a move operation accomplished by what is known as "drag-and-drop," rather than by "cut," then "paste." In the latter case the basic task of moving a screen object must be decomposed into several discrete operations. First the user selects the object, then copies it into a buffer and deletes it (usually one action), then copies it from the buffer into the new location. This seems more difficult and error prone than dragging and dropping in which the user selects the object and seemingly moves it with the mouse or pen to the new location.

Using Visual Design to Enhance GUI Usability

Good design can provide the "threads" that link screen content together into coherent and efficient action "phrases," to borrow Buxton's metaphor. The visual design of the form powerfully influences the perceived relationships among a number of small, varied tasks.

In considering issues such as the volume of information on the screen and its arrangement, thought must be given to preserving the visual autonomy of basic tasks, while indicating links between them. Overall cognitive load on the data collector (interviewer) may be decreased through an arrangement of groups or columns, use of frames, and similarity of controls used for similar functions. This kind of arrangement of subelements on a screen has been called a "graphic language" by Twyman (1979) and others.

The goal of good visual design is to use position, size, shading, color (if available), and other attributes of screen objects to guide the user to quickly understand the task(s) represented on the screen. In an interview situation these tasks include identifying and comprehending questions to be asked, discerning the proper order in which they should be asked, and determining how and where a response is to be entered. Finally the "graphic language" should tell the user what to do when finished with that screen.

Using GUIs to Solve CASIC Design Problems

GUIs and direct manipulation enable designers to solve certain longstanding problems in traditional CASIC applications. Some examples from our own experience are:

- Toolbar objects can replace command syntax. For instance, "irregular movement" can be guided by a jump menu accessed through a toolbar control. In a proposal which could extend these methods to a pen environment, Geisler (1995) has developed what he calls "*gedrics*" (GEsture-DRiven IConS). Gedrics are pen gestures or marks that can be tied to common, perhaps complex, application functions, thus freeing the user from having to target a particular screen control for tapping (see MacKenzie et al., 1994, and McQueen, 1995, for additional discussions of the use of pen gestures).

- Through "file folder" tabs, three to five individual screens can be combined into a single multipage; the user moves between pages by selecting "tabs" (see Figure 26.2, right edge of screen).

- Similarly the problem of quick movement between roster entities, along with visual feedback as to where one is in the roster, can be accomplished through the use of tabs which clearly identify each roster entity.

- Multiple items can be displayed on a single form under conditions of

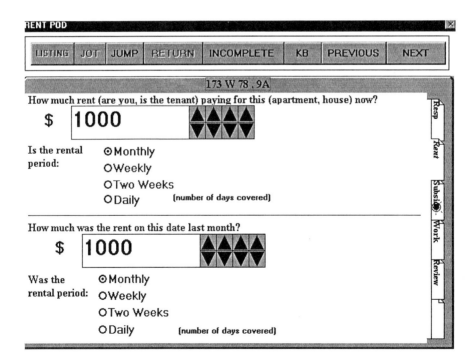

Figure 26.2. Using GUI tabs for instrument navigation.

simple branching by "graying" and disabling controls to render inappropriate items unavailable for use while maintaining their visibility. The same feature can be used to control default item order with multiple item forms.

26.4.2 Pen-Specific Design Issues

The goal of pen-centric design is not simply to copy desktop GUI designs but to take advantage of the specific characteristics of the pen as a unique input device. For example, if a text box is being used, the designer must decide how to give it "focus," that is, make the text field active. The basic choice is between having to tap the field with the pen to give focus, followed by writing to enter the value, or giving focus by the act of writing alone. The first method is less error prone; the second easier and quicker for the data collector under most circumstances. In either case the designer must decide whether there should be any particular visual feedback to indicate the "current" control or field.

The designer also must decide on the use of confirm sequences (for items or for entire forms). Should tapping a "complete" or "next" control be required to move to the next screen, or should completion of the last item on a screen automatically lead to forward movement?

Some design issues may be tied to specific hardware. For instance, some machines have border areas of the screen where pointing is inaccurately recognized and some pen events are not recognized at all. Thus, if a pen application is to perform well across hardware platforms, designers need to avoid interaction at the periphery of the screen.

Despite its ease of use, the pen—like any other device—is not ideally suited to every conceivable task. There may be times when other input or pointing techniques are desirable. Where both hands are fully occupied in the pen data-entry mode, any additional input must be provided by means such as sound (voice), nonmanual motor responses such as head movement, and the like. Studies by Oviatt and Olsen (1994) and Schmidt (1996) discuss conditions for the use of supplementary voice inputs with the pen.

26.4.3 Summary

CASIC use of forms-based, GUI instruments with pen input can provide new flexibility and more efficient handling of survey complexity. To take advantage of these benefits, however, the designer must come to grips with a new universe of issues, both in graphical screen design, and in the user's interaction with the computer via the pen. Only the most preliminary solutions now exist, but the first generation of pen instruments are showing that pen CASIC is a reality. We now turn to one such instrument, a pen-based CPI data collection tool developed at BLS, and our initial evaluation of its usability.

26.5 USABILITY TESTING FOR PEN CASIC

The two CPI housing instruments mentioned earlier ("housing listing" and "housing collection") have been refined on the basis of experienced CPI interviewers' informal comments and the results of two field tests. In addition the collection instrument has undergone a systematic usability evaluation. In this section we describe all of these efforts, focusing on the more systematic test.

26.5.1 Rationale for Usability Testing

Usability testing or usability engineering refers to a range of activities that provide evidence about how easily and accurately users can accomplish their tasks by interacting with an application. Usability testing is essential because it is an empirical alternative to developers' intuitions; because it is usually done by someone other than the developers, it can be relatively impartial. The expense and effort of usability testing are usually justified in terms of reduced cost derived from increased user satisfaction, faster learning, better retention, and more accurate use of the application (e.g., Nielsen and Levy, 1994).

The data from usability testing can include quantitative performance measures (e.g., the users' response times and error rates in performing particular tasks), satisfaction judgments, and impressions of where problems exist (as in a focus group). They do not always come from end users. "Usability inspection methods" (e.g., Nielsen and Mack, 1994) do not involve end users but can be valuable for answering certain types of usability questions. Our position is that any usability testing, whatever the method, almost always leads to a better product than if none is performed. However, methods differ in the kinds of usability problems they help researchers identify and some methods and approaches are more credible than others. Because our time and money were limited, we chose a usability inspection method—heuristic evaluation—that required relatively modest resources. We had found this method useful in detecting interface problems in other applications (Levi and Conrad, 1996).

Usability studies of pen computing are rare in the human computer interaction literature. Moreover usability testing is only gradually becoming a widely accepted part of the CASIC development process, despite at least one compelling argument to make the practice more routine (Couper, 1994). CASIC presents unique usability problems which are intensified by portable computers. For example, a user (interviewer) needs to interact with the respondent and the instrument simultaneously, possibly entering observed data about the respondent's environment (as in the CPI housing survey), while ignoring ambient noise and distractions. Therefore CASIC developers need to build applications that place minimal demands on the user's attention so that the user can devote more attention to the interview situation.

26.5.2 Informal Evaluation by End Users

Throughout development of the CPI housing listing and collection instruments, numerous users tried the pen platform and provided specific, though informal, feedback. These informal evaluations took several forms. In one form large groups of interviewers at a training session were asked to explore the instrument for about an hour. After this they discussed their experiences in a group format, moderated by a developer. In another, developers met with individual interviewers and talked to them as they interacted with the instrument. Finally a prototype review team that included two or three interviewers was formed. The team members conducted mock interviews in plausible field setting and summarized their reactions in a report.

There are several limits to the ways informal feedback can improve usability. First, such data are impressionistic; that is, they are the developers' impressions of where the users' problems lie and the users' impressions of what affects their performance. The developers derive some of these impressions by observing some interviewers interacting with the system, but more often they form impressions from talking to users about their experiences.

Second, our end users are expert at conducting interviews, not evaluating usability. While it is important to consider their intuitions about which features of the design are hard to use, their predictions that such features will hinder their performance must be viewed skeptically in the absence of behavioral data (e.g., times and error rates), and after a short period of use.

Third, the approach is limited by the fact that the feedback was gathered by the developers, who are hardly impartial. Developers recognize that modifying the interface to overcome usability problems takes time and effort. Therefore they may be inclined to recommend that users be trained to work with the existing system as an alternative to redesigning it. Fourth, developers are often not trained in the kind of experimental techniques needed to provide reliable data.

For all these reasons the data gained from informal user evaluations were not *systematically* incorporated into the design process. Nevertheless, the users' feedback led to numerous design changes throughout development. In addition the developers were able to articulate several broad pen CASIC design principles from the feedback. Some of the key principles follow.

One set of principles centers on using the pen for text entry. Because handwriting recognition still has fairly high error rates, text entry should be avoided if possible. When it cannot be avoided, the interface should highlight text entry fields and distinguish them from controls that are only sensitive to pen taps. The designers should make it as easy as possible for the users to write, for instance, by enabling the start of a writing action—bringing the pen into contact with the text field—to shift focus to that field. The idea is that this is easier for the user than requiring an explicit tap to bring focus to the text field before writing.

Another principle was derived from the users' depiction of scrolling as a clumsy navigational tool. Instead of vertical scrolling, users said they would prefer to move through pages of the instrument. The designers extended this idea so that users could move directly to sections of interest by using a set of notebooklike tabs in the collection instrument (Figure 26.2). The users indicated that horizontal scrolling was so demanding that the designers eliminated it entirely.

A further principle concerns positioning screen elements so that an element's location is commensurate with its importance. The principle maps an element's importance to its top-down, left-to-right position with the most important elements in the top left of the screen, and least important in the lower right.

Finally users' feedback led to an explicit consistency principle. Both the positioning of controls and the choice of screen elements used for common functions should be similar from one screen to the next.

On balance, we believe it was far superior to have informally collected impressionistic data than to have gathered no data from users. However, we also recognize that anecdotal data of this sort require skeptical interpretation. Although more systematic and structured evaluations were desirable, we did not have sufficient resources to test end users in a rigorous way. Given these constraints, we carried out a heuristic evaluation, a technique that is reputed to be faster and less expensive than systematic end-user testing (Nielsen, 1994). We focused on only the collection instrument, since the listing instrument seemed simpler and more usable and the results of the informal tests were consistent with this impression.

26.5.3 Heuristic Evaluation

One can divide popular usability testing methods into two major categories: traditional end user testing (e.g., Card, Moran, and Newell, 1983) and usability inspection (Nielsen and Mack, 1994). The first of these involves measuring the behavior of actual users interacting with the application, such as pauses, errors, and requests for help—and treating these as indications of usability problems. It relies on the methods and reasoning of experimental psychology. Although its results are highly credible, end user testing may require a formal laboratory and professional usability personnel to design the studies and interpret the results. The analysis can be time-consuming because it can involve detailed activities like coding video tapes of user sessions (see Hansen, Fuchs, and Couper, 1997, for an example of end user testing in the CASIC domain).

Usability inspection methods, in contrast, require relatively few resources and less background in experimental procedures. The inspection method we used, heuristic evaluation, involves recruiting three to five usability experts and asking them to inspect a software product on the basis of a set of usability principles or heuristics (Nielsen, 1994). They are asked to note violations of particular heuristics, and these violations become a list of usability problems. The problems on the list are then ordered by their rated severity. The entire inspection process generally takes about three days.

Where it is possible to do systematic end user testing, we advocate doing so rather than relying on experts' judgments. There is no substitute for end user performance data, like task completion time and accuracy, collected under both controlled and field conditions. In addition users' satisfaction judgments, collected with some sort of standardized instrument, can supplement user performance data for design decisions. Nonetheless, the heuristic evaluation provided valuable feedback to the developers.

Methodology

Heuristic evaluation, along with the other inspection methods, differs from conventional empirical usability testing in significant ways: (1) Evaluators are not drawn from the user community, (2) evaluations take less time, (3) evaluations are easier to set up and run, and (4) evaluations cost less.

Evaluating Interface Usability versus Instrument Ergonomics

CPI interviewers reported in the informal sessions that the ergonomics of pen computing in the interview situation may lead to instruments that are hard to use. Based on informal use of early versions of the listing and pricing instruments, they reported problems of glare on the screen from sunlight, fatigue from holding the pen computer while standing, visual obstruction of the screen by their clothing, belongings, and body parts, and so on. While acknowledging that this ultimately limits usability, we chose to look primarily at the interface's usability under relatively good conditions. For the usability inspection, evaluators were seated at a desk with the pen computer resting on the desk. There was uniform fluorescent lighting, no time pressure, no live respondents, no environmental distractions, and so on. Our reasoning was that if the interface is hard to use under such relatively ideal conditions, then it will be even more difficult to use in the field. The approach sought to improve the design first with further refinement as the demands of the field are imposed.

Identifying the Evaluators

In heuristic evaluation the evaluators are "usability experts." Evaluating a pen CASIC application requires two additional kinds of knowledge: familiarity with pen technology, so as not to confuse the differences between pen and desktop computing with usability problems, and some knowledge of the data collection task for which the application was designed—collecting housing data—including task-specific terms and concepts.

Not surprisingly these two types of knowledge (familiarity with pen computing and collecting CPI housing prices) rarely occur together. Our solution was to recruit evaluators with pen development experience and to provide them with a one-hour tutorial about specific data collection procedures. The standard recommendation (e.g., Nielsen, 1994) is that using three to five usability experts as evaluators is optimal: Fewer than that number will leave some usability problems undetected, and more will produce considerable duplication. We recruited two professional pen developers with "usability awareness," and one usability specialist without actual pen experience but with a general knowledge of the technology and familiarity with the CPI data collection process.

Developing Usability Heuristics

When used to evaluate a general purpose application, usability principles (heuristics) should be similarly general, for example, "Speak the user's language." However, the characteristics of pen CASIC impose different demands. For example, it is generally advisable that software products should be easy to learn to use: If novice users are frustrated, they may not persevere in trying to master the interface. However, in the case of CASIC, the typical users will repeatedly perform the same data collection task with a particular instrument and so become expert at that task relatively quickly when compared to the total amount of time they will use the instrument. Therefore a more appropriate principle for designing CASIC products should emphasize the needs of skilled users. Similarly fatigue due to moving a pen is potentially a more serious issue than fatigue due to moving a mouse because the pen is both the pointer and input device. Thus another pen (though not necessarily CASIC) design criterion is to minimize pen movement. We adapted a set of pen CASIC usability heuristics from Levi and Conrad (1996) and Nielsen (1994) and present it in Table 26.1. Some of the heuristics are general purpose, some are specific to pen computing, and some are specific to CASIC applications.

Procedure

The evaluators were first presented with an overview of their task and then given the CPI housing data collection tutorial. Next each evaluator explored the instrument in an individual session, accompanied by an experimenter and a CPI data collection instructor. The experimenter recorded the evaluator's problem reports, including the context in which the problem was encountered. The CPI instructor answered the evaluator's questions about the data collection procedure. The evaluators noted potential usability problems and for each one indicated the heuristic(s) that it violated.

After the individual sessions, all evaluators' reports were consolidated into a single list and formatted as a rating form. Evaluators were asked to rate the severity of each problem and return their ratings within a 24-hour period. They used the five-point rating scale developed by Nielsen (1994). The list of problems was then sorted by severity rating, location in the system, or both. This list was given to the developers within three days of the evaluation sessions.

Results

Many of the problems uncovered by the heuristic evaluation were not related to the pen CASIC implementation of the instrument. The same design decisions would have been flagged as potential problems in a desktop application in any application domain. We focus here on several problems that were directly related to pen CASIC usability.

Table 26.1. Pen CASIC Usability Heuristics

1. Speak the user's language

Use words, phrases, and concepts familiar to the user. Define new concepts the first time they are used.

2. Minimize user's memory load

Take advantage of recognition rather than recall. Do not force users to remember key information across tasks.

3. Minimize user fatigue
Minimize physical actions such as hand movements, and mental actions such as visual search or decisions. Design to facilitate pointing accuracy with stylus.

4. Design for skilled users

Support frequent repetition of a small set of well specified tasks. Provide proficient users with "short cuts" that do not violate data collection procedures. De-emphasize design for novices.

5. Provide sufficient guidance

Convey sufficient text or graphical information for the user to understand the forms-based task, but do not provide more information than users need to understand the task. Implicitly convey task instructions where possible through nonverbal cues, such as those provided by the spatial relationships among form elements on the screen.

6. Use visually functional design

Visually structure the user's task. Make it hard to confuse different tasks. User's eyes should be drawn to the correct place at the correct time, e.g., to actions to be performed, items to be remembered or referred to.

7. Use appropriately consistent design

Create consistent interfaces for tasks that are essentially the same. That is, use consistent formatting, phrasing, interface controls, task actions, etc., for tasks that closely resemble one another.

8. Design for easy navigation

Allow the user to move as necessary through the form, either forward or back to an earlier question. Enable an easy return from a temporary excursion to another portion of the survey. Enable user to determine current position.

First of all, even though the evaluators received a tutorial on the CPI collection task and had access to a CPI collection instructor, they felt that the way to use the interface was not always self-evident. These evaluators reported that heuristic 5 (provide sufficient guidance) was violated more than any other heuristic. This finding suggests that developers need to carefully consider what interface instructions they can reasonably expect a user to infer. If these can be made explicit without significant cost or compromising usability by experienced users, developers would be wise to do so. On the other hand, more knowledge of the domain can be assumed of CASIC users than many other user classes—at least for recurring surveys.

Second, the evaluators detected likely user fatigue in several cases that have their origins in the pen platform and stylus. Heuristic 3 (minimize user fatigue) was the second most frequently violated heuristic. In particular, important information was sometimes obscured by screen objects, especially the virtual keyboard. The screen on the pen computer is smaller than a standard laptop screen; on a larger screen overlapping objects could easily have been repositioned so that there was no overlap. The extra interface management required to move overlapping objects was a source of mental fatigue. The small size of screen controls like radio buttons required unreasonable precision in making contact with the screen, and this too was viewed as a likely source of mental fatigue.

Third, the evaluators expressed a preference for portrait orientation because of its similarity to the typical orientation of a pad of paper or notebook and because landscape orientation required too much scrolling (or paging) which they associated with mental fatigue. Unfortunately, portrait orientation was not an option for the developers because of operating system constraints.

Finally the evaluators felt that their inability to double tap as a "confirm" sequence led to undue fatigue. The alternative involves multiple, distinct pen actions, for example, a tap followed by pushing a confirmation button. In fact double tapping had been previously enabled, but in the informal evaluations, users indicated that (1) unintended pen contact followed by intended contact was sometimes misinterpreted as double tapping and (2) the hand movements involved in pen tapping required more concentration and motor precision than mouse clicking. As a result the developers intentionally disabled double tapping.

26.5.3 Relation between Informal User Feedback and Heuristic Evaluation Results

The list of potential problems that came out of the heuristic evaluation was useful to the developers but they were not surprised by any of the items on the list. Although these activities were not carried out to compare their results, we estimate that about 25 percent of the potential problems identified by the usability experts in the heuristic evaluation had also been detected in the informal end user studies. These involved positioning text on the screen, using the pen to enter text, choosing consistent controls to implement particular functionality, and providing visual feedback for user actions. It is hard to interpret this overlap figure because (1) the end user data were not systematically recorded and (2) the versions of the collection instrument tested in the heuristic evaluation had been extensively redesigned on the basis of the end user feedback; presumably these design changes had produced a more usable system.

In studies that compare the problems found with structured end user tests and heuristic evaluation, the overlap ranges from 10–15 percent (Jeffries, 1994) to 60 percent (Desurvire, Lawrence, and Atwood, 1991). Generally, end user tests identify more problems than heuristic evaluation, but Jeffries (1994) found the opposite to be true. While heuristic evaluation exposed more problems in the Jeffries study, the ratio of serious to minor problems was greater for end user testing.

This finding could be the result of asking evaluators in the heuristic study to inspect the design while instructing end users to perform a realistic task. It seems that the former approach could lead to an analysis of the interface—the choice and configuration of screen elements—while the latter may test the degree to which the application supports users in achieving their specific goals. Failure to support user goals is clearly more serious. This seemed to be the case in a heuristic evaluation of a web site (Levi and Conrad, 1996) which detected many interface problems but virtually no problems involving the structure of the site, organization of pages, and so on. In CASIC applications it would seem that heuristic evaluation would do a good job exposing problems such as inconsistencies in the way numerical data are entered on different screens but an end user study would better identify faulty logic in skip patterns.

The two methods can be used at different points in the development process. Heuristic evaluation can be done with a paper prototype and can be used early in the design process. End user testing generally requires a functional system and cannot be started as early. However, end users should be involved from the beginning in design reviews. In particular, developers and usability specialists should iteratively analyze the users' task(s) and solicit reactions to the analysis as a core design activity. As soon as there is a system with which users can interact—even if incomplete—they should be invited to use it in a test situation.

Both approaches have practical advantages and disadvantages for CASIC development. For example, heuristic evaluation requires identifying evaluators who are both expert in usability engineering and the survey domain. These "double experts" are rare if they exist at all. We tried to address the problem with a tutorial for usability specialists, but this was far from ideal. On the other hand, end user testing is difficult in CASIC because there is a limited pool of end users (interviewers), even in a large organization. Additionally at least some end user evaluations should take place in a wholly realistic interview situation. Unlike evaluating a standard desktop application, this kind of field study sacrifices some experimental control; on the other hand, a mock interview in the laboratory is bound to be artificial in some ways.

In sum, we advocate using multiple methods in complementary ways. Heuristic evaluation is a relatively fast and inexpensive alternative to end user testing. It can be used to refine the interface before end users are asked to interact with a system, and because it requires relatively few resources, it can be carried out frequently. End user testing identifies more severe problems at a greater cost. The chances of releasing a CASIC tool that successfully supports users are greatly increased if user performance data are collected (multiple times) during the development process. If heuristic evaluation or other inspection methods are the only options available to developers, then they should definitely be used.

26.6 DESIGN GUIDELINES FOR PEN INSTRUMENTS

Through these various forms of feedback and testing, we have established some preliminary design standards for instruments, some of which may have general applicability to forms-based instruments, regardless of input device. Some examples of these are shown in Table 26.2.

26.7 FUTURE OF PEN CASIC

While it has gained a firm foothold for simpler data collection tasks for largely commercial applications, pen computing is an "emerging technology" that is fast entering middle age without having seen substantial CASIC application.

Still, all the building blocks for successful implementation of complex CASIC surveys on the pen platform are now in place, in terms of sufficiently powerful hardware and visual programming environments. As the features available to designers grow more sophisticated, there will be increasingly successful use of alternatives to handwriting recognition, particularly in surveys in which numeric data predominate. Complex surveys will become easier to conduct on pen platforms as the full-featured CASIC packages release versions for Windows.

Regardless of the future of pen computing, the use of GUI offers to the CASIC designer new power and flexibility, and to the data collector the potential for greater ease of use of CASIC instruments (see Landauer, 1995, for a discussion of GUI usability). In the next several years most major CASIC studies will probably move to GUI platforms. As a result designers must learn a new "visual language" to effectively use the new power and recognize that the ease of our new design and programming tools does not guarantee that the resulting GUI will be easy and satisfying for data collectors to use. Usability testing thus becomes even more crucial as new CASIC instruments are developed. Fortunately the same graphical development tools that pose these sorts of problems also provide the support for new development strategies, such as rapid prototyping, which can facilitate more frequent testing.

The fielding of the first phase of CPI collection in 1997 constitutes a major "proof of concept" of pen CASIC. The demand for alternatives to keyboard input will grow as surveys that feature standing collection migrate

Table 26.2. Some Preliminary Pen Instrument Design Guidelines

---

I. From level and instrument momentum

1. Paging is superior to scrolling in nearly all instances. Vertical scrolling should be minimized; horizontal scrolling should be avoided at all costs.
2. Avoid wherever possible window movement or re-sizing as options to the user; maintain maximum designer control.
3. Wherever possible, restrict required items to the first or last items on a page. This may require breaking forms into multiple pages.
4. In general, if task sequences require jumping back and forth between two forms, attempt to take the related items from both forms and combine them into a third specific to that task.

II. Overall form design

1. Be careful not to make forms over crowded. Density should be substantially below what one would design for a paper form.
2. Use white space to mark off minor divisions, frames to mark off major divisions. Key required items should receive the same treatment as subgroups.
3. White space is not needed between screen edges and nontext controls, but text controls should not be placed at screen edges where possible.
4. For standing applications, portrait mode is superior to landscape due to the ergonomics of the machine.
5. In portrait mode, focus and thus the usability of the screen real estate declines as one moves to the bottom of the screen.
6. In part because of the greater focus at the top of the screen, major subform or window divisions should be accomplished vertically, even though this violates default order.
7. Grouped controls should be left-justified.
8. Where there is no vertical division of the screen, default order is top-left to bottom-right.
9. Place markers and functions that control the user surface (toolbars, other navigation controls) should be in a consistent position at the top of the screen.
10. User surface controls should be placed from right to left in order of increasing importance; the basic navigation controls (PREVIOUS, NEXT) should be at the upper right-hand corner of the control area.
11. When there are options for placement of textboxes versus nontext controls, text boxes should be placed higher on the screen. Text entry should never occur in the bottom $1\frac{1}{2}$ inch of the screen.
12. Text which is not intuitively attached or related to a control is less likely to be read.

III. Controls

1. Avoid text entry wherever possible—use lists and button controls. Use sliders or spin buttons for numeric entry options.
2. Where text entry is required, box or underline the text area. The method should indicate field character width. This also distinguishes textboxes from text displays and fills.
3. Separate textboxes to the extent that text is not entered in the wrong box (at least three characters apart).

Table 26.2. (*Continued*)

---

4. Use checkboxes for "all that apply" formats only; otherwise, use radio buttons.

5. For processing considerations, at all times minimize the use of actual controls per form.

6. Controls should be as large as possible; minimum size may be about $\frac{3}{4}$ inch square on a 9 inch diagonal screen.

7. Similar controls should be used for similar functions across the applications. Controls should generally take the same dimensions across the application.

8. For standard form confirm-and-move, a control with visual feedback is preferred to avoid multiple pushes.

9. When the label area is available as an active adjunct to a control, the label should be framed, unless this availability is consistent across all controls of the application or at least a prominent class of controls.

10. Avoid horizontal scrolling in list boxes wherever possible.

11. Whenever possible associate an edit box with a list box.

12. Always use gray shading to indicate disabled controls. In general, only use "gray shade" branching for branch-and-return sequences on the same form.

IV. Text and Font

1. Font size for most text elements should be 12 pt minimum as viewed on a 9 inch diagonal screen.

2. Except for emphasis (e.g., to set off instructions or observational items from interview items) do not use all caps.

---

to CAPI. But it may be that other new technologies—such as voice recognition—may provide many of the payoffs originally seen for pen. As the Census development project noted above indicates, the "mode" of the future may be some combination of these technologies—such as pen, audio-CAPI, and voice recognition working in tandem.

ACKNOWLEDGMENTS

We thank Irv Katz, for help in developing the Pen CASIC usability heuristics, and Vera Mitchell, for providing the CPI housing tutorial.

CHAPTER 27

Business Surveys of the Future: The World Wide Web as a Data Collection Methodology

Richard L. Clayton and George S. Werking
U.S. Bureau of Labor Statistics

27.1 INTRODUCTION

The Internet has become the symbol of the Information Age. This tool for accessing and sharing information is now extending throughout our culture. The Internet and the Universal Resource Locator (URL) of the World Wide Web (WWW) are linked to virtually all other types of communications media including newspapers, magazines, television, and radio. It is a technology that promises to supply any information needed in any format, to answer any question, and to replace much of print media. Governments are seeking ways of providing universal access to the Internet in an attempt to "level the playing field" for all citizens by offering equal access to information.

The promise of widespread and instant access via the Internet and the WWW has already spurred research into the potential for improving surveys. The early research will focus on business establishment surveys because Internet access has penetrated much further there than in households. This chapter focuses on using the Internet for business surveys. We briefly review the evolution of CASIC methods; then we review current Internet/WWW features relevant to data collection, identify considerations in the development of a data collection system based on the Web, profile the advantages of WWW collection against other CASIC methods in terms of quality, timeliness, and costs, and discuss issues relating to its future use for surveys. The particular frame of reference is research conducted at the U.S. Bureau of Labor Statistics (BLS) in the Current Employment Statistics (CES) program.

Computer Assisted Survey Information Collection, Edited by Mick P. Couper, Reginald P. Baker, Jelke Bethlehem, Cynthia Z. F. Clark, Jean Martin, William L. Nicholls II, and James M. O'Reilly. ISBN 0-471-17848-9 © 1998 John Wiley & Sons, Inc.

The specific features and capabilities of the WWW are evolving every day as new software is developed, as new uses are conceived for this new medium, and as we learn how to exploit its potential. The WWW as we know it now may be superseded by another yet-to-be-conceived information technology. The risk of writing about the WWW at this stage is one of being obsolete before the reader finishes these few pages. To postpone obsolescence, we provide only a brief reference to specific software used in CES research.

27.2 BACKGROUND

The availability of inexpensive computing power in an increasing number of forms has driven the evolution of CASIC methods. Each new technology, whether available to survey agencies or to our respondents, has offered new opportunities for improving some combination of timeliness, accuracy, or costs. The emerging WWW offers a new technology with many potential benefits and advantages over existing methods.

27.2.1 Evolution of CASIC Methods for Business Surveys

Over the last two decades alternative automated collection methods have developed rapidly, beginning with computer assisted telephone interviewing (CATI) in the 1970s. The availability of relatively inexpensive microcomputers later spawned research into computer assisted personal interviewing (CAPI), touchtone data entry (TDE), and voice recognition (VR). Each of these methods required little of the respondent except a touchtone telephone.

During the 1980s the telephone addressed many of the limitations inherent in personal interviews and in mail collection operations, especially high labor costs and slow turnaround. While in earlier years the telephone had been secondary, the 1980s saw CATI become a major collection mode for many surveys, including many large-scale government surveys (Clark, Martin, and Bates, Chapter 4). Methods also were developed for TDE systems for selected surveys and this was immediately followed by equivalent VR systems (Werking and Clayton, 1995). In the 13 years since its development, TDE has been used almost exclusively for a few select business surveys. TDE and VR methods are best suited for recurring surveys collecting a few numeric data items. Recently TDE/VR methodology has been creatively extended to capture very sensitive information in a household environment under the term telephone audio-CASI (Turner et al., Chapter 23). Subsequent research and discussions focused on timeliness of data, reducing direct labor costs for collection, hardware/software investment, and respondent acceptance. FAX transmission also emerged as a mode of collection. However, somewhat less than encouraging results were obtained from intelligent character recognition (ICR) systems that attempted to eliminate the inconvenience of the FAX paper and the subsequent key entry workload (Blom and Lyberg, Chapter 25).

Respondents' growing access to advanced microcomputers spurred the development of computerized self-administered questionnaires (CSAQ) in the 1980s (see Ramos, Sedivi, and Sweet, Chapter 20).

Over the past several years the availability of centralized databases has led to the emergence of electronic data interchange (EDI), also known as electronic commerce (EC). EDI allows large multiunit organizations to report large volumes of data in standardized formats, thereby reducing respondent burden and data collection costs. These data are transmitted electronically, making them instantly available for use in the survey (see Keller, 1994).

The primary goal of CASIC has been to improve the quality of data collected and edited at the source, while controlling error sources from interviewers through computer driven branching. These methods have offered improvements in data quality and timeliness and reduced costs (Nicholls, Baker, and Martin, 1997).

27.2.2   Internet and World Wide Web

The Internet is a global network of computers linked by a standard communications protocol. The World Wide Web (WWW) is a graphical interface to the Internet; it is at present probably the most popular application on the Internet.

27.2.3   Advantages of WWW for Data Collection

As in other CASIC methods, WWW offers many direct opportunities for addressing traditional survey problems: accuracy, timeliness, and costs. The unique features of the WWW also offer a range of other possible improvements in the survey process. For example, the WWW is a universal platform; that is, the systems work regardless of computer manufacturer or operating system. This is a distinct advantage over CSAQ which has suffered in needing costly parallel development for IBM, Apple, Unix, and other standards to cover the spectrum of computer users.

With the Internet and WWW providing an inexpensive and easy to use communications framework, and building on widespread availability of high-powered desktop computers, Web reporting is the next logical step in CASIC evolution. WWW data collection embodies all of the strengths of advanced telephone procedures of the 1980s, such as improved timeliness and on-line editing. It simultaneously eliminates many of the weaknesses, including the high costs of interviewers. It allows the user to enter data using a visually interesting interface. Links to other related sites can be provided, giving the respondent access to survey data products.

The WWW offers what may be the lowest-cost survey environment, especially for ongoing surveys (see Section 27.6.1). Cost reductions mirror those of TDE and CSAQ for on-line entry and editing, and far surpass any CASIC method for cost of transmissions. The unit cost per questionnaire is significantly reduced through the elimination of both postage charges and many labor-intensive activities of mail collection. Further savings are found in the reduction of telephone fees and labor costs for edit, prompting, and collection calls.

Some CASIC methods have limitations in their scope. For example, TDE and VR applications are usually restricted to the number of items for which a respondent is willing to push buttons and for the number of questions a respondent is willing to answer to a machine. CSAQ is limited by the number of diskettes to be mailed and platform sensitivity. WWW collection, however, has the potential of accommodating structured questionnaires of any form or length including "form layout" designs or traditional "question-by-question" designs. The respondent has the ability to refer to records as frequently as needed or to partially complete the questionnaire and return to it at a later time. WWW questionnaires may inexpensively offer individualized calculation spreadsheets, whereby the respondent can enter data that would automatically calculate the final response.

A major feature of this method is that the Web is truly on-line. In ways beyond CATI and CSAQ, WWW offers instantaneous response to virtually any request posed by a respondent. The improvements offered by automation and electronic communication will ultimately lead to simplified respondent reporting, more accurate data, more timely responses, and improved customer access to our survey products. Survey organizations have worked diligently to associate the utility and importance of the published information with data reporting by individual respondents. In the past these efforts have consisted of sending booklets, brochures, or press releases to respondents either as inducements for participation or as tokens of appreciation, with respondents and users waiting long periods for paper releases (i.e., press releases, periodicals, and bound volumes), calling or writing for specific tables, or purchasing specialized diskettes. Now users will have direct menu driven electronic access to our institutions' large, longitudinal public access databases. The WWW interface could take this effort to a new level by profiling the data provided by individual respondents against, for example, other data on their industry, their state, and the nation. Also the WWW's multimedia capabilities will enable survey organizations to provide on-line "clippings" showing the data in use, whether from the print media, radio, or television, further reinforcing to respondents the use and importance of their reporting efforts (see Section 27.7). This will significantly reduce the labor-intensive overhead associated with our information dissemination activities while providing improved services to users.

Using the WWW for surveys, while full of promise, does pose a variety of problems, namely limited penetration and inevitable mixed mode collection for years to come. Relatively spotty access also renders it useless for sampling (Fisher et al., 1995).

27.2.4   Current Employment Statistics Program

As was noted earlier, the principal lens through which this chapter views WWW data collection is the ongoing CES research program at BLS. The CES is an ongoing monthly survey of about 390,000 nonagricultural establishments. The CES preliminary estimates, released on the first Friday of each month, receive vast media coverage, are considered among the most influential data series for economic policy purposes, and are a driving force behind the financial markets. The CES is well-suited for CASIC methods because it has a very short collection period (between 2 to 2.5 weeks) and collects a small number (five or six) of numeric, commonly available payroll related data items. Since 1982 the CES staff has researched and implemented CATI, CAPI, TDE, VR, and EDI for traditional mail collection. At the end of 1997, over 240,000 respondents reported monthly using TDE with thousands more in transition. Some of the largest multiunit companies report through EDI. Under this array of automated methods, the average response rates for preliminary estimates have been raised by 20 percentage points and the average monthly revisions to the preliminary estimates have been reduced by 38 percent.

27.3   WWW METHODOLOGY: COMBINING TWO CASIC METHODS IN ONE

Developing a WWW survey methodology involves using tools developed under two other methods. These tools include the respondent contact procedures analogous to those used in TDE, and the automated self-interviewing techniques familiar from CSAQ. The combination of these features comprises the likely direction of WWW collection methodology.

In general, the WWW survey collection cycle begins with a sample control file containing the respondent's e-mail address in addition to the normal respondent contact information of name, address, and phone number. In the CES, as an ongoing panel survey, obtaining e-mail addresses from existing respondents may be relatively easy when compared to one-time surveys. The collection form is a standard "web page" containing an image of the questionnaire, survey instructions, definitions, and hypertext links to definitions. An e-mail address is provided for problem reporting and inquiries. As the collection cycle begins, the respondent opens his or her e-mail to find a reminder, points a browser to the survey homepage, accesses the data collection screen, and fills in the requested data (see Figure 27.1). The moment the respondent clicks the "submit data" icon, the data are transferred to the survey agency. The collection system electronically checks in the schedules, and at predetermined time periods, sends e-mail nonresponse reminder packages containing the full original information. During data collection the system conducts automated edits and notes failures on the screen. In the example in Figure 27.1, an edit failure is shown in the upper right portion of the screen and provides instructions on which data items to review and correct.

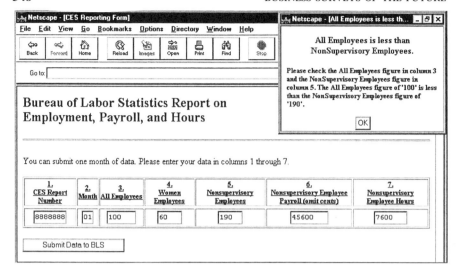

Figure 27.1. CES WWW data reporting page.

Existing TDE and VR methods largely eliminate labor-intensive activities for mail-out, mail-back, and data entry. However, neither method directly addresses another expensive activity: data editing and reconciliation. Current labor-intensive edit and reconciliation operations can also be directly handled under WWW collection, which allows the respondent, as in CSAQ, to directly review seeming edit failures and correct them as necessary. The WWW environment can implement both longitudinal and data integrity edits. Under Web methodology, most survey data collection operations can be fully automated and the overall process simplified for both the survey agency and the respondent.

27.3.1 Automated Self-Response Contact Methods

The CES is an ongoing survey where participating firms report data every month, most for at least a few years. In this traditional environment, mail surveys relied on the arrival of the form to spur the respondent's reply. If a response was not received by a certain time, a nonresponse prompt was issued. Often nonresponse prompts were wasted because there was no knowledge about whether or not the forms were in transit. The TDE methodology developed at BLS incorporates these same three message types in a carefully timed approach which can be compared to a WWW methodology (Harrell, Clayton, and Werking, 1995). The essential steps are:

- an advance notice message via postcard or FAX, which replaces the mailed form as a notice to take action,

- the self-initiated call to a toll free number for entering the numerical answers to the prerecorded questions, and
- a nonresponse prompt message by CATI or FAX.

Under WWW the advance notices and nonresponse prompt messages and timing are mirrored by e-mail messages, as shown in Table 27.1.

Under this view the entire methodology can be automated, from the timing and content of the e-mail messages to firm-specific nonresponse prompts based on the known availability of needed records. With all data entered and edited on-line and most or all messages handled automatically be e-mail, a truly "peopleless, paperless" methodology is possible (Werking, 1994).

27.3.2 WWW Instrument Design Issues

The WWW poses some difficult issues for designers. Maintenance of security, accounting for transmissions, and interface design are key issues. Federal surveys are required to maintain strict confidentiality about participants and their data, thereby adding requirements for encryption technologies, discussed in Section 27.6.3.

There are several issues regarding the design of the system as it appears to the respondent, including the approach used for screens, the design of each screen, and navigation within and among screens. The Shneiderman (1993) approach focuses on building systems that serve user needs. Good design retains and satisfies users to the benefit of the system. Well-designed interfaces should be simple and intuitive and provide some form of feedback or reward. Specifically for WWW development, Nielsen (1996) also focuses on the intuitiveness of the design. In most cases users should not have to read documentation.

Table 27.1. Comparison of Respondent Message Types: TDE versus WWW

Methodology Feature	TDE/Other Methods	World Wide Web	Fully Developed, Integrated WWW System
Monthly advance notice	Postcard or FAX	E-mail	Automatically generated
Data reporting	Call 800 number to TDE system	https:// www.ces.bls.gov	Incoming data are instantly received, edited, and stored
Nonresponse prompting	Phone call, FAX, or postcard	E-mail	Automatically generated according to monthly automated calendar

Scrolling versus "pages" is the first design consideration. Many WWW sites use scrolling capability, but these are mostly text-based sites. Nielsen emphasizes that the use of page-based sites is more intuitive and less confusing, since respondents may not realize that some questions are hidden. Page-based systems should be designed to follow branching as is routine in CATI, CAPI, and CSAQ. The trade-off may be the number of downloads, and the time these may take, versus the clarity and control of a page-based design. Research comparing both scroll and multiple page-based design indicates that the page approach offers the advantage of capturing important interview information, such as time spent on each screen, and does not increase total session time and overall completion rates (Vehovar and Batagelj, 1996).

Providing the respondent with the ability to navigate knowledgeably within and among screens is key to user satisfaction. The screen design must make it clear to the respondent how to move about within the questionnaire. Commonly used buttons, labeled for their exact purpose, aid in this design. Also familiar browser tools for "back" or "forward," or buttons with similar functionality, allow the user to move within the instrument and should be integrated to make such movement easy.

27.3.3 Total Design Method On-line

Many analogies have been made comparing traditional mail to e-mail/WWW. In many ways they are quite alike in their visual presentation and their "page-based" orientation, while the speed of delivery is an important difference. In view of their similarities, existing research into mail-based methods provides a strong basis for WWW system design and for attaining the highest possible response rates. One well-researched and documented framework is the Total Design Method (TDM) (see Dillman, 1978; Dillman, Treat, and Clark, 1994). The TDM states that response rates can be maximized when three basic tenets are incorporated into the survey design: (1) maximize respondent rewards, (2) minimize the costs of reporting, and (3) establish trust that rewards will actually be delivered.

First, TDM suggests that rewards are largely intangible, yet respondents want to be regarded positively and feel that their time, efforts, and comments are valued. These can be conveyed through personalized messages, and return postage should be a real stamp rather than metered postage. Second, Dillman suggested that respondent cost is measured primarily in time, both for the immediate response and any projected time. Lengthy, complicated questionnaires require substantial investments of time and should be presented as well-organized booklets, showing substantial white space with definitions included where they are needed. Of course there should be no monetary expense to the respondent. These can be minimized or eliminated by providing postage-paid envelopes for mail, or by extension, toll free numbers for telephone calls. The third tenet, establishing trust, can be encouraged by noting an affiliation with a respected organization and by diligently following through on any promise such as sending information or copies of survey results.

The WWW has four characteristics that fit this methodology. First, the page-based design of a WWW collection system, like a CATI instrument for interviewers, only shows the "pages" or screens needed based on previous interviews and prior answers. All other portions of the questionnaire remain unseen, limiting burden and crowding, so that the easy-to-complete screen encourages response.

Second, "the Internet is scaleable," says Bill Gates, founder and chairman of Microsoft Corporation, "in the sense that if something really catches your eye, you can be as educated and involved as you want to be" (see Kennedy, 1997). For surveys, this means that we can provide large amounts of information for those who want it. For example, by using visually highlighted hypertext links, definitions and related information can be hidden from view unless needed, and then instantly retrieved with a mouse click. For paper-based surveys, decisions on how much material to provide is both a cost and burden issue. However, since the marginal cost of adding information to the WWW server is negligible, we can inexpensively make available a large array of information. Examples include survey background, respondent-specific spreadsheets, published results in printed or graphic form, as well as video or audio clips showing the survey results being used in the media, by businesses, or by policy makers. This electronically stored information would be retrieved solely by respondents specifically requesting it. Research is needed on how to build in such features to appear helpful, and optional, without being perceived as burdensome.

Third, the WWW is by its very nature on-line and interactive, providing instant feedback—whether for editing, access to the survey-related background information, personalized messages of appreciation, or published output. If well-designed, the TDM trust concept is self-fulfilling, since the respondent instantly receives requested information from a menu of options linked to electronically stored responses.

Last, the WWW has the graphic flexibility to make the questionnaire as interesting as our skill and the subject matter allow, including audio and video cues. The challenge will be determining the appropriate uses of the WWW's interactive capabilities without being confusing or diverting attention from the questionnaire itself. Table 27.2 illustrates how some of the features of the TDM can be implemented on the WWW.

27.4 RESPONDENT ACCESS TO E-MAIL AND THE WWW

The tremendous potential for WWW collection and the enthusiasm for research currently should be tempered by the rather limited access. A number of estimates are available from both businesses and households. While varying considerably, they agree that WWW coverage is relatively small but growing rapidly. Inevitably mixed-mode collection will be needed for some time. Also WWW access somewhere within a business does not guarantee that a particular respondent needed for a particular survey has convenient access. The option of providing WWW access to all survey respondents, as in the telepanel

Table 27.2. Translating the Total Design Method to WWW Collection

Survey Function	TDM Recommendation	WWW Application
Correspondence	Personalized correspondence	Address advance notice and nonresponse prompt e-mail messages to a specific respondent. The questionnaire is always available on the web site.
Postage	Use actual stamp on return envelope	"Registered" e-mail could be used for the advance notice and nonresponse prompts.
Survey Form	Simplify form: white space, booklet approach, include definitions	Design the screen to leave the maximum amount of white space. Include definitions as hypertext links.
Question Lists	Prioritize questions, shorter, respondent friendly questionnaires, automatic branching	Use hypertext links to keep appearance of questionnaire simple and short.

approach (Saris, Chapter 21), would be prohibitively expensive for any single survey.

United States employers have responded to increasing international competitiveness pressures by downsizing and flattening their organizations, increasing their prductivity, and controlling their wage and price structures. However, perhaps more important, during the 1980s employers also responded by investing heavily in computing technology and communications to boost productivity, to link their national and international operations, and to provide instantaneous access to critical management information on inventories, personnel, and cash flow transactions. In 1991 companies for the first time spent more on computing and communications equipment than on industrial, mining, farm, and construction machines (Werking, 1994). These investments should be reflected in increasing availability of technology and Internet access for many worker groups.

In a 1995 survey of 404 randomly selected chief information officers (CIO) of Fortune 2000 companies (Spanski and Wickham, 1995), several key indicators of the current and future potential for e-mail were outlined. Significantly, 89 percent of the CIOs had e-mail within their companies. About half of the remainder expected e-mail access within the next two years. Also about 60 percent of their employees had e-mail access, and 44 percent had a link to the Internet. Most of these large businesses initially established e-mail linkages to improve internal communications and internal decision making. Those with Internet access point to it as a means for improving decision making, indicating that those without Internet access are likely to follow.

A 1996 Nielsen poll estimated that 30 percent of all working adults have Internet access (see Hoffman, Kalsbeek and Novak, 1996). Another survey, by the Bureau of National Affairs in January 1997, found about one-half of 494 questioned firms said selected workers have Internet access, with another 17 percent planning to give some employees access soon (*Wall Street Journal*, 1997).

An additional concern is the "churn rate," or changed e-mail addresses as people drop, add, or change Internet service. While the churn rate may tend to be lower for businesses than for households, keeping track of individual e-mail addresses will pose the same difficulties as telephone numbers when respondent turnover takes place.

These statistics, varying considerably and using different techniques, may be optimistic depending on the target respondent for a particular survey. Questionnaires targeted for CIOs may use WWW collection as the primary vehicle. However, these polls may be less promising for most large-scale, ongoing business survey respondents. In a 1995 review of 1300 respondents to the CES survey—typically payroll clerks—only 6 percent had e-mail access. In the 1996 WWW pilot test, 10.7 percent of existing CES respondents already reporting via TDE met the eligibility criteria of desktop Internet access, Netscape 2.0 browser or better (needed to support our screen design), and a willingness to participate in the pilot (see Table 27.3). On an optimistic note, an additional 10.5 percent of firms contacted had e-mail and WWW; however, either they had not extended availability to the CES respondent or the respondent has not used the Internet. Almost two-thirds of interviewed CES respondents, 62.5 percent, did not have e-mail or WWW access. The industries targeted for this test were thought likely to have high access rates, specifically computer and data processing services and state and local governments. There were no respondents meeting the criteria in other service industries.

A test of respondent preferences for transmitting data electronically to the annual Company Organization Survey at the Bureau of the Census (see Sweet and Russell, 1996) also found low availability. Only about 10 percent of those responding had Internet access and would use the Internet for data transmission. The low penetration of the Web, regardless of its growth, points directly to the inevitability of mixed-mode collection for some years to come. Developing and integrating two or more modes requires multiple systems for control and compounds tracking and integration difficulty.

27.5 RESULTS OF THE CES PILOT TEST

Beginning in March 1996, the CES launched a pilot test of WWW collection. The CES pilot was designed to build basic functions and to allow testing of important methodological features, rather than as an attempt to complete a production system while hardware and software are rapidly evolving. The pilot

Table 27.3. CES Respondent WWW Access and Conversion for Three Industry Groups: March–September 1996 (in %)

	Distribution of Units Contacted	Computer and Data Processing Services	Other Service Industries	State and Local Government
CES pilot units				
Compatible browser, e-mail, web on desktop	10.7	14.3	0.0	10.4
Ineligible units				
No e-mail or WWW access	62.5	49.4	79.1	74.9
Internal e-mail only	4.1	6.4	0.0	2.4
Company has capabilities that respondent does not use	10.5	12.8	7.0	8.7
Old or incompatible browser	4.5	6.4	5.8	0.6
E-mail and web access, but not on desktop	2.1	3.4	0.0	1.2
Prefer TDE	2.1	3.4	0.0	1.2
Out of business	3.5	3.9	8.1	0.6
Total	100.0	100.0	100.0	100.0
(Number of cases)	(698)	(313)	(121)	(264)

system uses hypertext markup language (HTML), JavaScript, and Perl to create the functional interface. HTML is a coding system used for text and images and to set up hyperlinks between documents. Javascript is an object-oriented scripting language that enhances the functionality of the Web browser; when embedded in an HTML document, it provides the ability to respond to the user without accessing the server. Perl is a scripting language that facilitates communication between the browser and the server.

The mail-based CES form collects five or six data items each month from a fixed panel of respondents where the number of items depends on the responding establishment's industry. The form is designed as a matrix that obtains data for an entire year, with a new row for each month. Column headings indicate the data item requested. The WWW pilot replicated this basic row of data for the current month only (see Figure 27.1), because of the difficulty of guaranteeing the security of previously reported data with current WWW security capabilities.

To enter the system, the respondent must enter a unique report number and password, and then proceed to a home page containing the Office of Management and Budget (OMB) mandated statements on confidentiality and respondent burden. The report number and password are also used to identify the correct industry form for each respondent. On screen, the respondent enters the number for the reference month and data for that month. The column headings appear as hypertext links to full definitions of the data items. Pressing a button labeled "send data to BLS" encrypts the data and transmits it to the BLS server. At this writing, prototype interactive editing has been developed and is now available to WWW respondents.

Currently 60 respondents are reporting their employment, payroll, and hours data by WWW. These respondents were formerly reporting via TDE and know the monthly reporting cycle. Each month just prior to the reporting cycle, BLS sends an e-mail message as an advance notice, paralleling the advance notice postcards and FAXes now sent to over 220,000 TDE respondents. Those who do not report within our specific time frame receive another e-mail as a nonresponse prompt.

In general, the WWW pilot test results mirrored TDE. This approach has yielded two critically important results. First, the basic response rates for our time critical preliminary estimates for these units are essentially the same under WWW collection (76%) as under TDE (78%). Thus we see no reason at this early juncture why WWW collection will not be able to match the same high response rates seen under similar telephone-based methods, TDE, and VR.

Second, e-mail messages seem to be as effective as other prompting methods. The proportion of WWW units needing e-mail nonresponse prompts (35%) is nearly the same for the overall TDE sample. This means that respondents can receive the advance notice messages through a variety of e-mail systems; they access their e-mail regularly enough for a time-critical monthly survey and they respond to e-mail messages. Thus the overall package of e-mail messages described in Table 27.1 is proving effective and is comparable to the proven TDE method. Based on this finding, more elaborate tests and refinements in message methodology, including the packaging of the e-mails with such features as voice files, graphics, or icon links to the Web site will be tested for user ease and acceptance.

The CES pilot test offered a few other interesting results. The length of the WWW session takes about two minutes, only eight seconds longer than the average TDE interview. Although the number of cases is very small, the distribution of the WWW sessions paralleled that seen for TDE, in that they occurred almost exclusively during business hours, and many sessions occurred closely after both advance notices and nonresponse prompts were sent. Thus the same workload peaks and capacity issues may be relevant (Werking and Clayton, 1995). Only one respondent has had trouble gaining access to the CES system, and this was because his WWW access provider was too busy to handle traffic out of New York City at midday. He was subsequently able to gain access earlier the next day. Several of our WWW respondents report for more than one location, in one case covering five separate locations, and have used the system without problem. There has been only one change in e-mail addresses for the 60 WWW respondents. Also we are slowly increasing the number of WWW respondents as individual TDE respondents request reporting via e-mail, suggesting that there is a market for this approach. Respondents have commented favorably on the ease of use of the interface and the display of the data in an array, providing an opportunity to check their data before submission.

27.6 IMPLEMENTATION ISSUES

Aside from the direct benefit of obtaining timely and accurate data, CASIC methods offer implementation opportunities to survey institutions, such as reduced costs, and some challenges, such as potentially profound organizational effects. The WWW demands security for the transmission and storage of information, as well as confidentiality of our respondents where law and institutions have yet to fully provide an umbrella for operations as exists for other electronic transmissions.

27.6.1 Costs

In most surveys the single most expensive survey function is data collection, ranging from one-third to one-half of total costs. One of the major thrusts of CASIC development is to change the cost structure by replacing the traditional labor-intensive processes with increasing portions of capital intensive factors, and varying degrees of labor costs. Under mail-based methods, clerical or semiprofessional staff produce forms, fold and stuff envelopes at mailout, and then open (often pre-edit and key-enter) the returned questionnaires. Thus mail is labor and time intensive, and error prone. Using CATI to improve data quality and timeliness reduces mail handling but adds new costs for computer and telephone technology and retains high labor costs for interviewers. The development of TDE as an automated self-response collection mode was spurred by the need to lower costs while retaining the timeliness of telephone collection. Under TDE, labor-intensive processes, except for editing, are replaced by respondent entry of data, and declining telephone costs reduce transmission costs. WWW collection offers the best of both CATI and TDE methods by transmitting data that are timely and edited at low costs.

Not only does WWW collection virtually eliminate the traditional labor-intensive activities of mail out, mail return, and data entry, but it continues to take advantage of significant and divergent cost trends. Labor is a continually rising cost, usually over one-half of any process, and rising at about 4 percent per year. Postage also averages a 5 percent annual increase. In contrast, the cost of computers has decreased by 22 percent annually over the last three years. Telephone costs have declined over 15 percent over the last decade and are likely to continue dropping. The costs to the telephone companies of a telephone call has dropped to one-thousandth its level in the late 1950s, while gross profit margins in the United States are over 40 percent (see *Economist*, November 1996). Thus telephone-based CASIC methods offer the likely prospect of continuing the acceleration of cost advantage.

For any CASIC method, systems development and maintenance is a significant new cost, and a cost that is also driving organizational change. Systems costs are very difficult to measure, primarily due to the continuous nature of improvements, enhancements, and incorporation of new technology.

CES unit costs of data transmission, hardware, and Internet access, assuming 10,000 monthly respondents, are shown in Table 27.4. This illustrates the dramatic cost-reducing opportunity offered by a switch from mail to TDE and WWW collection. Without including the labor savings, WWW operates at one-fifth the cost of mail collection. Also the WWW costs are not linear; the capacity included in this estimate for WWW could likely handle twice this workload without a significant increase in costs. For all but the smallest surveys, the growing disparity between postage and telephone costs will drive recurring surveys to automated self-response CASIC as transmission and hardware savings far exceed systems development and maintenance costs.

The drive for cost reductions at equal or higher quality is a constant force in any production function. Cost savings can be either redirected toward larger sample size or toward other quality-enhancing activities, or used to satisfy overall constraints of lower resource levels to prevent quality compromises.

27.6.2 Organization Effects

Survey organizations reflect the production requirements and the technology available at any given time. As these factors change, so does the organizational structure. Since the first computer was placed in the service of statistical operations, organizations and their staffs have had to evolve. A major automation advance such as computerized estimation causes a permanent reduction in staff, usually clerical, who had performed that function. Secondarily, computerized output, in its tabular and graphical form, demands greater sophistication of the remaining staff and provides the time to do previously prohibitive analysis and review.

Table 27.4. CES Monthly Unit Costs of Data Transmission (in U.S. $)

Messaging Function	Mail/Postage	TDE/FAX	WWW/E-mail
Outbound messages	0.32	0.07	0.00
Inbound messages (data)	0.32	0.14	0.00
Nonresponse Prompting	0.10	0.04	0.00
Total transmission costs	0.74	0.25	0.00
Hardware	0.01	0.03	0.01
High-speed communication lines	0.00	0.00	0.12
Total monthly unit cost	0.75	0.28	0.13

The large-scale implementation of CASIC methods, and WWW collection in particular, will result in the flow of clean, edited data directly to the survey organization. Thus both labor and nonlabor resources will be shifted among the remaining and new factors of production. The most obvious result is that postcollection data review and entry are virtually eliminated. The emphasis shifts instead toward computer systems development and toward the maintenance and operation of the new collection process.

The role of interviewers and collection staff depends on the specific CASIC method. For example, CATI and CAPI emphasize the interviewer's role in data capture and editing. TDE and VR rely on respondent data entry but retain a limited interviewer role for editing and prompting. CSAQ includes on-line entry and self-editing for further reductions in the interviewer's role. Yet all of these methods continue to require some labor costs for outbound mailing of forms, advance notices, and nonresponse prompts.

The view of WWW collection as a "peopleless, paperless" methodology (see Werking, 1994) has profound implications for survey organizations. The fully integrated WWW design using automated, outbound e-mail messaging and on-line entry and editing leaves virtually no role for clerical or professional interviewers.

Two other implications are important. First, there will be a need for new skills beyond those of the systems staff. Experts in computer human interfaces and graphical presentation will be critical to developing clear and consistent design and to obtaining meaningful results. Also the role of questionnaire designers will grow. Second, an interdisciplinary approach to solving survey problems will be increasingly critical. Barriers among various occupational and organizational groups must be reduced to ensure that staffs can interact freely.

27.6.3   Data Security

As any user of the Internet knows, the single most difficult aspect of commerce over the network is security. Security issues come in two basic forms: message transmission and the integrity of the individual sites in the network. Both of these pose serious difficulties for survey data collection. In addition, because security is the single largest roadblock to commercial uses of the Internet, there is intense interest in solving these problems as rapidly as possible.

The first set of security issues derives from the Internet's structure. It is a decentralized set of routers that directs traffic over the connecting telephone lines, with messages eventually finding their way from sender to receiver. This decentralization was initially conceived as a way to ensure that if one or more sites were destroyed or not working, the remaining structure could continue working without interruption. While the vast majority of messages flow without intervention, this structure allows persons at the routing points to, if necessary, view messages to complete the connection. These people therefore potentially have access to any passing message. This type of security issue can be handled primarily by encrypting the message so that routers cannot readily see their contents. The current security standard is Secure Sockets Layer (SSL) protocol.

The second set of security issues focuses on the hardware and software housing and running the survey instrument and the data used in it. The server holding the instrument must prevent intruders from interfering with its operation and must deny access to confidential files and systems critical to operations. There currently are numerous hardware and software solutions that can prevent interference with the system and its operation. Also the use of technological "firewalls," a detailed set of codes permitting access, is now routine for keeping intruders away from systems resources. Other security issues being addressed include authentication of the respondent, preventing eavesdropping and hijacking of the session, and protection of the survey Local Area Network against destruction or misuse of information. This field is rapidly changing, with new tools and software on the horizon for solving the barriers to both research and implementation.

The security issues often seem far more exhaustive for WWW collection than those for other CASIC methods. The differences lie in two areas: The Internet is currently open to technological eavesdropping or snooping, and the Internet is too new to have a body of case law restraining people's behavior.

The postal system has statutory laws prohibiting interference with the mail, a wide range of case law supporting those statutes, and even a special police force dedicated to catching and prosecuting criminals. The telephone system also has statutory and case law prohibiting wiretapping. For both the mail and telephone systems, the public at large is aware of these laws, and criminals are routinely publicized.

Laws already exist covering privacy on the Internet. The Electronic Communications Privacy Act of 1986 specifically covers the privacy of e-mail messages. In addition the Computer Fraud and Abuse Act of 1986 is cited in the Department of Defense WWW site as a legal protection of the site and its contents. Finally case law is being made, albeit slowly, that will help reduce the incidence of hacking.

27.7 THE FUTURE

While all students of CASIC methods have their own views of the future, the fascinating research of the next two decades is likely to reshape survey processes. This research will focus on reconciling the rapidly advancing technology with the increasingly specialized role of the interviewer.

Over the course of CASIC development, the role of the interviewer has been reshaped. Partly because of the technology and partly for cost reductions, interviewers' roles are forever changed. The CASIC challenge has been to combine technology and interviewer skills to maximum benefit given the specific survey environment. Under CATI and CAPI, computers relieve interviewers of tedious branching and manual editing workload, provide prompts to direct interviewers' work, and act to prevent errors. The result is viewed either as a refocusing of their skills toward the most meaningful and essential process of retrieving and interpreting information or as the gradual reduction of their contribution to mere data entry. Under TDE and VR, data entry is fully automated, leaving only postcollection edits for interviewer follow-up. CSAQ, in its various forms, and the WWW also seek to completely replace the interviewer, including editing, while allowing only interviewer follow-up and recontact for special cases.

Use of the WWW for surveys will no doubt begin with business establishments because of greater access to the technology. Household surveys will inevitably follow, although the time frame could be a long one given current levels of Internet access by households.

New technologies are becoming available that provide fascinating opportunities for survey instrument design in an on-line environment. For example, most personal computers now sold are packaged to include multimedia features, including sound cards, CD-ROM drives, and, in the near future, inexpensive cameras. It is already possible to download audio and video files for playback, increasing the power of our communications. For example, we could send e-mail message icons that allow for a direct link to the WWW reporting site, and with audio and video icons, as shown in Figure 27.2. The basic text message of the e-mail might vary by how long the respondent is delinquent, from "Please report your data by the 28th" to "We have not received your data for three months. It is highly important to the nation's employment report that we receive your data as soon as possible." Upon clicking one of the icons, the audio or video file would provide special

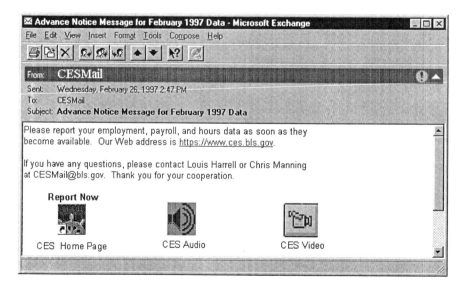

Figure 27.2. WWW screen for CES reporting voice and video prompts.

messages, radio or television clips discussing the importance and uses of the data, or short expressions of appreciation from agency heads or publicly recognized major users.

The world of telephone technology is rapidly advancing. For example, plans exist for merging the desktop PC and the telephone into a single unit, thereby reducing costs since much of the technology is duplicated. Another extension incorporates both telephone and computer into the "common" household television. Indeed "WebTV" is already commercially available. Whether the information is provided by the Internet or its successor, the survey features and possibilities are the same: to provide a clear and controlled set of questions able to elicit meaningful responses. When inexpensive cameras are added to PCs, interviewers in a remote CATI environment will be able to see their respondents in addition to using existing voice cues. Alternatively, surveys could use this camera technology for "on-line help desks." When a respondent spends too much time on a particular item, a pop-up screen could offer more information or examples of the desired data item. The pop-up screen could also display an icon which, when clicked, links to a live interviewer whose image and voice are carried to the respondent's screen. The interviewer would see the same screen and other historical information on the respondent and be able to talk respondents through a difficult item or teach them how to use the system. The results of such interviews would lead to improvements that would prevent a multitude of problems.

The CASIC challenge has been to merge the role of technology and interviewer in appropriate proportions for greatest quality and lowest cost. The WWW provides sufficient computing power and easy user interface to replace the interviewer and yet continue to offer the respondent all of the human interfacing that cyberspace will allow. Based on the concepts outlined above, it is possible to foresee the "virtual interviewer" built on prerecorded video/audio clips analogous to the interactive games now available. A portion of the screen would show the interviewer asking the questions, in any language, with the questions or form also shown. Response-driven branching by keyboard, touchscreen or voice recognition, would direct the presentation of visual and audio queues. The "virtual interviewer" vision could be field tested now using interactive CDs, like those our children are using for learning and video games.

This vision offers survey practitioners the best monitoring and questionnaire refinement capability imaginable. Monitoring can be conducted by keystroke capture or remotely by simply watching the self-response interview unfold. The virtual interviewer will have already requested and stored audio and video consent. Some researchers have found keystroke files useful in assessing interviewer competence in CAPI surveys (Couper, Hansen, and Sadosky, 1997); a WWW instrument could easily capture such files for assessing respondent understanding of the instrument.

The existing Internet, with its relatively limited bandwidth, will not fully support this vision. However, the next generation of the Internet is on the drawing board and discussions of its design focus on providing exactly the types of features that would make the virtual interviewer possible, real time multimedia applications capable of transferring and manipulating very large volumes of data (*Bulletin of Science Policy News*, 1996).

The development of strong, fully automated data collection via the WWW is inevitable. The very basic WWW research conducted to date, while slowed by the newness of the technology, is supported by the results of the previous two decades of CASIC research. The methodological issues, again supported by previous CASIC research and implementation, point directly to neutral or improved data quality, and improved timeliness compared to the traditional methods and equal to CATI and TDE at much reduced costs.

The Internet/WWW provides the greatest array of features of any CASIC method. However, until respondent access grows to near universal levels, mixed-mode operation will continue. Still the information superhighway has now opened up many new dynamic areas of electronic collection research for business surveys. The results from this research may ultimately position the World Wide Web as the most timely, accurate, and cost-effective approach to establishment and household surveys.

ACKNOWLEDGMENTS

The authors wish to recognize Louis Harrell and Chris Manning of BLS for their innovative development of the CES WWW prototype and analysis of the research results.

CHAPTER 28

Current and Future Technology Utilization in European Market Research

Bill Blyth
Taylor Nelson Sofres, United Kingdom

28.1   INTRODUCTION

The purpose of this chapter is to review current and future utilization of technology in survey and market research data collection in Europe. European market research is a diverse and highly vibrant sector where there is a great deal of technology application. In this discussion, I limit the detail to what I see as main European developments. Thus applications that are country specific—for example, by virtue of a local telecommunications supplier's activity—have been set aside. As in other industries new technologies often promise much but deliver little. I focus on those that, in my opinion, will not be merely passing fashions.

28.2   EUROPE AND EUROPEAN MARKET RESEARCH

The European Union's (EU) 15 member states have a population of over 370 million that spans north to south from Finland to Spain and west to east from Ireland to Austria. By the end of the millennium, there are plans to extend membership to the new Baltic states, to the old Eastern European states, and possibly to Turkey. In addition to the multitude of languages, there are substantial differences in population density, income per capita, education, and

Computer Assisted Survey Information Collection, Edited by Mick P. Couper, Reginald P. Baker, Jelke Bethlehem, Cynthia Z. F. Clark, Jean Martin, William L. Nicholls II, and James M. O'Reilly.
ISBN 0-471-17848-9   © 1998 John Wiley & Sons, Inc.

birth rates. These factors interact, affecting a technology's applicability to local conditions in different countries. Nevertheless, the EU's goal of a unified legal, cultural, and commercial entity requires consistent information across all member states. As a consequence there is a growing and healthy demand for market information.

The value of the worldwide market research industry increased from an estimated 5.5 billion European Currency Units (ECU) in 1990 to an estimated 7.9 billion ECU in 1995 (ESOMAR, 1996). Overall growth in Europe has been steady, over 40 percent for the period. Table 28.1 shows that this growth has also been relatively constant across all major regions of the world, reflecting the buoyancy of market research even in more slowly growing economies. Table 28.2 compares member states' shares of the EU by (1) market research industry, (2) advertising expenditures, (3) gross domestic product (GDP), and (4) population. These data show that among the top six countries, which account for over 85 percent of market research expenditures, there is little correlation with population size or gross domestic product. The U.K. has a relatively high share of research and Italy a somewhat low share. The reason for Italy's relatively low share is a matter of conjecture. The U.K. accounts for just under 40 percent of the European subcontracting business (i.e., the coordination of multinational research) which explains the disproportionately high U.K. share. Multinational research has been one of the fastest growing areas of research in Europe and will become increasingly important as management regards Europe as a single market.

Multinational research requires standardization and consistency to be of value to users. Technology is a potentially key component in ensuring standardization and consistency. In Europe the already high level of expenditure on market research and the sophisticated infrastructure of research suppliers provide additional impetus for technology application. Given that the individual member states vary greatly in income, technology absorption, and culture, it is yet undetermined the extent to which leading-edge technology can be used effectively and in a truly pan-European manner. This chapter focuses

Table 28.1. Regional Share of Worldwide Market Research, 1990 and 1995

	Regional Share	
	1990	1995
Europe	45%	45%
USA	35%	34%
Japan	9%	10%
Other	12%	11%
Total	100%	100%

Source: ESOMAR Annual Statistics (1996).

Table 28.2. National Shares of the Fifteen European Union Member States, 1995 (Percent)

	Market Research Expenditure	Advertising Expenditure	Gross Domestic Product	Population
Germany	27	32	29	22
United Kingdom	22	19	13	16
France	19	15	18	16
Italy	8	8	13	15
Spain	5	7	7	11
Netherlands	5	5	5	4
Sweden	4	3	3	2
Belgium	3	2	3	3
Austria	2	2	3	2
Denmark	2	2	2	1
Finland	1	2	1	1
Portugal	1	1	1	3
Greece	1	2	1	3
Ireland	1	1	1	1
Luxembourg	*	*	*	*

* Less than 0.5%.
Source: ESOMAR Annual Statistics (1996).

on technologies that are, or will be in the near future, suitable for use in surveys of the general public in most European countries. Thus, for example, applications for business-to-business research in companies with sophisticated information technology (IT) structures have been excluded.

28.3 TYPES OF DATA COLLECTION

Despite its relative maturity as a research market, Europe differs radically from the United States in preferred methods of data collection. In U.S. commercial market research almost all quantitative studies are conducted by telephone or mail. Personal interviewing in this context — particularly in-home interviewing — is comparatively rare. In Europe, broadly speaking, the in-home personal interview is the majority choice for ad hoc or customized quantitative surveys. Table 28.3 shows expenditure shares of data collection methods. Similarly omnibus services predominantly use in-home personal interviews, while telephone modes still lag behind. Self-completion — mainly mail — is a poor third, although it is in the area of self-completion that European researchers have made the highest investments in technology.

Table 28.3. Share of Ad-hoc Expenditure on Data Collection for Selected Countries, Plus European Total, 1995 (Percent)

Country	Face to face	Mode Mail	Telephone	Other	Total
France	51	5	34	10	100
Germany	67	6	22	3	100
Italy	69	3	25	3	100
Spain	61	7	25	7	100
U.K.	64	10	22	4	100
Total EU	60	9	27	4	100

Source: ESOMAR Annual Statistics (1996).

The mix of data collection methods depicted in Table 28.3 stems from the interaction of a number of peculiarly European factors. The use of personal interviewing emanates from the long-standing tradition of social research. Demographic conditions in the major industrial countries made personal interviewing feasible and cost effective for early social researchers. In the post war era, a large pool of skilled women were available for low-cost casual employment, which in turn contributed to the increased use of personal interviewing by commercial market researchers.

Monopolistic telephone utilities with high installation and per-call costs have inhibited the growth of telephone use throughout Europe. Only recently are new technologies and increasing competition beginning to reduce costs (although not in all countries). Low domestic telephone penetration also resulted in postwar generations who grew up without the telephone or who used it only in a limited way. Thus, the European market researcher never grew accustomed to the telephone as a data collection instrument, and for many agencies and clients this attitude persists. It is generally still a case of "special problems require telephone interviewing" rather than "is there any reason why we cannot do it by phone."

There seems to be an inhibition on the part of European market researchers to increase their use of the telephone. This attitude is not confined to researchers. For example, in 1995 draft EU telecommunications regulations proposed making unsolicited telephone calling for the purposes of survey research illegal on the grounds of its intrusive nature. Fortunately lobbying by both the market research industry and government statistical institutes prevented any enactment. However, the very presence of such a clause in centrally drafted legislation indicates that unsolicited calling is highly frowned upon. It should be emphasized that Europe has yet to experience the blitz of telemarketing that exists in the United States. If telemarketing ever reached similar levels in Europe, it would definitely worsen the public's perception of telephone survey activities. It is fully feasible that telephone research will grow faster in the future, and it is a matter of surmise what the European legislature's response will be.

Self-completion research performs poorly compared to either personal or telephone interviewing. Again the reasons are complex. By tradition Europe has never made great use of mail questionnaires (perhaps this is similar to the absence until recently on a large scale of that other great American postal service user: the mail order catalog). Mail response rates tend to be low. One exception is research users who have developed substantial expertise in using mail surveys often complemented by expertise in direct marketing. Such research is primarily carried out among existing customer bases. Other research fields have seen mail methods as being cheap and producing low-quality results, a perception that has inhibited the use of mail methods.

28.4 GENERAL BACKGROUND FACTORS

Before discussing current developments, it is necessary to treat a number of other factors that are likely to affect the use of technology in the next decade. At present the most important of these factors appears to be changing employment and tax laws as applied to field interviewers. Historically the status of interviewers within employment and tax law has been generally undefined across Europe. Definitions vary between countries as to whether interviewers are part-time, casual, or self-employed. The net effect has been that interviewers' entitlement to social security benefits and other employment rights has been limited. This situation is changing. While there is still not agreement at the EU level, it is increasingly apparent that their exclusion from rights enjoyed by the rest of the workforce is unlikely to continue in the future. At the time of writing, the Dutch market research industry is facing demands for social security payments for interviewers. A likely consequence is increased costs for personal interviewing. A cost increase may produce contradicting results. It is possible that the higher cost of personal interviewing compared to telephone interviewing will encourage a shift toward the latter method. Alternatively, if the European predilection for personal interviewing remains, it may encourage managers to value their interviewers more highly. Should this happen, one would expect the corps of interviewers to shrink, while those who remained would be employed on greatly improved terms and be equipped with first-rate technological tools. Samuels (1994) identified better terms, better conditions, and increased number of hours as a corollary of switching to CAPI. Thus the direction of the causation may be reversed, but the outcome would remain the same.

A second possible outcome, but for which the implications are much less clear, is the changing technology of the telephone, its connection/transmission media, and its ownership as these relate to the private citizen. It is a historic fact that the use of the telephone for data collection in surveys of domestic populations was predicated on a definable population of telephone installations which for all practical purposes was synonymous with the total domestic population. Implicit in this was the wire connection to the household and almost never more than one separate number per household, thus providing a sampling frame with calculable coverage and probabilities of selection. However, the advent of cellular or mobile telephones in all their variety is changing these fundamental axioms of telephone research.

In countries where a mature land-based telephone network was already in place prior to the introduction of mobile telephones, the mobile telephone has typically been an individual accessory for communications outside of the home. Thus for statistical sampling purposes the presence of such telephones could be almost entirely ignored. However, in those countries where mobile telephones have become available while the land-based telephone network is still immature, it is quite possible (particularly if state monopolies have inflexible pricing systems) that some individuals and households may become completely cellular in their phone possession and use. This coupled with the lack of up-to-date directories for cellular suppliers, the increasing growth of unlisted numbers, the lack of residence-related telephone numbering systems, and the incalculable probabilities of selection will seriously undermine the extent to which telephone interviewing in the future will be able to provide samples whose representativeness passes scrutiny.

The current scale of this problem is unknown. Administrative data on telephone ownership typically focuses on lines and tends to come from only the central telecom supplier. Survey-generated information can give some indication. However, since the most mobile proportion of the population is most likely to be nonrespondents and most differential in their behavior, the bias is likely to be extremely high. This issue can potentially have a considerable effect on the future use of telephone interviewing in Europe. As yet, however, there has not been any public discussion or consideration. It is noteworthy that accurate and up-to-date information on the characteristics of telephone ownership on a country-by-country basis across Europe is not publicly available. Until authoritative data are collected and published, this issue merits considerable concern.

The third and last major issue that can be identified as affecting the use of technology in European survey research is the availability of up-to-date optical fiber networks that are suitable for cable/interactive/high-speed data applications. In the absence of individual countries having centrally driven development plans, and the failure of the EU to yet announce any European-wide investment policy, any widespread application using this technology will not be practicable for a considerable number of years.

28.5 EMERGING EU TECHNOLOGIES

While substantial data exist and have been presented earlier about the overall shape of European market research, no data are collected either by ESOMAR or the major trade associations on the actual use of technology. Thus we must rely on a mix of desk research and personal conversations with key experts in the field — a classic business to business study. That has been the approach I have followed, and I am indebted to my colleagues in Europe for their knowledge and guidance. In the next three sections a broad categorization has been taken: (1) direct telephone applications, (2) field interview applications, and (3) self-completion applications.

28.6 DIRECT TELEPHONE APPLICATIONS

"Direct telephone" refers to applications where the respondent uses the telephone to provide responses in some way. Applications using other IT equipment connected to the telephone are dealt with in Section 28.6.3 below.

The use of the telephone for market research data collection in Europe did not start on a significant scale until the 1970s. Miln and Stewart-Hunter (1976) castigated researchers for not being more forward-looking. But by the late 1970s, the use of telephone in the U.K., France, and Germany for both domestic and business-to-business interviewing was becoming widespread, albeit still regarded with suspicion by a large section of the industry. The use of computer assisted telephone interviewing (CATI) occurred very early in the application of telephone research. The software was generally an extension of the analysis packages that were used increasingly by market research agencies as they moved away from the in-house developed packages that had been used in the 1960s. Perhaps surprisingly these software packages were European in origin, rather than from the United States, and still today almost all CASIC software used in Europe is homegrown.

Today CATI has become almost a standard for any major company offering telephone research. Over time CATI facilities have become more sophisticated, in line with general developments in CATI elsewhere in the world and in the social research sphere. Developments have generally been steady rather than spectacular — improved quality control features, improved scripting features, and better data handling. Recent developments of particular significance cover a number of applications.

28.6.1 Digital Recording

Increased computing power coupled with digital recording provide the ability to record the whole of an interview, index its constituent parts, and then directly access any chosen section (e.g., open-ended questions) for playback. The extent to which this is a benefit depends on the stance one takes with regard to listening secondhand to recorded interviews. Additionally there exists the question of whether recording without prior consent is either legal or ethically permitted in the country in which the fieldwork is conducted.

From a practical point of view, I doubt whether in an actual market research situation such recording is of much real value, since time and cost rarely permit the luxury of extensive analysis. However, providing verbatim quotes from quantitative research comparable to those provided by qualitative research will undoubtedly allow users to feel closer to the respondent. Such a benefit for clients is likely to be of significance in increasing the credibility of market research, and telephone research in particular. This will be particularly true for clients in countries where the concept of research is relatively new — for example, the former eastern bloc economies. For quality control purposes the benefits are likely to be negligible for CATI units because the existing ability to monitor centrally provides an instant and random facility that lacks comparability by any other data collection process. However, the application of digital recording within the CAPI process may provide major quality control advances, as described below.

28.6.2 Predictive Dialing

Until recently facilities such as predictive dialing and call handling were restricted to very large CATI units that could afford the expensive additional software routines that provide these applications. They are now becoming available as part of basic CATI packages. Generally, jobs must be of a certain minimum size before major economies of scale are realized, permitting even small units to compete effectively. Potentially they will allow some of the difficulties caused by directory noncoverage of cellular and unlisted numbers to be overcome by facilitating plus one dialing (a technique involving the random selection of listed numbers to whose last digit "1" is systematically added to provide samples of "all" numbers) where the number ranges are unknown.

My company has found that predictive dialing increases productivity up to 70 percent on very large samples with interview lengths of five minutes or less. The relative cost differential in favor of telephone research will be widened by the new technology. This might act as a catalyst to move telephone research up on the desirability scale in Europe.

One example of methodological improvements such efficiency can provide has been the introduction of telephone interviewing as an aid to continuous self-completion panels. In these applications respondents keep a diary about a defined area of purchasing or other behavior. Data are collected via a weekly phone call at a predesignated time — ideally by the same interviewer. The call acts as data collection, interactive edit correction, and stimulus to recall all relevant behavior. The technique is particularly applicable to high-frequency casual behavior such as payments, snacks and confectionery, and lottery ticket purchases. The use of the telephone in such circumstances is a valuable con-tributor to higher reporting continuity for teenagers and younger adults whose cooperation on such studies is otherwise notoriously difficult to maintain.

28.6.3 Multilingual Applications

The desire for comparable and consistent pan-European research has stimu-lated the conduct of multinational telephone research. Differences in base labor costs coupled with variation between European telecommunications companies in their international call tariffs has led to the growth of multilingual international interviewing from a single CATI installation. Technology has had little to do with this; rather one needs the existence of one or both of these factors in conjunction with a pool of relatively educated multilingual part-time workers. The U.K. and the Netherlands in particular satisfy these requirements, and a number of companies have built up this type of business.

One technology adjunct to this area has been the arrival of multilingual CATI software that enables interviewers to access whichever language version of the questionnaire they wish to, rather than doing, say, French on a particular shift. Subject to time zone differences, this functionality, coupled with predictive dialing and known interviewer linguistic abilities, enhances the productivity, price competitiveness, and speed of turnaround of the single-site multinational CATI facility.

28.6.4 Computer Telephone Integration

Emergent in the true sense of the word is computer telephone integration (CTI). The integration of computing — PCs in particular — and telephone potentially offers the ability to decentralize CATI and integrate it with a computer assisted personal interviewing (CAPI) facility on a single PC. This is achieved by the use of portable PCs complete with voice ports, integrated CAPI/CATI scripts, and data processing. In theory, one could disband a central CATI unit and employ a dispersed field force for both telephone and personal interviewing. Costs of the central overhead could be largely eliminated at the price of central location quality control. Digital recording could be used as a retrospective quality check, but it clearly lacks either the imperative or impact of the classic CATI unit system. Furthermore the extent to which a good personal interviewer possesses the same skills as a good telephone interviewer is debatable.

Just because a technology makes something feasible does not make it desirable. All too often methodologies are stretched to fit technologies rather than the reverse, and invariably such stretching and stressing end in the failure of the technology. In theory, the benefits of such integration are attractive. In reality, the differences between data sets arising from CATI and CAPI coupled with the different interviewing skills required render this technology of little lasting value to the user. This application is starting to be used in some central CATI units where the increased power and flexibility offer potential productivity benefits.

28.6.5 Touchtone Data Entry

Touchtone data entry (TDE) is seldom used in Europe. In large part this has probably been due to the slow arrival of touchtone telephones, and no great consumer incentive to use them. For example, at present in the U.K., approximately two-thirds of residential telephones can use touchtone, but of these only one-half (one-third of the total) actually use it. A further reason is the absence in Europe, until very recently, of that American phenomenon: the access panel. In these applications many tens of thousands or more homes and individuals are prescreened who then commit (in theory) to reply to questionnaires sent to them. This approach is starting to be marketed in the major European countries, and it seems a possible application for large-scale touchtone usage. One isolated example occurs in the U.K.: A quality of service measurement survey of a specially recruited panel. This survey measures the time lapse between the posting and receipt of mail.

28.6.6   Automatic Speech Recognition/Voice Recognition Entry

Blyth and Piper (1994) and Blyth (1995) have reported on the development of prototyped dialogues for interactive interviewing. Respondents with no special training were able to participate in interviews lasting up to three minutes. The interviews employed complex routing and the use of large vocabularies (over 200 words) for specific question answering. Agencies in the U.K., the Netherlands, and France are working on the implementation of this technology—for example, with small access panels—but at the time of writing no actual commercial applications exist.

The primary difficulty facing researchers developing applications is not the level of accuracy of the speech recognition. Rather, it is the absence of suitable dialogue design software that would enable survey researchers to integrate speech recognition into a free flowing interview format. EU funding to develop simple dialogue design implementation software has been granted to a consortium that includes NIPO of the Netherlands and Taylor Nelson AGB in the U.K. A number of different automatic speech recognition (ASR) applications are being tested during 1997–98. One of the objectives is to compare the similarity of the results of ASR and normal telephone research in more than one language.

In a recent survey of leading market research software suppliers, speech recognition was cited as the new technology likely to be the next major step in the application of technology. It is probable that within the next two years the leading European CATI and CAPI software suppliers will provide the facility to integrate some form of speech recognition into their repertoire. Given the power of the processor that is required, the effect and potential will be greatest for CATI systems. The options that are envisaged are essentially twofold: either the initiation of an interview by a human operator with speech recognition completing the main part (this would be particularly suitable for access panels) or the handling of inbound calls by speech recognition with referral to a human operator as a support. In the European context the potential of ASR to handle several languages and time zones from a single station is particularly appealing. Additionally ASR offers the opportunity for increased standardization of data collection.

Of all the potential technologies this is one of the most exciting—if not the most exciting—for widespread European application. The difficulty at this point is that no field trial data have been made available. The REWARD project (Real World Applications of Robust Dialogue) funded by the EU has committed to publish the results of its field trials during 1998–99, and these will include a variety of market research interviews in more than one language.

28.7 FIELD INTERVIEWER APPLICATIONS

Without doubt the emerging technology of the 1990s is CAPI. I focus the discussion of CAPI on three key questions. What have been the drivers of growth? How do the different CAPI technologies vary? Where is CAPI taking market research?

28.7.1 Historical Background

As an actual commercial application in Europe, CAPI goes back more than 10 years. In the Netherlands government survey research use of CAPI was followed by the first commercial service launched by NIPO (the largest Dutch research agency) which developed their own software in 1985. Initially the adoption of CAPI was slow. The first commercial use in the U.K. by Research Services Ltd. (RSL) was not until 1990 by which time Quantime, one of the leading market research software bureaus, was supplying CAPI software. Samuels (1994) reported that at that time, other than in the U.K. or the Netherlands, there was no other commercial use in Europe of CAPI. However, two years later the ESOMAR Directory of Research Organizations lists 65 research agencies within the EU claiming to offer CAPI interviewing. This probably is an underestimate, since several major agencies known to have CAPI do not mention it in their entry. Others have only recently announced their move to CAPI. Clearly the rate of growth of CAPI is staggering, and recent announcements indicate that it shows little sign of abating. This raises two questions: (1) Why was initial growth so slow? and (2) Why is it now so fast?

28.7.2 Why Was Initial Growth so Slow?

Most market research agencies in the 1980s and early 1990s were not intensive users of information technology. Editing, CATI, and tabulation were typically done using standard packages employing a few specialized specification writers and a relatively simple central processing facility. PC use was relatively uncommon, and only the very large companies had developed mainframe installations or specialized communications support of the sort required, for example, a TV meter panel. Thus the management skills were absent, the infrastructures were absent, and the specialist market research software houses had not gone beyond CATI to provide a basic product.

Second, CAPI requires a larger investment in hardware, software, and support staff than CATI. CATI originally used terminals with relatively low cost per unit software licenses. Even now a 50-screen CATI unit is relatively large requiring perhaps five or six times that number of staff to provide seven-day a week, full day cover. The cost of portable PCs suitable for CAPI interviewing is not low. Of the companies now using CAPI in the U.K., PC prices paid (as opposed to list price) range from £300 (about $480) per unit for relatively simple "tablets" to over £2000 (about $3200) per unit for sophisticated state-of-the-art systems complete with multimedia. Even where basic scripting software is available it only provides a very small part of the overall software requirement. The support software for fieldwork management, telecommunications, quality control, and data security cannot, by virtue of organizational structure and needs, be available off-the-shelf but generally has to be written from scratch. This is a lengthy, painstaking, arduous, and costly process which deters agencies.

Third, cost savings, while generally claimed for CAPI, are not easy to achieve until an organization has become almost totally CAPI. Maintaining paper-based field capabilities alongside electronic field capabilities increases cost and management aggravation. Unlike government agencies which typically carry out a small number of large and often complex surveys, the work pattern of commercial market research agencies is characterized by a high throughput of relatively small jobs. A medium to large agency may have between 15 and 30 personal interview surveys in the field at any one time. Software to despatch, progress control, and quality check work on this scale, with a field staff of five or six hundred CAPI interviewers, is complex and expensive to design, write, test, and support.

Fourth, CAPI removes a large proportion of variable direct costs — print, paper, postage, despatch, booking-in, data entry, and editing — and replaces them with a fixed overhead of equipment depreciation, interest, and software depreciation. A period of relatively poor growth in the late 1980s and early 1990s probably deterred many companies from changing their cost structure in such a fundamental way.

Fifth, and last, is the lack of commercial need. Experience first in the Netherlands, second the U.K., and third in France has shown that not until a major agency has taken the step, made it work, and experienced commercial success will other agencies follow suit. It is less the altruistic stimulus of quality, and more the commercial threat of slower delivery and lost revenue, that captures the research agency's corporate mind.

28.7.3 Why Is Growth Now so Fast?

A number of factors contribute to the rapid growth. First there is commercial anxiety. CAPI is rapidly moving from a gratification to a need. Survey requirements increasingly specify CAPI as the required data collection medium. For other service areas, such as an omnibus survey, CAPI is required to match the speed of competitive turnaround. In launching a pan-European omnibus using standardized interviewer terminals and software, the IPSOS Group (a European multinational research agency chain) has essentially made CAPI pan-European. Other international chains have to match the offering to stay in business.

Second, software suppliers have seen the large potential of this area. There will be many more CAPI units than CATI due to the relatively poor productivity of face-to-face interviewing. Few companies find it practical to transport CAPI equipment between interviewers. This potentially rich market is attracting increasing numbers of software suppliers. There are at least eight established companies in Europe offering off-the-shelf CAPI software at competitive prices.

Third, research agencies have become PC literate. "A screen for every desk" has become as European as "a chicken for every pot." Agencies have thus quickly developed the skills to support and manage PC networks both within and outside of the office, and this has increased CAPI's accessibility.

Finally there is the emotional and cultural aspect. European agencies feel happier with personal interviewing. CAPI inspires confidence and is easily accepted, all while maintaining the personal touch. The potential changes in European law mentioned earlier require a change in the structure of the industry's basic way of using labor. This is congruent with the business need of using CAPI equipment more intensively to get a greater return. Samuels (1994) reports experiments with bonus schemes to encourage interviewers to work more days and longer days to deal with this very point. This effect is common to agencies across all European countries.

In the short term CAPI's rapid growth will continue. Where it will peak is unclear. I would estimate that by the middle of 1997 there will be between 5000 and 6000 CAPI machines installed in the U.K. and over 15,000 across all of Europe. On the assumption that each machine is used $2\frac{1}{2}$ days per week for perhaps 48 weeks per annum, this represents a European CAPI capacity of 1.7 million field days a year. This is probably less than one-quarter of total European personal interviewing. As the advantages of CAPI are increasingly demonstrated, this figure will grow. It is likely, in my opinion, that it will double by the end of the millennium in terms of fieldwork days but less in terms of the number of machines. The apparent contradiction is accounted for by the increasing number of days per week that CAPI interviewers will have to work to justify the capital investment.

28.7.4 CAPI Technologies

CAPI is not a single technology but rather an acronym describing a range of technologies. Differences between the CAPI technologies, unlike the CATI technologies, are not obvious to the respondent. It follows therefore that each of the subtechnologies can produce different respondent, interviewer, and respondent-interviewer interaction effects. Within the commercial market re-search world there does not appear to be very high awareness of this as a factor, and this could undermine comparability between surveys. Little system-atic experimental work has been done or published on the subject. Four subtechnologies can be identified.

Bar Code Scanning Terminals

This inexpensive and robust technology lies outside the main development of CAPI. Bar code scanning preceded laptops and initially attracted the interest of companies doing a substantial amount of simple and repetitive fieldwork — often observational, for example in-store price checks. By issuing interviewers with precoded lists which include bar codes, it is possible for them to quickly enter substantial amounts of data using a mix of bar codes and a numeric keyboard. A few agencies went further and issued to interviewers precoded questionnaires which used bar codes. The advantage of this approach is that it reduces print, mail, and data-entry costs; for any questionnaire version only one hard copy per interviewer is required. The same hard copy could be scanned repeatedly for new respondents. Speed from the field was also the same as for conventional CAPI — at least for precoded parts.

There are limitations. First, show materials still have to be printed. Second, these scanning terminals were unsuitable for the collection of open-ended data, which still had to be written down and returned by conventional means. The associated delay, complexity of merging data sets, and lack of flexibility resulted in waning interest in this approach. While it has benefits for specific onetime applications, bar code scanning terminals can currently be regarded as a defunct technology for CAPI applications.

Conventional Laptop PCs

The conventional laptop, with a black and white screen, has become the everyday face of CAPI, accounting for an estimated 85 percent of CAPI installations. The independent observer might well ask what justification there is for this phenomenon. There is a generally held sentiment that CAPI improves data quality. For the work of commercial market research inter-viewers who work on many different surveys, usually without briefing, it is probable that CAPI will reduce the level of routing errors and contradictory answers. However, de Leeuw, Hox, and Snijkers (1995) found no evidence of general data quality improvements in their review of the literature. Similarly no case studies have been published that definitively address whether cost savings can be achieved.

For conventional market research interviewing, laptop CAPI suffers from a number of drawbacks. The first of these regards the siting of the interview. Laptop PCs have to be used sitting down. For long questionnaires, respon-dents may invite interviewers into the house, but this is not always the case, and less so for short questionnaires. Bosley, Conrad, and Uglow (Chapter 26) cite the advantages of pen computers for erect interviewing. However, I believe that regarding pen as the "outdoor" CAPI technique is unnecessarily limiting. This problem is accentuated by the increasing use of male interviewers in many countries. Male interviewers are less likely to be invited into homes. In these circumstances the laptop equipped interviewer cannot easily interview. This can decrease response rates and increase costs. The need for the interviewer to be sitting with access to the keyboard also inhibits the use of the screen for show materials, for it is difficult to keep showing the screen to respondents without giving them continuous access to the screen which is undesirable for many questions.

A second disadvantage concerns open-ended responses. Such responses must be entered through the keyboard, and except when the interviewer has specific keyboard skills training this is a relatively slow and sometimes clumsy process. Among a number of its effects, it reduces the questionnaire flow and diminishes the technological level of the process. It also reduces eye contact and rapport with the respondent. Notwithstanding these problems the laptop is *the* European CAPI technology at present.

Pen Computing

With regard to emergent technologies two strands are coming to the fore in CAPI. The first of these is the use of pen computers. A detailed review of pen technology is given by Bosley, Conrad, and Uglow (Chapter 26). The availability of reliable pen machines with screens of at least 20 cm diagonal and scripting software has made keyboard-free CAPI a reality. In Europe at least three companies now offer pen scripting. From the perspective of the researcher this largely overcomes the drawbacks associated with laptop PCs described in the preceding section. That is to say, doorstep interviewing is possible, open-ended responses can be written on the screen (for transmission only — recognition still lags some way behind the claims made); showcards can be put on screen together with rotations and other variations not possible with physical material. Pen also enables easy respondent use for self-completion of sensitive data and engenders a greater sense of confidentiality.

In my own company, interviewers with experience of bar code scanners, laptops, and pen overwhelmingly prefer pen. It has made their job easier and made them feel more professional. For standard interviewing, the only limitation to the future prospects of pen would appear to be the cost premium of around 50 percent that exists between a pen machine and a laptop with otherwise similar features. However, even if the research agency can justify the cost premium, there is a wider problem. Even at these prices there is a major concern as to whether there will be sufficient demand for the large screen pen to provide the necessary continuing availability of a range of robust, reliable equipment at competitive prices which are supported by a reputable manufacturer. It is improbable that the survey research market is sufficient to support the pen PC market. Currently there do not appear to be many other large-scale users of this technology.

Multimedia PCs

The enhanced power of PCs is resulting in a number of companies using enhanced materials — sound, moving image (video), and color. For specific applications, for example, advertising testing or tracking, such multimedia machines offer tremendous potential either for the in-home or small type operation. However, PC costs are relatively high and may not justify equipping all members of a fieldforce in this fashion.

28.7.5   The Future

The designer of a CAPI data collection instrument now has available a wide variety of facilities. To those already mentioned can be added digital recording (a reality), voice command (an emerging reality), and speech recognition (emerging). Meier (1997) has reported on experiments with pen computers that also integrated voice recording. These experiments found that the fullest data were obtained using recordings. Obviously for some applications this ability to provide essentially qualitative data will be extremely valuable. Other companies are experimenting with the integration of other software, such as maps to aid interviewers in their tasks.

Two broad conclusions can be drawn. The first is that we will see within CAPI a wider range of subtechnologies. Each will have advantages and disadvantages for specific aspects of the data collection process. An important point that is in danger of being ignored is that each of these technologies will produce their own bias structures. Without experimental work, the scale and effect of these on accuracy and comparability are unknown. Second, CAPI gives you what you pay for. That is to say, there is a strong correlation between cost per machine/software and the facilities available.

28.8   SELF-COMPLETION APPLICATIONS

In the European context the application of technology to self-completion has been led by dedicated consumer panels that provide either specific measurements (e.g., television viewing) or a number of related measurements (e.g., food and grocery purchasing) (see Saris, Chapter 21). In addition to these, there are two other main areas of the application of technology. First, the use of a panel to provide answers to ad hoc studies, and second, technology tailored to specific market research applications, for example, product testing.

28.8.1   Dedicated Panel Measurement

European research agencies pioneered the use of audience measurement using technology, first by mechanical means, and later, electronically. Such surveys require telecommunications, IT, engineering, and organizational skills that are far greater than those required for other forms of large-scale market research. As a consequence there exist within Europe a relatively large number of companies and individuals who have the experience to apply these skills across other survey areas and countries. These skills are required for the design, equipping, motivation, and collection of continuous data electronically from representative samples of the population.

Thus, for the collection of purchasing data on food, grocery, and toiletry, panels of many thousands of homes equipped with bar code scanners now exist in the U.K., France, Germany, Netherlands, Italy, and Switzerland. In this type of technology application there are close similarities between the United States and Europe.

As in the earlier discussion of future CAPI developments, the researcher has the facility to design equipment involving a range of features. The problem is one of cost. Consumer panels typically number several thousand members — either individuals or households. Developing and manufacturing task specific hardware and software on this scale requires a very large investment. This can typically be afforded by only a small number of syndicated or multiclient studies. Thus the trends in this technology will depend very much on which industries require that level of research — probably on a pan-European basis. One can surmise that the industries will be a mix of leisure/service industries such as the new electronic media, fast food chains, soft drinks, and other beverages. More complex and abstruse measurements such as time use, financial payments, and travel are also candidates. Such activity is typically individual rather than household, and it does not revolve around a simple focus — the TV set or the shopping basket.

For these types of measurement there exist two requirements. Electronic communication between a terminal and a central host or some form of local intelligent processor. A number of alternative technologies meet these requirements. They are the Internet, homes with a cable connection, homes with intelligent telephones (e.g., as with Minitel in France).

Much is promised for the Internet. For the major European countries (other than in the Netherlands) it is difficult to predict whether natural growth in Internet connections will be so high over the next five years as to provide naturally representative samples of the population. It is unlikely that the Internet will be used for panel development, since the cost of equipping and connecting a household for the Web will be too high.

Experiments are taking place in Germany to provide questionnaires to respondents' televisions via a spare channel and a set-top decoder. Depending on the cost of the signal transmission, such an approach offers a flexible and less expensive route. The difficulty is the fact that while the TV set is being used for one purpose, it cannot be used for another.

A telephone with some form of local processing intelligence, particularly if it includes a screen, offers the low-cost requirement and the potential for interactivity. Such systems for use in electronic ordering, electronic banking, and other applications are likely to become widely available in the next few years. They will also enable the direct use of speech recognition, which has a great potential for panel applications.

All of these systems will be capable of pan-European application. Of the three it is the intelligent telephone that offers the greatest potential for dedicated panel applications.

28.8.2 Panels for Ad hoc Research

Much of the foregoing applies equally well to panels that are recruited to answer ad hoc enquiries. For most ad hoc applications, speech recognition in a simulated CATI environment would per se be sufficient and have both the lowest large-scale entry and running costs.

The relative economics will depend entirely on the volume of questionnaires any one panel can bear and thus the trade-off between operating costs and setup costs for available alternatives. Ad hoc research does not appear to be a sector that requires or justifies a very high investment in technology.

28.8.3 Specific Ad hoc Applications

Of the technologies that have not been discussed, virtual reality offers a range of applications if it can be harnessed. This will not be a European phenomenon, rather it will probably emanate from the United States where such applications development is stronger. However, at the time of writing, no commercial applications of virtual reality have appeared in Europe. The use of multimedia PCs for advertising testing and tracking is already taking place and likely to expand. It is probable that multimedia will extend the range of research that can be carried out. For example, there is a growing demand for TV program research that tests alternative story lines and characters — interactive multimedia may offer a substantial potential there.

Other areas of technological application will be in the collection and analysis of observational data. Here the use of image recognition can already provide traffic counts at specific times — for example, in shops, banks, and railway stations. Such techniques are likely to increase in accuracy, giving more reliable data and allowing better projections and forecasts.

28.9 CONCLUSION

By virtue of the very nature of their business, market researchers have always been early innovators in the use of technology, be it to collect or analyze their data. The speed of innovation in data collection is not determined by the speed of technological development, but by the willingness of interviewers and respondents to accept and work with the technological advances.

The last 15 years have seen an amazing increase in the technology available to the professional market researcher and with effects that few could foresee. It has been, and continues to be, an exciting time for those working at the edge of technology applications. However, one must not forget that technology is a means to an end and not an end in itself.

The reality of the marketplace and the slow pace of technological diffusion among the general population result in a relatively slow adoption of new technology in applications that interface directly with the public. There is no reason why this should change in the future. It follows that the technologies that will be absorbed faster will be those that are obvious and natural to respondents or interfaced by interviewers. In this area extensions of CAPI and even speech recognition will be the primary contenders. Indeed speech recognition in all its forms is likely to be the hottest of the new technologies. The Internet will be extensively tested because it is there. However, its limited penetration and slowness of use makes it unsuitable for widespread ad hoc applications in the foreseeable future.

I do not expect any radical change within the next five years but rather evolution from the existing technology base. Applications of technology as previously stated provide savings in cost, savings in time, or arguably improvements in quality. For a major change to occur in the interim, we need a technology that will unarguably improve quality.

CHAPTER 29

The CASIC Future

Reginald P. Baker
Market Strategies, Inc.

29.1 INTRODUCTION

"Today mankind has the accelerator pedal of technology pressed hard to the floor." So writes James Martin in *Cybercorp: The New Business Revolution* (1996). Martin, long recognized as our most prolific and thoughtful writer on the topic of information technology strategy and management, goes on to describe the unprecedented rate of change in contemporary technological development, and the challenge that businesses and organizations of all kinds face in trying to adapt to it. This challenge is not optional; it must be taken up. Those who do so successfully will survive only to face new challenges. Those who fail will perish.

Survey research is almost completely a set of linked information processing tasks. A world of rapidly changing information technology presents us with both a challenge and an opportunity. The challenge is to find ways to capitalize on new technologies, while remaining true to the historical principles on which our industry and the survey profession rest. The opportunity is one that allows us to rethink and ultimately redefine the processes used to collect and process information. The evolution of survey research, especially over the last 20 years, has been intertwined with technological developments that now produce new technologies at an ever accelerating pace. As survey organizations absorb these new technologies, the processes of survey research, the structure of survey organizations, and the industry as a whole are changing.

This chapter considers expected future developments in information technology and speculates on their likely effect on the continued development of methods and technologies to support the multiple facets of what we have come to call computer assisted information collection (CASIC). It begins by looking

Computer Assisted Survey Information Collection, Edited by Mick P. Couper, Reginald P. Baker, Jelke Bethlehem, Cynthia Z. F. Clark, Jean Martin, William L. Nicholls II, and James M. O'Reilly. ISBN 0-471-17848-9 © 1998 John Wiley & Sons, Inc.

briefly at where CASIC has been, the underlying technological changes that produced the earliest CASIC technologies, and those that drive its continued development. Looking ahead, the chapter considers the future that technologists such as Martin and others see for us, and offers some thoughts on the likely effects on the practice of survey research.

29.2 UNDERLYING TECHNOLOGIES OF CASIC

Couper and Nicholls (Chapter 1) underscore the importance of the personal computer in the historical development of CASIC. Indeed many of the CASIC applications described in this volume were developed and run on microcomputer platforms. The invention and mass marketing of microprocessor-based technology has played a fundamental role in changing how we collect survey data, how we interact with our respondents, how we organize ourselves, and how we design our internal organizational processes.

Despite this reliance on the PC platform, the technological developments that serve as the basis for CASIC development — past, present, and future — are significantly more complex. Tapscott (1996) describes ten major technology shifts that he believes are revolutionizing our economy and society (Table 29.1). The shift from traditional semiconductor to microprocessor technology (i.e., the PC revolution) has been important, but other shifts such as those from host to client-server computing, from proprietary to open systems, or from dumb to intelligent networks have played key roles in the past, or will be major factors in the future. Some of these shifts have been underway for many years, while others are only now beginning on a broad scale. Virtually all of them have the capacity to create major changes in CASIC and therefore in survey research.

These shifts also serve as a lens through which we can view the evolution of CASIC, past and future. They define four broad phases of CASIC development.

Table 29.1. Tapscott's (1996) Ten Major Technology Shifts

Shift 1.	From analog to digital
Shift 2.	From traditional semiconductor to microprocessor technology
Shift 3.	From host to client-server computing
Shift 4.	From garden path bandwidth to the Information Highway
Shift 5.	From dumb access device to information appliance
Shift 6.	From separate data, text, voice, and image to multimedia
Shift 7.	From proprietary to open systems
Shift 8.	From dumb to intelligent networks
Shift 9.	From craft to object computing
Shift 10.	From GUIs to MUIs, MOLEs, MUDs, MOOs, and VR

The first two phases feature (1) the invention and widespread adoption of CATI followed by (2) the development and deployment of CAPI. The third stage, just now beginning, is likely to be still another wave of revolutionary technological change that will make possible a departure from our traditional reliance on interviewing as the primary mode of data collection, and move to broader use of self-administered methods. The fourth phase of CASIC development, which has not yet begun to unfold, may well result in a significantly reduced reliance on surveys to gather information for decision making. Under this scenario, organizations that today rely on survey data may choose to substitute databases developed through the mining of existing data resources. Although not described in this volume, the collection of point of sale (POS) data by market research firms (Ing and Mitchell, 1994) is perhaps a precursor of a much broader use of direct data collection methods in the future. It is easy to imagine others. Where surveys remain important information sources, the way we go about them, and the remaining role for interviewers will change dramatically.

29.3 TECHNOLOGY AND THE SURVEY ORGANIZATION

Porter (1985) has offered the concept of the value chain to describe the strategic use of information technology in the firm. Many of the competitive advantages described by Porter — shorter cycle times, lower costs, strategic links to suppliers and customers, and improved integration of internal processes — have also been clear goals of CASIC. More recently, Blattberg, Glazer, and Little (1994) proposed the information value chain (Figure 29.1) as a way of understanding how modern information technology is transforming market research. The information value chain describes the process by which "raw data" are collected and then transformed into information that can be used for decision making. Those decisions may concern which new products to bring to market or which public policies are likely to be most successful. While the information value chain concept was developed initially to describe information processing in market research organizations, it has applicability to information processing organizations generally, including government and

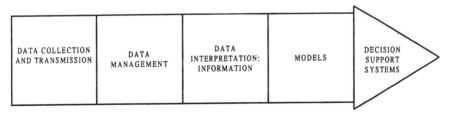

Figure 29.1. The information value chain. *Source*: Blattberg, Glazer, and Little (1994).

academic survey organizations. The framework is also useful for describing the functions of survey organizations regardless of the technologies used, and for understanding how those functions and the overall organizational processes and structures evolve with new technologies.

The information value chain consists of five stages. The first stage is the input stage, data collection. Any number of methods might be used including interviewing, respondent self-completion, or direct collection from on-line, machine-readable sources. All subsequent stages process these data with the objective of adding value. The second stage, data management, involves the processing and management of data into forms that are useful for processing by subsequent stages. The next stage translates raw data into information. Stage four is the statistical modeling of the information developed at stage three. Finally the output side of the chain relies on decision support systems to implement statistical models that provide actionable information for decision makers, whether they are concerned with marketing and sales, corporate strategy, or public policy development.

Survey organizations may perform one or all of these functions. Some organizations only do data collection, and then pass these raw data on to other organizations for additional processing. Other "full-service" firms not only collect data but also process, interpret, and model the resulting information.

Understanding how technological development affects the information value chain is key to understanding the evolution of survey organizations and the challenges they face as technology changes. A consistent theme in CASIC development is the capacity of technology to integrate discrete steps in a process into fewer steps. CATI, for example, integrated the formerly discrete steps of interviewing, editing, and data entry into the single step of interviewing. The same can be said for CAPI. As a result entire departments became obsolete, and survey organizations were restructured to accommodate and take advantage of this change. Looking to the future, should self-administered methods rise in popularity at the expense of interviewing, we would expect to see substantial restructuring of survey organizations that currently rely heavily on interviewing to acquire data.

29.4 EVOLUTION OF CASIC TO DATE

As conceived in this chapter, past and future CASIC development may be divided into four broad phases. Each of these phases has as its underpinnings one or more of the technology shifts cited by Tapscott and listed in Table 29.1. Phases I and II are now behind us. The CASIC technologies of those phases are described in detail throughout this volume. The following brief summary is intended only to place their development within the general theme of this chapter.

29.4.1 Phase I—Invention of CASIC

CATI was the first major application of information technology to survey research. We might call it the base CASIC technology. It is a kind of alpha point, not only because it was the first CASIC technology but also because its technological basis was first-generation computing. This is the generation of the first commercially available computers, a time before the technology shifts described by Tapscott were underway. The first CATI system relied on dumb terminals connected to a mainframe host computer. Proprietary software was run on proprietary hardware built on traditional semiconductor technology. Its user interface was basic and, by today's standards, crude.

The minicomputer soon offered significant computing power at a price low enough to bring CATI within the reach of smaller organizations (Baker, 1988). By the early 1980s survey organizations of all kinds began developing proprietary CATI systems, all for their own use and based in minicomputer technology. With the exception of their reliance on minicomputer platforms, the technological foundation for these systems was the same as the first mainframe systems. Yet these developments were the basis for a revolution in data collection, one in which CATI became not only the first CASIC technology but the primary means of data collection for much of the survey research conducted in the United States, especially by commercial organizations.

CATI also provided the basic paradigm for adopting the key elements of survey research to information technology. Paul Sheatsley (in Bogart, 1987, p. S187) once wrote that "the basic tools of opinion research are: a sample of people, a questionnaire, and an interviewer." With CATI, the problems associated with converting each of these tools from primarily manual processes to technology-based processes were solved (see Nicholls and Groves, 1985). Sample management was not simply automated, it became tightly integrated with data collection. The basic techniques for representing a questionnaire as a computer program were established. Methods for training and monitoring interviewers were developed, and the task of interviewing was redefined. These became the basic building blocks of CASIC, at least as developed thus far.

29.4.2 Phase II—Emergence of CAPI

Beginning in the mid-1980s, a number of important technological shifts set the stage for the development of the next major CASIC technology. Traditional semiconductor architectures gave way to the microprocessor, making it possible to design smaller yet more powerful as well as less expensive computers. Open systems began to emerge, gradually replacing vendor proprietary hardware architectures and software. The personal computer (PC) was at the center of both shifts, an open hardware architecture built around the microprocessor. With its introduction, not only was the computer hardware industry revolutionized but the modern software industry was born.

As the 1980s progressed, other shifts began. The old typology of the mainframe host accessed by dumb terminals gave way to client-server computing in which smart terminals (PCs) handle computing locally, communicating over a network to a central host to access information shared by a group of users working on similar tasks. Ease of use became an increasingly important design criterion for both hardware and software, as developing more intuitive user interfaces became a priority. New programming techniques, such as object-oriented programming, made it possible to develop more complex software much faster and more reliably than in the past.

The implications for survey research were many. Hardware to run CATI became affordable by almost all survey organizations. Literally dozens of software packages emerged (Carpenter, 1988) that could be run on a wide variety of manufacturers' computers, large and small. Software developers also took advantage of new techniques and more powerful computers to increase dramatically the functionality of CATI software, and software became easier to use. PC-based local area networks gradually replaced minicomputers.

But arguably the most significant shift in technology for survey researchers during this period was the freeing of computing from the tyranny of the centralized mainframe or minicomputer host. Instead of having to bring respondents to the computer (via telephone), it was now possible for interviewers to take the computer to respondents via lightweight, affordable, yet powerful laptops. CAPI was born, and the large, complex, in-person surveys that are the foundation of most government policy making and academic research were brought under the CASIC umbrella. Just as CATI became the default method for telephone interviewing in the mid- and late 1980s, the mid-1990s saw CAPI replace paper and pencil as the default method for face-to-face interviewing.

29.5 PHASE III — DECLINE OF INTERVIEWING AND RISE OF CASI AND CSAQ

The thrust of CASIC development under phases I and II has been overwhelmingly in the direction of computer assisted interviewing, that is, providing the survey interviewer with significantly better tools to produce better information, more quickly, and at a lower cost. There are, however, important exceptions.

One such exception is touchtone data entry (TDE), described by Clayton and Werking (Chapter 27). The development of TDE by the U.S. Bureau of Labor Statistics (BLS) and the subsequent deployment of voice recognition entry (VRE) at the same agency demonstrated the power of self-administration technologies for maintaining high levels of survey participation, lowering costs, reducing cycle times, and maintaining if not improving survey data quality. A similar argument can be made for computer assisted self-administered questionnaires (CSAQ) (see Ramos, Sedivi, and Sweet, Chapter 20), once again in the context of business surveys.

There have been important developments for household surveys as well. Saris (Chapter 21) describes the use of a computer assisted self-administered technology that is essentially an adaptation of CSAQ to household surveys. We also are seeing an increase in the deployment of audio computer assisted self-interviewing (ACASI) (see Turner et al., Chapter 23; Tourangeau and Smith, Chapter 22). One cannot help but be impressed by how successfully ACASI has been deployed and the ease with which respondents have learned and accepted it. The improvements in data quality are especially noteworthy. The major CASIC technologies — CATI and CAPI — have mostly been worthwhile process improvements with relatively modest increases in survey data quality (see Nicholls, Baker, and Martin, 1997). ACASI, on the other hand, has been shown to increase significantly the reporting of sensitive behaviors. By capitalizing on the strength of self-administered methods for evoking more truthful responses from respondents, and by using technology to compensate for the inherent weaknesses of these same methods, ACASI is arguably the most significant CASIC development since CATI.

Few in the industry believe these technologies will change the course of CASIC development. Rather, we tend to view them as niche technologies aimed at specific survey problems such as business panel data collection or sensitive behavior reporting. More likely we are in the early stages of a new round of technology shifts that will significantly alter our methods of data collection. Thus far we may have seen only hints of what will come over the next decade. The changes that lie ahead will be more dramatic than any we have seen thus far. The new emergent technologies will cause us to question and ultimately redefine the role of the interviewer in survey research.

Surveys without interviewers have at least two important advantages. The first and most obvious is their lower costs. The recruiting, training, deployment, and supervision of an interviewing staff are the most significant costs in the survey budget. Technologies that lower these costs are important, and survey organizations will adopt them. The second advantage to self-administration is the elimination of the interviewer as a source of error. Bradburn (1993) describes the role of the interviewer as "convey(ing) the meaning of the question and record(ing) the meaning of the answer." (Of course interviewers also gain cooperation, give assurances of confidentiality, and other things; more on that later.) Both of these tasks are subject to variation depending on the interviewer. Or, as Schaeffer (1991) puts it, "measurement-by-talk imposes constraints that require complicating theories of measurement." Elimination of these constraints ought to be a goal, and technology may gradually make that possible.

The principal barrier to elimination of the interviewer has been the mechanics of correctly administering a complex questionnaire and accurately recording the respondent's answers. Recently developed scanning technologies (see Dillman and Miller, Chapter 24; Blom and Lyberg, Chapter 25) have led to significant improvements in the cost and accuracy of mail surveys but have done nothing to overcome the inherent weaknesses of self-administration using paper forms. Mail survey respondents still skip items, have difficulty following even the simplest of skip patterns, and treat single-response questions as if they were multiple-response. Response rates still tend to be significantly lower than either telephone or face-to-face surveys. Creative forms design (see Jenkins and Dillman, 1997) can reduce some of this error but not eliminate it.

Technology is beginning to solve some of these problems. The same technical approaches used for questionnaire administration in CATI and CAPI are now successfully applied to self-administration, and we can expect to see this trend accelerate because of ongoing development in the key enabling technologies for self-administration. The emergence of the microprocessor, the shift from centralized host to client-server computing, and the evolution of the open systems approach to hardware and software development have all played key roles in the evolution of CASIC over the last two decades. Other key technology shifts now underway are likely to have an equally significant effect. These shifts are the dramatic change in user interfaces and the emergence of the Information Highway. There is little question that technology is evolving rapidly in directions that make it possible to conduct surveys without interviewers with all of the complexity and data integrity features of CATI and CAPI. The only question is how long it will be until the necessary technologies are adopted on a broad enough scale in the population at large to make such surveys scientifically viable.

29.5.1 Changes in the User Interface

Ease of use has become a key design element for both hardware and software. Such was not always the case, and the fact that computers often were hard to use was a major deterrent to their widespread adoption. This began to change in the 1980s with the development of the graphical user interface (GUI) and its first commercial application, the Apple Macintosh. Since then we have seen a host of improvements that make it increasingly easy to use and interact with computers. It is impossible to know how much of the growth in computer use by both businesses and households over, say, the last five years is due to the development of the Windows GUI. Would PC penetration in both of those markets be the same if we were still stuck in DOS? One suspects not. Bosley, Conrad, and Uglow (Chapter 26) provide an illuminating discussion of the potential advantages and implementation barriers of the GUI for interviewers. Surely there are similar advantages for self-administration, although few current CASIC applications realize them.

We are just beginning to see CSAQ and CASI applications migrate out of their current DOS standard to the easier to use and more powerful Windows environment. As this movement picks up speed, not only will we be able to design applications that respondents find easier to use, but we will also be able to deliver significantly more information. Sound is rapidly becoming standard; video is next. Prototype applications that integrate video into the interview situation are now being field tested (Baim, Frankel, and Arpin, 1996).

Outside of our industry we are seeing increasingly sophisticated multimedia applications that may be the precursors of our own future. Take the example of TurboTax, the best-selling software for preparing and filing U.S. income taxes. Delivered on CD-ROM, the software guides the user through an interview that elicits all of the information needed to complete the always difficult and sometimes byzantine U.S. tax forms. Context-sensitive help with hypertext links is available throughout to assist the user in navigating the complexities of the U.S. tax laws. Video clips offer advice from experts. There are frequent checks for consistency and completeness of information during the interview, and the user can navigate freely from section to section depending on the completeness and organization of his or her records, returning to unfinished sections later when the missing information has been found. When all forms have been completed correctly, they are filed electronically by modem or Internet.

Even more exciting changes in the user interface are just over the horizon. The next generation of user interfaces will be true multimedia user interfaces (MUI). The most obvious and important change will be the move to speech recognition. The point-and-click interface will give way to voice commands, and the simple voice recognition technology of, say, VRE will give way to continuous speech. Just as text only screen have given way to simple graphics, we will see even more frequent use of complex graphics, video, and sound throughout the user interface, not just in special applications.

There is a difference in opinion as to how soon this will happen. Tapscott (1996), Dertouzos (1997), Negroponte (1995), and a host of other futurists believe it could happen within the next five years. Others see a longer time horizon. Regardless of timing, when such interfaces become widespread, the implications for the general adoption of this technology by households and the usability of those technologies for survey research are substantial.

After the addition of voice as the prime interface, the most significant improvement in user interfaces is likely to be the move to virtual reality (VR). Over the next decade we can expect that VR, like speech recognition, will emerge as an essential component of the user interface. The mouse, trackballs, and touchpads will begin to disappear. In his especially compelling vision of future technology, Negroponte (1995) describes how today's icons on our computer screens will become holograms that the user can reach out to, touch, and manipulate.

Concurrent with and reinforcing the development of new, more intuitive interface will be a dramatic change in the devices we use to acquire and manage information of all kinds. The trend is one in which single-purpose "information appliances" (telephones, TVs, personal computers, electronic game systems, etc.) will give way to more generalized devices capable of a wide range of functions, a phenomenon frequently called "digital convergence." The functions of today's information appliances will be designed into a single, multipurpose device. Because of dramatic changes in user interfaces, interacting with these devices will become as natural as using the telephone, changing TV channels, or driving a car.

As these new technologies come to market and take hold, the range of what we might call "respondent-centric" applications will expand dramatically. As speech recognition becomes a truly reliable technology, we can expect to see it exploited in place of interviewer recording of answers. The potential of speech recognition will go well beyond the relatively simple applications of VRE. More advanced speech recognition systems currently in the feasibility testing stages (see Appel and Cole, 1994; Blyth, 1997; Rands, 1996) are an additional step forward, but they are still crude when compared with what will be possible. Sometime shortly after the turn of the century, we likely will be able to communicate with machines using normal, continuous speech, the speech of everyday conversation. And just as important, machines will be able to respond in kind. Respondents will be able to record their own answers as naturally as they now report them to an interviewer.

The next generation of user interfaces also will make interviewer and respondent tasks possible that thus far have eluded automation in the CASIC realm. For example, CASIC developers continue to struggle with the implementation of calendars to stimulate respondent recall and organize event series in meaningful ways. So-called show cards remain a feature even of CAPI interviews. Personalization or use of fills in ACASI scripts is still difficult. As powerful new user interfaces make it possible to deliver multimedia interviews that include these kinds of elements, the need for interviewers lessens.

29.5.2   Information Highway

We tend to think of the role of the interviewer as primarily that of conducting the actual interview. An equally important role is that of creating access to the respondent and getting him or her to agree to be interviewed. In CATI, interviewers dial the telephones and solicit cooperation. With CAPI they have the added logistical responsibilities of bringing the technology to the respondent. Even in a world where technology makes it possible to rely less on interviewers for asking the questions and recording the answers, the problems of access to respondents and delivery of the survey questionnaire remain.

The Information Highway is one potential solution. For now, we have the Internet. With its deployment comes the opportunity to connect millions. The PC is no longer isolated. What local area networks have made possible inside organizations, the Internet makes possible among organizations and households. We can now recruit respondents and elicit cooperation electronically. Respondents can complete questionnaires on-line or download them to be completed later. We can deliver multimedia material as part of the interview, although it is still difficult with current bandwidth constraints. Chat room software, one of the most popular Internet applications, can be used to create on-line focus groups.

Survey researchers are already taking advantage of this technology. For example, Clayton and Werking (Chapter 27) describe plans for transferring their groundbreaking TDE/VRE application to an on-line application on the Internet. There are at least a dozen Internet sites engaged in continual, on-line research, both standard surveys and focus groups. More and more organizations, especially market research firms, are beginning to use on-line methods in place of traditional mail, disk-by-mail, and even telephone surveys (e.g., CASRO, 1996; Green, 1996; MacElroy, 1996; Chen-Chi and Sung-Chi, 1996; Vehovar and Batagelj, 1996).

This is, of course, only the beginning. The full exploitation of the Information Highway by researchers awaits the development and deployment of a vast new information technology infrastructure, along with the information appliances to connect to it. The current Internet has only a fraction of the needed capacity. It is carrying mostly text and simple graphics, hardly the stuff of the complex survey applications we will require. As the level of bandwidth-intensive activities increases, the infrastructure must be strengthened. Even in its current state of relatively low penetration and modest throughput requirements, it is subject to frequent brownouts that diminish its utility.

The technologies already exist to solve these problems, as do the will and the resources. In the United States, opinions differ as to who will build and who will fund the next generation Internet, but few doubt that it will happen. Some see the traditional telecommunications companies as both architects and builders of the next phase; others look to cable TV companies. Regardless of who will do it, a good consensus estimate of when it will be done, that is, when it will be possible for virtually any U.S. household to connect directly to the Internet with sufficient bandwidth to run complex, multimedia applications, is toward the end of the first decade of the twenty-first century. For other parts of the world, such as Western Europe, the time line is likely to be even longer.

29.5.3 Will Respondents Go Along?

There is no question that the technology will exist to support a new generation of computer assisted self-administered methods. But the wild card, as in any survey, is the respondent. Will businesses and households acquire the technology to make it possible? The infrastructure will be built. Attractive and useful information appliances will be brought to market. But the ability of researchers to harness this technology depends on respondents' willingness to adopt it.

There is little question that businesses will do so, but households are another matter. The penetration of computers in American business now stands at near 90 percent; the comparable figure for households is only 40 percent. By the year 2000 the number of PCs bought for home use is expected to more than double. Already 40 percent of the PCs purchased every year in the United States are purchased for home use (Zwetchkenbaum, 1996). As new devices, such as sub-$1000 multimedia machines and so-called Internet boxes or "net-PCs" emerge, these numbers could change substantially.

Internet connectivity also is on the rise. A 1997 Harris Poll (*Business Week,* 1997) found that 21 percent of American adults use the Internet, about twice the number reported in 1996. CyberAtlas (1997), an on-line resource that tracks Internet usage, estimates that the Internet user population grew by 78 percent between September 1995 and January 1996, and an additional 31 percent from January to June of the same year. Whether usage will continue to grow at the pace suggested by these data is the subject of considerable debate, and it depends on a variety of factors, the most important being the services offered and the cost of going on-line.

In the early days of the PC revolution, there was frequent discussion of so-called killer apps. These were the software applications that stimulated the early adoption of PCs. Most observers at the time, for example, agreed that the first killer app for business was the electronic spreadsheet. Are there killer apps out there to drive the computer ownership and Information Highway access to the needed levels? Probably. Digital TV is one. Multimedia is another. But these are only manifestations of an underlying technology shift that could well give birth to a wide range of new killer apps. That shift is the conversion of information of all types from analog to digital, or as Negroponte (1995) describes it, from atoms to bits.

We still live in a world where the vast majority of information is stored in atoms: books, newspapers, magazines, television, film and video, telephone conversations, and even bills and checks. Slowly we are seeing information of all kinds converted to bits, that is, represented in electronic digital form instead of paper and analog transmissions. As more and more information is available in digital form, households and business will acquire the devices they need to access it.

Survey data suggest that Americans, at least, will be receptive to the lure of this new digital society. A 1994 survey by the Times Mirror Center for the People and the Press (Kohut, 1994) found that while only about one in three U.S. households contained a personal computer, over half of employed persons used a computer at work and 46 percent of teenagers had a home computer. Two-thirds of those polled said that they liked rather than disliked computers and technology, and among persons aged 18 to 29 that figure rises to 75 percent. Granted, these data do not describe a platform to be exploited by contemporary researchers. But they describe a U.S. population that gradually is becoming comfortable if not dependent on computers and technology. As the population ages and younger cohorts with an entirely different technology experience come of age, we can expect continued and dramatic increases in the ownership and use of computers or whatever information appliances evolve over the next decade.

The bad news in these numbers is the severe bias that currently exists in technology ownership and use. The Graphic, Visualization, and Usability Center at the Georgia Institute of Technology conducts regular on-line surveys of Internet users to measure general demographics, patterns of technology use, and attitudes on a wide variety of societal and technological topics. Their most recent survey (1996) describes a population that is younger, more frequently male than female, better educated, higher income, with more liberal social and political attitudes than the population at large. The survey also shows the dominance of the current Internet by U.S. users, and the heavy reliance on English as the unofficial language of the Web.

29.5.4 Transition

The transition to surveys without interviewers will begin first with business-to-business surveys. This is already happening. As Ramos, Sedivi, and Sweet (Chapter 20), and Clayton and Werking (Chapter 27) make clear, the business environment already has or is rapidly acquiring the needed technology and, just as important, is predisposed to respond to surveys electronically. This trend will continue.

On the household side, progress will be slower. The consumer mail panels currently maintained by many commercial organizations will gradually be brought on-line. The largest of these are already doing so. Where respondents are not equipped to respond electronically, survey organizations will supply the needed technology (see Saris, Chapter 21). Other household surveys will follow. For CATI surveys, respondents increasingly will be offered the opportunity to do the survey on-line at their own convenience rather than by telephone. Interviewing will gradually give way to recruiting respondents as a major activity in our telephone centers, but even that will fade as we gradually move from telephone to electronic mail as the means of initial contact on many surveys. Split samples and mixed-mode surveys will become common as we seek to control bias. Panels will increase in popularity both to help control recruiting costs and ensure that respondents have the needed technology. The already significant decline in face-to-face interviewing will accelerate. Interviewers will continue to visit respondent homes, but many will be there to set up equipment and facilitate its use, not to interview. In other forms of self-administration (e.g., Internet, surveys) human beings may be available to assist respondents with problems or clarify questions but have no direct role in the interview. Nicholls (1997, personal communication) describes this overall transition as one from "computer assisted personal interviewing to person assisted computer interviewing."

As we conduct more surveys without interviewers the whole issue of respondent cooperation will move center stage. The differences in response rates we generally see across the continuum from face-to-face to telephone to mail surveys are, to a large extent, a function of the human contact involved in each. Convincing a respondent to do a survey is somehow easier when you are standing face-to-face with that person, and hardest when you are trying to write an all purpose, concise cover letter for a mail survey. It seems clear that considerable research is going to be necessary to help us understand the dynamics of respondent cooperation with new forms of self-administered CASIC. It also seems likely that incentives will play a more significant role in the future than they do today.

In all of this, the commercial sector will lead the way. Less constrained by the purity of survey methodology, commercial firms have the resources and the economic incentive to experiment with new methods. Academic and government organizations will follow, but at a slower pace. The adoption of CATI in these organizations lagged that of commercial companies by almost a decade. The lag in the move to these new on-line methods probably will be shorter because the technology is more readily available and because survey organizations of all types are more adept today than 20 years ago at adopting new technologies.

Since its inception, survey research has stood on the three legs of sampling, interviewing, and questionnaire design. A decade from now one of these legs — interviewing — may well be disappearing in much, although not all, of the survey work we do. It already is possible to use the new cyber methods described here to survey certain populations without significant bias (engineers, MIS managers, home computer users, etc.). At a minimum we can expect that over the next decade the range of topics and populations for which we can effectively use self-administered CASIC methods will expand significantly. Many surveys, especially those relying on representative samples of the general population, will still require extensive use of interviewers to fill gaps in technology ownership and use. Nonetheless, the focus of CASIC development will have changed. The interviewer will no longer be at the center of the design of CASIC applications. The new challenge will be to develop systems and applications that respondents can use directly, without the assistance of an interviewer. It is a daunting challenge, but nowhere near as daunting as what could happen next.

29.6   PHASE IV—SURVEY RESEARCH IN THE TECHNOLOGICAL UTOPIA

In 1987, to commemorate its fiftieth anniversary, *The Public Opinion Quarterly* asked 16 well-known scholars and survey practitioners to offer their visions of the future of public opinion research (Bogart, 1987). In his response Harold Mendelsohn (p. S183) described how "the eventual 'computerizing' of the American home undoubtedly will contribute significantly to the speed, accuracy, and economy with which data will be gathered, analyzed, and readied for dissemination." James Beniger (p. S175) wrote that "a host of new technologies will...make possible the real time mass monitoring of individual behavior.....Survey research will increasingly give way to more direct measures of behavior made possible by new computer-based technologies." In a particularly chilling vision, and perhaps more tongue-in-cheek, Robert Worcester (p. S188) described how "...market researchers are close to their tactical ideal, the comprehensively wired micro model of segmented household 'norms' which can be conceptualized, pressurized, test marketed to, weighted (up and down), copy tested, product tested, studied, and, yes, manipulated by cables, satellites, and sensors (worn in rings, necklaces, earrings, or even — shades of George Orwell — implanted!)."

The foundation for these visions of the future is one that is increasingly popular among futurists such as Negroponte (1995), Tapscott (1996), Dertouzos (1997), Rawlins (1997), and Mitchell (1995). It has its beginnings in Weiner's seminal book, *Cybernetics* (1948). The vision is one of a technological utopia, a vision of a completely digital world in which everything is smart — home appliances, automobiles, clothes, money, buildings, and highways. In this utopia, computers will be omnipresent, and everything we use and touch as we go about our daily lives will be busy collecting, processing, and transmitting information. The implications for survey research are extraordinary.

A significant number of the surveys we conduct collect information about past behavior, whether by individuals or organizations. They seek information that often already exists as atoms, but because it is too difficult or too expensive to collect those atoms, we ask respondents to retrieve them for us. The questions we ask respondents are aimed at collecting information about behavior, the facts and figures that describe our lives, past and present. These surveys essentially ask us to report information that almost always exists somewhere else, but in a form or in a place that makes it inaccessible. The particulars of my birth, every school I ever attended, every class I ever took, every grade I received, and every degree I earned were meticulously recorded and stored somewhere. Every job I ever held, the pay I received and the taxes I paid, and any promotions I received are a matter of record. Every time I visit a doctor my reason for going, the procedures performed, the diagnoses made, the charge I pay, and how I pay are recorded. The problem is that the vast majority of this information was recorded on paper, stored as atoms. Or, if it has been converted to bits, it is stored in formats and in places where it is not easily retrieved. Most surveys retrieve this information, not by tracking down these atoms but by asking me to remember these events or, in some instances, search my own atoms or bits and pass that information on to an interviewer. Most business surveys do the same. Even electronic data interchange (EDI), as currently conceived, relies on the responding business providing the information in an agreed upon, standard format, rather than the survey organization retrieving it directly from the respondent's information system.

As the futurists see it, technology will change this. In the not too distant future all of this information will be stored as bits. We are about to enter an age in which the transactions of our daily lives will be entirely digital. As the transition from analog to digital, from atom to bits, becomes complete, virtually everything we do will be recorded and stored somewhere. It will be as if each of us were wired to some distant computer recording all of our actions without our particularly taking notice of it. There will be no sense of being a respondent or of providing information. Rather, we simply will go about our lives — being born, going to school, taking a job, getting married, buying a home, having children — and all of it, in the finest detail, will be stored as bits somewhere. No one will have to ask us questions about ourselves because everything already will be known.

Even with all of this information stored as bits, there is no guarantee it will be accessible. The technical problems of proprietary hardware and software architectures, limited networking, and inconsistent metadata standards will have to be solved. The key shifts that will solve these problems are the continued evolution and adoption of open systems based on industry standards and intelligent networking. These technical developments, already underway, are the same as those that provide the foundation for electronic commerce and the much heralded digital economy of the twenty-first century.

The technology, as always, is the easy part. As the technical barriers fall, new barriers may well be erected in the form of attempts to protect individual confidentiality in an age of free flowing information. Who will "own" all of this personal information that is constantly collected? How will research organizations get hold of it? These are key questions with no clear answers. And, of course, some of the information we will need, especially that about likely future behavior, attitudes, and opinions will not be in bits at all. How the survey industry deals with this dramatically altered social and technological landscape is an important issue. Equally important, however, will be the way in which the market for our product may change.

29.6.1 Open Systems

As discussed earlier in this chapter, the open systems movement began in earnest in the early 1980s and continues to gain momentum. Open systems implies not just standard data formats but standards in hardware, software systems, applications, database design, user interface, and communication protocols (Tapscott and Caston, 1993). Open systems eliminate all of the traditional technical barriers to free and easy access to information from outside of a specific system. It means that systems can communicate and exchange information transparently.

Inside organizations, the advantages of the open systems approach—reduced hardware and software costs, faster application development, lower IT maintenance costs, and internal integration of systems in multiple operating units—guarantee that the trend will only intensify. Increasingly the same approach is being used externally as organizations of all kinds find they must develop strategic partnerships with other organizations to be successful.

The need for these strategic partnerships will drive organizations of all kinds away from traditional, closed, proprietary systems to open systems based on industry standards. In the evolving business climate of the twenty-first century, alliances with supplies, customers, affinity organizations such as industry associations, labor unions, and even competitors will be essential to success. A key component of these alliances will be interconnected information systems that share a broad range of information. Data that today are inaccessible because of closed proprietary systems will become accessible. This will happen in all industries and all sectors of the economy, and it will be a global phenomenon.

29.6.2 Intelligent Networking

As these relationships among organizations evolve and their information infrastructures are interwoven, the effect will be to create a global network on a dramatically grander scale than today's Internet. This "Net" will be significantly beyond anything we experience today in terms of the number of users, capacity (i.e., bandwidth), and the information available. It will create what Tapscott (1996) calls, "The Age of Networked Intelligence," an era in which the technologies of computing, communications, and content will create a world in which businesses, governments, organizations, households, individuals, even whole societies, will be on-line. Retrieving information on this network will be the job of intelligent agents or "knowbots," computer programs that will prowl the Net collecting the information we need, whether to plan a vacation or maintain a statistical model.

29.6.3 Implications for Survey Research

If the future unfolds as most futurists believe, survey research as we know it today will change significantly. Under one possible scenario it might evolve into something that is more akin to today's discipline of information systems design, or even disappear altogether. We have already seen how the role of the interviewer is likely to diminish over the next decade as user interfaces become more intuitive and information technology use becomes more widespread. With the above described developments in open systems and networking, and with the ongoing development of massive databases and information systems that are constantly fed by the activities of everyday life, we also may see sampling and questionnaire design, at least as we conceive of them today, change dramatically.

A basic premise of survey research and the foundation for much of its methodology is the impracticality of interviewing an entire population; thus we draw a sample. In the future it may be possible, at least in some applications, to collect information on entire populations. We may not need to contact respondents and solicit interviews because much of what we want to know will already be stored as bits in a system whose information is arranged in ways we can understand and access. Under these conditions the marginal cost of collecting data on 50 million people is not substantially different from that of five thousand. The investment in infrastructure will be substantial, and the temptation to forget the lessons of *The Literary Digest* will be extreme, but the speed with which the information can be collected and the analytical power of these huge databases will be compelling.

Questionnaire design may also decline in importance if not become unnecessary altogether. Because of problems of comprehension, recall, and bias, we put great effort into writing questions and designing questionnaires that we hope will elicit "truth." Since in many instances we will not collect information directly from respondents, we will not need questionnaires. In their place we might design and build intelligent agents that prowl the vast Information Highway collecting information. They will do essentially what today's interviewers do, only in cyberspace.

These changes will present a formidable set of challenges to traditional survey organizations, their composition, and structure. The information value chain described earlier will look much the same, although the methods of data acquisition likely will be completely different from what they are today. More important, the users of those data, that is, our customers and the organizations relying on the decision support systems to which the data and information flow, will also have undergone substantial transformation. It will be the age of Martin's (1996) "cybercorp," an organization designed using the principles of cybernetics, one that constantly senses changes in its environment and reacts to them, an organization "designed for fast change, which can learn, evolve, and transform itself rapidly." In the digital economy of the twenty-first century, organizations of all kinds will depend on knowledge and information to achieve and sustain competitive advantage. Haeckel (1994) argues that eventually we will see the development of "enterprise models" that institutionalize an organization's know-how and decision making into "formal (if not entirely mathematical) representation." These models will be fed with information that is continually collected, in real time. As organizations evolve in this direction, all processes in the information value chain become core business processes and therefore essential functions of the organization. One potential result is that more and more organizations — private corporations, public agencies, and even academic institutions — may find it desirable, if not necessary, to internalize all of the functions in the information value chain, and technology will make that possible. As a consequence there may no longer be a need to develop and maintain a corps of trained interviewers; nor will there be a need for the specialized skills of the survey designer. What role remains for survey organizations is unclear at best.

Put another way, survey organizations may face what Porter (1985) calls "the threat of substitution." Buyers of survey data, whether in the private or public sectors, will have an increasingly broader array of alternatives on which to rely for decision making. Perceptions of the value of those alternative sources versus the value of survey data compared to the relative price of each will determine whether the survey industry continues to flourish or becomes a casualty of the Post-Information Age.

Fortunately there is no shortage of reasons why the above scenario might not come to pass. At the top of the list are deep doubts as to whether the vision of the futurists will become reality in anywhere near the time frame they envision. Schnaars (1989) has cataloged the failed predictions of the recent past to show just how poorly futurists have predicted the pace of technological change and its societal impact. Surely some of what Negroponte and others predict will happen, that is, become technically feasible, but how deeply it will penetrate American society, let alone global society, is less certain. The uneven adoption of the enabling technologies will make it possible for a significant amount of research currently done as surveys to be done by these advanced cyber methods. As the data warehousing phenomenon becomes more wide-spread, open systems move to new levels of standardization, and a new and more powerful Internet is deployed, it will become possible to retrieve and model data that today are only obtainable through surveys. Nevertheless, it seems unlikely that these methods will totally replace surveys over the next 30 years.

A second significant issue is privacy and confidentiality of information. How will they be safeguarded in this brave new world, and how will such constraints affect the evolution of the survey industry? There already is considerable suspicion in the general public that with the growth of technologies, especially the Information Highway, privacy is increasingly threatened. We are seeing rising public hostility to surveys as evidenced by both falling response rates and hundreds of legislative initiatives introduced in state legislatures every year aimed at restricting telephone solicitations of all kinds, including surveys. We are seeing restricted access to some traditional sources of information by survey researchers both in the private and public domains. The political climate in both the U.S. and Europe virtually guarantees that this trend will continue. Unfortunately, the challenge to traditional survey research is every bit as daunting as that posed to the newer methods of the twenty-first century.

Conceptually at least, the problem of protecting confidentiality has a technical solution. It rests on the use of encryption algorithms and personal keys that guarantee individuals will continue to control their own information, releasing it only to persons they designate and under agreed-upon conditions. Of course, how these concepts are implemented is another matter. As Rawlins (1996) says, "Today's encryption technology could, if used widely enough, make us the last generation ever to have to fear for our privacy ... if misused, it could make us the last generation with any notion of personal privacy at all."

The third objection to this future for survey research is that it does not encompass the full breadth of survey topics we pursue. Two in particular come to mind: socially undesirable behavior and attitudinal measures. Government and academic interest in the former, including drug and alcohol abuse and sexual behavior, has been the primary impetus for the development of ACASI. The extent to which government policy making continues in the decades ahead to be concerned with these issues of private behavior will have some effect on the overall shape and development of information gathering technologies for public policy decision making. Attitudes and opinions, on the other hand, whether about product quality, public policy issues, or political candidates, are likely to continue to be a major component of the survey industry for the foreseeable future. These studies of future behavior will no doubt benefit from more sophisticated predictive modeling techniques, but in all likelihood we will still want to ask people questions about their intentions for some time to come.

29.6.4 Alternative Scenario: Survey Research in the Information Marketplace

Dertouzos (1997) offers a somewhat different vision of the future but with equally significant implications for survey research. He calls it the Information Marketplace and defines it as "the collection of people, computers, communications, software, and services that will be engaged in the intra-organizational and interpersonal informational transactions of the future." His inspiration was the Athens flea market with its bustling trade in virtually everything—the constant buying, selling, and trading of all kinds of goods. Like other futurists Dertouzos sees a future where computers are ubiquitous and interconnected, fostering unprecedented levels of communication and information sharing. The foundation on which the Information Marketplace rests is the same technological infrastructure described by Tapscott for the digital economy, one that could produce the world without surveys described in the previous scenario. But Dertouzos's vision is a far more romantic model for electronic commerce, one with a more obvious human face. His vision is that of "a twenty-first century village marketplace where people and computers buy, sell, and freely exchange information and information services."

The key to managing transactions in the Information Marketplace will be something called the electronic form or "e-form," an agreed-upon standard among common interest groups for specifying routine and recurrent transactions among its members. For example, there might be an e-form standard for buying a car. I fill in the e-form on my computer specifying my requirements and send it off to a variety of automakers' computers. Those computers read and understand the e-form, compare my requirements against their inventories, and send an electronic offer to sell me a car if they can meet my specifications.

The Information Marketplace is not unlike a model for surveys in the future proposed by Beniger (1997). He too sees a future where computers are everywhere. He further sees interviewers evolving into information specialists who contact respondents electronically to solicit their participation and negotiate (perhaps even buy) information from them. In Beniger's vision, surveys become transactions in the Information Marketplace like any other, where survey organizations either secure information directly from respondents (perhaps by sending them electronic questionnaires or even accessing their personal information systems) or get the necessary permissions to secure it from third parties. One can easily imagine the development of e-form standards among survey organizations, respondents, and information providers that automate these transactions.

For those in the survey profession who worry about technology undermining the historical standards and traditions of survey research, Beniger's vision is an appealing one. It is easy to see it as a logical evolution from where we are today with CATI, for example, to emerging or still undeveloped data collection technologies. While it is not interviewing per se, it maintains the human side of survey research, redefining the relationship between interviewer (or whatever he or she will be called) and the respondent in a world where electronic communication is the norm. It recognizes the fundamental role of technology in everyday life yet retains a role for human judgment and for building the trust that many in our profession believe is essential to learning the truth.

29.7 CONCLUSION

The challenge of CASIC so far has been one of making creative use of technology to do the things that we have always done in surveys, but to do them better and faster. It has caused us to look inward, to make significant changes in how we do surveys, while leaving the basic underpinnings of our discipline pretty much intact. The promise of re-engineering, "the *fundamental* rethinking and *radical* redesign of business processes" (Hammer and Champy, 1994) within the survey profession has yet to be fulfilled.

The future will present more fundamental challenges than anything we have faced thus far. The challenge will be to look outward as well as inward. The defining issue will not be how effectively we adopt succeeding generations of technology to do what we have always done, but how we change the things we do to accommodate dramatic changes in the role of technology in society at large. Until now we have been concerned with the technology inside our organizations. In the twenty-first century we will need to be equally concerned with the technology that is out there, in the hands of our respondents, an integral part of their daily lives and businesses. In the future envisioned by the futurists, shifting our paradigm is no longer an option.

Nor can we avoid confronting and resolving, at least within the survey profession, what Dertouzos (1997) calls "the humie-techie split." He traces this split back to the eighteenth century and the Enlightenment, to a time when science split off from religion and morality; reason separated from faith. Science became unfettered, free to take us wherever it could. It flourished of course, bringing us first the Industrial Revolution, and then the Information Revolution, and soon the Post-Information Age. Over the centuries the split has deepened. The humies remain steadfast in their belief that technology and science should only be pursued in the service of a clear human purpose. The techies believe just as strongly that science should be pursued for its own sake, and that we should go wherever our collective minds and technical know-how take us. Conflicts and controversies are frequent (nuclear power, cloning, etc.). There seems to be little progress in healing the split.

What is true for society at large is also true for the survey profession. The humie-techie split is part of our culture as well. Its resolution within our profession is a key challenge for the twenty-first century. How will we reconcile the use of an overwhelmingly powerful and omnipresent (some might say invasive) information technology with the standards of a discipline whose main goal is to increase our understanding of human behavior? Will we heal the split and establish survey research as an enduring activity of human society in the Post-Information Age? Or will one side triumph over the other, condemning our discipline to either the bloodless exercise of the most advanced cyber techniques or a quaint anachronism, a cottage industry from another age?

For now we might take some solace in the knowledge that predictions about technology and our future are almost always overblown if not just plain wrong. But there are exceptions, and as Schnaars (1989) points out, "the only industry where . . . dazzling predictions have consistently come to fruition is computers." In 1980 IBM estimated the worldwide market for PCs for the decade of the 1980s would be 275,000 units. When the decade closed, 60 million PCs had been manufactured and sold.

These are interesting times for survey research, and CASIC is at the center of it. Whatever happens in our industry in the decades ahead will be more strongly driven by technology than ever before. The accelerator pedal is indeed pushed to the floor, and we are in for what surely will be a very exciting ride.

References

ACSF Investigators (1992), "Analysis of Sexual Behavior in France: A Comparison between Two Modes of Investigation—Telephone Survey and Face-to-Face Survey. *AIDS*, **6**, pp. 315–323.

Addelman, S. (1962), Symmetrical and Asymetrical Fractional Factorial Plans. *Technometrics*, **4**, pp. 47–58.

Allard, B., Brisebois, F., Dufour, J., and Simard, M. (1996), "How Do Interviewers Do Their Job? A Look at New Data Quality Measures for the Canadian Labour Force Survey," paper presented at the International Conference on Computer Assisted Survey Information Collection, San Antonio, TX.

Allen, D. F. (1987), "Computer versus Scanners: An Experiment in Nontraditional Forms of Survey Administration," *Journal of College Student Personnel*, 28, pp. 266–73.

Ambler, C., and Mesenbourg, T. M. (1992), "EDI—Reporting Standard of the Future," *Proceedings of the Annual Research Conference*, U.S. Bureau of the Census, pp. 289–297.

Anderson, S., Bethel, J., Tourangeau, K., Vincent, C., Waksberg, J., and Ward, P. (1994), "Final Field Test Methodology Report, National Medical Expenditure Survey, IPC Feasibility study," unpublished Report, Rockville, MD: Agency for Health Care Policy and Research.

Andersson, R., and Lyberg, L. (1983), "Automated Coding at Statistics Sweden," *Proceedings of the Section on Survey Research Methods, American Statistical Association*, pp. 41–50.

Andrews, F. M. (1984), "Construct Validity and Error Components of Survey Measures: A Structural Modeling Approach," *Public Opinion Quarterly*, **48**, pp. 409–422.

Appel, M. V. (1987), "Automated Industry and Occupation Coding," paper presented at the Seminar on Development of Statistical Expert Systems, Luxembourg.

Appel, M. V. and Cole, R. (1994), "Spoken Language Recognition for the Year 2000 Census Questionnaire," paper presented at the American Association for Public Opinion Research Annual Conference, Danvers, MA, May.

Appel, M. V. and Hellerman, E. (1983), "Census Bureau Experience with Automated Industry and Occupation Coding," *Proceedings of the Survey Research Methods Section, American Statistical Association*, pp. 32–40.

Appel, M. V., Petunias, T. F., and Russell, C. E. (1994), *Experiences with Fax Data Reporting and Questionnaire Distribution*. Washington, DC: U.S. Bureau of the Census.

Aquilino, W. (1993), "Effects of Spouse Presence During the Interview on Survey Responses Concerning Marriage," *Public Opinion Quarterly*, **57**, pp. 358–376.

Aquilino, W. (1994), "Interview Mode Effects in Surveys of Drug and Alcohol Use," *Public Opinion Quarterly*, **58**, pp. 210–240.

Aquilino, W. and LoSciuto, L. (1990), "Effect of Interview Mode on Self-reported Drug Use," *Public Opinion Quarterly*, **54**, pp. 362–395.

Archer, D. and Scott, D. (1995), "Use of Imaging in the 1996 New Zealand Census of Population and Dwellings," *Proceedings of the 1995 Seminar on New Techniques and Technologies for Statistics*, Bonn, November.

Attewell, P. (1994), "Information Technology and the Productivity Paradox," in D. P. Harris (ed.), *Organizational Linkages; Understanding the Productivity Paradox*, Washington, DC: National Academy Press.

Australian Bureau of Statistics (1993), *Data Editing*. Belconnen: Australian Bureau of Statistics.

Bailar, B. A. (1975), "The Effect of Rotation Group Bias on Estimates from Panel Surveys," *Journal of the American Statistical Association*, **70**, pp. 23–30.

Baim, J., Frankel, M., and Arpin, D. (1996), "Audio-Visual CASI (AV-CASI) in Personal Interviews," paper presented at the International Conference on Computer-Assisted Survey Information Collection, San Antonio, TX, December.

Baines, R. (1995), "Getting Information to Everyone's Home," *Electronic Engineering Times*, October 2, pp. 48–49.

Baird, D. and Walker, G. (1995). "A Data Collection Strategy for UK Business Statistics," paper presented at the 1995 International Conference on New Techniques and Technologies for Statistics (NTTS-95), Bonn, Germany, November.

Baker, R. P. (1988), "Computer-Based Question Banks: Why and How," *Proceedings of Symposium 88, The Impact of High Technology on Survey Taking*, Ottawa: Statistics Canada, pp. 141–154.

Baker, R. P. (1992), "New Technology in Survey Research: Computer-Assisted Personal Interviewing," *Social Science Computer Review*, **10**, pp. 145–157.

Baker, R. P. (1994), "Managing Information Technology in Survey Organizations," *Proceedings of the 1994 Annual Research Conference and CASIC Technologies Interchange,* Washington, DC: U.S. Bureau of the Census, pp. 637–646.

Baker, R. P. and Bradburn, N. M. (1991), "CAPI: Impacts on Data Quality and Survey Costs," paper presented at the 1991 Public Health Conference on Records and Statistics, Washington, DC.

Baker, R. P. and Lefes, W. L. (1988), "The Design of CATI Systems: A Review of Current Practice," in R. M. Groves, P. Biemer, L. E. Lyberg, J. Massey, W. L. Nicholls II, and J. Waksberg (eds.), *Telephone Survey Methodology*, New York: Wiley.

Baker, R. P., Bradburn, N. M., and Johnson, R. (1995), "Computer Assisted Personal Interviewing: An Experimental Evaluation of Data Quality and Survey Costs," *Journal of Official Statistics*, **11**, pp. 415–434.

Bankier, M., Luc, M., Nadeau, C., and Newcombe, P. (1995), "Imputing Numeric and

Qualitative Variables Simultaneously," unpublished report, Ottawa: Statistics Canada.

Barcaroli, G. and Venturi, M. (1995), "DAISY (Design Analysis and Imputation System): Structure, Methodology and First Applications," paper presented at the United Nations Work Session on Statistical Editing, Athens, November.

Bassi, F., Torelli, N., and Trivellato, U. (1996), "Latent Class Analysis for Estimating Gross Flows Affected by Classification Errors in SIPP," *Proceedings of the International Conference on Survey Measurement and Process Quality*, Alexandria, VA: American Statistical Association, pp. 121–126.

Bates, B. and Gregory, D. (1995), *Voice and Data Communications Handbook*. New York: McGraw-Hill.

Bemelmans-Spork, M. and Sikkel, D. (1985), "Data Collection with Handheld Computers," *Proceedings of the International Statistical Institute*, 45 Session, vol. 3, topic 18.3. Voorburg, The Netherlands: International Statistical Institute.

Beniger, J. R. (1997), "Comments made in panel discussion on CASIC: Brave New World or Death Knell for Survey Research," at the annual meeting of the American Association for Public Opinion Research, Norfolk, May.

Bergman, L. R., Kristiansson, K. E., Olofsson A., and Säfström, M. (1994), Decentralised CATI versus Paper and Pencil Interviewing: Effects on the Results in the Swedish Labour Force Surveys," *Journal of Official Statistics*, **10**, pp. 181–195.

Bernard, C. (1990), "Survey Data Collection Using Laptop Computers," INSEE Report 01/C520, Institut National de la Statistique et des Etudes Economiques, Paris.

Berry, S. H. and O'Rourke, D. (1988), "Administrative Designs for Centralized Telephone Survey Centers: Implications of the Transition to CATI," in R. M. Groves, P. P. Biemer, L. E. Lyberg, J. Massey, W. L. Nicholls II, and J. Waksberg (eds.), *Telephone Survey Methodology*, New York: Wiley.

Bethlehem, J. G. (1997), "Integrated Control Systems for Survey Processing," in L. Lyberg, P. Biemer, M. Collins, E de Leeuw, C. Dippo, N. Schwarz, and D. Trewin (eds.), *Survey Measurement and Process Quality*, New York: Wiley.

Bethlehem, J. G. and Hofman, L. P. M. B. (1995), "Macro-Editing with Blaise III," in V. Kuusela (ed.), *Essays on Blaise, Proceedings of the Third Blaise Users' Conference*, Helsinki: Statistics Finland, pp. 1–22.

Bethlehem, J. G., Hundepool, A. J., Schuerhoff, M. H., and Vermeulen, L. F. M. (1989), *Blaise 2.0, CAPI/CATI Gebruikershandleiding*. Voorburg, The Netherlands: Centraal Bureau voor de Statistiek.

Bethlehem, J. G. and Kersten, H. M. P. (1986), "Werken met Non-respons," *Statistische Onderzoekingen M30*, The Hague: Staatsuitgeverij.

Bethlehem, J. G. and Schuerhoff, M. H. (1994), "Managing Computer Assisted Survey Processing," in *Proceedings of the Bureau of the Census 1994 Annual Research Conference and CASIC Technologies Interchange*. Washington, DC: U.S. Bureau of the Census, pp. 697–714.

Biemer, P. and Caspar, R. (1994), "Continuous Quality Improvement for Survey Operations: Some General Principles and Applications," *Journal of Official Statistics*, **10**, pp. 307–326.

Billiet, J. and Loosveldt, G. (1988), "Improvements in the Quality of Responses to Factual Survey Questions by Interviewer Training," *Public Opinion Quarterly*, **52**, pp. 190–211.

Bittner, D. and Gill, B. (1996), "Training Interviewers at Home on CAPI: Measuring the Effectiveness of Westat's On-Line Tutorial CAPITRAIN as a Home Study Training Tool," paper presented at the Annual Conference of the American Association for Public Opinion Research, Salt Lake City, UT, May.

Blattberg, R. C., Glazer, R., and Little, D. C. (eds.) (1994), *The Marketing Information Revolution*, Boston: Harvard Business School Press.

Blom, E. (1993), "New Technologies in Data Collection at Statistics Sweden," paper presented at the 1993 Field Directors/Technologies Conference, Chicago, May.

Blom, E. (1994), "Building Integrated Systems of CASIC Technologies at Statistics Sweden," *Proceedings of the Bureau of the Census Annual Research Conference and CASIC Technologies Interchange*, Washington, DC: U.S. Bureau of the Census, pp. 623–636.

Blom, E. and Friberg, R. (1995), "The Use of Scanning at Statistics Sweden," *Proceedings of the International Conference on Survey Measurement and Process Quality*, Alexandria, VA: American Statistical Association, pp. 52–63.

Blum, O. (1995), "Editing Definition and Operation in the Optical Data Entry System (ODE) of the 1995 Census of Population in Israel," Statistical Commission and Economic Commission for Europe, Conference of European Statisticians, Work Session on Statistical Data Editing, Athens, Greece, November, Working Paper 48.

Blum, O. and Ben-Moshe, E. (1996), "Automated Record Linkage and Editing: Essential Supporting Components in Data Capture Process," Israeli Central Bureau of Statistics, Census of Population and Housing.

Blyth, W. G. (1997), "Developing a Speech Recognition Application for Survey Research," in L. Lyberg, P. Biemer, M. Collins, E. deLeeuw, C. Dippo, N. Schwarz, and D. Trewin (eds.), *Survey Measurement and Process Quality*, New York: Wiley, pp. 249–266.

Blyth, W. G. and Piper, H. (1994), "Speech Recognition—A New Dimension in Survey Research." *Journal of the Market Research Society*, **36**, pp. 183–204.

Bobbitt, L. and Carroll, C. D. (1993), "Coding Major Field of Study," *Proceedings of the Survey Research Methods Section, American Statistical Association*, pp. 177–182.

Boekeloo, B., Schiavo, L., Rabin, D., Conlon, R., Jordan, C., and Mundt, D. (1994), "Self-reports of HIV Risk Factors at a Sexually Transmitted Disease Clinic: Audio vs Written Questionnaires," *American Journal of Public Health*, **84**, pp. 754–760.

Bogart, L. (1987), "The Future of Public Opinion: A Symposium," *Public Opinion Quarterly*, **51**, pp. S173–S191.

Bon, E. and van Doorn, L. (1988), "Some Results and Experiences of a New Consumer Panel Technique: Tele-Interview," in: *ESOMAR Seminar on Improving the Use of Consumer Panels for Marketing Decisions*, Amsterdam: ESOMAR, pp. 85–101.

Boucher, L. (1991), "Micro-Editing for the Annual Survey of Manufactures: What Is the Value Added?" *Proceedings of the Bureau of the Census' Annual Research Conference*, Washington, DC: U.S. Bureau of the Census, pp. 765–781.

Brackstone, G. (1985), "Discussion of New Technologies of Data Collection and Capture," *Proceedings of the 45th Session, International Statistical Institute*, book III, topic 18.2, pp. 157–158.

Bradburn, N. M. (1983), "Response Effects," in P. Rossi, J. Wright, and A. Anderson (eds.), *Handbook of Survey Research*, New York: Academic Press, pp. 289–328.

Bradburn, N. M. and Sudman, S. (1989), *Polls and Surveys: Understanding What They Tell Us.* San Francisco: Jossey-Bass.

Bradburn, N. M., Sudman, S., and Schwarz, N. (1996), *Thinking about Answers.* San Francisco: Jossey-Bass.

Brakenhoff, W. J., Remmerswaal, P. W. M., and Sikkel, D. (1987), "Integration of Computer Assisted Survey Research," *Automation in Survey Processing,* CBS Select 4, Voorburg, The Netherlands: CBS Publications.

Brent, E. E., Jr. and Anderson, R. E. (1990), *Computer Applications in the Social Sciences,* Philadelphia: Temple University Press.

Brick, M., Allen, B., Cunningham, P., and Maklan, D. (1996), "Outcomes of a Calling Protocol in a Telephone Survey," *Proceedings of the Survey Research Methods Section, American Statistical Association,* pp. 142–149.

Briere, D. and Heckart, C. (1996), "What's All This about ADSL? *Network World,* March 11, p. 20.

Brooks, F. P. (1982), *The Mythical Man-Month, Essays in Software Engineering,* Reading, MA: Addison-Wesley.

Brown, A. and Veevers, R. (1996), "Quality Assurance for the Self-sufficiency Project," *Proceedings of the International Conference on Survey Measurement and Process Quality.* Alexandia, VA: American Statistical Association, pp. 76–80.

Buck, S. (1987), "Television Audience Measurement Research: Yesterday and Tomorrow," unpublished paper, London: AGB Research.

Bulletin of Science Policy News (1996), American Institute of Physics, no. 145, October 11.

Burgess, R. D. (1989), "Major Issues and Implications of Tracing Survey Respondents," in D. Kasprzyk, G. Duncan, G. Kalton, and M. P. Singh (eds.), *Panel Surveys,* New York: Wiley.

Burkhead, D. and Coder, J. (1985), "Gross Changes in Income Recipiency from the Survey of Income and Program Participation," *Proceedings of the Section on Social Statistics, American Statistical Association,* pp. 351–355.

Burt, C. (1994), "Results of a Pilot Test of a Pen-Based Data Collection System," paper presented at the International Field Directors and Technologies Conference, Chicago, May.

Bushnell, D. (1996), "Computer Assisted Occupation Coding," *Proceedings of the Second ASC International Conference,* Chesham, UK: Association for Survey Computing, pp. 165–173.

Business Week, (1997), "A Census in Cyberspace." May 5.

Buxton, W. (1986), "Chunking and Phrasing and the Design of Human-Computer Dialogues," in H. J. Kugler (ed.), *Proceedings of the IFIP 10th World Computer Conference—Information Processing '86,* Amsterdam: Elsevier Science, pp. 475–480.

Cain, A. (1995), "Security, Authentication and Privacy on the World Wide Web," in: *Proceedings of the 4th International WWW Conference,* Boston, December.

Camburn, D., Cynamon, D., and Harel, Y. (1991), "The Use of Audio Tapes and Written Questionnaires to Ask Sensitive Questions during Household Interviews," paper presented at the National Field Technologies Conference, San Diego, CA.

Campanelli, P., Thomson, K., Moon, N., and Staples, T. (1997), "The Quality of

Occupational Coding in the United Kingdom," in: L. Lyberg, P. P. Biemer, M. Collins, E. de Leeuw, C. Dippo, N. Schwarz, and D. Trewin (eds.), *Survey Measurement and Process Quality*. New York: Wiley.

Cannell, C. F., Fowler, F. J., and Marquis, K. (1968), *The Influence of Interviewer and Respondent Psychological and Behavioral Variables on the Reporting in Household Interviews*, Vital and Health Statistics, series 2, no. 26, Washington, DC: Government Printing Office.

Cannell, C. F. and Robison, S. (1971), "Analysis on Individual Questions," in: L. Lansing, S. Withey, and A. Wolfe (eds.), *Working Papers on Survey Research in Poverty Areas*. Ann Arbor: Institute for Social Research, University of Michigan.

Card, S., Moran, T., and Newell, A. (1980), *The Psychology of Human Computer Interaction*. Hillsdale, NJ: Lawrence Erlbaum Associates, Inc.

Carley, K. (1993), "Coding Choices for Textual Analysis: A Comparison of Content Analysis and Map Analysis," *Sociological Methodology*, **93**, pp. 75–126.

Carlson, J. E. and Dillman, D. A. (1988), "The Influence of Farmers' Mechanical Skill on the Development and Adoption of New Agricultural Practice," *Rural Sociology*, **53**, pp. 235–245.

Carpenter, E. H. (1988), "Software Tools for Data Collection: Microcomputer Assisted Interviewing," *Social Science Computer Review*, **6**, pp. 353–368.

Carroll, J. M. (ed.) (1991), *Designing Interaction: Psychology at the Human-Computer Interface*, New York: Cambridge University Press.

CASRO (1996), *Survey Research on the Internet: Trends and Practices among Major Companies Operating in the United States*. Port Jefferson, NY: Council of American Survey Research Organizations.

Catania, J. A., Coates, T. J., Stall, R., Turner, H., Peterson, J., Hearst, N., Dolcini, M. M., Hudes, E., Gagnon, J., Wiley, J. et al. (1992), "Prevalence of AIDS-Related Risk Factors and Condom Use in the United States," *Science*, **258**, pp. 1101–1106.

Catania, J., Stall, R., Binson, D., Dolcini, M., et al. (1994), "Family of AIDS Behavioral Surveys," unpublished proposal for funded NIH grant RO1-MH51523, San Francisco: University of California, San Francisco.

Catlin, G. and Ingram, S. (1988), "The Effects of CATI on Costs and Data Quality: A Comparison of CATI and Paper Methods in Centralized Interviewing," in: R. M. Groves, P. P. Biemer, L. E. Lyberg, J. Massey, W. L. Nicholls II, and J. Waksberg (eds.), *Telephone Survey Methodology*, New York: Wiley.

Cattin, P. and Wittink, D. R. (1982), "Commercial Use of Conjoint Analysis, a Survey," *Journal of Marketing*, **46**, pp. 44–53.

Chambers, J. M., Cleveland, W. S., Kleiner, B., and Tukey, P. A. (1983), *Graphical Methods for Data Analysis*. Boston: Duxbury Press.

Chen, B. C., Creecy, R. H., and Appel, M. V. (1993), "Error Control of Automated Industry and Occupation Coding," *Journal of Official Statistics*, **9**, pp. 729–745.

Chen-Chi, S. and Chen-Chi, C. (1996), "A Web-Based Intelligent Survey Tool, paper presented at the International Conference on Computer-Assisted Survey Information Collection, San Antonio, TX, December.

Christie, A., Illingworth, M. M., and Lange, L. (1995), "One World?" *Information Week*, October 2, pp. 52–58.

Clayton, R. L. and Harrell, Jr., L. J. (1989), "Developing a Cost Model of Alternative

Data Collection Methods: Mail, CATI, and TDE," *Proceedings of the Section on Survey Research Methods, American Statistical Association*, pp. 264–269.

Clayton, R. L. and Werking, G. S. (1994), "Integrating CASIC into the Current Employment Statistics Survey," *Proceedings of the Bureau of the Census Annual Research Conference and CASIC Technologies Interchange*, Washington, DC: U.S. Bureau of the Census, pp. 738–750.

Clayton, R. L. and Winter, D. L. S. (1992), "Speech Data Entry: Results of a Test of Voice Recognition for Survey Data Collection," *Journal of Official Statistics*, **8**(3), pp. 377–388.

Clemens, J. (1984), "The Use of View Data Panels for Data Collection," in: *ESOMAR Seminar on Are Interviewers Obsolete? Drastic Changes in Data Collection and Data Presentation*, Amsterdam: ESOMAR, pp. 47–65.

Colledge, M., Wensing, F., and Brinkley, E. (1996), "Integrating Metadata with Survey Development in a CAI Environment," *Proceedings of the Bureau of the Census Annual Research Conference and Technology Exchange*, pp. 1078–1100.

Collins, M. (1971), "Market Segmentation—The Realities of Buyer Behavior," *Journal of The Market Research Society*, 13, p. 3.

Collins, M. (1983), "Computer Assisted Telephone Interviewing in the U.K.," *Proceedings of the Section on Survey Methods Research, American Statistical Association*, pp. 636–641.

Connett, W. E. (1996), "Computer Aided Interviewing: Has It Ever, and Will It Still Work?" *Survey and Statistical Computing 1996, Proceedings of the Second ASC International Conference*, pp. 29–38.

Connett, W. E., Blackburn, Z., Gebler, N. Greenwell, M., Hansen, S. E., and Price, P. (1990), "A Report on the Evaluation of Three CATI Systems," unpublished report, Ann Arbor: Survey Research Center, University of Michigan.

Connett, W. E., Mockovak, W., and Uglow, D. (1994), "CAI Systems—The Users Perspective," *Proceedings of the Bureau of the Census Annual Research Conference and CASIC Technologies Interchange*, Washington, DC: U.S. Bureau of the Census.

Converse, J. M. (1987), *Survey Research in the United States: Roots and Emergence, 1890–1960*. Berkeley: University of California Press.

Converse, J. M. and Presser, S. (1986), *Survey Questions: Handcrafting the Standardized Questionnaire*. Beverly Hills: Sage.

Cooley, P. C. and Turner, C. F. (in press), "Implementing Audio CASI on Windows' Platforms," *Computers in Human Behavior*, in press.

Cooley, P. C., Turner, C. F., Allen, D. R., O'Reilly, J., Hamill, D. N., and Paddock, R. E. (1996), "Audio-CASI: Hardware and Software Considerations in Adding Sound to a Computer-Assisted Interviewing System," *Social Science Computer Review*, **14**, pp. 197–204.

Corbett, J. P. (1972), "Encoding from Free Word Descriptions," unpublished report, Washington DC: U.S. Bureau of the Census.

Couper, M. P. (1994), "What Can CAI Learn from HCI?" paper presented at the COPAFS seminar on New Directions in Statistical Methodology. Bethesda, MD.

Couper, M. P. (1998), "The Application of Cognitive Science to Computer Assisted Interviewing," in: M. G. Sirken, D. J. Hermann, and S. Schechter (eds.), *Cognition and Survey Research*. New York: Wiley.

Couper, M. P. and Burt, G. (1994), "Interviewer Attitudes toward Computer-Assisted Personal Interviewing (CAPI)", *Social Science Computer Review*, **12**(1), pp. 38–54.

Couper, M. P. and Groves, R. M. (1990), "Interviewer Expectations Regarding CAPI: Results of Laboratory Tests II," unpublished paper, Ann Arbor: Survey Research Center, University of Michigan.

Couper, M. P. and Groves, R. M. (1992), "Interviewer Reactions to Alternative Hardware for Computer-Assisted Personal Interviewing," *Journal of Official Statistics*, **8**, pp. 201–210.

Couper, M. P., Groves, R. M., and Kosary, C. (1989), "Methodological Issues in CATI," *Proceedings of the Survey Research Methods Section*, Alexandria, VA: American Statistical Association, pp. 349–354.

Couper, M. P., Hansen, S. E., and Sadosky, S. A. (1997), "Evaluating Interviewer Use of CAPI Technology," in: L. Lyberg, P. Biemer, M. Collins, E. De Leeuw, C. Dippo, N. Schwarz, and D. Trewin (eds.), *Survey Measurement and Process Quality*, New York: Wiley.

Couper, M. P. and Rowe, B. (1996), "Evaluation of a Computer-Assisted Self-Interview (CASI) Component in a CAPI Survey," *Public Opinion Quarterly*, **60**, pp. 89–105.

Couper, M. P., Sadosky, S., and Hansen, S. E. (1994), "Measuring Interviewer Behavior Using CAPI," *Proceedings of the Section on Survey Research Methods, American Statistical Association*, pp. 845–850.

Creecy, R. H., Masand, B. M., Smith, S. J., and Waltz, D. L. (1992), "Trading MIPS and Memory for Knowledge Engineering," *Communications of the ACM*, **35**, pp. 48–63.

Curry, J. (1987), "Computer-Assisted Telephone Interviewing: Technology and Organizational Management," Sawtooth Software.

Curry, J. (1990a), "Future Developments in Computer Assisted Personal Interviewing," paper presented at the 1990 AAPOR/WAPOR Conference, Lancaster, PA.

Curry, J. (1990b), "Interviewing by PC: What We Could Not Do Before," *Applied Marketing Research: A Journal for Practitioners*, **30**, pp. 30–37.

Cyberatlas (1997), ⟨http://www.cyberatlas.com⟩.

Daata, R. and Wojcik, M. (1994), "Sample Creation in the Field—A CAPI Application," presented at the Field Directors and Technologies Conference, Boston, May.

Dalton, P. and Keogh, G. (1996), "Automatic Coding of Occupations the Irish Experience," unpublished report, Dublin: Central Statistics Office.

Danielsson, L. and Maarstad, P. A. (1982), *Statistical Data Collection with Hand-Held Computers: A Consumer Price Index*. Orebro, Sweden: Statistics Sweden.

deBie, S. E., Ineke, A. L. S., and de Vries, K. L. M. (1989), *CAI Software: An Evaluation of Software for Computer Assisted Interviewing*. Leiden: VOI, Association of Social Research Institutes.

Degerdal, H., Hoel, T., and Thirud, T. (1995), "The CAI System of Statistics Norway," *Essays on Blaise 1995, Third International Blaise User's Conference*, Statistics Finland, pp. 35–44.

De Heer, W. F. (1991), "The Use of Handheld Computers in Social Surveys of The Netherlands Central Bureau of Statistics," *The Statistician*, **40**, pp. 125–138.

Dekker, F. and Dorn, P. (1984), "Computer Assisted Telephone Interviewing: A

Research Project in the Netherlands," paper presented at the Conference of the Institute of British Geographers, Durham, U.K.

De Leeuw, E. D. (1993), "Mode Effects in Survey Research: A Comparison of Mail, Telephone, and Face to Face Surveys," *Bulletin de Méthodologie Sociologique*, **41**, pp. 3–15.

De Leeuw, E. D. (1994), *Computer Assisted Data Collection Data Quality and Costs: A Taxonomy and Annotated Bibliography*, Methods and Statistics Series Publication 55, University of Amsterdam, Department of Education.

De Leeuw, E. and Collins, M. (1997), "Data Collection Methods and Survey Quality," in: L. Lyberg, P. Biemer, M. Collins, E. De Leeuw, C. Dippo, N. Schwarz, and D. Trewin (eds.), *Survey Measurement and Process Quality*, New York: Wiley.

De Leeuw, E. D. and Hox, J. J. (1995), "Computer Assisted Data Collection, Data Quality and Costs: An Annotated Bibliography," *Survey Statistician*, **32**, July, pp. 5–10.

De Leeuw, E. D., Hox, J. J., and Snijkers, G. (1995), "The Effect of Computer-Assisted Interviewing on Data Quality," *Journal of the Market Research Society*, **37**, pp. 325–344.

De Leeuw, E. D. and Nicholls, W. L. (1996), "Technological Innovations in Data Collection: Acceptance, Data Quality and Costs," *Sociological Research Online*, vol. 1.

De Leeuw, E. D. and van der Zouwen, J. (1988), "Data Quality in Telephone and Face to Face Surveys: A Comparative Meta-Analysis," in: R. Groves, P. Biemer, L. Lyberg, J. Massey, W. Nicholls, and J. Waksberg (eds.), *Telephone Survey Methodology*, New York: Wiley.

DeMaio, T. J. and Rothgeb, J. M. (1996), "Cognitive Interviewing Techniques: In the Lab and in the Field," in: N. Schwarz and S. Sudman (eds.), *Answering Questions*, San Francisco: Jossey-Bass.

Deming, W. E. (1986), *Out of the Crisis*, Cambridge University Press: Cambridge.

Den Boon, A. (1996), "From People Meter to Decoder: Revolution on the TV Screen," in: *Media Data for Market Profit*, ESOMAR series, 205, pp. 293–311.

Denteneer, D., Bethlehem, J. G., Hundepool, A. J., and Keller, W. J. (1987), "The Blaise System for Computer-Assisted Survey Processing," *Proceedings of the Bureau of the Census Annual Research Conference*, Washington, DC: U.S. Bureau of the Census, pp. 112–127.

DePijper, W. M. and Saris, W. E. (1986a), "Computer Assisted Interviewing Using Home Computers," *European Research*, **14**, pp. 144–152.

De Pijper, W. M. and Saris, W. E. (1986b), "The Formulation of Interviews Using the Program INTERV," Amsterdam: Sociometric Research Foundation.

Dertouzos, M. L. (1997), *What Will Be: How the New World of Information Will Change Our Lives*. New York: Harper Collins.

Desurvire, H. W., Lawrence, D., and Atwood, M. (1994), Empiricism versus Judgment: Comparing User Interface Evaluation Methods on a New Telephone-Based Interface," *ACM SIGCHI Bulletin*, **23**(4), pp. 58–59.

de Waal, A. G. (1996), "CHERRYPI: A Computer Program for Automatic Edit and Imputation," paper presented at the United Nations Working Session on Statistical Data Editing. Voorburg, The Netherlands, November.

Dibbs, R., Hale, A., Loverock, R., and Michaud, S. (1995), "Some Effects of Computer Assisted Interviewing on the Data Quality of the Survey of Labour and Income Dynamics," *Proceedings of the International Conference on Survey Measurement and Process Quality*. Alexandria, VA: American Statistical Association, pp. 174–177.

Dielman, L. and Couper, M. P. (1995), "Data Quality in a CAPI Survey: Keying Errors," *Journal of Official Statistics*, **11**(2), pp. 141–146.

Dillman, D. A. (1978), *Mail and Telephone Surveys: The Total Design Method*, New York: Wiley.

Dillman, D. A. (1991), "The Design and Administration of Mail Surveys," *Annual Review of Sociology*, **17**, pp. 225–249.

Dillman, D. A. (1995), "Image Optimization Test: Summary of 15 Taped Interviews," Technical Report 95-40, Social and Economic Sciences Research Center, Washington State University, Pullman, WA.

Dillman, D. A. (1996), "Why Innovation is Difficult in Government Surveys," *Journal of Official Statistics*, **12**, pp. 113–124.

Dillman, D. A. (1997), "Token Finanical Incentives and the Reduction of Nonresponse Error in Mail Surveys," *Proceedings of the government survey section, American Statistical Association*.

Dillman, D. A. and Tarnai, J. (1988), "Administrative Issues in Mixed Mode Surveys," in: R. M. Groves, P. Biemer, L. Lyberg, J. Massey, W. L. Nicholls II, and J. Waksberg (eds.), *Telephone Survey Methodology*, New York: Wiley.

Dillman, D. A., Clark, J. R., and Treat, J. (1994), "The Influence of 13 Design Factors on Response Rates to Census Surveys," *Proceedings of the Bureau of the Census' Annual Research Conference Proceedings*, Washington, DC: U.S. Bureau of the Census, pp. 137–159.

Dillman, D. A., Jenkins, C., Martin, E., and DeMaio, T. (1996), *Cognitive and Motivational Properties of Three Proposed Decennial Census Forms*. Technical Report 96-29 of the Social and Economic Sciences Research Center, Washington State University, Pullman, WA (also released by the Center for Survey Methods Research, U.S. Bureau of the Census, Washington, DC).

Dillman, D. A., Sinclair, M. D., and Clark, J. R. (1993), "Effects of Questionnaire Length, Respondent-Friendly Design, and a Difficult Question on Response Rates for Occupant-Addressed Census Mail Surveys," *Public Opinion Quarterly*, **57**, pp. 289–304.

Dippo, C., Polivka, A., Creighton, K., Kostanich, D., and Rothgeb, J. (1992), "Redesigning a Questionnaire for Computer-Assisted Data Collection: The Current Population Survey Experience," paper presented at the Field Directors and Technologies Conference, St. Petersburg, FL, May.

Downes-LeGuin, T. and Hoo, B. S. (1994), "Disk-by-Mail Data Collection for Professional Populations," paper presented at the American Association for Public Opinion Research Annual Conference, Danvers, MA, May.

Driessen, J. H. W., Keller, W. J., Mokken, R. J., Vrancken, E. H. J., de Vries, W. F. M., and Wit, J. W. W. A. (eds.) (1987), *Automation in Survey Processing*. Voorburg/Heerlen, The Netherlands: Centraal Bureau voor de Statistiek.

Dubnoff, S., Kiesler, S., and Turner, C. F. (1986), "Silicon Surveys: Computer Administered Questionnaires," unpublished proposal for unfunded NIH grant under SBIR program. Boston: Circle Systems.

Duffy, J. C. and Waterton, J. J. (1984), "Under-reporting of Alcohol Consumption in Sample Surveys: The Effect of Computer Interviewing in Fieldwork," *British Journal of Addiction*, **79**, pp. 303–308.

Duggan, J. A. (1996), "Electronic Imaging in Support of the 1996 Census of Agriculture. Statistics Canada," paper presented at the International Conference on Computer Assisted Survey Information Collection, San Antonio, TX, December.

Dumas, J. S. and Redish, J. C. (1994), *A Practical Guide to Usability Testing*. Norwood, NJ: Ablex.

Dumicic, S. and Dumicic, K. (1994), Optical Reading and Automatic Coding in the Census '91 in Croatia," Statistical Commission and Economic Commission for Europe, Conference of European Statisticians, Work Session on Statistical Data Editing. Cork, Ireland, October, Working Paper 2.

Duncan, G. J. (1992), "Household Panel Studies: Prospects and Problems," paper presented at the International Conference on Social Science Methodology, Trento, Italy.

Duncan, G. J. and Kalton, G. (1987), "Issues of Design and Analysis of Surveys across Time," *International Statistical Review*, **55**, pp. 97–117.

Dussert, F. and Luciani, G. (1995), CAPI plus et Blaise III. Une Organization général pour les enquêtes de l'INSEE," *Essays on Blaise, Third International Blaise User's Conference, Statistics Finland*, pp. 45–52.

Dutka, S. and Frankel, L. (1980), "Sequential Survey Design through the Use of Computer Assisted Telephone Interviewing," *Proceedings of the Business and Economic Statistics Section*, Alexandria, VA: American Statistical Association, pp. 73–76.

Dykema, J., Lepkowski, J. M., and Blixt, S. (1997), "The Impact of Interviewer and Respondent Behavior on Data Quality: Analysis of Interaction Coding in a Validation Study," in L. Lyberg, P. Biemer, M. Collins, E. De Leeuw, C. Dippo, N. Schwarz, and D. Trewin (eds), *Survey Measurement and Process Quality*, New York: Wiley.

Ecklund, B. (1991), "CAPI Applications at NASS," paper presented at the International Field Directors and Technologies Conference, San Diego, CA.

Economist (1996), "November 9–15th, Economist Newspaper Limited, pp. 20, 71.

Edwards, B., Bittner, D., Edwards, W. S., and Sperry, S. (1993), "CAPI Effects on Interviewers: A Report from Two Major Surveys," in: *Proceedings of the Bureau of the Census' Annual Research Conference*, pp. 411–428.

Edwards, B., Sperry, S., and Schaeffer, N. C. (1995), "CAPI Design Techniques for Improving Data Quality," *Proceedings of the International Conference on Survey Measurement and Process Quality*, pp. 168–173.

Egan, M., Lemaître, G., Michaud, S., and Murray T. S. (1990), "Invisible Seams? The Experiences with the Canadian Labour Market Activity Survey," *Proceedings of the Bureau of the Census' Annual Research Conference*, Washington, DC: U.S. Bureau of the Census, pp. 715–729.

Ehrenberg, A. S. C. (1969), "Towards an Integrated Theory of Consumer Behavior," *Journal of The Market Research Society*, **11**, p. 3.

Eiderbrandt, G. and Lyberg, L. (1974), "Control of Optical Character Recognition," Research Report KEX2007, Stockholm: Statistics Sweden.

Eikvil, L. (1993), "OCR: Optical Character Recognition," Norwegian Computing Center (Norsk Regnesentral), Report, Publikasjonsnr: 876, December.

Elder, S. and McAleese, I. (1996), "Application of Document Scanning. Automated Data Recognition and Image Retrieval to Paper Self-completion Questionnaire," *Survey and Statistical Computing 1996*, Proceedings of the Second ASC International Conference, London, U.K., September.

Elias, P. (1996), "Automatic Coding of Occupational Information for the 2001 Census of Population: A Feasibility Study," unpublished report, London: Institute for Employment Research.

Engström, P. and Ängsved, C. (1994), "A Description of a Geographical Macro-Editing Application," paper presented at the Conference of European Statisticians, Work Session on Data Editing, Cork, Ireland.

Esposito, R., Fox, J. K., Lin, D., and Tidemann, K. (1994), "ARIES: A Visual Path in the Investigation of Statistical Data," *Journal of Computational and Graphical Statistics*, **3**, pp. 113–125.

Esposito, R. and Lin, D. (1993), "The ARIES System in the BLS Current Employment Statistics Program," *Proceedings of the International Conference on Establishment Surveys*, Alexandria, VA: American Statistical Association.

ESOMAR (1984), *Are Interviewers Obsolete; Drastic Changes in Data Collection and Data Presentation*, Amsterdam: ESOMAR.

ESOMAR (1996), *Annual Statistics*, Amsterdam: ESOMAR.

Evan, W. M. and Miller, J. R. (1969), "Differential Effects of Computer vs. Conventional Administration of a Social Science Questionnaire: An Exploratory Methodological Experiment," *Behavioral Science*, **14**, p. 216–227.

Falthzik, A. M. (1972), "When to Make Telephone Interviews," *Journal of Marketing Research*, **9**, pp. 451–452.

Falzon, P. (ed.) (1990), *Cognitive Ergonomics: Understanding, Learning, and Designing Human-Computer Interaction*, San Diego: Academic Press.

Fay, R. E., Turner, C. F., Klassen, A. D., and Gagnon, J. H. (1989), "Prevalence and Patterns of Same-Gender Sexual Contact among Men," *Science*, **243**, pp. 338–348.

Federal Committee on Statistical Methodology (1984), *The Role of Telephone, Mail, and Personal Interviews in Federal Statistics*. Statistical Policy Working Paper 12, Washington DC: Statistical Policy Office, Office of Management and Budget.

Federal Committee on Statistical Methodology (1990a), *Data Editing in Statistical Agencies*. Statistical Policy Working Paper 18, Washington, DC: Statistical Policy Office, Office of Management and Budget.

Federal Committee on Statistical Methodology (1990b), *Computer-Assisted Survey Information Collection*. Statistical Policy Working Paper 19, Washington, DC: Statistical Policy Office, Office of Management and Budget.

Federal Committee on Statistical Methodology (1991), *Seminar on Quality of Federal Data*. Statistical Policy Working Paper 20, Washington, DC: Statistical Policy Office, Office of Management and Budget.

Fellegi, I. P. and Holt, D. (1976), a Systematic Approach to Automatic Edit and Imputation," *Journal of the American Statistical Association*, **71**, pp. 17–35.

Fendrich, M. and Vaughn, C. (1994), "Substance Abuse Underreporting," *Public Opinion Quarterly*, **58**, pp. 96–123.

Finegan, J. E., and Allen, N. J. (1994), "Computerized and Written Questionnaires: Are They Equivalent," *Computers in Human Behavior*, **10**, pp. 483–496.

Fink, J. C. (1981), "Quality Improvement and Time Savings Attributed to CATI—Reflections on 11 Years of Experience," paper presented at the Annual Conference of the American Association for Public Opinion Research, Buck Hill Falls, PA, May.

Fink, J. C. (1983), "CATI's First Decade: The Chilton Experience," *Sociological Methods and Research*, **12**, pp. 153–168.

Fisher, B., Margolis, M., and Resnick, D. (1995), "A Study of Civic Life on the Internet," paper presented at the American Association for Public Opinion Research Annual Conference, Fort Lauderdale, FL, May.

Fisher, B., Resnick, D., Margolis, M., and Bishop, G. (1995), "Survey Research in Cyberspace: Breaking Ground on the Virtual Frontier," *Proceedings of the American Association of Public Opinion Research*, pp. 178–183.

Fitti, J. E. (1979), "Some Results from the Telephone Health Interview System," *Proceedings of the Section on Survey Research Methods, American Statistical Association*, pp. 63–72.

Forsyth, B. H. and Lessler, J. T. (1991), "Cognitive Laboratory Methods: a Taxonomy," in P. P. Biemer, R. M. Groves, L. E. Lyberg, N. A. Mathiowetz, and S. Sudman (eds.), *Measurement Errors in Surveys*, New York: Wiley.

Foster, D. and Snow, R. W. (1995), "An Assessment of the Use of Hand-Held Computers during Demographic Surveys in Developing Countries. *Survey Methodology*, **21**(2), pp. 179–184.

Fowler, F. J. Jr., (1995), *Improving Survey Questions—Design and Evaluation*, Thousand Oaks, CA: Sage.

Fowler, F. J., Jr., and Mangione, T. W. (1990), *Standardized Survey Interviewing*. Newberry Park, CA: Sage.

Frankish, C., Hull, R., and Morgan, P. (1995), "Recognition Accuracy and User Acceptance of Pen Interfaces," *Proceedings of CHI'95 Conference on Human Factors in Computing Systems*, New York: ACM, pp. 503–510·.

Frankovic, K. A., Ramnath, B., and Arnedt, C. M. (1994), "Interactive Polling and Americans' Comfort Level with Technology," paper presented at meetings of the American Association for Public Opinion Research, Danvers, MA, May.

Franzosi, R. (1990), "Computer-Assisted Coding of Textual Data. An Application to Semantic Grammars," *Sociological Methods and Research*, **19**, pp. 225–257.

Freeman, H. E. (1983), "Research Opportunities Related to CATI," *Sociological Methods and Research*, **12**, pp. 143–152.

Frese, M. and Altmann, A. (1989), "The Treatment of Errors in Learning and Training," in L. Bainbridge and S. A. Ruiz Quintanilla (eds.), *Developing Skills with Information Technology*, Chichester: Wiley, pp. 65–86.

Frey, J. (1989), *Survey Research by Telephone*. 2nd ed., Beverly Hills: Sage.

Friedman, D., Thornton, A., Camburn, D., Alwin, D., and Young-DeMarco, L. (1988), "The Life History Calendar: a Technique for Collecting Retrospective Data," in C. C. Clogg (ed.) *Sociological Methodology*, **18**, San Francisco: Jossey Bass, pp. 37–68.

Friedman, W. F. (1993), "Memory for the Time of Past Events," *Psychological Bulletin*, **113**, pp. 44–66.

Galbraith, G., Strauss, M., Jordon-Viola, E., and Cross, H. (1974), "Social Desirability Ratings from Males and Females: a Sexual Item Pool," *Journal of Consulting and Clinical Psychology*, **42**, pp. 909–910.

Gambino, J. (1996), "The New Design of the Canadian Labour Force Survey," *Proceedings of the Government Statistics Section, American Statistical Association*, pp. 216–220.

Gates, B. (1995), *The Road Ahead*. New York: Viking.

Gautier, J. M. (1995), "Structuration et modélisation statistique des données de consommation receillies automatiquement," paper presented at the International Statistical Institute meeting, Beijing, August.

Gavrilov, A. J. (1988), "Computer Assisted Interviewing in the USSR," paper presented at the International Methodology Conference, Moscow, November.

Gawen, P. (1996), "The Emerging Market for Performance Scanners," *Document World*, July/August.

Geisler, J. (1995), "Gedrics: The Next Generation of Icons," *Proceedings of the 5th International Conference on Human-Computer Interaction (INTERACT'95)*, Lillehamer, Norway, pp. 73–78.

Geist, J., Wilkinson, R. A., Janet, S., Brother, P. J., Hammond, B., Larsen, N. W., Clear, R. M. S., Matsqui, M. J., Burges, C. J. C., Creecy, R., Hull, J. J., Vogl, T. P., and Wilson, C. L. (1994), *The Second Census Optical Character Recognition Systems Conference: A Report of the National Institute for Standards and Technology*, NIST IR 5452, U.S. Department of Commerce.

Geldrop, M. (1993), *Proef op de Steekproef: Een Evaluatie van de Kwaliteit van De Samenstelling van het Telepanel*, Amsterdam: NIMMO.

Gershenfeld, S., Atherton, T., Ben-Akiva, M., and Musetti, L. (1989), "Context-specific Choice Experiments for Multi-featured products: a Disk-by-Mail Survey Application," *Proceedings of the Sawtooth Software Conference, Gaining a Competitive Advantage through PC-Based Interviewing and Analysis*, vol. 1, Sun Valley, ID: Sawtooth Software, pp. 19–24.

Gfroerer, J. C. and Hughes, A. L. (1992), "Collecting Data on Illicit Drug Use by Phone," in C. F. Turner, J. T. Lessler, and J. D. Gfroerer (eds.), *Survey Measurement of Drug Use: Methodological Issues*, DHHS Pub. No. 92-1929. Washington, DC: Government Printing Office.

Giangrasso, D. (1995), "Mobile Strategies," *Pen Computing Magazine*, **6**, August/September, pp. 54–55.

Gibbs, M. (1993), "Handwriting Recognition: A Comprehensive Comparison," *Pen Computing Magazine*, March/April, pp. 31–35.

Gillman, D. and Appel, M. V. (1994), "Automated Coding Research at the Census Bureau," unpublished report, Washington, DC: Bureau of the Census, Statistical Research Report Series (No. RR94/04).

Glasser, N. (1978), "Computer Assisted Telephone Interviewing," a paper distributed by the New York University Graduate School of Business Administration, Marketing Division.

Goldstein, H. (1987), "Computer Surveys by Mail," *Proceedings of the Sawtooth Software Conference on Perceptual Mapping, Conjoint Analysis, and Computer Interviewing*, Sun Valley, ID: Sawtooth Software, pp. 55–59.

Goodger, C. (1995), "Training Interviewers in the Use of Blaise," *Essays on Blaise, Third International Blaise User's Conference, Statistics Finland*, pp. 53–59.

Goodyear, J. (1995), in *Financial Systems*, Winter 1995 issue.

Gooselin, J.-F., Chinnappa, B. N., Ghangurde, P. D., and Tourigny, J. (1978), *A Compendium of Methods of Error Evaluation in Censuses and Surveys*, Ottawa: Statistics Canada.

Granquist, L. (1984), "Data Editing and Its Impact on the Further Processing of Statistical Data," paper presented at the Workshop on Statistical Computing, Budapest.

Granquist, L. (1990), "A Review of Some Macro-Editing Methods for Rationalizing the Editing Process," *Proceedings of the Statistics Canada Symposium*, Ottawa: Statistica Canada, pp. 225–234.

Granquist, L. (1994), "Macro-Editing—A Review of Methods for Rationalizing the Editing of Survey Data," in *Statistical Data Editing: Methods and Techniques*, Vol. 1, Geneva: United Nations.

Granquist, L. (1995), "Improving the Traditional Editing Process," in B. Cox, D. Binder, A. Christianson, M. Colledge, and P. Kott (eds.), *Business Survey Methods*, New York: Wiley, pp. 385–401.

Graphic, Visualization, and Usability Center (1996), *GVU's 5th WWW User Survey*, Atlanta: Georgia Tech Research Corporation.

Gray, J. (1994), "Laptop Support Systems for Multiple Survey Management," presented at the Field Directors and Field Technologies Conference, Boston.

Gray, J. (1995), "An Object Based Approach for the Handling of Survey Data," *Essays on Blaise, Third International Blaise User's Conference, Statistics Finland*, pp. 61–70.

Gray, J. and Anderson, S. (1996), "The Data Pipeline—Processing Survey Data on a Flow Basis," *Survey and Statistical Computing 1996, Proceedings of the Second ASC International Conference*, pp. 389–398.

Green, P. E. (1974), "On the Design of Choice Experiments Involving Multifactor Alternatives," *Journal of Consumer Research*, 1, pp. 61–68.

Green, T. M. (1996), "An Investigation of Response Effects for Responders and Refusers in an On-Line Organizational Survey," paper presented at the International Conference on Computer Assisted Survey Information Collection, San Antonio, TX, December.

Greenberg, B. S. and Stokes, S. L. (1990), "Developing an Optimal Call Scheduling Strategy for a Telephone Survey," *Journal of Official Statistics*, 6, pp. 421–435.

Greist, J. H., Klein, M. H., and Erdman, H. P. (1976), "Routine On-line Psychiatric Diagnosis by Computer," *American Journal of Psychiatry*, 12, pp. 1405–1408.

Grilley, N., Kean, Y., and Nichols, B. (1996), "ACASI: A Practical Analysis," unpublished report, Chicago: National Opinion Research Center.

Grondin, C. and Michaud, S. (1994), "Data Quality of Income Data Using Computer Assisted Interview: The Experience of the Canadian Survey of Labour and Income Dynamics," *Proceedings of the Section on Survey Research Methods*, Alexandria, VA: American Statistical Association, pp. 839–844.

Groves, R. M. (1983), "Implications of CATI: Costs, Errors, and Organization of Telephone Survey Research," *Sociological Methods and Research*, 12, pp. 199–215.

Groves, R. M. (1989), *Survey Errors and Survey Costs*, New York: Wiley.

Groves, R. M. (1994), "Challenges to Methodological Innovation in Government Statistical Agencies," in Z. Kennessy (ed.), *The Future of Statistics, An International Perspective*, Voorburg, The Netherlands: ISI.

Groves, R. M. and Kahn, R. L. (1979), *Surveys by Telephone: A National Comparison with Personal Interviews*, New York: Academic Press.

Groves, R. M. and Magilavy, L. (1980), "Estimates of Interviewer Variance in Telephone Surveys," *Proceedings of the Section on Survey Research Methods, American Statistical Association*, pp. 622–627.

Groves, R. M. and Mathiowetz, N. A. (1984), "Computer Assisted Telephone Interviewing: Effect on Interviewers and Respondents," *Public Opinion Quarterly*, **48**, pp. 356–359.

Groves, R. M. and Nicholls, W. L. II (1986), "The Status of Computer-Assisted Telephone Interviewing: Part II—Data Quality Issues," *Journal of Official Statistics*, **2**(2), pp. 117–134.

Grudin, J. (1989), "The Case against User Interface Consistency," *Communications of the ACM*, **32**, pp. 1164–1173.

Gum, G. S. (1989), "Using Ci2 and ACA to Obtain Complex Pricing Information," *1989 Sawtooth Software Conference Proceedings, Gaining a Competitive Advantage Through PC-Based Interviewing and Analysis*, vol. 1, Sun Valley, ID: Sawtooth Software, pp. 65–69.

Haeckel, S. H. (1994), "Managing the Information-Intensive Firm of 2001," in R. C. Blattberg, R. Glazer, and D. C. Little (eds.), *The Marketing Information Revolution*, Boston: Harvard Business School Press, pp. 328–354.

Hale, A. (1988), "Computer Assisted Industry and Occupation Coding in the Canadian Labour Force Survey," *Proceedings of the Bureau of the Census' Fourth Annual Research Conference*, pp. 387–395.

Hale, A. and Dibbs, R. (1993), "Questionnaire Design in a Paperless Society. *Proceedings of the Bureau of the Census' Annual Research Conference*, Washington, DC: U.S. Bureau of the Census, pp. 353–361.

Hale, A. and Michaud, S. (1995), "Dependent Interviewing: Impact on Recall and on Labour Market Transitions," *Proceedings of the Bureau of the Census' Annual Research Conference*, Washington, DC: U.S. Bureau of the Census, pp. 467–474.

Hammer, M. and Champy, J. (1994), *Reengineering the Corporation*, New York: Harper Business.

Harlow, B. L., Rosenthal, J. F., and Ziegler, R. G. (1985), "A Comparison of Computer-Assisted and Hard Copy Telephone Interviewing," *American Journal of Epidemiology*, **122**(2), pp. 335–340.

Harmon, D. (1991), "How Effective is Suffixing?" *Journal of the American Society for Information Science*, **42**, pp. 7–15.

Harrell, L. J., Jr., Clayton, R. L., and Werking, G. S. (1996), "TDE and Beyond: Data Collection on the World Wide Web," in *Proceedings of the Section of Survey Research Methods, American Statistical Association*, pp. 768–773.

Hartman, H., Saris, W. E., Gallhofer, I. N., Leeuwin, J., Verwey, J., and Lemmens, N. (1991), *Data Collection on Expenditures*, Amsterdam: Sociometric Research Foundation (SRF).

Harvell, A. (1990), "Summary of the Final Debriefing for the Pittsburgh CAPI Test," unpublished memorandum, Washington, DC: Automation Implementation Branch, U.S. Bureau of the Census.

Helgadóttir, H. (1995), "News from Statistics Iceland," *Internatonal Blaise User Group Newsletter*, December.

Heller, J.-L. (1993), "The Use of CAPI and BLAISE in the French Labour Force Survey," *Proceedings of the Second International BLAISE Users Conference*, London.

Henden, M., Wensing, F., Smith, K., and Georgopolous, M. (1997), "An Office Management System in Blaise III," *Proceedings of the Fourth International Blaise Users Conference*, May 5–7, Paris.

Hendershot, T. P., Thornberry, J., Rogers, S. M., Miller, H. G., and Turner, C. F. (1996), "Multilingual Audio-CASI: Using English-Speaking Field Interviewers to Survey Elderly Korean Households," in R. Warnecke (ed.), *Health Survey Research Methods*. Hyattsville, MD: National Center for Health Statistics.

Henne, J. (1993), "Evaluating Existing Systems: Computer-Aided Telephone Interviewing," *Proceedings From Sawtooth Software's Conference for Small- and Medium-Sized CATI Facilities*, May, Evanston, IL, pp. 13–31.

Herek, G. M. and Capitanio, J. P. (1993), "Public Reaction to AIDS in the United States: A Second Decade of Stigma," *American Journal of Public Health*, **83**, pp. 574–577.

Herold, E. and Way, L. (1988), "Sexual Self-disclosure among University Women." *Journal of Sex Research*, **24**, pp. 1–14.

Heuvelmans, F., Kerssemakers, F., and Winkels, J. (1997), "Integrating Surveys with Blaise III," in *Actes de la 4 Conference Internationale des Utilisateurs de Blaise 1997*, Paris: INSEE, pp. 123–133.

Hidiroglou, M. A. and Berthelot, J.-M. (1986), "Statistical Editing and Imputation for Periodic Business Surveys," *Survey Methodology*, **12**, pp. 73–83.

Higgins, C. A., Dimnick, T. P., and Greenwood, H. P. (1987), "The DISKQ Survey Method," *Journal of Market Research Society*, **29**, pp. 437–445.

Hochstim, J. (1967), "A Critical Comparison of Three Strategies of Collecting Data from Households," *Journal of the American Statistical Association*, **62**, pp. 976–989.

Hoffman, D. L., Kalsbeek, W. D., and Novak, T. (1996), "Internet Use in the United States: 1995 Baseline Estimates and Preliminary Market Segments," Project 2000 Working Paper, Owen Graduate School of Management, Vanderbilt University.

Hofman, L. P. M. and Keller, W. J. (1991), "Design and Management of Computer Assisted Interviews in the Netherlands," *Proceedings of the Section on Social Statistics*, American Statistical Association, pp. 306–311.

Hofman, L. P. M. and Keller, W. J. (1993), "Management of Computer Assisted Interviews in the Netherlands," *Journal of Official Statistics*, **9**, pp. 765–782.

Hogendoorn, A. and Sikkel, D. (1994), "Optimizing Response Burden in Panels," *Kwantitatiewe Methoden*, **47**, pp. 47–67.

Holt, D., McDonald, J. W., and Skinner, C. J. (1991), "The Effect of Measurement Error on Event History Analysis," in P. P. Biemer, R. M. Groves, L. E. Lyberg, N. Mathiowetz, and S. Sudman (eds.), *Measurement Errors in Surveys*, New York: Wiley, pp. 665–687.

House, C. C. (1985), "Questionnaire Design with Computer Assisted Telephone Interviewing," *Journal of Official Statistics*, **1**, pp. 209–219.

House, C. C. and Nicholls, W. L. II (1988), "Questionnaire Design for CATI: Design Objectives and Methods," in R. M. Groves, P. P. Biemer, L. Lyberg, J. T. Massey, W. L. Nicholls II, and J. Waksberg (eds.), *Telephone Survey Methodology*, New York: Wiley, pp. 421–436.

Houston, G. and Bruce, A. G. (1993), "Geographical Editing for Business and Economic Surveys," *Journal of Official Statistics*, **9**, pp. 81–90.

Hox, J., De Leeuw, E. D., and Snijkers, G. (1993), "Het Effect van Computergestuurd Interviewen op de Data Kwaliteit; Een Vergelijking van Computergestuurde en Papier-en-Pen Methoden voor Dataverzameling," in P. Debets, E. D. De Leeuw, F. Schelbergen, D. Sikkel, and G. Snijkers (eds.), *De Computer als Veldwerker*, Amsterdam: Informatiseringscentrum.

Hsu, F., Anantharaman, T., Campbell, M., and Nowatzyk, A. (1990), "A Grandmaster Chess Machine," *Scientific American*, **263**, pp. 44–50.

Human Resources Development Canada and Statistics Canada, (1996), *Growing up in Canada, the National Longitudinal Survey of Children and Youth (Canada)*, *Technical Appendix*, Catalogue no. 89-550-MPE, no. 1, Human Resources Development Canada and Statistics Canada.

Hyman, H. H., Cobb, W. J., Feldman, J. J., Hart, C. W., and Stember, C. H. (1954), *Interviewing in Social Research*, Chicago: University of Chicago Press.

Informatica Comunidad de Madrid (1993), *Lince, Sistema de Validaciòn e Imputation Automatica de Datos Estadisticos; Manual de Usario*, Madrid: ICM.

Ing, D. and Mitchell, A. A. (1994), "Point-of-Sale Data in Consumer Goods Marketing: Transforming the Art of Marketing into the Science of Marketing," in R. C. Blattberg, R. Glazer, and D. C. Little (eds.), *The Marketing Information Revolution*, Boston: Harvard Business School Press.

Iverson, S. (1992), "Test Reports on Computer Assisted Personal Interviewing (CAPI): New Technologies and Methods," unpublished report, Washington, DC: NICHD.

Jabine, T. B. (1985), "Flow Charts: A Tool for Developing and Understanding Survey Questionnaires," *Journal of Official Statistics*, **1**(2), pp. 189–207.

Jabine, T. B., Straf, M. L., Tanur, J. M., and Tourangeau, R. (eds.) (1984), *Cognitive Aspects of Survey Methodology: Building a Bridge Between Disciplines*, Washington, DC: National Academy Press.

Jacobs, L. C. (1986), "Effect of the Use of Optical-Scan Sheets on Survey Response Rate," paper presented at Annual Meeting of the American Educational Research Association, San Francisco, April.

Jacobson, I. (1992), *Object-Oriented Software Engineering: A Use Case Drive Approach*, Reading, MA: Addison-Wesley.

Jeffries, R. (1994), "Usability Problem Reports: Helping Evaluators Communicate Effectively with Developers," in J. Nielsen, and R. Mack (eds.), *Usability Inspection Methods*, New York: Wiley.

Jenkins, C. R. and Dillman, D. A. (1995), "The Language of Self-administered Questionnaires as Seen through the Eyes of Respondents," in *Statistical Policy Working Paper 23: New Directions in Statistical Methodology,* vol. 3, Washington, DC: U.S. Office of Management and Budget, pp. 470–516.

Jenkins, C. R. and Dillman, D. A. (1997), Toward a Theory of Self-administered Questionnaire Design," in Lyberg, L., Biemer, P., Collins, M., De Leeuw, E., Dippo, C., Schwarz, N., and Trewin, D. (eds.), *Survey Measurement and Process Quality*, New York: Wiley, pp. 165–196.

Jobe, J. B., Pratt, W. F., Tourangeau, R., Baldwin, A., and Rasinski, K. (1997), "Effects of Interview Mode on Sensitive Questions in a Fertility Survey," in L. Lyberg, P. Biemer, M. Collins, E. DeLeeuw, C. Dippo, N. Schwarz, and D. Trewin (eds.), *Survey Measurement and Process Quality*, New York: Wiley, pp. 311–329.

Johnson, R. M. (1987), "Adaptive Conjoint Analysis," *Sawtooth Software Conference Proceedings*, Sun Valley, ID: Sawtooth Software, pp. 253–265.

Johnson, R. M. (1992), "Ci3: Introduction and Evolution," *Sawtooth Software Conference Proceedings*, Sun Valley, ID: Sawtooth Software, pp. 91–102.

Johnston, J. and Walton, C. (1995), "Reducing Response Effects for Sensitive Questions: A Computer-Assisted Self Interview with Audio," *Social Science Computer Review*, **13**(3), pp. 304–319.

Jones, E. and Forrest, J. (1992), "Underreporting of Abortions in Surveys of U.S. Women: 1976 to 1988," *Demography*, **29**, pp. 113–126.

Jones, E. E. and Sigall, H. (1971), "The Bogus Pipeline: A New Paradigm for Measuring Affect and Attitude," *Psychological Bulletin*, **76**, pp. 349–364.

Kalfs, N. (1993), *Hour By Hour: Effects of the Data Collection Mode in Time Use Research*, Amsterdam: NIMMO.

Kalton, G., Kasprzyk, D., and McMillen, D. (1989), "Nonsampling Errors in Panel Surveys," in D. Kasprzyk, G. Duncan, G. Kalton, and M. P. Singh (eds.), *Panel Surveys*, New York: Wiley.

Kaper, E. and Saris, W. E. (1996), "Problems of Continuous Registration," paper presented at the Sixth Biennial Conference on Panel Data, Amsterdam, June.

Katz, I. R. (1996), "Toward Development of a Questionnaire Design Environment for Survey Professionals: Final Project Report," unpublished report, Princeton: Educational Testing Service.

Katz, I. R. and Conrad, F. G. (1996), "Questionnaire Designer: Software Tools for Specification of CASIC Instruments," demonstration presented at the International Conference on Computer Assisted Survey Information Collection, San Antonio, TX, December.

Katz, I. R., Conrad, F. G., and Stinson, L. L. (1996), "Questionnaire Designers versus Instrument Authors: An Investigation of the Development of CASIC Instruments at the U.S. Bureaus of Labor Statistics and of the Census," paper presented at the International Conference on Computer Assisted Survey Information Collection, San Antonio, TX, December.

Kaushal, R. and Laniel, N. (1993), "Computer Assisted Interviewing Data Quality Test," *Proceedings of the Annual Research Conference*, U.S. Bureau of the Census, pp. 513–524.

Keller, W. J. (1993), "Trends in Survey Data Processing," *Journal of the Market Research Society*. **35**, pp. 211–219.

Keller, W. J. (1994), "Changes in Statistical Technology," in Z. Kennessy (ed.), *The Future of Statistics, An International Perspective*, Voorburg, The Netherlands: ISI.

Keller, W. J., Bethlehem, J. G., and Metz, K. G. (1990), "The Impact of Microcomputers on Survey Processing at the Netherlands Central Bureau of Statistics," *Proceedings of the Bureau of the Census' Annual Research Conference*, Washington, DC: U.S. Bureau of the Census, pp. 637–645.

Kelley, K. L. and Charness, N. (1995), "Issues in Training Older Adults to Use Computers," *Behaviour and Information Technology*, **14**(2), pp. 107–120.

Kemsley, W. F. F. (1961), "The Household Expenditure Enquiry of the Ministry of Labor," *Applied Statistics*, 10, **3**, pp. 117–135.

Kennedy, J. F. Jr. (1997), "Bill Gates Man and Mouse," *George*, February 1997, pp. 20.

Kennedy, J. M., Lengacher, J. E., and Demerath, L. (1990), "Interviewer Entry Error in CATI Interviews," paper presented at the International Conference on Measurement Errors in Surveys, Tucson, AZ.

Kerin, R. A. and Peterson, R. A. (1983), "Scheduling Telephone Interviews: Lessons from 250,000 Dialings," *Journal of Advertising Research*, **23**, pp. 97–112.

Kiesler, S. and Sproull, L. S. (1986), "Response Effects in the Electronic Survey," *Public Opinion Quarterly*, **50**, pp. 402–413.

Kindel, C. B. (1992), "Electronic Data Collection at the National Center for Education Statistics: Successes in the Collection of Library Data," *Journal of Official Statistics*, **10**, pp. 93–102.

Kinder, D. R. and Sanders, L. M. (1990), "Mimicking Political Debate with Survey Questions: The Case of White Opinion on Affirmative Action for Blacks," *Social Cognition*, **8**, pp. 73–103.

Kinsey, A., Pomeroy, W., Martin, C., and Gebhard, P. (1953), *Sexual Behavior in the Human Female*, Philadelphia: Saunders.

Kinsey, S. H. (1994), "CAPI Questionnaires: Common Problems and the Optimal Approach to Development and Testing," paper presented at the Field Directors and Field Technologies Conference, Boston, May.

Kinsey, S. H., Thornberry, J. S., Carson, C. P., and Duffer, A. P. (1995), Respondent Preferences Toward Audio-CASI and How That Affects Data Quality," paper presented at the 50th Annual Conference of the American Associaton for Public Opinion Research, Fort Lauderdale, FL.

Klassen, A., Williams, C., and Levitt, E. (1989), *Sex and Morality in the U.S.: An Empirical Enquiry under the Auspices of the Kinsey Institute*, Middletown, CT: Wesleyan University Press.

Klehn, P. (1993), "Managing a CATI Installation on Limited Resources," *Computer-Aided Telephone Interviewing, Proceedings from Sawtooth Software's Conference for Small- and Medium-Sized CATI Facilities*, May, Evanston, IL, pp. 87–93.

Klose, A. and Ball, A. D. (1995), "Using Optical Mark Read Surveys: An Analysis of Response Rate and Quality," *Journal of the Market Research Society*, **37**, 3, pp. 269–279.

Knaus, R. (1987), "Methods and Problems in Coding Natural Language Survey Data," *Journal of Official Statistics*, **3**, pp. 51–60.

Kohut, A. (1994), *Technology in the American Household*, Washington, DC: Times Mirror Center for The People and The Press.

Kojetin, B. A., Kurlander, J., and Rope, D. (1994), "Exploring Longitudinal Nonresponse in the New Current Population Survey," paper presented at the Fifth International Workshop on Household Survey Nonresponse, Ottawa, Canada.

Kovar, J. G. and Whitridge, P. (1990), "Generalized Edit and Imputation System: Overview and Applications," *Revista Brasileira Estatistica*, **51**, pp. 85–100.

Kovar, M. G. (1990), "Computer Assisted Personal Interviewing: Lessons from Experience," *Proceedings of the Section on Survey Research Methods, American Statistical Association*, pp. 378–381.

Kuijlen, A. A. A. (1993), *De Scenariobenadering. Een Onderzoek naar Complexe Consumentenbeslissingen met Behulp van Computergestuurd Enqueteren*, Amsterdam: Postbank.

Kulka, R. A. and Weeks, M. F. (1988), "Toward the Development of Optimal Calling Protocols for Telephone Surveys: A Conditional Probabilities Approach," *Journal of Official Statistics*, **4**, pp. 319–332.

Kuklinski, J. H., Sniderman, P. M., Knight, K., Piazza, T., Tetlock, P. E., Lawrence, G. R., and Mellers, B. (1997), "Racial Prejudice and Attitudes toward Affirmative Action," *American Journal of Political Science*, **41**, pp. 402–419.

Kuusela, V. (1995), "Interviewer Interface of the CAPI-System of Statistics Finland," *Essays on Blaise, Third International Blaise User's Conference, Statistics Finland*, pp. 89–100.

Laflamme, F., Barrett, C., Johnson, W., and Ramsay, L. (1996), "Experiences in Re-engineering the Approach to Editing and Imputing Canadian Imports Data," *Proceedings of the Bureau of the Census' Annual Research Conference and Technology Interchange*. Washington, DC: U.S. Bureau of the Census, pp. 1025–1037.

LaLomia, M. J. (1994), "User Acceptance of Handwritten Recognition Accuracy," *Companion Proceedings of the CHI'94 Conference on Human Factors in Computing Systems,* New York: ACM, p. 107.

Landauer, T. K. (1995), *The Trouble with Computers: Usefulness, Usability and Productivity*. Cambridge: MIT Press.

Latouche, M. and Berthelot, J.-M. (1992), "Use of a Score Function to Prioritize and Limit Recontacts in Business Surveys," *Journal of Official Statistics*, **8**, pp. 389–400.

Laumann, E., Gagnon, J., Michael, R., and Michaels, S. (1994), *Social Organization of Sexuality*, Chicago: University of Chicago Press.

Laurie, H. and Moon, N. (1996), "Converting to CAPI in a Longitudinal Panel Survey," paper presented at the International Conference on Computer Assisted Survey Information Collection, San Antonio, TX, December.

Lawrence, P. N. (1992), "The Census Dataset Demonstration Reports," unpublished report, Austin, TX: Lawrence Technologies.

Lebow, I. (1995), *Information Highways and Byways: From the Telegraph to the 21st Century*. New York: IEEE Press.

Legum, S. E. (1996), "A Computer Assisted Coding and Editing System for Nonnumeric Educational Transcript Data," *Proceedings of the Data Editing Workshop and Exposition*. Statistical Policy Working Paper 25, Subcommittee, Federal Committee on Statistical Methodology, Statistical Policy Office, Office of Management and Budget.

Lemaître, G. (1992), "Dealing with the Seam Problem for the Survey of Labour and Income Dynamics," SLID Research Paper No. 92-05, Statistics Canada.

Lenat, D. B. (1984), "Computer Software for Intelligent Systems," *Scientific American*, **251**, pp. 204–212.

Lepkowski, J. M., Sadosky, S. A., Couper, M. P., Carn, L., Chardoul, S., and Scott, L. J. (1995), "Exploring Mode Differences in Interviewer Entry Errors," *Proceedings of the Section on Survey Research Methods, American Statistical Association*, pp. 521–526.

Lessler, J. T. and Kalsbeek, W. D. (1992), *Nonsampling Errors in Surveys.* New York: Wiley.

Lessler, J. T. and O'Reilly, J. M. (1995), "Literacy Limitations and Solution for Self-administered Questionnaires to Enhance Privacy," *Seminar on New Directions in Statistical Methodology*, Statistical Policy Working Paper 23, U.S. Office of Management and Budget, Part 2, pp. 453–469.

Levi, M. D. and Conrad, F. G. (1996), "A Heuristic Evaluation of a World Wide Web Prototype," *Interactions*, **3**, July/August, pp. 50–61.

Liepins, G. E., Garfinkel, R. S., and Kunnathur, A. S. (1982), "Error Localization for Erroneous Data: A Survey," *TIMS/Studies in the Management Sciences*, **19**, pp. 205–219.

Lindell, K. (1994), "Evaluation of the Editing Process of the Salary Statistics for Employees in Country Councils," paper presented at the United Nations Congress on Data Editing, Cork, Ireland.

Linton, M. (1975), "Memory for Real-World Events," in D. A. Norman and D. E. Rumelhart (eds.), *Explorations in Cognition*, San Francisco: Freeman.

Little, R. J. A. and Smith, P. J. (1987), "Editing and Imputation for Quantitative Survey Data," *Journal of the American Statistical Association*, **82**, pp. 58–68.

Loftus, E. F. and Marburger, W. (1983), "Since the Eruption of Mt. St. Helens, Has Anyone Beaten You Up? Improving the Accuracy of Retrospective Reports with Landmark Events," *Memory and Cognition*, **11**, pp. 14–120.

London, K. and Williams, L. (1990), "A Comparison of Abortion Underreporting in an In-Person Interview and Self-administered Questonnaire," paper presented at the Annual Meeting of the Population Association of America, Toronto, Canada, May.

Lorigny, J. (1988), "QUID, A General Automatic Coding Method," *Survey Methodology*, **14**, pp. 289–298.

Lucas, R. W., Mullin, P. J., Luna, C. B., and McInray, D. C. (1977), "Psychiatrists and a Computer as Interrogators of Patients with Alcohol-Related Illnesses: A Comparison," *British Journal of Psychiatry*, **131**, pp. 160–167.

Luce, R. D. and Tukey, J. W. (1964), "Simultaneous Conjoint Measurement: A New Type of Fundamental Measurement," *Journal of Mathematical Psychology*, **1**, pp. 1–27.

Lyberg, L. E. (1985), "Plans for Computer-Assisted Data Collection at Statistics Sweden," *Proceedings of the 45th Session, International Statistical Institute*, book III, topic 18.2, pp. 1–11.

Lyberg, L. E. and Dean, P. (1990), "International Review of Approaches to Automated Coding," paper presented at the Conference on Advanced Computing for the Social Sciences, Williamsburg, VA.

Lyberg, L. E. and Kasprzyk, D. (1991), "Data Collection Methods and Measurement Error: An Overview," in P. P. Biemer, R. M. Groves, L. E. Lyberg, N. A. Mathiowetz, and S. Sudman (eds.) *Measurement Errors in Surveys.* New York: Wiley, pp. 237–257.

Lyberg, L. E., Biemer, P. P., Collins, M., De Leeuw, E., Dippo, C., Schwarz, N., and Trewin, D. (eds.) (1997), *Survey Measurement and Process Quality*, New York: Wiley.

Maartens, M. (1995), *Praktisch Bekeken: Het Casip-Project Geëvalueerd*. Amsterdam: NIMMO.

MacBride, J. N. and Johnson, R. M. (1980), "Respondent Reaction to Computer-Interactive Interviewing Techniques," paper presented at the 1990 ESOMAR Conference, Montecarlo.

MacElroy, W. (1996), "Comparative Results between Computer-Aided Data Collection Methods: Internet Web Survey versus Disk-by-Mail," paper presented at the International Conference on Computer-Assisted Survey Information Collection, San Antonio, TX, December.

Machrone, B. (1992), "User Groups Go High(er) Tech," *PC Magazine*, June 30, vol. 11, p. 87.

MacNeill, D. (1995), "Outstanding in Their Field—The Benefits of Pen," *Pen Computing Magazine*, **6**, August/September, pp. 28–31.

MacNeill, D., and Giangrasso, D. (1996), "Pen Solutions in Transportation," *Pen Computing Magazine*, **8**, May/June, pp. 18–25.

Mahajan, V. and Peterson, R. A. (1985), *Models for Innovation and Diffusion*, Beverly Hills: Sage.

Manners, T. (1992), "New Developments in Computer Assisted Survey Methodology (CASM) for the British Labour Force Survey and Other OPCS Surveys," *Proceedings of the Bureau of the Census Annual Research Conference*, Washington, DC: U.S. Bureau of the Census, pp. 491–500.

Manners, T. and Deacon, K. (1997), "An Integrated Household Survey in the UK: The Role of Blaise III in an Experimental Development by the Office for National Statistics," in *Actes de la 4 Conference Internationale des Utilisateurs de Blaise 1997*, Paris: INSEE, pp. 197–214.

Marick, B. (1995), *The Craft of Software Testing: Subsystem Testing Including Object-Based and Object-Oriented Testing*, Englewood Cliffs, NJ: Prentice Hall.

Martin, J. (1996), *Cybercorp: The New Business Revolution*, New York: American Management Association.

Martin, J. (1993), "PAPI to CAPI: The OPCS Experience," *Essays on Blaise 1993*, *Proceedings of the Second International Blaise User's Conference*, pp. 96–117.

Martin, J. (1995), "The CAPI Revolution: Experiences at OPCS," paper presented at the International Conference on New Techniques and Technologies for Statistics, Bonn.

Martin, J. and Manners, T. (1995), "Computer Assisted Personal Interviewing in Survey Research," in R. M. Lee (ed.), *Information Technology for the Social Scientist*, London: UCL Press, pp. 51–72.

Martin, J., Matchett, S., Baker, R., and Jamieson, R. (1994), "Organization Impacts of Moving to CAPI," panel discussion at the Field Directors and Technologies Conference, Boston.

Martin, J., O'Muircheartaigh, C., and Curtice, J. (1993), "The Use of CAPI for Attitude Surveys: An Experimental Comparison with Traditional Methods," *Journal of Official Statistics*, **9**, 3, pp. 641–661.

Martini, A. (1989), "Seam Effect, Recall Bias, and the Estimation of Labour Force Transition Rates from SIPP," *Proceedings of the Section on Survey Research Methods, American Statistical Association*, pp. 387–392.

Mathiowetz, N. A. and Cannell, C. F. (1980), "Coding Interviewer Behavior as a Method of Evaluating Performance," *Proceedings of the Section on Survey Research Methods, American Statistical Association*, pp. 525–528.

May, R., Anderson, R., and Blower, S. (1989), "The Epidemiology and Transmission Dynamics of HIV-AIDS," *Daedalus*, **118**, pp. 163–201.

McCarthy, V. (1995), "The Web: Open for Business," *Datamation*, December 1.

McGarr, M. S. (1995), "At DISA's EC/EDI '95 the Message was Clear," *EDI World*, June, p. 48.

McKay, R. B. and Robinson, E. L. (1994), "Touchtone Data Entry for CPS Sample Expansion," *Proceedings of the Section on Survey Research Methods*, American Statistical Association, pp. 509–511.

McKenna, J. F. (1993), "RISC-y Business: Technophobia," *Industry Week*, December 20.

McNulty, F. L. (1994), "Security on the Internet," National Institute of Standards and Technology, U.S. Department of Commerce statement presented before the Subcommittee on Science, Committee on Science, Space, and Technology, U.S. House of Representatives, Washington, DC, on March 22, 1994.

McQueen, C., MacKenzie, I. S., Nonnecke, B., Riddersma, S., and Meltz, M. (1994), "A Comparison of Four Methods of Numeric Entry on Pen-Based Computers," *Proceedings of Graphics Interface '94*, Banff, Alberta. Toronto: Canadian Information Processing Society, pp. 75–82.

McQueen, C., MacKenzie, I. S. and Zhang, S. X. (1995), "An Extended Study of Numeric Entry on Pen-Based Computers," *Proceedings of Graphics Interface '95*, Toronto: Canadian Information Processing Society, pp. 215–222.

Meeker, M. and DePuy, C. (1996), *The Internet Report*, New York: Harper Business.

Meier, G. (1997), "A New Generation of CAPI with New Input Techniques and Multimedia Features," In *Proceedings of the ESOMAR Seminar on Technology*, January.

Mersch, M., Gbur, P., and Russell, C. (1993), "Statistical Process Control In Decennial Census Industry and Occupation Coding," *Proceedings of the Section on Survey Research Methods, American Statistical Association*, pp. 58–65.

Messinger, N. L. (1989), "Maintaining Quality in Large Computer Interactive Interviewing Projects," *Proceedings of the Sawtooth Software Conference, Gaining a Competitive Advantage through PC-Based Interviewing and Analysis*, vol. 1, Sun Valley, ID: Sawtooth Software, pp. 19–24.

Meurs, H., van Wissen, L., and Visser, J. (1989), "Measurement Biases In Panel Data," *Transportation Research*, **16**, pp. 175–194.

Meyer, A. (1995), "Pen Computing: A Technology Overview and a Vision," *ACM SIGCHI Bulletin*, **27**, (3), pp. 46–90.

Meyers, G. J. (1979), *The Art of Software Testing*, New York: Wiley.

Miller, H. G., Turner, C. F., and Moses, L. E. (eds.) (1990), *AIDS: The Second Decade*, Washington, DC: National Academy Press.

Miller, H. G., Gribble, J. N., Rogers, S. M., and Turner, C. F. (in press), "Abortion and

Breast Cancer Risk: Fact or Artifact?" in A. Stone (ed.), *Science of Self Report*, Mahwah, NJ: Lawrence Erlbaum Associates.

Miller, K. J. (1996), "The Influence of Different Techniques on Response Rates and Nonresponse Error in Mail Surveys," unpublished masters thesis, Western Washington University, Bellingham, WA.

Miln, D. and Stewart-Hunter, D. (1976), "The Case for Telephone Research," in *Proceedings of the Annual Conference of the Market Research Society*, pp. 41–54.

Mitchell, W. J. (1995), *City of Bits: Space, Place, and the Infobahn*, Cambridge: MIT Press.

Morrison, R. (1988), "Disks-by-Mail," paper presented at the Sawtooth Software Conference on Perceptual Mapping, Conjoint Analysis, and Computer Interviewing, Lancaster, PA, May.

Mosher, W. D. and Duffer, A. P. (1994), "Experiments in Survey Data Collection: The National Survey of Family Growth Pretest," paper presented at the annual meeting of the Population Association of America, Miami, May.

Mosher, W. D. and Duffer, A. P. (1995), "Innovations in the 1995 National Survey of Family Growth," paper presented at the annual meeting of the Population Association of America, San Francisco.

Mott, F. (1985), *Evaluation of Fertility Data and Preliminary Analytic Results from the 1983 Survey of the National Longitudinal Surveys of Work Experience of Youth*, a report to the National Institute of Child Health and Human Development by the Center for Human Resources Research, January.

Mudryk, W., Burgess, J., and Xiao, P. (1996), "A Quality Control Approach to CATI Operations In Statistics Canada," paper presented at the International Conference on Computer Assisted Survey Information Collection, San Antonio, TX, December.

Muller, N. (1996), *Network Planning, Procurement, and Management*, New York: McGraw-Hill.

Murray, D., O'Connell, C., Schmid, L., and Perry, C. (1987), "The Validity of Smoking Self-reports by Adolescents: A Reexamination of the Bogus Pipeline Procedure," *Addictive Behaviors*, **12**, pp. 7–15.

Nabseth, L. and Ray, G. F. (1974), *The Diffusion of New Industrial Processes: An International Study*. London: Cambridge University Press.

Nathan, G. and Givol, I. (1996), "The ODE (Optical Data Entry) Experience in Israel," paper presented at the International Conference on Computer Assisted Survey Information Collection, San Antonio, TX, December.

National Agricultural Statistics Service (1992), *Criteria for the Evaluation of Interactive Survey Software*, Report of the Interactive Survey Software Committee, Washington, DC: U.S. Department of Agriculture.

National Center for Health Statistics and Bureau of the Census (1988), *Report of the 1987 Automated National Health Interview Survey Feasibility Study*, Working Paper no. 32, Hyattsville, MD: National Center for Health Statistics.

Negroponte, N. (1995), *Being Digital*, New York: Alfred A. Knopf.

Nelson, R. O., Peyton, B. L., and Bortner, B. Z. (1972), "Use of an On-line Interactive System: Its Effects on Speed, Accuracy, and Cost of Survey Results," paper presented at the 18th ARF Conference, New York City, November.

Nicholls, W. L. II (1978), "Experiences with CATI in a Large-Scale Survey," *Proceedings of the Section on Survey Research Methods*, American Statistical Association, pp. 9–17.

Nicholls, W. L. II (1983), "CATI Research and Development at the U.S. Census Bureau," *Sociological Methods & Research*, **12**(2), pp. 191–197.

Nicholls, W. L. II (1988), "Computer Assisted Telephone Interviewing: A General Introduction," In R. M. Groves, P. P. Biemer, L. E. Lyberg, J. T. Massey, W. L. Nicholls II, and J. Wasksberg (eds.), *Telephone Survey Methodology*, New York: Wiley.

Nicholls, W. L. II (1997), "The Meanings of Data Quality In Assessments of New Data Collection Technologies," paper presented at the International Statistical Institute, Istanbul, Turkey, August.

Nicholls, W. L. II, and Appel, M. V. (1994), "New CASIC Technologies at the U.S. Bureau of the Census," *Proceedings of the Section of Survey Research Methods*, American Statistical Association, pp. 757–762.

Nicholls, W. L. II, Baker, R. P., and Martin, J. (1997), "The Effect of New Data Collection Technologies on Survey Data Quality," in L. Lyberg, P. Biemer, M. Collins, E. DeLeeuw, C. Dippo, N. Schwarz, and D. Trewin (eds.), *Survey Measurement and Process Quality*, New York: Wiley, pp. 221–248.

Nicholls, W. L. II, and DeLeeuw, E. (1996), "Factors In Acceptance of Computer-Assisted Interviewing Methods: A Conceptual and Historical Review," *Proceedings of the Section of Survey Research Methods*, American Statistical Association, pp. 758–763.

Nicholls, W. L. II, and Groves, R. M. (1986), "The Status of Computer-Assisted Telephone Interviewing: Part I—Introduction and Impact on Cost and Timeliness of Survey Data," *Journal of Official Statistics*, **2**, pp. 93–115.

Nicholls, W. L. II, and House, C. C. (1987), "Designing Questionnaires for Computer Assisted Interviewing: A Focus on Program Correctness," *Proceeding of the Third Annual Research Conference*, Washington, DC: U.S. Bureau of the Census, pp. 95–111.

Nicholls, W. L. II, and Kindel, K. K. (1993), "Case Management and Communications for Computer Assisted Personal Interviewing," *Journal of Official Statistics*, **9**, pp. 623–639.

Nicholls, W. L. II, and Matchett, S. D. (1992), "CAI Issues at the Census Bureau as Seen by Members of Outside Panels," *Proceedings of the Bureau of the Census Annual Research Conference*, Washington, DC: U.S. Bureau of the Census, pp. 371–394.

Nicholls, W. L. II, West, B., Williams, B., and Baker, R. (1994), "Recovery from CAPI Failures," panel discussion at the Field Directors and Technologies Conference, Boston.

Nielsen, J. (1994), "Heuristic Evaluation," in J. Nielsen, and R. Mack (eds.), *Usability Inspection Methods*, New York: Wiley.

Nielsen, J. (1996), "Usability Testing of WWW Designs," http://www.sun.com/sun-on-net/udesign/usabilitytest.html.

Nielsen, J. and Levy, J. (1994), "Measuring Usability; Preference vs. Performance," *Communications of the ACM*, 37, pp. 66–75.

Nielsen, J. and Mack, R. (eds.) (1994), *Usability Inspection Methods*, New York: Wiley.

Norman, D. A. (1988), *The Psychology of Everyday Things*, New York: Basic Books.

Norman, K. L. (1991), *The Psychology of Menu Selection: Designing Cognitive Control of the Human/Computer Interface.* Norwood, NJ: Ablex Publishing.

Nusser, S. M., Thompson, D. M., and DeLozier, G. S. (1996a), "Using Personal Digital Assistants to Collect Survey Data," paper presented at the Annual Conference of the American Statistical Association, Chicago, May.

Nusser, S. M., Thompson, D. M., and Delozier, G. S. (1996b), "Conducting Surveys with Personal Digital Assistants," paper presented at the International Conference on Computer Assisted Survey Information Collection, San Antonio, TX, December.

O'Brien, T. and Dugdale, V. (1978), "Questionnaire Administration by Computer," *Journal of the Market Research Society*, **20**, pp. 228–237.

Ogden Government Services (1993), "U.S. Bureau of the Census Technology Assessment of Data Collection Technologies for the Year 2000: Final Technology Assessment Report," a report prepared for the U.S. Bureau of the Census Year 2000 Staff.

Ohly, H. P. (1993), "Knowledge-Based Systems: Another Data Approach for Social Scientists?" *Social Science Computer Review*, **11**, (1), pp. 84–94.

Olivier, A. J. (1987), *Het Samenstellen en Beheer van Gegevens als Onderdeel van een Beslissing-Ondersteunend System ten Behoeve van het Marketing Management: Een Case Study*, Ph.D. Dissertation, Groningen: Universiteit van Groningen.

Olsen, R. J. (1991), "Mode Effects on Data Quality—CAPI versus Pencil and Paper," unpublished paper, Ohio State University.

Olsen, R. J. (1992), "The Effects of Computer-Assisted Interviewing on Data Quality," in *Working Papers of the European Scientific Network on Household Panel Studies*, Paper 36, Colchester: University of Essex.

Olson, L. and Schneiderman, M. (1995), "Physicians' Participation in a Disk-by-Mail Survey," paper presented at the American Association for Public Opinion Research Annual Conference, Fort Lauderdale, FL, May.

O'Muircheartaigh, C. and Murphy, M. (1991), *Evaluation of Computer Assisted Survey Systems. Report 2: Overview of Computer Assisted Interviewing Software.* Working Paper Series 5, London: Joint Centre for Survey Methods.

Oppenhuisen, J. (1994), *Panels Ins en Outs*, Amsterdam: NIMMO.

Oppermann, M. (1995), "E-mail Surveys—Potentials and Pitfalls," *Marketing Research*, 7, 3, pp. 29–33.

O'Reagan, R. T. (1972), "Computer-Assigned Codes from Verbal Responses," *Communications of the ACM*, **15**, pp. 455–459.

O'Reilly, J. M., Weeks, M., Deloach, D., Dewitt, D. S., and Batts, J. (1989), "Management of a Field Survey Using Laptop Computers," *Proceedings of the Fifth Annual Research Conference*, U.S. Bureau of the Census, pp. 357–365.

O'Reilly, J. M., Hubbard, M., Lessler, J., Biemer, P. P., and Turner, C. F. (1994), "Audio Computer Assisted Self-Interviewing: New Technology for Data Collection on Sensitive Issues and Special Populations," *Journal of Official Statistics*, **10**, pp. 197–214.

Orfali, R., Harkey, D., and Edwards, J. (1994), *Essential Client/Server Survival Guide*, New York: Wiley.

O'Rourke, D., Sudman, S., and Ryan, M. (1996), "The Growth of Academic and Not-for-Profit Survey Research Organizations," *Survey Research*, **27**(1–2), pp. 1–5.

Oviatt, S. L. and Olsen, E. (1994), "Integration Themes in Multimodal Human-Computer Interaction," *Proceedings of the 1994 International Conference on Spoken Language Processing, Acoustical Society of Japan*, **2**, pp. 551–554.

Padgett, T. (1996), "Fortune 500s Seek Multimedia Training," *Black Enterprise*, May, pp. 34–36.

Palit, C. and Sharp, H. (1983), "Microcomputer-Assisted Telephone Interviewing," *Sociological Methods and Research*, **12**, pp. 169–189.

Paxson, M. C. (1992), "Unpublished Data: Response Rates for 183 Studies," Pullman, WA: Department of Hotel and Restaurant Administration, Washington State University.

Pellerud, T. S. (1994), "Optical Character Recognition and Document Image Processing, Statistics Norway," paper presented at the 8th International Roundtable on Business Survey Frames, Heerlen, the Netherlands, May.

Perloff, B., Anand, J., Ingwersen, L., and LaComb, R. (1996), "USDA's Experience with Computer Assisted Food Coding in its 1994 Nationwide Food Survey," *Proceedings of the Bureau of the Census' Annual Research Conference and Technology Interchange*, pp. 1127–1133.

Perron, S., Berthelot, J. M., and Blakeney, R. D. (1991), "New Technologies in Data Collection for Business Surveys," *Proceedings of the American Statistical Association, Survey Research Methods Section*, pp. 707–712.

Phipps, P. A. and Tupek, A. R. (1991), "Assessing Measurement Errors in a Touchtone Recognition Survey," *Survey Methodology*, **17**, (1), pp. 15–26.

Piazza, T., Sniderman, P. M., and Tetlock, P. E. (1989), "Analysis of the Dynamics of Political Reasoning: A General-Purpose Computer-Assisted Methodology," in J. A. Stimson (ed.), *Political Analysis*, vol. 1, Ann Arbor: University of Michigan Press.

Pierzchala, M. (1990), "A Review of the State of the Art in Automated Data Editing and Imputation," *Journal of Official Statistics*, **6**, pp. 355–377.

Pierzchala, M. (1992), "Generating Multiple Versions of Questionnaires," in *Essays on Blaise. Proceedings of the First International Blaise Users Meeting*, Voorburg, The Netherlands: CBS, pp. 131–145.

Pierzchala, M. (1993), "Computer Generation of Mega-Version Instruments for Data Collection and Interactive Editing of Survey Data," *Proceedings of the Bureau of the Census Annual Research Conference*, Washington, DC: U.S. Bureau of the Census, pp. 362–374.

Pierzchala, M. (1994), "Optimal Edit Writing," in *International BLAISE User Group Newsletter*, **3**, pp. 10–13.

Pierzchala, M. (1995), "The 1995 June Area Frame Experience," in *Essays on Blaise 1995. Proceedings of the Third International Blaise Users Conference*, Helsinki: Statistics Finland, pp. 143–148.

Pilon, T. L. and Craig, N. C. (1988), "Disks-by-Mail: A New Survey Modality," *Proceedings of the 1988 Sawtooth Software Conference on Perpetual Mapping, Conjoint Analysis and Computer Interviewing*, Sun Valley, ID: Sawtooth Software.

Pitkow, J. and Recker, M. (1994), "Results from the First World-Wide Web User Survey," *Computer Networks and ISDN Systems*, **27**(2), pp. 243–254.

Platek, R. (1985), "Some Important Issues in Questionnaire Development," *Journal of Official Statistics*, **1**, (2), pp. 119–136.

Polivka, A. E. and Rothgeb, J. M. (1993), "Overhauling the Current Population Survey: Redesigning the Questionnaire," *Monthly Labor Review*, **116**(9), pp. 10–28.

Popper, K. (1957), *The Poverty of Historicism*, London: Routledge and Keagan Paul, Preface, pp. v–vi.

Porst, R., Schneid, M., and van Brouwershaven, J. W. (1994), "Computer Assisted Interviewing in Social and Market Research," in I. Borg, and P. Mohler (eds.), *Trends and Perspectives in Empirical Social Research*, Berlin: Gruyter.

Porter, M. E. (1985), *Competitive Advantage*, New York: Free Press.

Potter, D. E. B. (in press), *Design and Methods for the 1996 Medical Expenditure Panel Survey, Nursing Home Component*, Rockville, MD: Agency for Health Care Policy and Research, U.S. Public Health Service.

Presser, S. and Blair, J. (1994), "Survey Pretesting: Do Different Methods Produce Different Results?" in P. V. Marsden, (ed.), *Sociological Methodology*, vol. 24, Washington, DC: American Sociological Association.

Quarterman, J. S. (1994), "Preliminary Results of the Second TIC/MIDS Internet Demographic Survey," *Matrix News*, **4**, 12, December ⟨http//www.tic.com⟩.

Quirk's Marketing Research Review (1987), "Conjoint Analysis Enhances Computer-Based Interviews," *Quirk's Marketing Research Review*, March, 1987, 1(4), pp. 22–23.

Ramos, M. and Sweet, E. (1995), "Results from 1993 Company Organization Survey (COS) Computerized Self-Administered Questionnaire (CSAQ) Pilot Test," Economic Statistical Methods Report Series ESM-9501, U.S. Department of Commerce, Bureau of the Census.

Rands, R. (1996), "CfMC Voice Technology," paper presented at the International Conference on Computer-Assisted Survey Information Collection, San Antonio, TX, December.

Rasmussen, J., Duncan, K., and Leplat, J. (eds.) (1987), *New Technology and Human Error*, New York: Wiley.

Rauscher, T. G. and Ott, L. M. (1987), *Software Development and Management for Microprocessor-Based Systems*, Englewood Cliffs, NJ: Prentice Hall.

Rawlins, G. J. E. (1996), *Slaves of the Machine: The Quickening of Computer Technology*, Cambridge: MIT Press.

Rawlins, G. J. E. (1997), *Moths to the Flame: The Seductions of Computer Technology*, Cambridge: MIT Press.

Reason, J. (1990), *Human Error*, Cambridge: Cambridge University Press.

Reinhardt, A. (1995), "New Ways to Learn," *Byte*, March, pp. 50–71.

Robinson, R. and West, R. (1992), "A Comparison of Computer and Questionnaire Methods of History-Taking in a Genito-Urinary Clinic," *Psychology and Health*, **6**, pp. 77–84.

Rogers, E. M. (1995), *Diffusion of Innovations*, 4th ed., New York: Free Press.

Rogers, E. M. and Shoemaker, F. F. (1971), *Communication of Innovations: A Cross Cultural Approach*, London: Collier Macmillan.

Rogers, S. M. and Turner, C. (1991), "Patterns of Same-Gender Sexual Contact among Men in the U.S.A.: 1970–1990," *Journal of Sex Research*, **28**, pp. 491–519.

Rogers, S. M., Miller, H. G., Forsyth, B. H., Smith, T. K., and Turner, C. F. (1996), "Audio-CASI: The Impact of Operational Characteristics on Data Quality," *Proceedings of the American Association for Public Opinion Research/American Statistical Association, Survey Methods Research Section.*

Rogers, T. F. (1976), "Interviews by Telephone and in Person," *Public Opinion Quarterly,* **40**, pp. 51–65.

Rokeach, M. (1973), *The Nature of Human Values,* New York: Free Press.

Roshwald, I. (1984), "CATI and the Dynamics of Research," ESOMAR Seminar "Are Interviewers Obsolete," *Proceedings of the ESOMAR Congress,* Sophia Antipons, France, November 7–10.

Rossi, P. H. and Nock, S. L. (eds.) (1982), *Measuring Social Judgments: The Factorial Survey Approach,* Beverly Hills: Sage.

Rowe, E. and Appel, M. A. (1994), "Image Processing of Facsimile Data Reporting. Initial Technical Assessment," U.S. Bureau of the Census Statistical Report, Series No 2.

Rowland, J. H. and Kinack, M. D. (1994), "A Comparison of Two Methods of Automated Industry Coding," *Proceedings of the Survey Research Methods Section, American Statistical Association,* pp. 1130–1133.

Rothgeb, J. M., Polivka, A. E., Creighton, K. P., and Cohany, S. R. (1991), "Development of the Proposed Revised Current Population Survey Questionnaire," *Proceedings of the Survey Research Methods Sections, American Statistical Association,* pp. 649–654.

Rothschild, B. B. and Wilson, L. B. (1988), "Nationwide Food Consumption Survey 1987: A Landmark Person Interview Surveys Using Laptop Computers," *Proceedings of the Fourth Annual Research Conference,* U.S. Bureau of the Census, pp. 341–356.

Rowe, E. and Appel, M. V. (1993), "Image Processing of Facsimile Data Reporting: Initial Technical Assessment," a report to the CASIC Committee on Technology Testing, U.S. Bureau of the Census.

Rudolph, B. A. and Greenberg, A. G. (1994), *Surveying of Public Opinion: The Changing Shape of an Industry,* Chicago: National Opinion Research Center: Report to the Office of Technology Assessment.

Rustemeyer, A. (1977), "Measuring Interviewer Performance in Mock Interviews," *Proceedings of the American Statistical Association, Social Statistics Section,* pp. 341–346.

Salant, P. and Dillman, D. A. (1994), *How To Conduct Your Own Survey,* New York: Wiley.

Saltzman, A. (1992), "Improving Response Rates in Disk-by-Mail Surveys," *1992 Sawtooth Software Conference Proceedings,* Sun Valley, ID: Sawtooth Software, pp. 27–38.

Samuels, J. (1994), "From CAPI to HAPPI: A Scenario for the Future and Its Implications for Research," in *Proceedings of the Annual Conference of ESOMAR,* pp. 327–352.

Saris, W. E. (1988), "A Full Automatic Procedure for Data Collection," in F. Faulbaum and H. M. Uehlinger (eds.), *Fortschritte der Statistik-Software 1,* Stuttgart: Fischer Verlag, pp. 93–105.

Saris, W. E. (1989), "A Technological Revolution in Data Collection," *Quality and Quantity*, **23**, pp. 333–349.

Saris, W. E. (1991), *Computer Assisted Interviewing*, Newbury Park, CA: Sage

Saris, W. E. (1995), "Telepanel," paper presented at the Conference on Longitudinal Econometrics, Rotterdam, February.

Saris, W. E. and Andrews, F. M. (1991), "Evaluation of Measurement Instruments Using a Structural Modeling Approach," in P. P. Biemer, R. M. Groves, L. E. Lyberg, N. Mathiowetz, and S. Sudman (eds.), *Measurement Errors In Surveys*, New York: Wiley, pp. 575–599.

Saris, W. E. and de Pijper, M. (1986), "Computer Assisted Interviewing Using Home Computers", *European Research*, **14**, pp. 144–152.

Saris, W. E. and van Meurs, A. (1990), *Evaluation of Measurement Instruments by Meta-Analysis of Multitrait-Multimethods Studies*, Amsterdam: North Holland.

Saris, W. E., Prastacos, P., and Marti-Recober, M. (1995), *CASIP: A Complete Automated System for Information Processing*, Amsterdam: NIMMO.

Sawtooth Software (1989), "Disks by Mail," *Sawtooth News*, **5**, (2), pp. 4–5.

Schaadt, P. (1992), "Character Recognition in Document Image Applications," IMC Document Imaging 92, International Conference Center Berlin, September.

Schaeffer, N. C. (1991), "Conversation with a Purpose—or Conversation? Interaction in the Standardized Interview," in P. P. Biemer, R. M. Groves, L. E. Lyberg, N. A. Mathiowetz, and S. Sudman (eds.), *Measurement Errors In Surveys*, New York: Wiley.

Scherpenzeel, A. C. (1995a), *A Question of Quality: Evaluating Survey Questions in Multi-Trait-Multimethod Studies*, Leidschendam, Royal PTT, The Netherlands.

Scherpenzeel, A. C. (1995b), "Meta-Analysis of a European Comparative Study," in W. E. Saris and A. Münnich (eds.), *The Multitrait-Multimethod Approach to Evaluate Measurement Instruments*, Budapest: Eötvös University Press, pp. 225–243.

Scherpenzeel, A. C. and Saris, W. E. (1995), "Effects of Data Collection Technique on the Quality of Survey Data. An Evaluation of Interviewer and Self-Administrated Computer Assisted Data Collection Techniques," in A. Scherpenzeel, *A Question of Quality: Evaluating Survey Questions by Multitrait-Multimethod Studies*, Amsterdam: University of Amsterdam.

Scherpenzeel, A. C. and Saris, W. E. (1997), "The Validity and Reliability of Survey Questions: A Meta-Analysis of MTMM Studies," *Sociological Methods and Research*, **25**, pp. 341–383.

Schiopu-Kratina, I. and Kovar, J. (1989), "Use of Chernikova's Algorithm in the Generalized Edit and Imputation System," Working Paper, BSMD-89-001E, Ottawa: Statistics Canada.

Schmidt, K. E. (1996), "A Pilot Study of Speech and Pen User Interface for Graphical Editing," Technical Report WUCS-96-17, Department of Computer Science, Washington University–St. Louis.

Schnaars, S. (1989), *Megamistakes: Forecasting and the Myth of Rapid Technological Change*, New York: Free Press.

Schneid, M. (1991), "Einsatz Computergestutzter Befragungssysteme in der Bundes-republik Deutschland," *ZUMA-Arbeitsbericht*, 91/20.

Schneid, M. (1995), "The Use of Computer Assisted Interviewing Systems in South America, Asia, Africa and Australia (A Fax Survey)," *ZUMA-Arbeitsbericht*, 95/03 (in German).

Schober, S., Caces, M. F., Pergamit, M., and Branden, L. (1992), "Effects of Mode of Administration on Reporting of Drug Use in the National Longitudinal Survey," in C. F. Turner, J. T. Lessler, and J. C. Gfroerer (eds.), *Survey Measurement of Drug Use: Methodological Studies*, Rockville, MD: National Institute on Drug Abuse, pp. 267–276.

Schou, R. and Pierzchala, M. (1993), "Standard Multi-Survey Shells in NASS," in *Essays on Blaise 1993. Proceedings of the Second International Blaise Users Conference,* London: OPCS, pp. 133–142.

Schuhl, P. (1996), "SICORE, The INSEE Automatic Coding System," *Proceedings of the Bureau of the Census' Annual Research Conference and Technology Interchange*, pp. 810–820.

Schuldt, B. A. and Totten, J. W. (1994), "Electronic Mail vs. Mail Survey Response Rates," *Marketing Research, A Magazine for Management and Applications*, **6**, pp. 36–139.

Schuman, H. and Bobo, L. (1988), "An Experimental Approach to Surveys of Racial Attitudes," in H. J. O'Gorman (ed.), *Surveying Social Life: Papers in Honor of Herbert H. Hyman*, Middletown, CT: Wesleyan University Press.

Schwarz, N. and Hippler, H. (1987), "What Response Scales May Tell Your Respondents: Informative Functions of Response Alternatives," in H. Hippler, N. Schwarz, and S. Sudman (eds.), *Social Information Processing and Survey Methodology*, New York: Springer-Verlag.

Schwarz, N., Hippler, H., Deutsch, B., and Strack, F. (1985), "Response Categories: Effects on Behavioral Reports and Comparative Judgments," *Public Opinion Quarterly*, **49**, pp. 388–395.

Schwarz, N. and Sudman, S. (eds.) (1996), *Answering Questions; Methodology for Determining Cognitive and Communicative Processes in Survey Research*, San Francisco: Jossey-Bass.

Scott, D. (1995), "Imaging in the 1996 New Zealand Population Census," Statistics New Zealand and XII Conference of Commonwealth Statisticians.

Sebestik, J., Zelon, H., DeWitt, D., O'Reilly, J. M., and McGowan, K. (1988), "Initial Experiences with CAPI," *Proceedings of the Bureau of the Census' Annual Research Conference*, Washington, DC: U.S. Bureau of the Census, pp. 357–365.

Sedivi, B. and Rowe, E. (1993), "Computerized Self-Administered Questionnaire, Mail or Modem, Initial Technical Assessment," prepared for the CASIC Committee on Technology Testing, U.S. Bureau of the Census.

Semmer, N. and Pfäfflin, M. (1978), *Interaktionstraining. Ein Handlungs-Theoretische Ansatz zum Training Sozialer Fertigkeiten*, Basel: Weinheim.

Shangraw, R. F., Jr. (1986), "Telephone Surveying with Computers: Administrative, Methodological and Research Issues," *Evaluation and Program Planning*, **19**, pp. 107–111.

Shanks, J. M. (1983), "The Current Status of Computer-Assisted Telephone Interviewing: Recent Progress and Future Prospects," *Sociological Methods & Research*, **12**, (2), pp. 119–142.

Shanks, J. M. (1989), "Information Technology and Survey Research: Where Do We Go From Here?" *Journal of Official Statistics*, **5**, pp. 3–21.

Shanks, J. M., Nicholls, W. L. II, and Freeman, H. E. (1981), "The California Disability Survey: Design and Execution of a Computer-Assisted Telephone Study," *Sociological Methods and Research*, **10**, (2), pp. 123–140.

Shanks, J. M. and Tortora, R. (1985), "Beyond CATI: Generalized and Distributed Systems for Computer-Assisted Surveys," *Proceedings of the Bureau of the Census First Annual Research Conference*, pp. 358–371.

Sheldon, T. (1994), *LAN TIMES Encyclopedia of Networking*, New York: McGraw-Hill.

Shepherd, J., Hill, D., Bristor, J., and Montalvan, P. (1996), "Converting an Ongoing Health Study to CAPI: Findings from the National Health and Nutrition Examination Survey III," In *Proceedings of the Health Survey Research Methods Conference*, pp. 159–164.

Shneiderman, B. (1992), *Designing the User Interface: Strategies for Effective Human-Computer Interaction*, 2nd ed., Reading, MA: Addison-Wesley.

Shneiderman, B. (1997), *Designing the User Interface: Strategies for Effective Human-Computer Interaction*, 3rd ed., Reading, MA: Addison-Wesley.

Sikkel, D. (1987), "Computer-Assisted Data Collection," *Mens en Maatschappij*, **62**, pp. 289–301 (in Dutch).

Sikkel, D. (1994), "Event History Analysis: Een Techniek voor het Analyseren van Tijdsduren," *Onderzoek*, **23**, pp. 22–25.

Silberstein, A. S. and Scott, S. (1991), "Expenditure Diary Surveys and their Associated Errors," In P. P. Biemer, R. M. Groves, L. E. Lyberg, N. Mathiowetz, and S. Sudman (eds.), *Measurement Errors in Surveys*, New York: Wiley, pp. 303–327.

Simpson, H. (1985), *Design of User-Friendly Programs for Small Computers*, New York: McGraw-Hill.

Simpson, J. B. Jr., Pratt, D. J., Burkheimer, G. J., and Bethke, A. D. (1995), "Dictionary Based Development and Automatic Software Generation for Computer Assisted Data Collection," *Proceedings of the Annual Research Conference*, Washington, DC: U.S. Bureau of the Census, pp. 575–592.

Smith, E. and Behringer, R. (1992), "Survey Non-Response and Bias as a Function of Paper, Disk and Phone Formats," *1992 Sawtooth Conference Proceedings*, Sun Valley, ID: Sawtooth Software, pp. 45–53.

Smith, E. and Squire, P. (1990), "The Effects of Prestige Names in Question Wording," *Public Opinion Quarterly*, **54**, pp. 97–116.

Smith, J. (1995), "From Products to Systems—Addressing the Needs of CAI Surveys at Westat," *Essays On Blaise, Third International Blaise User's Conference*, Statistics Finland, pp. 175–180.

Smith, J. E. and Bayless, D. L. (1994), "Improving the Quality of Survey Data: Westat's Computer Assisted Coding and Editing (CACE) System," *Proceedings of the SAS User's Group International Conference*, pp. 427–430.

Smith, R. and Smith, R. (1980), "Evaluation and Enhancements for Computer Controlled Telephone Interviewing," *Proceedings of the Section On Survey Research Methods*, American Statistical Association, pp. 513–515.

Smith, T. (1984), *A Comparison of Telephone and Personal Interviewing*, GSS Methodological Report no. 28, Chicago: NORC.

Smith, T. (1992), "Discrepancies between Men and Women in Reporting Number of Sexual Partners: A Summary from Four Countries," *Social Biology*, **39**, pp. 203–211.

Sniderman, P. M., Brody, R. A., and Tetlock, P. E. (1991), *Reasoning and Choice: Explorations in Political Psychology*, Cambridge: Cambridge University Press.

Sniderman, P. M. and Grob, D. B. (1996), "Innovations in Experimental Design in General Population Attitude Surveys," *Annual Review of Sociology*, **23**, pp. 377–399.

Sniderman, P. M. and Piazza, T. (1993), *The Scar of Race*, Cambridge: Belknap/Harvard University Press.

Sniderman, P. M., Piazza, T., Tetlock, P. E., and Kendrick, A. (1991), "The New Racism," *American Journal of Political Science*, **35**, pp. 423–447.

Somer, E. P. and Murphy, D. J. (1989), "Computer Interviewing Applications in the Navy," *1989 Sawtooth Software Conference Proceedings, Gaining a Competitive Advantage through PC-Based Interviewing and Analysis,* vol. 1, Sun Valley, ID: Sawtooth Software, pp. 91–95.

Sonenstein, F. L., Pleck, J. H., and Ku, L. C. (1991), "Levels of Sexual Activity among Adolescent Males in the United States," *Family Planning Perspectives*, **23**, pp. 162–167.

Spaeth, M. A. (1987), "CATI Facilities at Survey Research Organizations," *Survey Research*, Summer–Fall.

Spanski, R. and Wickham, L. (1995), *Connecting the Workplace: Electronic Commerce in Business and Government*, Study 944013, Louis Harris and Associates, Inc.

Speizer, H. and Dougherty, D. (1991), "Automating Data Transmission and Case Management Functions for a Nationwide CAPI Survey," *Proceedings of the Seventh Annual Research Conference*. U.S. Bureau of the Census, pp. 375–388.

Spencer, L., Faulkner, A., and Keegan, J. (1988), *Talking about Sex*, Pub. no. P.5997, London: Social and Community Planning Research.

Sperry, S., Bittner, D., and Branden, L. (1991), "Computer-Assisted Personal Interviewing on the Current Beneficiary Survey," paper presented at the annual conference of the American Association for Public Opinion Research, Phoenix, AZ, May.

Springett, P. (1994), "A Dent in the Bureaucrat's Armour," *Government Computing*, November, U.K.

Stallings, W. and Van Slyke, R. (1994), *Business Data Communications*, Englewood Cliffs, NJ: Prentice Hall.

Stanley, J. and Erth, S. G. (1995), "Development of a Field Monitoring Program for Telephone Interview and Instrument Quality Evaluation," *Proceedings of the International Conference On Survey Measurement and Process Quality*, Alexandria, VA: American Statistical Association, pp. 24–28.

Statistics Netherlands (1996), *Blaise Developers Guide*, Voorburg: Statistical Informatics Department.

Statistics Sweden (1989), *Computer Assisted Data Collection in the Labour Force Surveys, Report of Technical Tests*, Stockholm: Statistics Sweden.

Stoker, L. (1998), "Understanding Whites' Resistance to Affirmative Action: the Role of Principled Commitments and Racial Prejudice," in J. Hurwitz, and M. Peffley (eds.), *Perception and Prejudice: Race and Politics in the United States*, New Haven: Yale University Press.

Stokes, S. L. and Greenberg, B. S. (1990), "A Priority System to Improve Callback Success in Telephone Surveys," *Proceedings of the Survey Research Methods Section, American Statistical Association*, pp. 742–747.

Stoudt, N. (1995), "Standardized Occupation and Industry Coding Software," paper presented at the Annual Meeting of the American Public Health Association, San Diego, CA.

Stratton, A. K. and Hardy, A. M. (1996), "The Need for Questionnaire Specifications in a CAPI Environment," paper presented at the International Conference on Computer Assisted Survey Information Collection, San Antonio, TX, December.

Strube, G. (1987), "Answering Survey Questions: The Role of Memory," in H. J. Hippler, N. Schwartz, and S. Sudman (eds.), *Social Information Processing and Survey Methodology*, New York: Springer-Verlag.

Sudman, S. (1983), "Survey Research and Technological Change," *Sociological Methods and Research*, **12**(2), pp. 217–230.

Sudman, S. and Ferber, R. (1979), *Consumer Panels*, Chicago: American Marketing Association.

Sudman, S., Blair, E., Bradburn, N. M., and Stocking, C. (1977), "Estimates of Threatening Behavior Based on Reports of Friends," *Public Opinion Quarterly*, **41**, pp. 261–264.

Sudman, S., Bradburn, N. M., and Schwarz, N. (1996), *Thinking about Answers: The Applications of Cognitive Processes to Survey Methodology*, San Francisco: Jossey-Bass.

Sweet, E. and Ramos, M. (1995), "Evaluation Results from a Pilot Test of a Computerized Self-Administered Questionnaire (CSAQ) for the 1994 Industrial Research and Development (R&D) Survey," Economic Statistical Methods Report Series ESM-9503, U.S. Department of Commerce, Bureau of the Census.

Sweet, E. and Russell, C. (1996), "A Discussion of Data Collection via the Internet," *Proceedings of the Section on Survey Research Methods, American Statistical Association*, pp. 774–779.

Sykes, W. and Collins, M. (1987), "Comparing Telephone and Face-to-Face Interviewing in the U.K.," *Survey Methodology*, **13**(1), pp. 15–28.

Synodinos, N. E. and Brennan, J. M. (1988), "Computer Interactive Interviewing in Survey Research," *Psychology and Marketing*, **5**, pp. 117–137.

Synodinos, N. E., Papacostas, C. S., and Okimoto, G. M. (1994), "Computer-Administered versus Paper-and-Pencil Surveys and the Effect of Sample Selection," *Behavior Research Methods, Instruments, and Computers*, **26**(4), pp. 395–401.

Tanfer, K. (1993), "National Survey of Men: Design and Execution," *Family Planning Perspectives*, **25**, pp. 83–86.

Tapscott, D. (1996), *The Digital Economy: Promise and Peril in the Age of Networked Intelligence*, New York: McGraw-Hill.

Tapscott, D. and Caston, A. (1993), *Paradigm Shift: The New Promise of Information Technology*, New York: McGraw-Hill.

Taylor, F. W. (1911), *The Principles of Scientific Management*, New York: Norton.

Thomas, P. (1994), "Optical Character Recognition and Document Image Processing," U.K. Employment Department, 8th International Round Table on Business Survey Frames, Heerlen, The Netherlands.

Thornberry, O., Rowe, B., and Biggar, R. (1991), "Use of CAPI with the U.S. National Health Interview Survey," *Bulletin de Méthodologie Sociologique*, **30**, pp. 27–43.

Tinari, R. (1988), "Discussion of Werking, Tupek and Clayton, and Catlin, Ingram and Hunter Papers," *Proceedings of the Bureau of the Census Annual Research Conference*. Washington, DC: U.S. Bureau of the Census, pp. 300–304.

Tortora, R. D. (1985), "CATI in an Agricultural Statistical Agency," *Journal of Official Statistics*, **1**, pp. 301–314.

Tortora, R. and Faulkenberry, D. (1979), "A Study of Measurement Error Suitable for a Classroom Example," *The American Statistician*, **33**, 19–22.

Tourangeau, R., Jobe, J. B., Pratt, W. F., Smith, T., and Rasinski, K. (1997), "Design and Results of the Women's Health Study," in L. Harrison and A. Hughes (eds.), *The Validity of Self-reported Drug Use: Improving the Accuracy of Survey Estimates*, Rockville, MD: National Institute on Drug Abuse, pp. 344–365.

Tourangeau, R. and Smith, T. (1996), "Asking Sensitive Questions: The Impact of Data Collection Mode, Question Format, and Question Context," *Public Opinion Quarterly*, **60**, pp. 275–304.

Tourigny, J. Y. and Moloney, J. (1992), "The 1991 Canadian Census of Population Experience with Automated Coding," unpublished report, Ottawa, Ontario: Statistics Canada.

Tozer, C. and Jaensch, B. (1994), "CASIC Technologies Interchange, Use of OCR Technology in the Capture of Business Survey Data," in *Proceedings of the U.S. Bureau of the Census Annual Research Conference*, Arlington, VA.

Trumble, T. L., Cushing, B. D., Kindred, P., Filson, K. A., and Monoghan, C. J. (1995), *NSF EPSCoR Automated Collection System (ADCS) Users Guide*. Bethesda, MD: Quantum Research Corporation.

Tukey, J. W. (1977), *Exploratory Data Analysis*, Reading, MA: Addison-Wesley.

Turner, C. F. (1981), "Surveys of Subjective Phenomena: A Working Paper," in D. Johnston (ed.), *Measurement of Subjective Phenomena*, Washington, DC: Government Printing Office.

Turner, C. F. (1989), "Research On Sexual Behaviors That Transmit HIV: Progress and Problems." *AIDS*, **3**, pp. S63–S71.

Turner, C. F. (1991), "Voice-Administered CAPI: Memorandum to the Subcommittee on Measurement for the Multisite Trail of Behavioral Strategies to Prevent the Spread of HIV," unpublished memorandum, May 23.

Turner, C. F., Danella, R., and Rogers, S. M. (1995), "Sexual Behavior in the United States: 1930–1990: Trends and Methodological Problems," *Sexually Transmitted Diseases*, **22**, pp. 173–190.

Turner, C. F., Lessler, J. T., and Devore, J. (1992), "Effects of Mode of Administration and Wording on Reporting of Drug Use," in C. F. Turner, J. T. Lessler, and J. D. Gfroerer (eds.), *Survey Measurement of Drug Use: Methodological Issues*. DHHS Pub. no. 92-1929. Washington, DC: Government Printing Office.

Turner, C. F., Lessler, J. T., and Gfroerer, J. D. (1992a), "Improving Measurements of Drug Use: Future Directions for Research and Practice," in C. F. Turner, J. T.

Lessler, and J. D. Gfroerer (eds.), *Survey Measurement of Drug Use: Methodological Issues*, Washington, DC: Government Printing Office.

Turner, C. F., Lessler, J. T., and Gfroerer, J. D. (eds.) (1992b). *Survey Measurement of Drug Use: Methodological Issues*. Washington, DC: Government Printing Office.

Turner, C. F., Miller, H. F., and Catania, J. A. (1995), "Supplement for Methodological Augmentation of the Gay Urban Men's Survey (GUMS) Using a New Survey Technology: Telephone Audio-CASI," unpublished proposal for a funded competitive supplement to NIH grant R01-MH54320 (J. Catania, P. I.), Center for AIDS Prevention Studies, University of California, San Francisco, and Research Triangle Institute.

Turner, C. F., Miller, H. G., and Moses, L. E. (eds.) (1989), *AIDS, Sexual Behavior, and Intravenous Drug Use*. Washington, D.C.: National Academy Press.

Turner, C. F., Miller, H. G., and Rogers, S. M. (in press), "Survey Measurement of Sexual Behaviors: Problems and Progress," in J. Bancroft (ed.), *Researching Sexual Behavior*. Bloomington: Indiana University Press.

Turner, C. F., Ku, L., Sonenstein, F. L., and Pleck, J. H. (1996a), "Impact of Audio-CASI on Bias in Reporting of Male-Male Sexual Contacts," in R. Warnecke (ed.), *Health Survey Research Methods*, Hyattsville, MD: National Center for Health Statistics.

Turner, C. F., Rogers, S. M., Hendershot, T. P., Miller, H. G., and Thornberry, J. P. (1996b), "Improving Representation of Linguistic Minorities in Health Surveys: A Preliminary Test of Multilingual Audio-CASI," *Public Health Reports*, **111**, pp. 276–279.

Turner, C. F., Miller, H. G., Smith, T. K., Cooley, P. C., and Rogers, S. M. (1996c), "Telephone Audio Computer-Assisted Self-Interviewing (T-ACASI) and Survey Measurements of Sensitive Behaviors: Preliminary Results," in R. Banks, J. Fairgrieve, L. Gerrard et al. (eds.), *Survey and Statistical Computing 1996: Proceedings of the Second ASC International Conference*, Chesham, Bucks, U.K.: Association for Survey Computing.

Twyman, M. (1979), "A Schema for the Study of Graphic Language," in P. A. Kolers, M. E. Wrolstad, and H. Bouma (eds.), *Processing of Visible Language*, vol. 1, New York: Plenum, pp. 117–149.

Udry, J. R. (1995), "Prospective Longitudinal Study of Adolescent Health," NIH-CRISP abstract for funded proposal P01-HD31921, Bethesda, MD: National Institutes of Health.

Uglow, D. (1991), "Design Issues in Forms-Based CAPI Instruments," paper presented at the International Field Directors and Technologies Conference, San Diego, May.

Uglow, D. (1992), "Pen-Based Computers: Remodeling Computer-Assisted Data Collection?" paper presented at the International Field Directors and Technologies Conference, St. Petersburg, FL, May.

United Nations (1994), *Statistical Data Editing: Methods and Techniques*, vol. 1, Statistical Standards and Studies, no. 44, Geneva: United Nations Statistical Commission and Economic Commission for Europe.

U.S. Department of Commerce (1965), "United States Censuses of Population and Housing 1960," *Quality Control of Preparatory Operations, Microfilming, and Coding*. Reports on Methodology of the 18th Decennial Census.

U.S. Department of Energy (1992), "EIA's Petroleum Supply Survey Respondents

Transmit Data Electronically," Energy Information Administration, Washington, DC. unpublished report.

U.S. Energy Information Administration (1989), *PEDRO—Respondent User Guide to the Petroleum Electronic Data Reporting Option, Version 3.0.* Washington, DC: U.S. Department of Energy.

Van Bastelaer, A., Kerssemakers, F., and Sikkel, D. (1987), "A Test of the Continuous Labour Force Survey with Hand-Held Computers, Interviewer Behaviour and Data Quality," in *CBS Select 4, Automation in Survey Processing.* Voorburg: Statistics Netherlands, pp. 37–54.

Van Bastelaer, A., Kerssemakers, F., and Sikkel, D. (1988), "A Test of the Netherlands' Continuous Labour Force Survey with Hand-Held Computers: Contributions to Questionnaire Design," *Journal of Official Statistics*, **4**, pp. 141–154.

van den Oord, E., and Saris, W. E. (1994), *Bestedingen aan Vlees en Vleeswaren*, Amsterdam: STP.

Van de Pol, F., and Diederen, B., (1996), "A Priority Index for Macro-Editing the Netherlands Foreign Trade Survey," in *Proceedings of the Data Editing Workshop and Exposition*, Washington, DC: U.S. Bureau of Labor Statistics.

Van de Pol, F., and Molenaar, W., (1995), "Selective and Automatic Editing with CADI-Applications," in V. Kuusela, (ed.), *Essays on Blaise 1995, Proceedings of the Third International Blaise Users' Conference,* Helsinki: Statistics Finland, pp. 159–168.

Van der Spiegel, J. (1995), "New Information Technologies and Changes in Work," in A. Howard (ed.), *The Changing Nature of Work*, San Francisco: Jossey-Bass.

Vehovar, V. and Batagelj, Z. (1996), "The Methodological Issues in WWW Surveys," paper presented at the International Conference on Computer Assisted Survey Information Collection, San Antonio, TX, December.

Verhallen, T. M. M. (1988), *Psychologisch Marktonderzoek*, Tilburg: Katholieke Universiteit Brabant.

Verwey, M., Saris, W. E., Mosselman, K., and van Doorn, L. (1989), "Tele-interviewing in Practice: Household Incomes Show Large Monthly Fluctuations," *Marketing and Research Today*, **17**, pp. 230–240.

Verwey, J. (1992), *De Kwaliteit van Inkomensdata: Een Analyse van Vier Maanden Inkomensdata*, Amsterdam: NIMMO.

Vézina, S. M. (1994), "Imaging and Intelligent Character Recognition," pilot project, prepared by Operations and Integration Division, Statistics Canada.

Vigerhous, G. (1981), "Scheduling Telephone Interviews: A Study of Seasonal Patterns," *Public Opinion Quarterly*, **45**, pp. 250–259.

Vis, C. and Wouters, E. (1996), "Embedded Measurement of Life Histories," paper presented at the NOSMO Methodology Day, Utrecht, December.

Wagenaar, W. A. (1986), "My Memory: A Study of Autobiographical Memory over Six Years. *Cognitive Psychology*, **18**, pp. 225–252.

Walker, S. and Jones, R. M. (1987), "Improving Subject Retrieval in Online Catalogues," British Library Research Paper 24.

Wall Street Journal (1996), "U.S. Households with Internet Access Doubled to 14.7 Million in Past Year," October 11, p. B11.

Wall Street Journal (1997), "Work Week," February 18, vol. 229, no. 33, p.1.

Walker, C., Brown, A., and Veevers, R. (1996), "Data Validation in the CAI World," paper presented at the Joint Statistical Meetings of the American Statistical Association, Chicago, August.

Walsh, J. P., Kiesler, S., Sproull, L. S., and Hesse, B. W. (1992), "Self-selected and Randomly Selected Respondents in a Computer Network Survey," *Public Opinion Quarterly*, **56**, pp. 241–244.

Warde, W. (1986), *Problems with Telephone Surveys*, NASS Staff report no. SRB-NERS-86-01, Washington, DC: U.S. Department of Agriculture.

Waterton, J. J. and Duffy, J. C. (1984), "A Comparison of Computer Interviewing Techniques and Traditional Methods in the Collection of Self-report Alcohol Consumption Data in a Field Survey," *International Statistical Review*, **52**, pp. 173–182.

Weeks, M. F. (1988), "Call Scheduling with CATI: Current Capabilities and Methods," in R. M. Groves, P. P. Biemer, L. E. Lyberg, J. Massey, W. L. Nicholls II, and J. Waksberg (eds.), *Telephone Survey Methodology*, New York: Wiley, pp. 403–420.

Weeks, M. F. (1992), "Computer-Assisted Survey Information Collection: A Review of CASIC Methods and Their Implications for Survey Operations," *Journal of Official Statistics*, **8**, pp. 445–465.

Weeks, M. F., Kulka, R. A., and Pierson, S. A. (1987), "Optimal Call Scheduling for a Telephone Survey," *Public Opinion Quarterly*, **51**, pp. 540–549.

Weiner, N. (1948), *Cyberspace*, New York: Wiley.

Weise, E. (1996), "Mike and Terry's Dreadful Adventure," April 8, 1996, Associated Press. http://www.dhp.com/~sloppy/files/text_about_hackers/purpcon.txt.

Wensing, F. (1995), "Update from 'Downunder': History, Plans and Functions We've Built for CAI and Blaise in Australia," *Essays On Blaise*, *Third International Blaise User's Conference, Statistics Finland*, pp. 197–206.

Wenzowski, M. J. (1988), "ACTR: A Generalized Automated Coding System," *Survey Methodology*, **14**, pp. 299–307.

Wenzowski, M. J. (1996), "Advances in Automated and Computer Assisted Coding Software at Statistics Canada," *Proceedings of the Bureau of the Census' Annual Research Conference and Technology Interchange*, Washington, DC: U.S. Bureau of the Census, pp. 1117–1126.

Werking, G. S. (1994), "Establishment Surveys: Designing the Survey Operations of the Future," *Proceedings of the Section on Survey Research Methods, American Statistical Association*, pp. 163–169.

Werking, G. S. and Clayton, R. L. (1991), "Enhancing Data Quality through the Use of a Mixed Mode Collection," *Survey Methodology*, **17**, (1), pp. 3–14.

Werking, G. S. and Clayton, R. L. (1995), "Automated Telephone Methods for Business Surveys," in B. Cox, D. Binder, A. Christianson, M. Colledge, and P. Kott, (eds.), *Business Survey Methods*, New York: Wiley.

Werking, G., Tupek, A., and Clayton, R. (1988), "CATI and Touchtone Self-response Applications for Establishment Surveys," *Journal of Official Statistics*, pp. 349–362.

Werner, J., Maisel, R., and Robinson, K. (1995), "The Prodigy Experiment in Using E-mail for Tracking Public Opinion," paper presented at the American Association for Public Opinion Research Annual Conference, Fort Lauderdale, FL, May.

Westat, Inc. (1996), *WesVarPC: A User's Guide to WesVarPC*, Rockville, MD: Westat, Inc.

Whelan, J., Karlsen, R., and Yost, P. (1995), "Computer-Assisted Data Collection: Concepts in User-Friendly Menus, Coding, and Modular Application Design for PC Laptop Data Collection," in *Proceedings of the Bureau of the Census Annual Research Conference*, Washington, DC: U.S. Bureau of the Census, pp. 593–603.

Wilson, B. (1989), "Disk-by-Mail Surveys: Three Years' Experience," *1989 Sawtooth Software Conference Proceedings, Gaining a Competitive Advantage through PC-Based Interviewing and Analysis,* vol. 1, Sun Valley, ID: Sawtooth Software, pp. 1–4.

Wilson, C. L., Geist, J., Garris, M. D., and Rama, C. (1996), *Design, Integration, and Evaluation of Form-Based Handprint and OCR Systems*, NISTIR Report 5932, Washington, DC: U.S. National Institute of Standards and Technologies.

Winkler, W. and Draper, L. R. (1994), "Application of the SPEER Edit System," Washington, DC: U.S. Bureau of the Census, Unpublished Paper.

Wiseman, F. and McDonald, P. (1979), "Noncontact and Refusal Rates in Consumer Telephone Surveys," *Journal of Marketing Research,* **16**, pp. 478–484.

Witt, K. J. and Bernstein, S. (1992), "Best Practices in Disk-by-Mail Surveys," *1992 Sawtooth Software Conference Proceedings,* Sun Valley, ID: Sawtooth Software, pp. 1–22.

Wittink, D. R. and Cattin, P. (1989), "Commercial Use of Conjoint Analysis: An Update," *Journal of Marketing,* **53**, pp. 91–96.

Woelfel, J. (1993), "Artificial Neural Networks in Policy Research: A Current Assessment," *Journal of Communication,* **43**, pp. 63–80.

Wojcik, M. S. and Baker, R. P. (1992), "Interviewer and Respondent Acceptance of CAPI," in *Proceedings of the Bureau of the Census 1992 Annual Research Conference*, pp. 613–621.

Wojcik, M. S., Bard, S., and Hunt, E. (1991), "Training Field Interviewers to Use Computers: A Successful CAPI Training Program," paper presented at the Annual Conference of the American Association for Public Opinion Research, St. Petersburg, FL, May.

Wojcik, M. S. and Hunt, E. (1994), "Using Self-Study to Train CAPI Interviewers," paper presented at the Annual Conference of the American Association for Public Opinion Research, St. Charles, IL, May.

Woltman, H. F., Turner, A. G., and Bushery, J. M. (1980), "A Comparison of Three Mixed-Mode Interviewing Procedures in the National Crime Survey," *Journal of the American Statistical Association,* **75**, (371), pp. 534–543.

Young, N. (1989), "Wave-Seam Effects in the SIPP," *Proceedings of the Section on Survey Research Methods, American Statistical Association*, pp. 393–398.

Zandan, P. and Frost, L. (1989), "Customer Satisfaction Research using Disk-by-Mail," *1989 Sawtooth Software Conference Proceedings, Gaining a Competitive Advantage through PC-Based Interviewing and Analysis,* vol. 1, Sun Valley, ID: Sawtooth Software, pp. 5–17.

Zwetchkenbaum, R. (1996), "Developments in Home Computing," in *1996–97 Information Industry and Technology Update.* Framingham, MA: International Data Corporation.

Glossary of CASIC Acronyms

ACASI, audio-CASI Audio computer assisted self-interviewing. Respondent self-administers the survey on a laptop computer, listening to questions on audio headphones and entering responses into computer. The response categories are usually displayed on the screen. The questions may or may not also be simultaneously displayed on the screen. See also T-ACASI and VCASI or video-CASI.

ASR Automatic speech recognition. Respondents listen to an automated recording of the questions and speak their responses into the telephone. System recognizes and codes response. See also VRE.

CADE Computer assisted data entry.

CADI Computer assisted data input. Automated tools to facilitate the postcollection keying and editing of paper forms or questionnaires. The acronym CADE has also been used to refer to computer assisted data editing, the postcapture or postkeying review, and edit of survey data.

CAI Computer assisted interviewing. The subset of CASIC methods that are interviewer-administered. Mostly used to refer to computer assisted telephone interviewing (CATI) and computer assisted personal interviewing (CAPI).

CAPI Computer assisted personal interviewing. Face-to-face or personal visit survey in which the interviewer carries a laptop or palmtop computer to the site of the interview and administers the survey.

CASI Computer assisted self-interviewing. The family of methods in which the respondent self-administers the instrument. CASI usually refers to methods where an interviewer is present or delivers the computer, and often occurs in combination with CAPI. CASI is sometimes referred to as video-CASI (or VCASI) to contrast it with audio-CASI.

CASIC Computer assisted survey information collection. Defined here as use of computers for survey data collection, data capture, and data preparation and for related activities preparing for, supporting, managing, and coordinating those tasks with each other and with later stages of the survey process.

CATI Computer assisted telephone interviewing. Telephone interviewing with the interviewers using the computer. May be centralized (in a central facility with desktop computers connected to a LAN) or decentralized (interviewers using laptop or desktop computers to conduct interviews from their homes).

CSAQ   Computerized self-administered questionnaire. Self-administered data collection in which an interviewer is usually not present. Includes methods such as disk-by-mail (DBM), electronic mail surveys (EMS), touchtone data entry (TDE), voice recognition entry (VRE), World Wide Web or Internet surveys, etc.

DBM   Disk-by-mail. Computerized self-administered questionnaire (CSAQ) mailed on a disk to the respondents. Respondents install and complete the instrument on their own computers. Responses are saved on disk and mailed back to the survey organization.

DIP   Document image processing. Family of technologies related to scanning, OMR, OCR, and ICR.

EDI   Electronic data interchange. The automatic exchange of information between a respondents' database and that of the survey organization. Used predominantly in establishment surveys.

EMS   Electronic mail surveys. Computerized self-administered questionnaire (CSAQ) transmitted to respondents and returned via electronic mail, or delivered via the Internet or World Wide Web (WWW).

IAQ   Interviewer-administered questionnaire. A term sometimes used as a contrast to self-administered questionnaires (SAQs). Computer assisted IAQ methods include CATI and CAPI.

ICR   Intelligent character recognition. Automated system for reading and interpreting freeform and unconstrained handwriting. See also OCR.

OCR   Optical character recognition. Automated system for reading and interpreting handwritten characters (numbers or letters). Sometimes called intelligent character recognition (ICR), especially when referring to the recognition of cursive handwriting.

OMR   Optical mark recognition. Automated system for reading and interpreting simple marks, such as checks and filled circles.

PAPI   Paper and pencil interviewing. Used as a constrast to CATI and CAPI. Also referred to as paper and pencil (P&P) interviewing.

SAQ   Self-administered questionnaire. Usually used to refer to a paper and pencil mode of data collection, in contrast to CSAQ.

T-ACASI, telephone ACASI   Telephone audio computer assisted self-interviewing. Respondents listen to an automated voice recording of the questions and enter their responses using the keypad on a touchtone telephone. Uses similar technology to VRE, but is viewed as an extension of ACASI, in that it is usually part of an interviewer-administered survey.

TDE   Touchtone data entry. Respondents listen to an automated voice recording of the questions and enter their responses using the keypad on a touchtone telephone.

VCASI, video-CASI   Video computer assisted self-interviewing. Same as CASI, but sometimes used as an explicit contrast to audio-CASI.

VRE   Voice recognition entry. Respondents listen to an automated recording of the questions and speak their responses into the telephone. System recognizes and codes responses. VRE is also sometimes referred to as VR or ASR.

Index

abortion, 431–433, 437–439, 441–442, 458–460

academic survey organizations, 8–9, 12–13, 16, 18, 45, 49–50, 56, 92, 168–169, 286, 390, 399, 481, 586, 596, 600

access control, 259, 266, 281, 395

acquiescence, 181–182

adaptive conjoint analysis, 158–159

adolescents, 448, 450, 458, 460, 471

affirmative action, 172, 178–180

AIDS, 170, 176, 431, 456, 458, 462

analysis of variance, 172, 302, 442

Asymmetric Digital Subscriber Line (ADSL), 324–325, 328

asynchronous transfer mode (ATM), 322

audio-CASI, 14, 19, 47–48, 165, 331, 334, 347, 431–452, 455–473, 525, 541, 589, 592, 601

audio computer assisted self interviewing (ACASI, audio-CASI), 14, 19, 47, 165, 432, 435, 443–448, 451, 455–473, 525, 589, 601

Australian Bureau of Statistics (ABS), 62–83, 277, 510–513, 515

authentication, 326, 395, 559

authoring systems, 111–112, 114, 342, 392, 525, 527

authors, instrument, 70, 75, 87, 89, 96, 107, 114–116, 121–123, 404

autodialing, 289–291, 303. *See also* predictive autodialing

automated call scheduling, 98, 246, 273, 285–306. *See also* call scheduling

automatic speech recognition (ASR), 572

backing up data, 87, 91, 98, 260–262, 270–271, 317, 326, 519

backing up in instrument, 352–357, 362–363, 466

bar code recognition (BCR), 506

bar codes, 12, 428, 500, 505, 509, 513

barcode scanning, 576, 577, 579

behavior codes, 374, 387, 388

bilingual, 288, 298

Blaise, 65, 70, 76, 81, 162, 212, 219, 278, 282, 354, 392, 469, 472

bogus pipeline, 438, 448–449

branching, 8, 20, 47, 57, 70, 96, 97, 111, 390, 393, 422, 457, 530, 543, 550, 559, 561

Bureau of Justice Statistics, 64

Bureau of Labor Statistics (BLS), 12, 64, 298, 391, 462, 521, 543, 588

Bureau of the Census, 9, 16, 39, 48, 62–83, 106, 111, 129, 135, 165, 209, 225, 228–231, 252, 289, 337, 341, 344, 351, 391, 397, 400, 404, 476, 481, 484, 510, 516, 525, 553

business survey, 16, 18, 62, 64–67, 72, 79–81, 129, 210, 392, 482, 543–562, 588, 595, 597

cable, 324, 328, 568, 579, 593, 597

computer assisted personal interviewing (CAPI), 10–12, 17–20, 23–43, 48, 50–51, 55–58, 62–82, 85, 118, 119, 136, 147, 169, 247, 249, 253–254, 263–283, 307–329, 331–349, 351–365, 369, 432, 443–448, 521–522, 558–559, 573–579, 587–588, 592

WILEY SERIES IN PROBABILITY AND STATISTICS
ESTABLISHED BY WALTER A. SHEWHART AND SAMUEL S. WILKS

Editors
Vic Barnett, Ralph A. Bradley, Noel A. C. Cressie, Nicholas I. Fisher,
Iain M. Johnstone, J. B. Kadane, David G. Kendall, David W. Scott,
Bernard W. Silverman, Adrian F. M. Smith, Jozef L. Teugels;
J. Stuart Hunter, Emeritus

Probability and Statistics Section

*ANDERSON · The Statistical Analysis of Time Series
ARNOLD, BALAKRISHNAN, and NAGARAJA · A First Course in Order Statistics
ARNOLD, BALAKRISHNAN, and NAGARAJA · Records
BACCELLI, COHEN, OLSDER, and QUADRAT · Synchronization and Linearity:
   An Algebra for Discrete Event Systems
BASILEVSKY · Statistical Factor Analysis and Related Methods: Theory and
   Applications
BERNARDO and SMITH · Bayesian Statistical Concepts and Theory
BILLINGSLEY · Convergence of Probability Measures
BOROVKOV · Asymptotic Methods in Queuing Theory
BRANDT, FRANKEN, and LISEK · Stationary Stochastic Models
CAINES · Linear Stochastic Systems
CAIROLI and DALANG · Sequential Stochastic Optimization
CONSTANTINE · Combinatorial Theory and Statistical Design
COOK · Regression Graphics
COVER and THOMAS · Elements of Information Theory
CSÖRGŐ and HORVÁTH · Weighted Approximations in Probability Statistics
CSÖRGŐ and HORVÁTH · Limit Theorems in Change Point Analysis
DETTE and STUDDEN · The Theory of Canonical Moments with Applications in
   Statistics, Probability, and Analysis
*DOOB · Stochastic Processes
DRYDEN and MARDIA · Statistical Analysis of Shape
DUPUIS and ELLIS · A Weak Convergence Approach to the Theory of Large Deviations
ETHIER and KURTZ · Markov Processes: Characterization and Convergence
FELLER · An Introduction to Probability Theory and Its Applications, Volume 1,
   Third Edition, Revised; Volume II, *Second Edition*
FULLER · Introduction to Statistical Time Series, *Second Edition*
FULLER · Measurement Error Models
GELFAND and SMITH · Bayesian Computation
GHOSH, MUKHOPADHYAY, and SEN · Sequential Estimation
GIFI · Nonlinear Multivariate Analysis
GUTTORP · Statistical Inference for Branching Processes
HALL · Introduction to the Theory of Coverage Processes
HAMPEL · Robust Statistics: The Approach Based on Influence Functions
HANNAN and DEISTLER · The Statistical Theory of Linear Systems
HUBER · Robust Statistics
IMAN and CONOVER · A Modern Approach to Statistics
JUREK and MASON · Operator-Limit Distributions in Probability Theory
KASS and VOS · Geometrical Foundations of Asymptotic Inference
KAUFMAN and ROUSSEEUW · Finding Groups in Data: An Introduction to Cluster
   Analysis

*Now available in a lower priced paperback edition in the Wiley Classics Library.

*Now available in a lower priced paperback edition in the Wiley Classics Library.

*Now available in a lower priced paperback edition in the Wiley Classics Library.

*Now available in a lower priced paperback edition in the Wiley Classics Library.

*Now available in a lower priced paperback edition in the Wiley Classics Library.

Texts and References Section

AGRESTI · An Introduction to Categorical Data Analysis

ANDERSON · An Introduction to Multivariate Statistical Analysis, *Second Edition*

ANDERSON and LOYNES · The Teaching of Practical Statistics

ARMITAGE and COLTON · Encyclopedia of Biostatistics: Volumes 1 to 6 with Index

BARTOSZYNSKI and NIEWIADOMSKA-BUGAJ · Probability and Statistical Inference

BERRY, CHALONER, and GEWEKE · Bayesian Analysis in Statistics and
    Econometrics: Essays in Honor of Arnold Zellner

BHATTACHARYA and JOHNSON · Statistical Concepts and Methods

BILLINGSLEY · Probability and Measure, *Second Edition*

BOX · R. A. Fisher, the Life of a Scientist

BOX, HUNTER, and HUNTER · Statistics for Experimenters: An Introduction to
    Design, Data Analysis, and Model Building

BOX and LUCEÑO · Statistical Control by Monitoring and Feedback Adjustment

BROWN and HOLLANDER · Statistics: A Biomedical Introduction

CHATTERJEE and PRICE · Regression Analysis by Example, *Second Edition*

COOK and WEISBERG · An Introduction to Regression Graphics

COX · A Handbook of Introductory Statistical Methods

DILLON and GOLDSTEIN · Multivariate Analysis: Methods and Applications

DODGE and ROMIG · Sampling Inspection Tables, *Second Edition*

DRAPER and SMITH · Applied Regression Analysis, *Third Edition*

DUDEWICZ and MISHRA · Modern Mathematical Statistics

DUNN · Basic Statistics: A Primer for the Biomedical Sciences, *Second Edition*

FISHER and VAN BELLE · Biostatistics: A Methodology for the Health Sciences

FREEMAN and SMITH · Aspects of Uncertainty: A Tribute to D. V. Lindley

GROSS and HARRIS · Fundamentals of Queueing Theory, *Third Edition*

HALD · A History of Probability and Statistics and their Applications Before 1750

HALD · A History of Mathematical Statistics from 1750 to 1930

HELLER · MACSYMA for Statisticians

HOEL · Introduction to Mathematical Statistics, *Fifth Edition*

JOHNSON and BALAKRISHNAN · Advances in the Theory and Practice of Statistics: A
    Volume in Honor of Samuel Kotz

JOHNSON and KOTZ (editors) · Leading Personalities in Statistical Sciences: From the
    Seventeenth Century to the Present

JUDGE, GRIFFITHS, HILL, LÜTKEPOHL, and LEE · The Theory and Practice of
    Econometrics, *Second Edition*

KHURI · Advanced Calculus with Applications in Statistics

KOTZ and JOHNSON (editors) · Encyclopedia of Statistical Sciences: Volumes 1 to 9
    wtih Index

KOTZ and JOHNSON (editors) · Encyclopedia of Statistical Sciences: Supplement
        Volume

KOTZ, REED, and BANKS (editors) · Encyclopedia of Statistical Sciences: Update
        Volume 1

KOTZ, REED, and BANKS (editors) · Encyclopedia of Statistical Sciences: Update
        Volume 2

LAMPERTI · Probability: A Survey of the Mathematical Theory, *Second Edition*

LARSON · Introduction to Probability Theory and Statistical Inference, *Third Edition*

LE · Applied Categorical Data Analysis

LE · Applied Survival Analysis

MALLOWS · Design, Data, and Analysis by Some Friends of Cuthbert Daniel

MARDIA · The Art of Statistical Science: A Tribute to G. S. Watson

MASON, GUNST, and HESS · Statistical Design and Analysis of Experiments with
    Applications to Engineering and Science

*Now available in a lower priced paperback edition in the Wiley Classics Library.

Texts and References (Continued)
MURRAY · X-STAT 2.0 Statistical Experimentation, Design Data Analysis, and Nonlinear Optimization
PURI, VILAPLANA, and WERTZ · New Perspectives in Theoretical and Applied Statistics
RENCHER · Methods of Multivariate Analysis
RENCHER · Multivariate Statistical Inference with Applications
ROSS · Introduction to Probability and Statistics for Engineers and Scientists
ROHATGI · An Introduction to Probability Theory and Mathematical Statistics
RYAN · Modern Regression Methods
SCHOTT · Matrix Analysis for Statistics
SEARLE · Matrix Algebra Useful for Statistics
STYAN · The Collected Papers of T. W. Anderson: 1943–1985
TIERNEY · LISP-STAT: An Object-Oriented Environment for Statistical Computing and Dynamic Graphics
WONNACOTT and WONNACOTT · Econometrics, *Second Edition*

WILEY SERIES IN PROBABILITY AND STATISTICS

ESTABLISHED BY WALTER A. SHEWHART AND SAMUEL S. WILKS

Editors
Robert M. Groves, Graham Kalton, J. N. K. Rao, Norbert Schwarz, Christopher Skinner

Survey Methodology Section

BIEMER, GROVES, LYBERG, MATHIOWETZ, and SUDMAN · Measurement Errors in Surveys
COCHRAN · Sampling Techniques, *Third Edition*
COUPER, BAKER, BETHLEHEM, CLARK, MARTIN, NICHOLLS, and O'REILLY (editors) · Computer Assisted Survey Information Collection
COX, BINDER, CHINNAPPA, CHRISTIANSON, COLLEDGE, and KOTT (editors) · Business Survey Methods
*DEMING · Sample Design in Business Research
DILLMAN · Mail and Telephone Surveys: The Total Design Method
GROVES and COUPER · Nonresponse in Household Interview Surveys
GROVES · Survey Errors and Survey Costs
GROVES, BIEMER, LYBERG, MASSEY, NICHOLLS, and WAKSBERG · Telephone Survey Methodology
*HANSEN, HURWITZ, and MADOW · Sample Survey Methods and Theory, Volume 1: Methods and Applications
*HANSEN, HURWITZ, and MADOW · Sample Survey Methods and Theory, Volume II: Theory
KASPRZYK, DUNCAN, KALTON, and SINGH · Panel Surveys
KISH · Statistical Design for Research
*KISH · Survey Sampling
LESSLER and KALSBEEK · Nonsampling Error in Surveys
LEVY and LEMESHOW · Sampling of Populations: Methods and Applications
LYBERG, BIEMER, COLLINS, de LEEUW, DIPPO, SCHWARZ, TREWIN (editors) · Survey Measurement and Process Quality
SKINNER, HOLT, and SMITH · Analysis of Complex Surveys

*Now available in a lower priced paperback edition in the Wiley Classics Library.